Advanced Production, Processing and Characterization of Industrial Materials

Advanced Production, Processing and Characterization of Industrial Materials

Guest Editors

Jozef Maščenik
Tibor Krenicky

MDPI

Basel • Beijing • Wuhan • Barcelona • Belgrade • Novi Sad • Cluj • Manchester

Guest Editors

Jozef Maščenik
Department of Design and Monitoring of Technical Systems
Technical University of Kosice
Presov
Slovakia

Tibor Krenicky
Department of Design and Monitoring of Technical Systems
Technical University of Kosice
Presov
Slovakia

Editorial Office

MDPI AG
Grosspeteranlage 5
4052 Basel, Switzerland

This is a reprint of the Special Issue, published open access by the journal *Materials* (ISSN 1996-1944), freely accessible at: https://www.mdpi.com/journal/materials/special_issues/APPCIM.

For citation purposes, cite each article independently as indicated on the article page online and as indicated below:

Lastname, A.A.; Lastname, B.B. Article Title. *Journal Name* **Year**, *Volume Number*, Page Range.

ISBN 978-3-7258-7098-1 (Hbk)
ISBN 978-3-7258-7099-8 (PDF)
https://doi.org/10.3390/books978-3-7258-7099-8

Contents

About the Editors

Jozef Maščenik

Jozef Maščenik is an Associate Professor at the Faculty of Manufacturing Technologies in Prešov, Technical University of Kosice, Slovakia. His professional profile focuses on the design of specific engineering solutions incorporating systems for monitoring operating conditions and parameters. His scientific work in mechanical transmissions has contributed to research on optimizing gear production using unconventional technologies such as laser, plasma, and water-jet machining. His research activities further extend to the design, control, and testing of belt transmissions and other machine components. He disseminates his results through utility models and publications in domestic and international journals indexed in the Web of Science and Current Contents databases, and he also serves on the Editorial Boards of international scientific conferences and journals. Additionally, he is a reviewer for several international scientific journals.

Tibor Krenicky

Tibor Krenicky is an Associate Professor at the Faculty of Manufacturing Technologies in Prešov, Technical University of Kosice, Slovakia. His research focuses on the simulation, determination, and evaluation of operating parameters and technical condition of engineering systems through multiparametric monitoring and virtual instrumentation, as well as the design and implementation of experimental diagnostic devices. He collaborates extensively with national and international research projects. He is also an experienced Editorial Board Member, Guest Editor, and reviewer for several international scientific journals.

materials

Editorial

Advanced Production, Processing and Characterization of Industrial Materials

Jozef Mascenik [1] **and Tibor Krenicky** [2,*]

1 Department of Computer Aided Manufacturing Technologies, Faculty of Manufacturing Technologies, Technical University of Kosice, Bayerova 1, 080 01 Presov, Slovakia; jozef.mascenik@tuke.sk
2 Department of Manufacturing Process Control, Faculty of Manufacturing Technologies, Technical University of Kosice, Bayerova 1, 080 01 Presov, Slovakia
* Correspondence: tibor.krenicky@tuke.sk

Abstract: This Special Issue presents recent advances in the production, modelling, processing, and characterization of advanced industrial materials, highlighting the diversity and sophistication of contemporary research discussing metallic, polymeric, composite, and nano-structured systems. The collected contributions address key challenges in materials science, ranging from surface quality control, the development of novel machining and fabrication tools, and optimization of thermoplastic composite consolidation, to provide fundamental insights into additive manufacturing, rheology, and constitutive modelling. The showcased studies introduce innovative approaches to metrology, including advanced optical, fluorescence, and X-ray scattering techniques for characterizing nano-particles, microstructures, and thermal properties. The presented research also features investigations into the welding of dissimilar steels, binder jetting of stainless steel, and the influence of heat treatment on functional steel performance, alongside environmentally oriented research on natural-fibre energy devices and bio-based polymer composites. Further research topics include defect structures in doped crystals, low-temperature synthesis of oxide films, and mechanical behaviour of steels under extreme conditions. Collectively, these articles demonstrate the strong synergy between experimental methods, computational modelling, and industrial applications, underscoring the continued progress in materials reliability, surface engineering, and advanced manufacturing technologies. This Special Issue therefore provides a comprehensive overview of current trends and emerging directions, offering valuable methodological and conceptual insights in the field.

Keywords: production technologies; thermal processing; structure modelling; material surface; composites; material design; surface treatment; material quality test

1. Introduction

The rapid development of advanced industrial materials over the past two decades has significantly transformed modern manufacturing, structural engineering, and high-performance applications. Contemporary materials engineering is increasingly marked by the convergence of innovative production technologies, multi-scale computational modelling, and high-precision characterization tools, all aimed at enabling materials with tailored mechanical and functional properties [1–4]. The aerospace, automotive, biomedical engineering, and renewable energy sectors are increasingly relying on materials optimized for lightweight design, thermal stability, and reliability under complex service conditions [5,6].

 https://doi.org/10.3390/ma18235366

Among recent technological advances, additive manufacturing (AM) represents a fundamental shift in production methodologies. Processes such as selective laser melting, electron beam melting, binder jetting, fused deposition modelling, and directed energy deposition allow near-net-shape fabrication of complex structures with controlled microstructural features [4–7]. Significant progress has been achieved in understanding melt pool dynamics, powder–laser interactions, solidification behaviour, and defect prevention, supported by advances in computational modelling and high-speed diagnostics [6–9]. Complementary research in AM process monitoring has introduced real-time optical, thermal, and acoustic approaches capable of detecting pores, spatter, and melt pool instabilities, laying the foundation for closed-loop control strategies [8–12].

Composite materials have likewise seen rapid development. Fibre-reinforced polymers, hybrid laminates, natural-fibre composites, and thermoplastic matrices are now widely used across modern engineering disciplines due to their high specific strength, corrosion resistance, and fatigue performance [13–16]. Thermoplastic composites in particular have benefited from new forming, consolidation, and automated placement technologies, supported by integrated thermal–mechanical modelling and improved control of crystallization and interlaminar bonding mechanisms [14–17].

Advances in surface engineering and high-resolution metrology have played a central role in enhancing material performance and reliability. Optical interferometry, confocal and laser profilometry, and interference microscopy provide increasingly detailed insight into topography, texture, and deformation at the micro- and nano-scale [12,18]. Progress in profilometric analysis has led to improved detection of wear, abrasion, and subsurface changes, particularly for advanced industrial alloys and tribological systems [19].

At smaller scales, nanomaterials research has been strengthened by powerful experimental tools for structural and chemical analysis, including small-angle X-ray scattering, advanced electron microscopy, and hybrid simulation approaches, which, together, enable precise characterization of nano-particle assemblies and defect structures [20,21]. In parallel, modern constitutive modelling frameworks—including viscoelastic, viscoplastic, and multi-physics formulations—support improved prediction of deformation, creep, and damage across a range of materials from polymers to metals [22,23].

Sustainability has grown into a central theme in materials research. Efforts to reduce environmental impact have accelerated the development of low-energy coating technologies, new corrosion-resistant systems, and environmentally resilient industrial materials [24,25]. At the same time, natural-fibre composites and bio-based modifiers have attracted increasing attention as renewable and recyclable alternatives for large-volume applications [16,26].

Artificial intelligence and data-driven approaches have become essential tools across manufacturing and materials science. AI-augmented modelling supports process–structure–property prediction, accelerates optimization, and enhances quality control, while enabling new forms of adaptivity in additive manufacturing and biomaterials development [27,28].

The current state of the art in advanced production, processing, and characterization of industrial materials is defined by stronger integration of modelling and experimentation, increased emphasis on in situ monitoring, the adoption of sustainable material systems, and the emergence of intelligent, data-driven design and manufacturing strategies. The significant engagement of the scientific community on this topic is evidenced by the numerous views on the articles, but more importantly by the rich citation response. To date, the articles in this Special Issue have been cited a total of 143 times, with an average of 8.4 citations per article, and this figure is continuing to rise, although the distribution of citations is considerably uneven, with review articles [29,30] standing out significantly so far.

2. Overview of Published Articles

This Special Issue presents a comprehensive view of current advances in industrial materials research. The articles included within collectively expand the knowledge on additive manufacturing, composite processing, advanced surface characterization, modelling, and sustainable materials engineering. Each study contributes to filling existing gaps—whether related to process optimization, joining of dissimilar materials, nano-scale metrology, or microstructural control. Simultaneously, this Special Issue outlines challenges that must be addressed to facilitate future research on advanced materials production, processing, and characterization.

2.1. Additive Manufacturing and Advanced Processing of Metallic Materials

In their study, Král' et al. (Contribution 8) applied binder jetting technology to 316L stainless steel, examining how different printing strategies affected both mechanical strength and dimensional accuracy. Their work emphasized the importance of post-processing and powder–binder interactions in determining final component performance. Similarly, Gel'atko et al. (Contribution 9) investigated electron beam welding of dissimilar stainless and maraging steel joints, focusing on how welding parameters influenced microstructure, hardness distribution, and joint integrity. Their findings underscored the challenges of joining hybrid AM–conventional steel systems under industrial conditions. In a complementary theoretical contribution, Stejskal et al. (Contribution 17) developed a virial-theorem-based model to predict stiffness–density relationships in additively manufactured lattice structures. Their approach highlighted the potential of physics-based modelling to support design optimization in AM.

2.2. Composite Materials, Thermoplastic Processing, and Machining Technologies

Campos et al. (Contribution 3) analysed process parameters governing the continuous consolidation of thermoplastic composites in their statistical investigation, identifying optimal conditions for achieving consistent thickness and quality. Their results demonstrated how temperature, pressure, and speed interact in high-rate composite processing. Nomura et al. (Contribution 2) developed an electrodeposited wire-mesh grinding wheel for machining CFRP materials, emphasizing improved chip evacuation and reduced delamination during cutting. Their study highlighted the need for specialized tooling in advanced composite manufacturing. Acierno and Patti (Contribution 4) synthesize findings on material flow, thermal effects, and processing parameters to outline key challenges and opportunities in thermoplastic-based additive manufacturing using the principles and technological factors governing fused deposition modelling (FDM) of thermoplastic filaments, with particular emphasis on how rheological behaviour influences printability, process stability, and final part quality. Kulikov et al. (Contribution 15) explored the influence of epoxidized soybean oil on wood–polymer composites, demonstrating improvements in impact strength and dimensional stability, while emphasizing the potential of bio-based modifiers in sustainable composite development. In a related metallurgical context, Dudda (Contribution 16) evaluated the torsional behaviour of two steel grades at elevated temperatures, offering insights into thermomechanical stability under complex loading.

2.3. Surface Metrology, Materials Modelling, and Advanced Characterization

In his work on surface integrity, Ružbarský (Contribution 1) employed a laser profilometer to assess the surface roughness of AWJ-cut materials, highlighting the advantages of non-contact optical techniques for rapid, precise topography measurements. Complementing this, Staško et al. (Contribution 7) experimentally determined the emissivity of polished steel strips in continuous annealing lines, stressing the relevance of accurate thermal radiation data for non-contact temperature control in industrial heat treatment. Doveiko et al. (Contribution 6) advanced nano-particle metrology by combining dye fluorescence anisotropy, SAXS, and molecular dynamics to characterize silicate clusters, underscoring the need for multi-method approaches when dealing with nano-scale reactive systems. In a crystallographic study, Sidorov et al. (Contribution 12) examined homogeneous and gradient-doped $LiNbO_3$:Er^{3+} crystals, revealing how defect structures manifest in IR transmission spectra and influence proton-related conductivity. At the modelling level, Guedes and Morais (Contribution 10) compared nonlinear constitutive models for epoxy resin behaviour, demonstrating that even minimally calibrated models can maintain predictive capability across creep, relaxation, and constant strain conditions. Nandipati et al. (Contribution 5), in their article, examine the convergence of nanomanufacturing and artificial intelligence, highlighting how data-driven and machine-learning methods enhance nano-scale fabrication, process optimization, defect prediction, and material behaviour modelling, framing AI as an enabling tool for next-generation nanomanufacturing systems and intelligent material processing workflows.

2.4. Environmentally Oriented Materials and Sustainable Processing

Barbaccia et al. (Contribution 13), in their contribution to sustainable surface engineering, describe a low-temperature method for forming copper oxide films under basic conditions, emphasizing cost- and energy-efficient processing routes. Varga et al. (Contribution 14) investigated a gravitational vortex turbine manufactured from natural fibres, analysing its hydrodynamic performance and identifying manufacturing-induced imbalances as key efficiency limitations. Their work demonstrated the challenges and opportunities associated with natural-fibre-reinforced components in small-scale renewable-energy technologies. The environmentally oriented composite study of Kulikov et al. (Contribution 15) (also referenced above) similarly highlighted the role of green additives in improving material performance without compromising sustainability.

2.5. Microstructure–Property Relations in Specialized Materials

In their microstructural analysis, Jia et al. (Contribution 11) studied SWP-B music steel wire subjected to various annealing treatments, revealing how cementite spheroidization and phase orientation influence both mechanical strength and acoustic performance. Their work demonstrated the close coupling of microstructure and functional properties in specialty steels. Additionally, the modelling study by Guedes and Morais (Contribution 10) provided valuable insight into the predictive reliability of time-dependent constitutive formulations for polymers, reinforcing the role of modelling in guiding material design.

3. Discussion and Conclusions

Despite the advancements described above, several important knowledge gaps remain that limit the full realization of advanced material technologies in industrial practice. Addressing these gaps is essential for improving reliability, sustainability, and design predictability in the field of industrial materials. For example, there is a need for more accurate multi-scale process–microstructure–property models. Although significant progress has been made, current models often fail to capture complex thermal–field interactions, phase transformations, porosity evolution, or fibre–matrix interfacial behaviour under dynamic processing conditions. For example, melt pool simulations in AM still struggle with capturing transient convection effects, while composite consolidation models inadequately predict interlaminar bonding at high processing speeds. Fully integrated multi-physics, multi-scale frameworks remain largely underdeveloped.

Another need concerns the implementation of real-time, standardized in-process monitoring. While advances in acoustic emission, thermography, pyrometry, digital image correlation, and AI-driven anomaly classification are encouraging, industrial adoption is limited. Many manufacturing processes still rely on post-manufacture inspection rather than continuous monitoring. This increases scrap rates, reduces manufacturing efficiency, and prevents the use of adaptive control strategies capable of correcting process deviations in real time.

In constitutive modelling, a major constraint is the limited availability of robust datasets needed for calibrating sophisticated models. Long-term behaviour, such as viscoelastic–viscoplastic coupling, environmental degradation, fatigue under variable loading, and moisture-induced changes, is still insufficiently understood, particularly for polymers and hybrid materials. The integration of machine learning helps alleviate data scarcity, but risks creating models with limited interpretability or restricted generalizability.

A persistent challenge also exists in the area of sustainable materials, where the variability of natural fibres, limited thermal stability, and incomplete understanding of degradation mechanisms hinder broader industrial deployment. Low-temperature or energy-efficient processes show promise but remain underexplored at scale. Lifecycle assessments for emerging materials are often incomplete, preventing informed decision-making regarding environmental trade-offs.

Surface integrity and defect characterization present another unresolved issue. As materials become increasingly complex—incorporating nano-modifiers, multi-phase microstructures, or functional gradients—traditional measurement techniques may lack the spatial resolution, environmental resistance, or quantitative reliability required to characterize subtle or multi-scale defects. High-temperature, in-line, or chemically reactive environments present additional challenges. A system-level gap also lies in the lack of comprehensive integrated design frameworks. Materials science, mechanical engineering, process optimization, sustainability analysis, and lifecycle evaluation often remain siloed disciplines. Without unified methodologies, materials may perform well in laboratory conditions but fail under real-world constraints related to manufacturability, durability, or environmental exposure.

Overall, while notable progress has been achieved, future work requires deeper interdisciplinarity and broader technological integration. Advances in modelling, sensing, the development of sustainable materials, and process standardization will be essential to address these gaps and to fully exploit the potential of next-generation industrial materials.

Author Contributions: Conceptualization, J.M. and T.K.; formal analysis, J.M. and T.K.; writing—original draft preparation, J.M. and T.K.; writing—review and editing, J.M. and T.K.; supervision, T.K.; project administration, J.M. and T.K.; funding acquisition, J.M. and T.K. All authors have read and agreed to the published version of the manuscript.

Funding: This work was funded by the Ministry of Education, Science, Research, and Sport of the Slovak Republic within projects VEGA 1/0834/25, KEGA 012TUKE-4/2025 and APVV-18-0316. This research was also supported by the EU NextGenerationEU through the Recovery and Resilience Plan for Slovakia under the projects No. 09I03-03-V03-00075 and No. 09I05-03-V02-00042.

Institutional Review Board Statement: Not applicable.

Informed Consent Statement: Not applicable.

Acknowledgments: The Guest Editors wish to extend their gratitude to all authors whose contributions were accepted, as well as to those whose contributions were not included in this Special Issue. The latter were declined not as a result of a lack of scientific rigour, but rather because they did not strongly align with the Special Issue focus. The authors who demonstrated a commendable response rate, effectively implementing the constructive feedback provided by reviewers and editors, are deserving of the deepest gratitude. Their cooperative and conscientious efforts were instrumental in the advancement of the articles, thereby making a substantial contribution to the field. In addition, gratitude is extended to all reviewers who have assumed this significant responsibility and have contributed to the success of this Special Issue. Finally, the *Materials* Editorial Team merits particular recognition for their indispensable assistance in the successful realization of this Special Issue.

Conflicts of Interest: The authors declare no conflicts of interest.

List of Contributions:

1. Ružbarský, J. Roughness Control of Surfaces Using a Laser Profilometer with the Selected Material Cutting Technology. *Materials* **2023**, *16*, 4109. https://doi.org/10.3390/ma16114109.
2. Nomura, M.; Kurashige, S.; Ito, Y.; Fukuhara, Y.; Sasahara, H. Development of Electrodeposited Wire Mesh Grinding Wheel for Cutoff and Grooving Carbon Fiber Reinforced Plastic. *Materials* **2023**, *16*, 5247. https://doi.org/10.3390/ma16155247.
3. Campos, D.; Maimí, P.; Martín, A. Statistical Study of the Process Parameters for Achieving Continuous Consolidation of a Thermoplastic Composite. *Materials* **2023**, *16*, 6723. https://doi.org/10.3390/ma16206723.
4. Acierno, D.; Patti, A. Fused Deposition Modelling (FDM) of Thermoplastic-Based Filaments: Process and Rheological Properties—An Overview. *Materials* **2023**, *16*, 7664. https://doi.org/10.3390/ma16247664.
5. Nandipati, M.; Fatoki, O.; Desai, S. Bridging Nanomanufacturing and Artificial Intelligence—A Comprehensive Review. *Materials* **2024**, *17*, 1621. https://doi.org/10.3390/ma17071621.
6. Doveiko, D.; Martin, A.R.G.; Vyshemirsky, V.; Stebbing, S.; Kubiak-Ossowska, K.; Rolinski, O.; Birch, D.J.S.; Chen, Y. Nanoparticle Metrology of Silicates Using Time-Resolved Multiplexed Dye Fluorescence Anisotropy, Small Angle X-ray Scattering, and Molecular Dynamics Simulations. *Materials* **2024**, *17*, 1686. https://doi.org/10.3390/ma17071686.
7. Staško, Š.; Jablonský, G.; Varga, A.; Dzurňák, R.; Kizek, J. Experimental Measurement of Emissivity of Polished Steel Strips from a Continuous Annealing Line. *Materials* **2024**, *17*, 3084. https://doi.org/10.3390/ma17133084.
8. Kráľ, J.; Dzuro, T.; Debski, H. Applying Binder Jetting Technology to 316L Stainless Steel Materials and Testing Its Mechanical and Dimensional Properties Depending on the Printing Method. *Materials* **2024**, *17*, 4400. https://doi.org/10.3390/ma17174400.
9. Geľatko, M.; Vandžura, R.; Botko, F.; Hatala, M. Electron Beam Welding of Dissimilar Stainless Steel and Maraging Steel Joints. *Materials* **2024**, *17*, 5769. https://doi.org/10.3390/ma17235769.
10. Jia, X.; Li, Q.; Li, F.; Fang, X.; You, J.; Zhao, Q.; Wang, X.; Lu, J. Influence of Microstructure on Music Properties of SWP-B Music Steel Wire Under Different Annealing Treatments. *Materials* **2025**, *18*, 440. https://doi.org/10.3390/ma18020440.
11. Guedes, R.M.; Morais, J.L. Comparison of the Performance of Nonlinear Time-Dependent Constitutive Models Calibrated with Minimal Test Data Applied to an Epoxy Resin. *Materials* **2025**, *18*, 404. https://doi.org/10.3390/ma18020404.
12. Sidorov, N.; Pyatyshev, A.; Stroganova, E.; Galutskiy, V.; Bushunov, A.; Tarabrin, M. Features of the Defect Structure of the Compositionally Homogeneous Crystal $LiNbO_3$:Er^{3+}(3.1 wt%) and the Gradient Crystal $LiNbO_3$:Er^{3+} and Their Manifestation in the IR Transmission Spectra in the Region of Stretching Vibrations of Hydrogen Atoms of OH^--Groups. *Materials* **2025**, *18*, 579. https://doi.org/10.3390/ma18030579.
13. Barbaccia, F.I.; de Caro, T.; Federici, F.; Mezzi, A.; Sansone, L.; Giordano, M.; Macchia, A. Preparation of Copper Oxide Film at Low Temperature in Basic Conditions on a Copper Substrate. *Materials* **2025**, *18*, 1487. https://doi.org/10.3390/ma18071487.
14. Varga, M.; Velásquez, L.; Rubio-Clemente, A.; Ramón Valencia, B.; Chica, E. Experimental Analysis of Gravitational Vortex Turbine Made from Natural Fibers. *Materials* **2025**, *18*, 2352. https://doi.org/10.3390/ma18102352.

15. Kulikov, A.; Kryvolapov, D.; Sukhyy, K.; Yeromin, O.; Fedak, M.; Prokopenko, O.; Sukha, I.; Musaio, A.; Hrebik, T. Study of the Impact of Epoxidized Soybean Oil on the Characteristics of Wood-Polymer Composites. *Materials* **2025**, *18*, 2455. https://doi.org/10.3390/ma18112455.
16. Dudda, W. Mechanical Characteristics of 26H2MF and St12T Steels Under Torsion at Elevated Temperatures. *Materials* **2025**, *18*, 3204. https://doi.org/10.3390/ma18133204.
17. Stejskal, T.; Maláková, S.; Lascsáková, M.; Frankovský, P. Stiffness and Density Relationships in Additively Manufactured Structures: A Virial Theorem-Based Approach. *Materials* **2025**, *18*, 3432. https://doi.org/10.3390/ma18153432.

References

1. Kalpakjian, S.; Schmid, S.R. *Manufacturing Engineering and Technology*, 8th ed.; Pearson: London, UK, 2020.
2. Callister, W.D.; Rethwisch, D.G. *Materials Science and Engineering: An Introduction*, 10th ed.; Wiley: Hoboken, NJ, USA, 2018.
3. Ashby, M.F. *Materials Selection in Mechanical Design*, 5th ed.; Butterworth-Heinemann: Oxford, UK, 2016.
4. Gibson, I.; Rosen, D.W.; Stucker, B.; Khorasani, M. *Additive Manufacturing Technologies*, 3rd ed.; Springer: Cham, Switzerland, 2021. [CrossRef]
5. Tofail, S.A.M.; Koumoulos, E.P.; Bandyopadhyay, A.; Bose, S.; O'Donoghue, L.; Charitidis, C. Additive manufacturing: Scientific and technological challenges, market uptake and opportunities. *Mater. Today* **2017**, *21*, 22–37. [CrossRef]
6. DebRoy, T.; Wei, H.L.; Zuback, J.S.; Mukherjee, T.; Elmer, J.W.; Milewski, J.O.; Beese, A.M.; Wilson-Heid, A.; De, A.; Zhang, W. Additive manufacturing of metallic components—Process, structure and properties. *Prog. Mater. Sci.* **2018**, *92*, 112–224. [CrossRef]
7. Herzog, D.; Seyda, V.; Wycisk, E.; Emmelmann, C. Additive manufacturing of metals. *Acta Mater.* **2016**, *117*, 371–392. [CrossRef]
8. Tapia, G.; Elwany, A.H. A review on process monitoring and control in metal-based additive manufacturing. *J. Manuf. Sci. Eng.* **2014**, *136*, 060801. [CrossRef]
9. Lee, J.; Park, H.J.; Chai, S. Review on Quality control Method in metal additive manufacturing. *Appl. Sci.* **2021**, *11*, 1966. [CrossRef]
10. Cai, Y.; Xiong, J.; Chen, H.; Zhang, G. A review of in-situ monitoring and process control system in metal-based laser additive manufacturing. *J. Manuf. Syst.* **2023**, *70*, 309–326. [CrossRef]
11. Peng, X.; Kong, L.; An, H.; Dong, G. A Review of In Situ Defect Detection and Monitoring Technologies in Selective Laser Melting. *3D Print. Addit. Manuf.* **2023**, *10*, 438–466. [CrossRef] [PubMed]
12. Aydogan, B.; Chou, K. Review of in situ detection and ex situ characterization of porosity in laser powder bed fusion metal additive manufacturing. *Metals* **2024**, *14*, 669. [CrossRef]
13. Mazumdar, S.K. *Composites Manufacturing: Materials, Product, and Process Engineering*; CRC Press: Boca Raton, FL, USA, 2001.
14. Gay, D. *Composite Materials: Design and Applications*, 4th ed.; CRC Press: Boca Raton, FL, USA, 2022.
15. Soutis, C. Fibre reinforced composites in aircraft construction. *Prog. Aerosp. Sci.* **2005**, *41*, 143–151. [CrossRef]
16. Pickering, K.L.; Efendy, M.G.A.; Le, T.M. A review of recent developments in natural fibre composites and their mechanical performance. *Compos. Part A* **2016**, *83*, 98–112. [CrossRef]
17. Zhao, T. Advanced manufacturing technologies of thermoplastic composites. *Materials* **2024**, *17*, 5564. [CrossRef] [PubMed]
18. Chen, S.; Lu, W.; Guo, J.; Zhai, D.; Chen, W. Flexible and high-resolution surface metrology based on stitching interference microscopy. *Opt. Lasers Eng.* **2022**, *151*, 106915. [CrossRef]
19. Pawlus, P.; Reizer, R. Profilometric measurement of low wear: A review. *Wear* **2023**, *532–533*, 205102. [CrossRef]
20. Glatter, G.; Kratky, O. (Eds.) *Small-Angle X-Ray Scattering*; Academic Press: London, UK, 1982.
21. Egerton, R.F. *Physical Principles of Electron Microscopy: An Introduction to TEM, SEM, and AEM*, 2nd ed.; Springer: Cham, Switzerland, 2016. [CrossRef]
22. Simo, J.C.; Hughes, T.J.R. *Computational Inelasticity*; Springer: New York, NY, USA, 1998. [CrossRef]
23. Ward, I.M.; Sweeney, J. *Mechanical Properties of Solid Polymers*, 3rd ed.; Wiley: Chichester, UK, 2012.
24. Kutz, M. *Handbook of Environmental Degradation of Materials*, 3rd ed.; Elsevier: Amsterdam, The Netherlands, 2018.
25. Łatka, L.; Pawłowski, L.; Winnicki, M.; Sokołowski, P.; Małachowska, A.; Kozerski, S. Review of functionally graded thermal sprayed coatings. *Appl. Sci.* **2020**, *10*, 5153. [CrossRef]
26. Peças, P.; Carvalho, H.; Salman, H.; Leite, M. Natural fibre composites and their applications: A review. *J. Compos. Sci.* **2018**, *2*, 66. [CrossRef]
27. Sani, A.R.; Zolfagharian, A.; Kouzani, A.Z. Artificial intelligence-augmented additive manufacturing: Insights on Closed-Loop 3D Printing. *Adv. Intell. Syst.* **2024**, *6*, 2400102. [CrossRef]
28. Martelli, A.; Bellucci, D.; Cannillo, V. The role of artificial intelligence in biomaterials science: A review. *Polymers* **2025**, *17*, 2668. [CrossRef] [PubMed]

29. Acierno, D.; Patti, A. Fused Deposition Modelling (FDM) of Thermoplastic-Based Filaments: Process and Rheological Properties—An Overview. *Materials* **2023**, *16*, 7664. [CrossRef]
30. Nandipati, M.; Fatoki, O.; Desai, S. Bridging Nanomanufacturing and Artificial Intelligence—A Comprehensive Review. *Materials* **2024**, *17*, 1621. [CrossRef]

materials

MDPI

Article

Roughness Control of Surfaces Using a Laser Profilometer with the Selected Material Cutting Technology

Juraj Ružbarský

Faculty of Manufacturing Technologies, Technical University of Kosice, Sturova 31, 080 01 Presov, Slovakia; juraj.ruzbarsky@tuke.sk

Abstract: The article aims to assess the roughness of parting surfaces in the context of abrasive water jet technology for various materials. The evaluation is based on the feed speed of the cutting head, which is adjusted to achieve the desired final roughness, taking into consideration the stiffness of the material being cut. We used non-contact and contact methods to measure selected parameters of the roughness of the dividing surfaces. The study included two materials—namely, structural steel material S235JRG1 and aluminum alloy AW 5754. In addition to the above, the study involved using a cutting head with varying feed rates to achieve different surface roughness levels required by customers. The roughness parameters Ra and Rz of the cut surfaces were measured using a laser profilometer (laser profilometer). To ensure the accuracy of the laser profilometer, a control roughness measurement was conducted using a contact roughness gauge. The roughness values obtained for Ra and Rz from both measurement methods were plotted on a graph to illustrate their dependencies and were subsequently evaluated and compared. By measuring the roughness parameters Ra and Rz, the study was able to provide insights into the effectiveness of the cutting head's feed rates in achieving the desired roughness levels. Additionally, by comparing the results of the laser profilometer and contact roughness gauge, the accuracy of the measurement non-contact method used in the study was verified.

Keywords: quality; cutting; profilometer; roughness; parameter

1. Introduction

The quality of machined surfaces is constantly growing in importance and becoming increasingly demanding. Surface quality has a major impact on the functionality of the entire equipment [1].

Fine micro geometric surface irregularities deviating from the ideal surface are referred to as surface roughness, caused by the tool used in machining and finishing [2].

The surface texture of the material, which can take many forms (i.e., fine grooves, grooved structures created by machining, etc.), is gaining importance, especially in the field of surface engineering [3].

The parameter of roughness is one of many factors that define the quality of machining, the effect of which is increased productivity. In terms of mechanical vibrations, higher values of roughness are undesirable because they cause noise and dynamic stress. These result in fatigue, malfunction of the mechanical structure or its functional parts, loss of energy, or loss of performance [4,5].

Experimental analysis of surface roughness in the process of machining is important in several ways. Unwanted surface roughness may indicate damage to the functional parts of the machine tool, wear of the tools, the workpiece, the cutting head, or other parts of the machine tool. These processes then initiate damage to the machine's structure, machine stiffness, bearing components, or other machining and cutting parts of the machine [6].

By evaluating the roughness parameters, it is possible to predict the sequence of operations necessary to achieve the required surface quality and also to optimize the process of its formation [7].

The value of surface roughness is often a critical factor for products that come into contact with each other [8]. It also has a significant impact on the durability, reliability, and operation of technical equipment [9,10].

The emerging surface roughness is not only a carrier of partial information, but above all, an image of the technology used to create it [11].

With the development of science, technology, and the application of the results of this development, the importance of surface quality of components also increases, with their durability and reliability largely affected [12,13].

Surface roughness is a technical value that has a critical influence on friction in contact surfaces. The surface roughness is assessed according to the size of protrusions and depressions in the surface. At present, contact profilometers and profile graphs are mainly used for the quantitative evaluation of the surface structure. Recently, however, new methods have emerged that bring several benefits. These digital instruments analyze the surface very precisely and evaluate the surface roughness parameters based on the data obtained. These are determined by optical vision, magnifying glass, microscope, light interference, or electrical measurement [2].

One of the most recently developed approaches to measuring surface roughness is 3D profilometry using a laser, while pulsed thermography has been used to combine the effect of surface roughness with thermal radiometry [14].

Contactless measurement methods replace contact measurements of surfaces sensitive to mechanical damage, soft materials, etc. Contactless measurement avoids damaging the measured surface. The inspected surface is monitored by a focused measuring head, whose program-controlled adjustment is quick and easy [2].

The setup of most experimental assemblies involving optical 3D methods is very similar in principle. There is always an optical (light) source, a measured object, and an optical detector in the assembly. Light sources are projected via the so-called optical probe, which is an optical structure, falling on the surface of the object measured. Individual methods differ from each other in the type of this optical structure and in the way of evaluating its interaction with the measured surface [12].

Optical profilometric systems are described as systems for real-time image reading, digital signal processing, and full-color 3D profile display. Ongoing discussions address the relevant operating principles, strengths, and weaknesses of these devices, as well as the main constraints and future challenges in the field of high-speed optical profilometry [15,16].

Optical measuring devices in which the scanning tip is replaced by an electromagnetic radiation beam avoid damaging the surface of the measured part. Radiation reaching the surface can be reflected in three different ways: reflection, scattering, or a combination of the two. The interaction between the radiation and the measured sample depends primarily on the surface roughness and the radiation wavelength used [2,15].

The main advantages of these devices are contactlessness, non-destructiveness, the possibility of continuous checking, and measuring a larger part of the surface than contact methods make possible. Their disadvantage is that the measurement of surface topography is indirect. This results in the need to compare experimentally obtained results from these devices with the values obtained by the contact profilometer [17].

Lu, Yi, Liu, Wang, and Ao [18] designed a similar method of measuring surface roughness to prove that the roughness measurement process is largely influenced by a laser light source that contains two colors—i.e., red, and green—creating a relationship between the overlap index and roughness.

According to C. Kang and H.X. Liu [19] and C. Lu [20], the surface profile and roughness of a product are crucial qualitative features that often indicate technical requirements for structural components. Ensuring the desired surface quality is essential for the proper functioning of the component. Moreover, the significance of the machined surface quality and the associated demands are continually rising [21,22].

The type and degree of roughness depend on the method of machining, the physical and mechanical properties of the material being machined, the cutting conditions, and, in particular, the size of the tool feed and the cutting speed [2]. The evaluated surface roughness parameters using laser profilometry were Ra, the mean arithmetic deviation of the profile, and Rz, the greatest height of the profile unevenness. The Ra and Rz parameters are the most monitored roughness parameters in practice and can provide sufficient information about the quality of the measured surface [23]. The feed rate of the cutting head is one of the most important and technologically most easily regulated technological parameters affecting the quality parameters of Ra and Rz [2].

Laser profilometry is a form of non-contact optical profilometric measurement of the surface of a scanned object, enabling a subsequent creation of a three-dimensional model of that object. The composition of such a profilometric system (Figure 1) is based on the presence of a laser beam emitted by an optical source of electromagnetic radiation and an image-reading device in the form of a camera, usually based on a CCD or CMOS sensing chip [12].

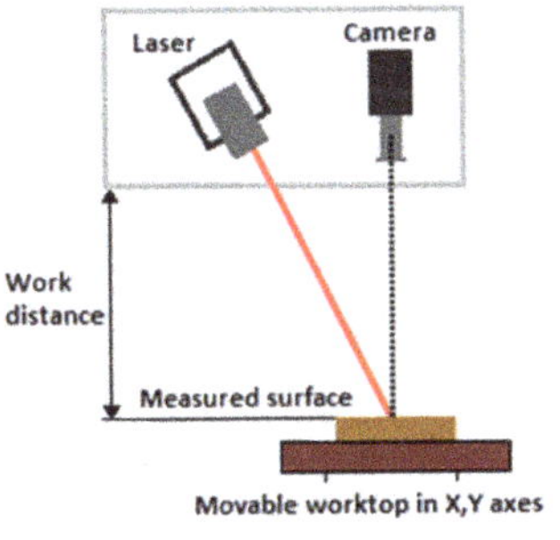

Figure 1. Principle of measuring with a laser profilometer.

Laser profilometry can evaluate not only roughness parameters such as Ra and Rz, but also other surface characteristics such as Rp, Rq, and Rv. These parameters can provide additional information about the nature of individual parts of the surface. For example, Rp can indicate the height of the highest profile protrusion, Rq can give the average quadratic deviation of the profile, and Rv can reveal the depth of the largest depression in the profile. Therefore, using a laser profilometer for surface characterization can be beneficial in understanding the surface properties and optimizing the cutting process for specific materials [24].

The objective of the experiment is to compare different methods of measuring roughness on parting surfaces of samples from different materials. The samples included structural steel S235JRG1 and aluminum alloy AW 5754, which were cut using waterjet (AWJ) technology in a manufacturing plant.

Abrasive waterjet (AWJ) technology leaves visible scoring on the cut surface, which can significantly affect the dimensional accuracy of workpieces and the quality of the finished surface. The surface after cutting consists of two different regions: a smooth zone in the top part of the cut, which is the result of cutting wear, and a rough, grooved zone in the lower part of the cut, which arises because of deformational wear during water jet cutting [25–27]. A relatively smooth area, in the top part of the cut, is the result of the cutting wear zone, and the second grooved area, in the lower part of the cut, arises because of the deformational wear during waterjet cutting [25,26]. Most research works [28,29] qualify the surface condition through roughness parameters depending on the set cutting parameters. According to Chao, Zhou, Leu, and Geskin [27], the smooth region is characterized as homogeneous with a random profile and a slightly isotropic texture. The roughness parameters Ra and Rq are commonly used to quantify the surface condition, and they show a weak dependence on the cutting speed and are almost independent of the depth of cut for the smooth zone. However, for the grooved zone, the roughness characteristics—Ra and Rq—increase sharply with the increase in the depth of cut and the speed of the water jet. The findings suggest that the machined surface is smooth with inconspicuous grooves or waviness up to a certain depth of cut, while below this depth, the surface striation is pronounced, and the cut zone is visibly divided into two parts, smooth and grooved [28,30].

In many laboratory or industrial measurement applications, it is necessary to measure and evaluate the spatial shape of objects through non-contact 3D profile measurement. Laser profilometry is intended precisely for the quick and efficient solution of these tasks. The use of spatial control of the surface offers improvement of the properties of functional surfaces and progress in production processes, e.g., energy savings, etc. Therefore, it is in the interest of the technical development of production and control technologies to create all prerequisites for the comprehensive use of spatial surface assessment in normal production and metrological practice. For the reasons mentioned above, I focused the measurement experiment on the measurements of specific parameters of the roughness of the dividing surfaces and the analysis of the relationship between the material dividing speed and the type of material that is used most often in practice. To ensure accuracy, the roughness measurement experiment was performed using a non-contact laser profilometer, which was compared to a control roughness measurement using a contact roughness meter. The advantages of the spatial evaluation of the surface structure show that these are the metrological methods of the future. The evaluated spatial parameters of the 3D texture are based on a significantly larger amount of measured data and have a much higher reliability. The real image of the 3D surface obtained in this way contains a large amount of information that can be interpreted by an experienced observer. When using the spatial assessment of the surface structure, the dangerous neglect of the influence of some of the parameters from important surface properties is also significantly reduced. The processed data from individual experimental measurements were displayed in graphical dependencies and evaluated at the end of the paper. Classic contact measurement of roughness is still at the forefront in practice due to the standardized and clear surface roughness parameters. The practical management of the transition from 2D to 3D will represent a long-term process, both with the surface measurement itself, but also with the processing of the results and with the effective use of the obtained results. The success of enforcing the spatial quality control of the surface in metrological practice will undoubtedly be decided not only by technical but also by economic considerations. The practical benefit of the experiment is, based on the analysis of the sample surfaces, a view of whether the input parameters of the machine for cutting the material are set correctly. The comparison of different roughness measurement methods made it possible to determine their advantages and disadvantages, which are described in the final part of the paper. This information can be useful for determining the most suitable roughness measurement method for a particular type of material. Additionally, the results of the experiment can provide useful information for improving the water jet cutting process and optimizing the selection of cutting parameters for these materials.

2. Materials and Methods

The conducted experiments were aimed at monitoring the differences in the roughness of the cut surface of different types of materials produced under the same conditions of material separation, but at different feed rates of the cutting head adapted to the required resulting roughness and stiffness of the material. The tested samples were made of structural steel and aluminum alloy, and the roughness of the cut surface was measured in a non-contact and contact manner. When applying both methods, the cut areas measured were the same. The roughness parameters of the samples were measured and evaluated using both a contactless laser profilometer manufactured by KVANT and a contact roughness meter Surftest SJ-400 by the company Mitutoyo. The use of two different methods for roughness measurement allowed for a comparison of the results and helped to ensure the accuracy of the measurements.

The actual production of samples consisted of cutting the material using the AWJ technology, which was done directly in the production plant of DRC s.r.o where the cutting machine WJ4020 1Z-C0-P60 COBRA (Figure 2) was used to separate the material, together with the high-pressure pump from FLOW SYSTEMS.

Figure 2. Cutting machine WJ4020 1Z-C0-P60 COBRA.

Table 1 is a technical specification table that shows the parameters of the cutting machine used to prepare the experimental samples.

Table 1. Specification of the cutting machine WJ4020 1Z-C0-P60 COBRA.

Parameter	Value
Working speed (max)	20,000 mm/min
Accuracy	+/−0.04 mm/300 mm
Repeatable accuracy	+/−0.03 mm
Speed (max)	30,000 mm/min
Movement in the Z-axis	500–700 mm
Cutting heads (No)	1–4
Portal width (max)	4 m
Length of linear guide (max)	30,000 mm

These parameters play an important role in determining the quality of the cut surfaces and can also affect the roughness of the cut surfaces.

The analyzed materials (structural steel S235JRG1 and aluminum alloy AW 5754) were selected based on several factors, including their extensive usability, availability in the industry, and the variety of properties they exhibit.

Demonstration of the (nomenclature) marking:

- FS050—structural steel S235JRG1—cutting head feed rate (50 mm/min)
- FS100—structural steel S235JRG1—cutting head feed rate (100 mm/min)
- FS150—structural steel S235JRG1—cutting head feed rate (150 mm/min)
- AA120—aluminum alloy AW 5754—cutting head feed rate (120 mm/min)
- AA220—aluminum alloy AW 5754—cutting head feed rate (220 mm/min)
- AA370—aluminum alloy AW 5754—cutting head feed rate (370 mm/min)

Six samples were produced using the AWJ technology, as shown in Figure 3. Three of the samples were made of structural steel (FS050, FS100, FS150), and the other three were made of aluminum alloy (AA120, AA220, AA370). Each sample was produced using a different feed rate of the cutting head during the material separation process.

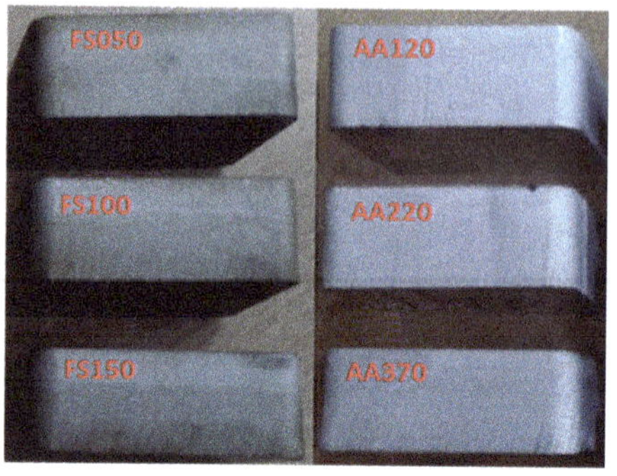

Figure 3. Marking of used samples from different materials.

It is common for production companies to provide roughness specifications for parts or surfaces that they manufacture using various cutting or machining processes. These specifications provide customers with an indication of the quality of the finished product and help to ensure consistency in the manufacturing process. It is important to note, however, that the actual roughness values obtained during the manufacturing process can depend on a variety of factors the material properties of the workpiece and the accuracy of the cutting parameters used. The production company guarantees the roughness values (6.3; 12.5; 25 μm) for the cut surfaces only by changing the feed speed of the cutting head for different types of materials using the AWJ technology.

By providing roughness guarantees for parts produced using this technology, the production company can assure their customers of the quality of the finished product and demonstrate their commitment to meeting specific performance requirements.

Table 2 shows the input parameters set for cutting the material by the AWJ technology, constant for all samples produced as opposed to the changing speed of the cutting head.

Table 2. Input parameter setting of the material cutting machine using the AWJ technology.

Parameter	Value
Type of abrasive	Australian Grenade
Amount of abrasive (g/min)	430
Diameter of the guide tube (mm)	0.889
Diameter of the water jet (mm)	0.406
Abrasive grain size	MESH 80
Pressure (MPa)	379.21

The laser profilometer assembly is composed of both basic and additional parts, as shown in Figure 4. The basic parts consist of a sturdy aluminum support structure made up of components that allow for the vertical positioning of the measuring tube. The assembly also includes a worktop mounted on stepper motors that provide movement in the X and Y axes, a laser beam source, a lens, and a camera equipped with a CMOS (complementary metal oxide semiconductor) sensor and backlight. Additional parts that make up the assembly include a computer that runs the laser profilometer View control and evaluation software, as well as an image splitter.

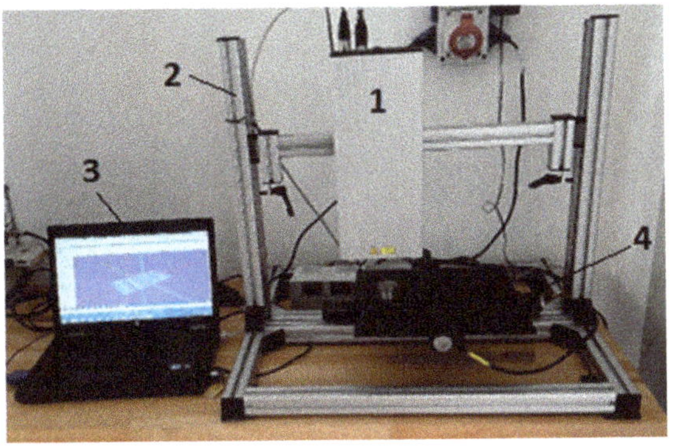

Figure 4. The laser profilometer setup. 1—Tube with laser and camera; 2—laser profilometer frame; 3—PC with operating and evaluation software; 4—worktop with step motors with movable in the X and Y axes.

The X- and *Y*-axis automated micrometric horizontal feed of the sample worktop was ensured by 2 Standa 8MT16-30 stepper motors with a maximum limit of up to 300 mm on each axis. The feed in the Z axis was resolved by the mechanical lifting of the entire profilometric tube using locking pins. The optical part of the laser profilometer system consists of an AVT Marlin 131B camera and a Tamron 23F50SP 50 mm lens with a visible area of 154 mm^2 (22 × 7 mm). The frame design of the system allows the measurement of samples weighing up to a maximum of 8 kg with a positioning accuracy of 2.5 μm per step, with each step consisting of 8 μm. The camera sensor resolution is 0.02 mm/pixel.

When measured by the laser profilometer non-contact system, the experimental samples on the worktop were first placed in an anti-vibration mass. The first reason was that the machined surface of the sample was imperfectly fitted to the laser profilometer worktop and the sample on the laser profilometer worktop shifted during the movement of the stepper motors during the measurement. The second reason was to eliminate the influence of ambient vibrations, which can affect the quality of the measured data.

The experimental laser profilometer system allows for the measurement and evaluation of surface roughness parameters of various samples. The roughness parameters that can be measured using this system include Rq, Rv, Rz, Ra, and Rp. The roughness data collected from this system can be exported in CSV format, which is highly suitable for further processing of experiments in the form of raw data.

The Matrox Triple Head 2Go—Digital Edition image splitter ensures high-quality processing and display of images that include measured data. This device allows for the connection of up to three monitors to a single graphical output from a computer. In addition, by using the enhanced desktop option and an additional graphical output, up to four independent monitors can be obtained.

The main principle of the laser profilometer system consists of using the well-known so-called triangulation principle, where the light source, together with the camera sensor and the examined object, forms the so-called triangulation triangle (Figure 5). A thin laser line is projected onto the measured profile; the image of the laser line is captured at an angle by a digital camera placed perpendicularly above the scanned surface; the profile of the object in the cross-section specified by the laser line is then evaluated from the scanned image. In this way, it is possible to create a real 3D profile from each scanned image. This contactless profilometer cooperates with a classic computer via the laser profilometer View profilometric software LPMView with the user environment. Laser profilometer View is used for communication with a linear laser profilometer, evaluating the data captured by the camera sensor, which is then processed in the desired form. FireWire [2] is required for communication.

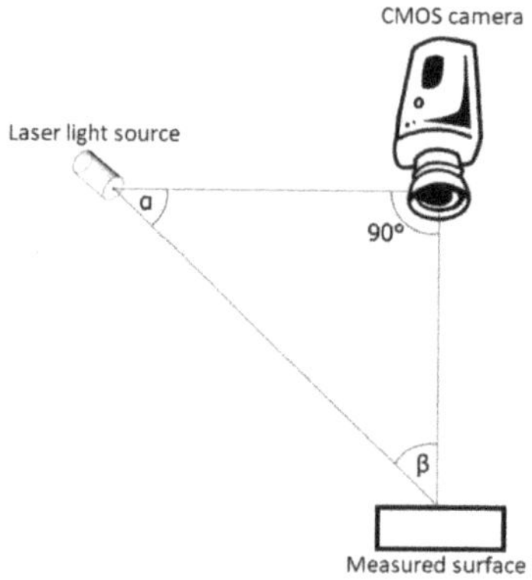

Figure 5. The principle of active triangulation.

As mentioned, to verify the correctness and compare the measured values with the contactless laser profilometer, the measurement of samples was also carried out in a contact way, using the Mitutoyo Surftest SJ-400 roughness meter (Figure 6). Table 3 is the listed parameters of the contact roughness meter.

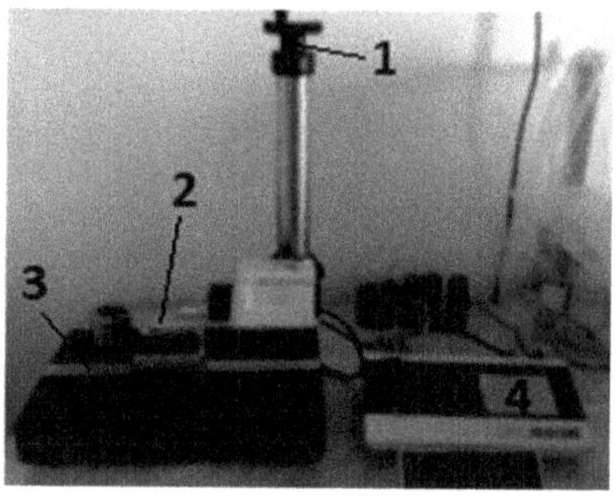

Figure 6. Surftest SJ-400 contact roughness meter. 1—Adjusting the standoff distance of the measuring tip; 2—measuring tip; 3—worktop; 4—touch display with the printer.

Table 3. Technical parameters of the Surftest SJ-400 roughness meter.

Parameters	Value
Speed of return (mm/s)	0.5; 1.0; 2.0
Positioning	±1.5° (inclination); 10 mm (up/down)
Evaluated parameters	0.889
Measuring speed (mm/s)	0.05; 0.1; 0.5; 1.0
Measurement range/resolution (μm)	P (primary); R (roughness); W (filtered waviness)
Cutoff length (mm)	0.08; 0.25; 1.8; 2.5; 8

For each sample, the same section of the measured surface was chosen for the application of both measurement methods. The materials used for the samples were defined by the relevant standards as structural steel S235JRG1 and aluminum alloy EN AW 5754.

Experimental Samples Measurement

Measurement of the roughness of the cut surfaces, created in the process of material separation by water jet technology (AWJ), was carried out on two types of materials. The measured samples were of the same thickness; however, they were made at different feed rates of the cutting head. Each material was used to create three samples.

The measurement of the roughness parameters Ra and Rz of the cut surface of the sample, when measured with the laser profilometer, was determined on each sample at ten lines of the cut surface measurement (Figure 7, right-hand side)—specifically, at the measurement lines 0.11; 2.31; 4.51; 6.71; 8.91; 11.11; 13.31; 15.51; 17.71; 19.91 mm from the bottom of the sample. The thickness of the sample material measured with both methods was the same.

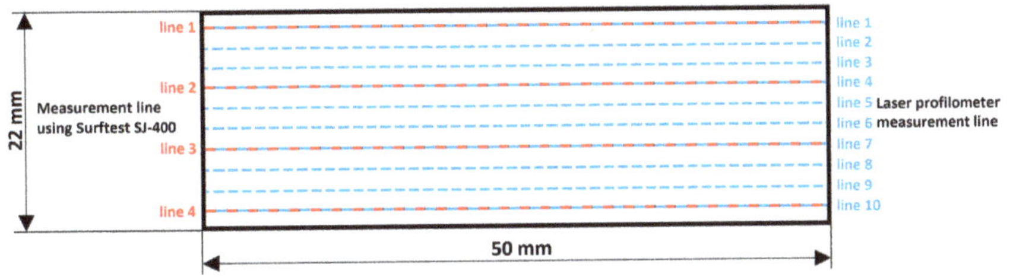

Figure 7. Lines of measurement are specified on individual samples.

Using the contact method—i.e., the Surftest SJ-400—the roughness parameters of the cut surface were determined in four measurement lines of the cut surface (Figure 7, left-hand side)—specifically, at the measurement lines of 0.11; 6.71; 13.31; 19.91 mm from the bottom of the sample. The number of measurements was reduced due to this method's complexity and difficulty of measurement.

Table 4 displays the values measured with the Surftest SJ-400. These values serve as a comparison to the values measured with the laser profilometer and to verify the accuracy of the measurements.

Table 4. Measured surface roughness parameters of the cut surface using Surftest SJ-400 in individual measurement lines.

Declared Roughness by the Company	Measurement Line	Structural Steel S235JRG1		Aluminum Alloy AW 5754	
		Ra [μm]	Rz [μm]	Ra [μm]	Rz [μm]
6.3	1	2.61	16.4	2.18	13.9
	2	2.88	15.5	3.35	16.2
	3	3.95	17.5	4.12	17.2
	4	4.36	19.5	5.74	25.8
12.5	1	2.53	14.7	3.59	18.9
	2	3.32	20.5	5.27	23.9
	3	5.96	33.2	8.71	39.2
	4	9.42	42.2	9.38	41.2
25	1	3.45	19.4	6.25	20.7
	2	4.25	28.3	8.72	41.5
	3	10.89	56.4	14.12	64.9
	4	15.73	59.4	21.34	74.8

3. Results and Discussion

When assessing the roughness of the surface, the Ra and Rz parameters were used, representing the arithmetic average roughness and the maximum height of the roughness profile, respectively. The surface of the samples was measured non-contacting in 181 steps with a step size of 0.11 mm at 10 measurement lines, with a step shift of 2 mm from the top of the sample. The Gain mode, which amplifies the image signal in the camera, was set to 1. The cutoff time, which determines the time for scanning the surface, was set to 19.520 ms. The input parameters of the laser profilometer were adjusted based on previous test measurements to provide the purest image preview and the least noise in the image.

In this section of the paper, the roughness data obtained from the surface of a sample that wares produced using the AWJ technology were evaluated using the laser profilometer.

The data obtained for the sample material, varying with the feed rate of the cutting head, were plotted concerning the roughness parameters, Ra and Rz. The surface roughness was measured at ten different lines on three structural steel samples and three aluminum alloy samples. The input parameters used for setting up the laser profilometer to measure surface roughness are presented in Table 5. These parameters were consistent with those observed in previous studies by Ružbarský [2], Srivastava et al. [31], and Valicek et al. [32].

Table 5. Input Parameters for Sample Measurement with Laser Profilometer Setting.

Scan Distance—Axis Y	Scan Line—Axis X
Scan lines count 200 µm Scan lines step 110 µm	4000 µm

The measured surface roughness values of the individual samples were processed, resulting in the dependence of the mean arithmetic deviation of the Ra profile unevenness on the measurement line on the sample h at the feed rate of the cutting head FS050 mm/min, FS100 mm/min, FS150 mm/min, AA120 mm/min, AA220 mm/min, and AA370 mm/min, and the dependence of the greatest elevation of the unevenness of the Rz profile on the measurement line on the sample h at the feed rate of the cutting head FS050 mm/min, FS100 mm/min, FS150 mm/min, AA120 mm/min, AA220 mm/min, and AA370 mm/min. The results are also confirmed by the conclusions reached by Azhari et al. [33] and Jurko et al. [34].

The plotted dependence of the parameter Ra (Figure 8) shows the surface roughness at the measurement lines of the sample at the feed rates of the cutting head 50, 100, and 150 mm/min for structural steel, and 120, 220, and 370 mm/min for aluminum alloy, respectively. These results are also confirmed by the conclusions reached by Azhari et al. [35] and Foldyna et al. [36].

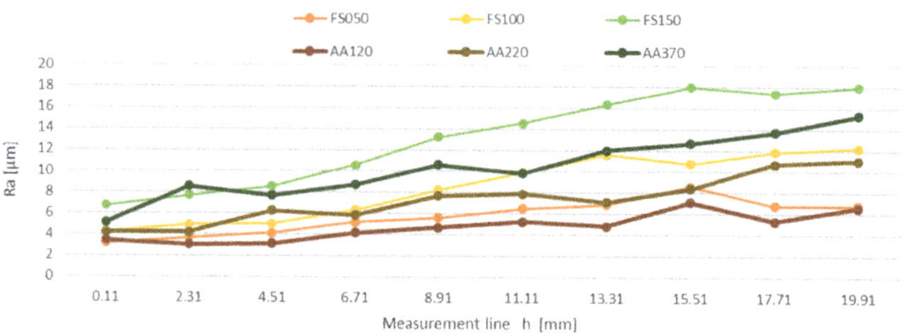

Figure 8. Plotted dependence of Ra on the measurement line for the materials FS050, FS100, FS150, and for the materials AA120, AA220, and AA370.

The plotted dependence of the Ra parameter (Figure 9) shows the surface roughness at the measurement lines of the sample at the feed rates of the cutting head 50, 100, and 150 mm/min for structural steel, and 120, 220, 370 mm/min for aluminum alloy. These results are also confirmed by the conclusions reached by Akkurt [37], Hreha et al. [38], and Boud et al. [39].

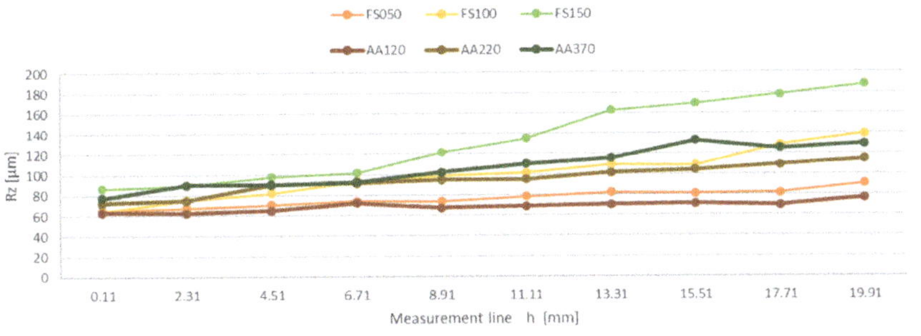

Figure 9. Plotted dependence of Rz on the measurement line for the materials FS050, FS100, FS150, and for the materials AA120, AA220, and AA370.

The curves of the monitored parameters Ra and Rz show an increasing trend on all samples depending on the line of measurement.

Figure 10 shows 3D models of scanned samples made of structural steel S235JRG1, which were created by placing a series of measured profiles side by side via a laser profilometer. The models show the 3D surface of the cut material of experimental samples captured by the laser profilometer. Figures show a clear impact of the cutting head speed on the surface quality. A greater difficulty of the cutting water jet passing through the material is visible on experimental samples (FS050, FS120, FS150), and the water jet left more depressions and unevenness on the surface of the cut areas. The comparison of the plotted dependencies of the parameters Ra and Rz (Figures 8 and 9) for structural steel shows that the roughness values on the cut surface are more stepped, which reflects the greater strength and lower toughness of the evaluated material. The surface roughness of the cut areas in these experimental samples is greater across the entire cut surface than that of the aluminum alloy. These results are also confirmed by the conclusions reached by Tomas et al. [40], Kunaporn et al. [41], and Hascalik et al. [42].

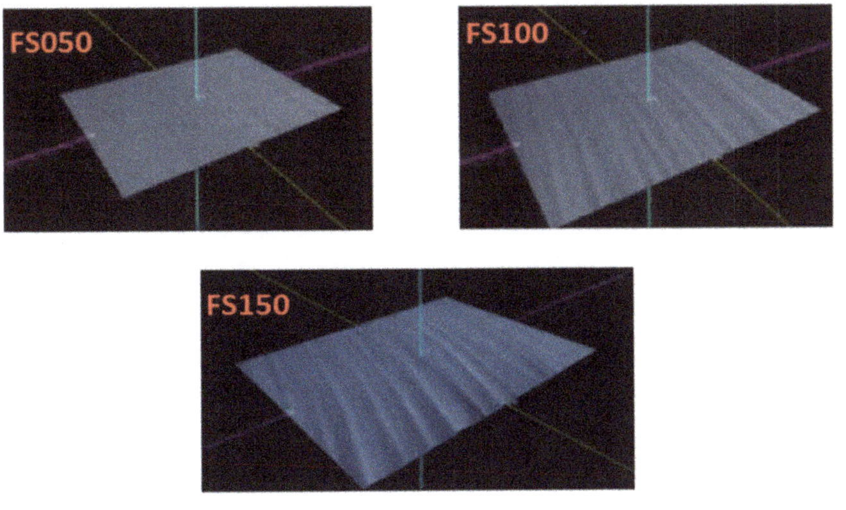

Figure 10. 3D model of scanned S235JRG1 samples.

Figure 11 shows 3D models of scanned samples made of AW 5754 aluminum alloy, which were created by placing a series of measured profiles side by side via a laser profilometer. The figures also show the effect of the cutting head speed on the surface quality of the aluminum alloy. A comparison of the created 3D models (Figures 10 and 11) shows a clear difference in the surface quality of the samples of structural steel and aluminum alloy. Samples made of aluminum alloy (AA120, AA220, AA370) show a smoother cut across the entire profile of the cut surface, where the passage of the water jet through the material did not leave deep depressions on the surface of the cut areas, unlike on samples that were made of structural steel. The comparison of plotted dependencies of the parameters Ra and Rz (Figures 8 and 9) for aluminum alloy shows that the roughness values of the cut surface are less stepped, which reflects the lower strength and greater toughness of the materials evaluated. These results are also confirmed by the conclusions reached by Akkurt et al. [43] and Monno et al. [44].

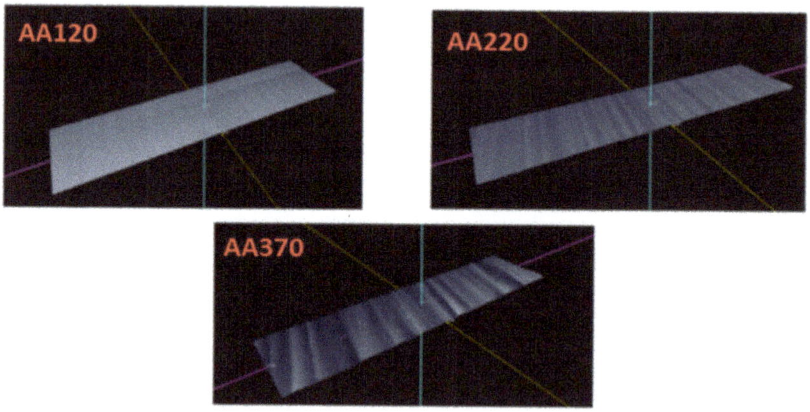

Figure 11. 3D model of scanned EN 5754 samples.

In the non-contact measurement method using laser profilometry, the surface area of the cut sample was divided into ten measurement lines, and for the contact measurement method using Surftest SJ-400, the surface area of the cut sample was divided into four measurement lines (Figure 7). To verify the correctness and compare the values measured with the laser profilometer, all samples were first measured with the contact Surftest SJ-400 roughness meter (Table 4). The mutual comparison of the measured data shows that the values of the Ra parameter measured on the laser profilometer are within the range of values that were measured with the contact Surftest SJ-400 roughness meter. The values of the Rz parameter that were measured with the contact Surftest SJ-400 roughness meter are two-to-three times smaller than the values measured with the laser profilometer. Similar values and conclusions were reached by the authors Krenicky [1] and Hlavac et al. [11]. Krenicky [1], Mital et al. [17], and Ružbarský [45] assumed the reason might be that the contact roughness meter's tip diameter could not enter every surface recess (the top of the measuring tip is of a diameter larger than some surface recesses) and partly due to the image noise during measurements with the laser profilometer, where in some measurement lines there was a reflection of laser light from the surface of the sample to the CMOS camera, which resulted in a sharp increase in Rz values and thus a measurement error. This measurement error can be corrected by removing such distorted values from the exported data.

4. Conclusions

Roughness control using a laser profilometer can provide valuable information about the surface quality of the material after the cutting process. The conducted experiments suggest that in praxis, crucial factors that affect the surface roughness parameters Ra and Rz is the AWJ cutting technology, the feed rate of the cutting head, and the type of material. The results of the experiment can be summarized in the following points:

- All measured dependencies show an increasing trend of the selected roughness parameter values, and the surface roughness varies linearly, with the depth of the cut done by the AWJ cutting technology.
- The smooth zone occurs mainly in the first part of the cut of the separated area. This is also confirmed by the plotted dependencies Ra and Rz, where we measured the smallest initial surface roughness.
- In the smooth zone of the separated surface, the measured values of the Ra and Rz parameters are closest to each other, but even at these values, there is an obvious difference in the roughness of the surface of the cut area depending on the speed of the cutting head.
- With a decrease in the feed rate of the cutting head, the size of the smooth zone (characterized by lower roughness parameters) increases.
- The linearly increasing roughness of the cut surfaces applies to the smooth and medium-smooth zones of the cut surfaces. An exponential increase in the roughness of the cut surfaces occurs in the middle and end zones of the cut surfaces.
- The occurrence of visible reliefs on the cut surfaces was less common in aluminum samples than in structural steel samples. It follows from the above that to achieve the same surface roughness in different materials, a different cutting head speed must be used.
- This non-contact method of measuring surface roughness enables checks in real-time, which is a very important advantage in terms of overall production efficiency.

The results of the experiment indicated that the surface roughness of the samples produced by the DRC company did not exceed the declared roughness of any of the samples used—i.e., the input parameters of the material cutting machine are set correctly. However, the Surftest SJ-400 contact roughness meter used to measure the roughness parameters did not provide a complete picture of the roughness on the surface being assessed or the roughness in different parts of the surface under evaluation.

To summarize, the laser profilometer provides a set of values that describe the surface areas at each step of measurement, allowing for the evaluation of multiple parameters and creating a 3D image of the surface being measured for better visualization of the surface characteristics. Overall, the experiment provides valuable insights into the comparison of different roughness measurement methods and their suitability for specific applications.

A certain disadvantage is glossy or highly reflective surfaces, which can cause issues when measuring surface roughness parameters using contact-free methods. The laser light can reflect off the surface, creating unwanted reflections that may be captured by the CMOS camera and these reflections can lead to inaccurate measurements of surface roughness parameters. In such cases, contact methods may be more appropriate for obtaining accurate measurements. In summary, while contact-free methods, such as laser profilometry, can provide accurate measurements of surface roughness parameters on many surfaces, highly reflective surfaces can pose challenges.

Funding: This research was funded by the Cultural and Educational Grant Agency of the Ministry of Education, Science, Research and Sport of the Slovak Republic (KEGA), grant number 015TUKE-4/2022.

Institutional Review Board Statement: Not applicable.

Informed Consent Statement: Not applicable.

Data Availability Statement: Not applicable.

Acknowledgments: The author would like to thank the KEGA grant agency for supporting research work within the project 015TUKE-4/2022.

Conflicts of Interest: The authors declare no conflict of interest. The funders had no role in the design of the study; in the collection, analyses, or interpretation of data; in the writing of the manuscript; or in the decision to publish the results.

References

1. Krenický, T. Non-contact Study of Surfaces Created Using the AWJ Technology. *Manuf. Technol.* **2015**, *15*, 61–64. [CrossRef]
2. Ružbarský, J. *Contactless System for Measurement and Evaluation of Machined Surfaces*; Springer Nature: Cham, Switzerland, 2022; p. 97. ISBN 978-3-031-08980-0.
3. Bennett, J.M. Review Article: Recent Developments in Surface Roughness Characterization. *J. Meas. Sci. Technol.* **1992**, *3*, 1119–1127. [CrossRef]
4. Bozek, P.; Nikitin, Y.; Krenicky, T. Model systems for diagnostics of mechatronic. In *Diagnostics of Mechanical Systems*; Studies in Systems, Decision and Control; Springer: Cham, Switzerland, 2021; Volume 345, pp. 27–61. ISBN 978-3-030-67055-9.
5. Gajdoš, I.; Spišák, E.; Kaščák, L.; Krasinskyi, V. Surface Finish Techniques for FDM Parts. *Mater. Sci. Forum* **2015**, *818*, 45–48. [CrossRef]
6. Panda, A.; Olejárová, Š.; Valíček, J.; Harničárová, M. Monitoring of the condition of turning machine bearing housing through vibrations. *Int. J. Adv. Manuf. Technol.* **2018**, *97*, 401–411. [CrossRef]
7. Vikram, G.; Babu, N.R. Modelling and analysis of abrasive water jet cut surface topography. *Int. J. Mach. Tools Manuf.* **2002**, *42*, 1345–1354. [CrossRef]
8. Yashchuk, V.V.; Tyurin, Y.N.; Tyurina, A.Y. Modeling of surface metrology of state-of-the-art x-ray mirrors as a result of stochastic polishing process. *Opt. Eng.* **2016**, *55*, 74106. [CrossRef]
9. Wang, W.; Zhang, S.; Li, Y. Surface microstructure profilometry based on laser confocal feedback. *Rev. Sci. Instrum.* **2015**, *86*, 103108. [CrossRef] [PubMed]
10. Chen, L.C.; Nguyen, X.L.; Huang, H.S.; Chen, J.L. Dynamic Surface Profilometry and Resonant-Mode Detection for Microstructure Characterization Using Nonconventional Stroboscopic Interferometry. *IEEE Trans. Ind. Electron.* **2010**, *57*, 1120–1126. [CrossRef]
11. Hlaváč, L.M.; Kocich, R.; Gembalová, L.; Jonšta, P.; Hlaváčová, I.M. AWJ cutting of copper processed by ECAP. *Int. J. Adv. Manuf. Technol.* **2016**, *86*, 885–894. [CrossRef]
12. Ružbarský, J.; Mital, G. Diagnostics of selected surface characteristics with laser profilometry. *MM Sci. J.* **2018**, *2018*, 2140–2143. [CrossRef]
13. Jeyapoovan, T.; Murugan, M. Surface roughness classification using image processing. *Measurement* **2013**, *46*, 2065–2072. [CrossRef]
14. Stemp, W.; Stemp, M. UBM Laser Profilometry and Lithic Use-Wear Analysis: A Variable Length Scale Investigation of Surface Topography. *J. Archaeol. Sci.* **2001**, *28*, 81–88. [CrossRef]
15. Van der Jeught, S.; Dirckx, J.J. Real-time structured light profilometry: A review. *Opt. Lasers Eng.* **2016**, *87*, 18–31. [CrossRef]
16. Akram, M.; Bashir, S.; Rafique, M.S.; Hayat, A.; Mahmood, K. Laser Induced Surface Morphology of Molybdenum Correlated with Breakdown Spectroscopy. *Plasma Chem. Plasma Process.* **2016**, *37*, 287–304. [CrossRef]
17. Mital, G.; Spisak, E.; Mulidran, P.; Kascak, L.; Jezny, T. 3D Evaluation of the topography of the surface by abrasive water jet machining technology. *MM Sci. J.* **2021**, *2021*, 4847–4852. [CrossRef]
18. Liu, J.; Lu, E.; Yi, H.; Wang, M.; Ao, P. A new surface roughness measurement method based on a color distribution statistical matrix. *Measurement* **2017**, *103*, 165–178. [CrossRef]
19. Kang, C.; Liu, H. Small-Scale Morphological Features on a Solid Surface Processed by High-Pressure Abrasive Water Jet. *Materials* **2013**, *6*, 3514–3529. [CrossRef]
20. Lu, C. Study on prediction of surface quality in machining process. *J. Mater. Process. Technol.* **2008**, *205*, 439–450. [CrossRef]
21. Kumar, B.M.; Ratnam, M.M. Machine vision method for non-contact measurement of surface roughness of a rotating workpiece. *Sens. Rev.* **2015**, *35*, 10–19. [CrossRef]
22. Jin, G.-C.; Bao, N.-K. Surface detection and 3D profilometry for microstructure using optical metrology. *Opt. Lasers Eng.* **2001**, *36*, 1–9. [CrossRef]
23. Mital, G.; Dobránsky, J.; Ružbarský, J.; Olejárová, Š. Application of Laser Profilometry to Evaluation of the Surface of the Workpiece Machined by Abrasive Waterjet Technology. *Appl. Sci.* **2019**, *9*, 2134. [CrossRef]
24. Hirabayashi, A.; Ogawa, H.; Kitagawa, K. Fast surface profiler by white-light interferometry by use of a new algorithm based on sampling theory. *Appl. Opt.* **2002**, *41*, 4876–4883. [CrossRef]
25. Dadvandipour, S. *Water Jet Cutting*; Interné materiály Bay-Zoltán Foundation for Applied Research: Miskolc-Egyetemváros, Hungary, 1996.
26. Hashish, M. Material Properties in Abrasive-Waterjet Machining. *J. Eng. Ind.* **1995**, *117*, 578–583. [CrossRef]
27. Chao, J.; Zhou, G.; Leu, M.C.; Geskin, E. Characteristics of Abrasive Waterjet Generated Surfaces and Effects of Cutting Parameters and Structure Vibration. *ASME J. Eng. Ind.* **1995**, *117*, 516–525. [CrossRef]
28. Momber, A. Aktuelle Probleme der Abrasiv-Wasserstrahl-Bearbeitung. In *wt spezial*; Warnecke, H.J., Ed.; Springer: Berlin/Heidelberg, Germany, 1991; pp. 437–441.
29. Warnecke, H.J.; Przyklenk, K.; Schaltter, M. *Eintgraten durch Hochdruckwasserstrahlen*; Springer: Berlin/Heidelberg, Germany, 1985; Volume 75, pp. 425–429.
30. Kovacevic, R.; Hashish, M.; Mohan, R.; Ramulu, M.; Kim, T.J.; Geskin, E.S. State of the Art of Research and Development in Abrasive Waterjet Machining. *ASME J. Manuf. Sci. Eng.* **1997**, *119*, 776–784. [CrossRef]

31. Srivastava, M.; Hloch, S.; Krejci, L.; Chattopadhyaya, S.; Dixit, A.R.; Foldyna, J. Residual stress and surface properties of stainless steel welded joints induced by ultrasonic pulsed water jet peening. *Measurement* **2018**, *127*, 453–462. [CrossRef]
32. Valíček, J.; Harničárová, M.; Öchsner, A.; Hutyrová, Z.; Kušnerová, M.; Tozan, H.; Michenka, V.; Sepelak, V.; Mital, D.; Zajac, J. Quantifying the Mechanical Properties of Materials and the Process of Elastic-Plastic Deformation under External Stress on Material. *Materials* **2015**, *8*, 7401–7422. [CrossRef] [PubMed]
33. Azhari, A.; Schindler, C.; Kerscher, E.; Grad, P. Improving surface hardness of austenitic stainless steel using waterjet peening process. *Int. J. Adv. Manuf. Technol.* **2012**, *63*, 1035–1046. [CrossRef]
34. Jurko, J.; Panda, A.; Valicek, J.; Harnicarova, M.; Pandaova, I. Study on Cone Roller Bearing Surface Roughness Improvement and the E_ect of Surface Roughness on Tapered Roller Bearing Service Life. *Int. J. Adv. Manuf. Technol.* **2016**, *82*, 1099–1106. [CrossRef]
35. Azhari, A.; Schindler, C.; Li, B. Effect of waterjet peening on aluminum alloy 5005. *Int. J. Adv. Manuf. Technol.* **2012**, *67*, 785–795. [CrossRef]
36. Foldyna, J.; Sitek, L.; Scucka, J.; Martinec, P.; Valicek, J.; Palenikova, K. E_ects of Pulsating Water Jet Impact on Aluminium Surface. *J. Mater. Process Technol.* **2009**, *209*, 6174–6180. [CrossRef]
37. Akkurt, A. Surface Properties of the Cut Face Obtained by Di_erent Cutting Methods from AISI 304 Stainless Steel Materials. *Indian J. Eng. Mater. Sci.* **2009**, *16*, 373–384.
38. Hreha, P.; Radvanska, A.; Knapcikova, L.; Królczyk, G.M.; Legutko, S.; Królczyk, J.B.; Hloch, S.; Monka, P. Roughness Parameters Calculation By Means Of On-Line Vibration Monitoring Emerging From AWJ Interaction With Material. *Metrol. Meas. Syst.* **2015**, *22*, 315–326. [CrossRef]
39. Boud, F.; Carpenter, C.; Folkes, J.; Shipway, P. Abrasive waterjet cutting of a titanium alloy: The influence of abrasive morphology and mechanical properties on workpiece grit embedment and cut quality. *J. Mater. Process. Technol.* **2010**, *210*, 2197–2205. [CrossRef]
40. Tomáš, M.; Džupon, M.; Evin, E.; Spišák, E. Surface Analysis of Uncoated and PVD Coated Punch at the Hole-Flanging Process. *Metals* **2018**, *8*, 218. [CrossRef]
41. Kunaporn, S.; Chillman, A.; Ramulu, M.; Hashish, M. Eject of Waterjet Formation on Surface Preparation and Profiling of Aluminum Alloy. *Wear* **2008**, *265*, 176–185. [CrossRef]
42. Hascalik, A.; Caydas, U.; Gurun, H. E_ect of Traverse Speed on Abrasive Waterjet Machining of Ti-6Al-4V Alloy. *Mater. Des.* **2007**, *28*, 1953–1957. [CrossRef]
43. Akkurt, A.; Kulekci, M.K.; Seker, U.; Ercan, F. Eject of Feed Rate on Surface Roughness in Abrasive Waterjet Cutting Appli-cations. *J. Mater. Process Technol.* **2004**, *147*, 389–396. [CrossRef]
44. Monno, M.; Ravasio, C. The E_ect of Cutting Head Vibrations on the Surfaces Generated by Waterjet Cutting. *Int. J. Mach. Tools Manuf.* **2005**, *45*, 355–363. [CrossRef]
45. Ružbarský, J. The Difficulty of Measuring the Roughness of Glossy Surfaces Using the Triangulation Principle. *Appl. Sci.* **2023**, *13*, 5155. [CrossRef]

materials

MDPI

Article

Development of Electrodeposited Wire Mesh Grinding Wheel for Cutoff and Grooving Carbon Fiber Reinforced Plastic

Mamoru Nomura [1,2], Shuji Kurashige [1], Yukio Ito [2], Yoshiya Fukuhara [3] and Hiroyuki Sasahara [1,*]

[1] Department of Mechanical Systems Engineering, Tokyo University of Agriculture and Technology, Tokyo 184-8588, Japan; mamoru.nomura@ibarakiseito.com (M.N.); kurashu004427@gmail.com (S.K.)

[2] IBARAKI GRINDING WHEEL Co., Ltd., Ibaraki 300-2521, Japan; jms007@qc.commufa.jp

[3] Mitsubishi Heavy Industry, Hyogo 676-0008, Japan; yoshiya.fukuhara.25@mhi.com

* Correspondence: sasahara@cc.tuat.ac.jp; Tel.: +81-42-388-7417

Abstract: Carbon fiber reinforced plastic (CFRP) is used in various industries because of its high specific strength, but it is well known as a difficult material to cut. In this study, we developed a disc-shaped electrodeposited diamond wire mesh grinding wheel as a new method for cutoff and grooving with a large aspect ratio for CFRP. We confirmed that this tool could be used for machining at a feed rate of 1000 mm/min, equivalent to that of an abrasive waterjet. This tool discharges generated chips through the spaces in the wire mesh, preventing clogging and thereby enabling the suppression of machining temperature. No burrs or delamination were observed on the surface machined with the wire mesh grinding wheel, and the surface roughness was $Ra = 2.76$ μm. However, the groove width was larger than the wheel thickness due to the runout of the wheel. Additionally, the moderate elasticity and durability of the tool suggest that it might extend tool life by avoiding the crushing of abrasive grains.

Keywords: CFRP; wire mesh grinding wheel; clogging; temperature; surface roughness; diamond abrasive grain

1. Introduction

Carbon fiber reinforced plastic (CFRP) is widely used in various industries due to its high specific strength and high specific modulus. Demand for CFRP is increasing, especially in the aerospace industry [1,2]. However, CFRP is known as a difficult material to cut, and aircraft components require a lot of cutoff and grooving operations. Cutting with an end mill, which is a common machining method, has the following problems: machining efficiency cannot be increased to prevent delamination of CFRP fibers [3–7], and tool life is short due to high wear rates [8–12]. Abrasive waterjet processing has also been used [13–15], but it has the problem of high initial investment, maintenance, and running costs [16]. In addition, most of these previous studies focused on the cutting or edge-trimming process; there have been few studies on grooving with a large aspect ratio. In this study, we investigate a grooving method with long tool life and low initial investment.

Grinding is widely regarded as a promising machining process for achieving high-quality results when working with CFRP. Hu et al. [17,18] dry-ground CFRP using a resinoid grinding wheel with WA abrasive grains and showed the relationship between surface roughness and the angle of the fibers relative to the grinding direction. The most pronounced surface roughness was observed when the fiber orientation resulted in an obtuse angle between the grain and the cutting direction. This can be attributed to the cutting of fibers as they are lifted during the process.

Diamond abrasives are generally used for grinding CFRP due to their high thermal conductivity and hardness. Gao et al. [19] constructed a mechanical model of diamond single-particle grinding and verified it experimentally, considering the differences in the coefficient of friction between abrasive grains and CFRP under different lubrication conditions. The results showed that minimum quantity lubrication (MQL) grinding can reduce the grinding force compared to dry grinding. Both Wang and Ning et al. [20,21] showed that when rotary ultrasonic machining (RUM) is applied to a metal-bonded diamond grinding tool, the frictional force decreases, and the cutting force and torque decrease because of the shorter contact time between the workpiece and the tool with increasing ultrasonic power. Yuan et al. [22] showed that it is possible to cut CFRP at high cutting speeds (76 m/s, 25,000 rpm) using a diamond-abrasive ultrathin dicing blade. However, the serious vibration of the spindle in the horizontal direction limited the rotational speed to increase further. In addition, the depth of the cut they performed was 0.6 mm, which was rather small for high-efficiency machining. In contrast, our proposed tool can cut and groove with a cut depth of more than 7 mm.

While grinding is expected to produce high-quality workpieces, problems related to clogging are difficult to cut. Clogging problems should be considered because they increase grinding force and generate abnormal temperature increases. Soo et al. [23] showed that tool wear, grinding resistance, and machined surface roughness were lower when diamond abrasive was used in end-face grinding of CFRP than when CBN was used. They also observed that excessive clogging shortened tool life.

Suzuki et al. [24] developed a tool with diamonds electrodeposited on a cylindrically rolled wire mesh as a method to prevent clogging. The tool was applied to the drilling of CFRP, revealing that it could achieve a high-efficiency equivalent to a twist drill. In this study, we took advantage of this feature to develop a tool with diamond abrasive grains electrodeposited on a disc-shaped wire mesh. This tool is characterized by its ability to machine grooves with large aspect ratios and by its long life. In this paper, we show that clogging can be avoided and that low stiffness is effective in extending the life of grinding wheels.

2. Wire Mesh Grinding Wheel Features

2.1. Appearance and Advantages of Grinding Wheels

Figures 1 and 2 show the side view of the proposed tool and its outer circumference, respectively. Diamond abrasive grains were electrodeposited around the wire mesh corresponding to the base metal. The diameter was 150 mm, the protrusion from the base disk was 10 mm, and the thickness was 1 mm. The wire is austenitic stainless steel (SS304), 0.5 mm in diameter, with a space of 0.5 × 0.5 mm. The grain size was #100 (average grain size 149 μm). The mesh size depends almost entirely on the wire diameter. In addition, a space must remain after the abrasive grains are electrodeposited. After considering these factors comprehensively, a space size of 0.5 mm × 0.5 mm was selected.

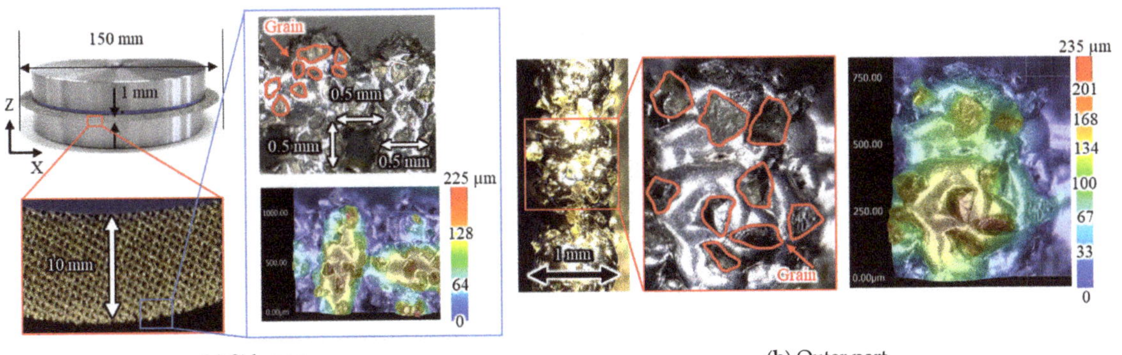

(a) Side part (b) Outer part

Figure 1. Diamond electrodeposited wire mesh grinding wheel.

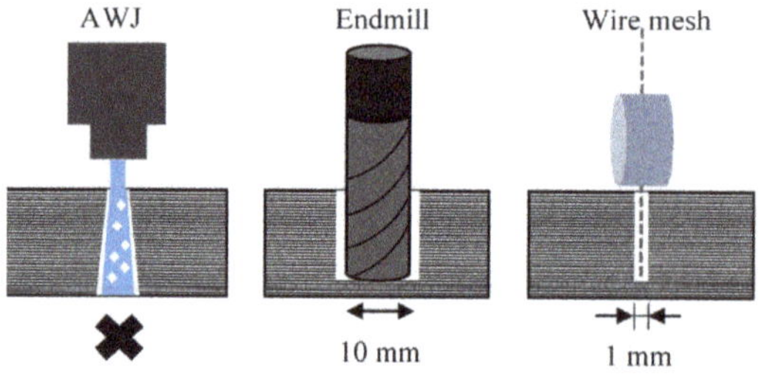

Figure 2. Comparison of grooving methods.

In addition, the wire mesh grinding wheel had peripheral runout, which is difficult to remove. One of the reasons for this is the low roundness of the wire mesh when it is cut into disk shapes. Nevertheless, as will be described later, cutting is possible without any problem, even if there is a large runout.

Figure 2 shows the difference in groove width when machining CFRP using an abrasive waterjet (AWJ), an end mill, and a wire mesh grinding wheel. AWJ is not suitable for grooving because it only allows machining through penetration. A wire mesh grinding wheel can be used to machine thin and deep grooves with larger aspect ratios than end mills. Although such grooving is possible with general electroplated cutters, the difficulty in removing chips results in frequent tool changes due to clogging. In the case of a wire mesh grinding wheel, the chip discharge effect shown in Figure 3 can prevent clogging. In the part of the grinding wheel involved in machining, chips are temporarily held in the wire mesh space, as shown in Figure 3a. In the part of the grinding wheel not involved in machining, chips are discharged by centrifugal force due to rotation, as shown in Figure 3b. The pre-processed part of the grinding wheel remains free of clogging, as shown in Figure 3c.

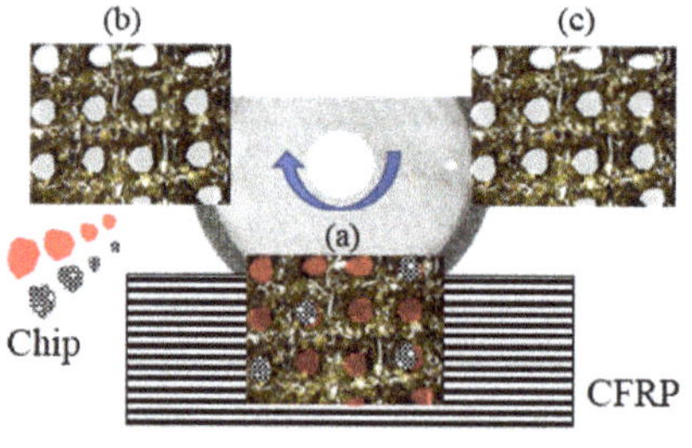

Figure 3. Schematic diagram of chip discharge process. (**a**) Chips are temporarily held in the wire mesh space, (**b**) Chips are discharged, (**c**) Free of clogging before engaging.

2.2. *Stiffness of Grinding Wheel*

The base metal of the proposed tool was wire mesh, which is less rigid than a typical solid base metal, even considering that the electrodeposition layer strengthens the wires by bonding them to each other. We first investigated the stiffness of this wire mesh electrodeposited grinding wheel. Figure 4 shows the method of testing for out-of-plane bending stiffness. A wheel was fixed on a dynamometer (KISTLER 9257B), a pin was pressed down 72 mm from the center against a wheel with a diameter of 150 mm (radius 75 mm), and displacement and load were measured. In addition to the electrodeposited diamond mesh wheel (hereafter referred to as the wire mesh wheel), the bending stiffnesses of two other wheels were measured for comparison: a resinoid wheel (grain: GC, grain size: #100, bond: N) and an electrodeposited diamond solid wheel (base metal: SUS304, grain: diamond, grain size: #100, hereafter referred to as the solid wheel). Figure 5 shows these three types of grinding wheels.

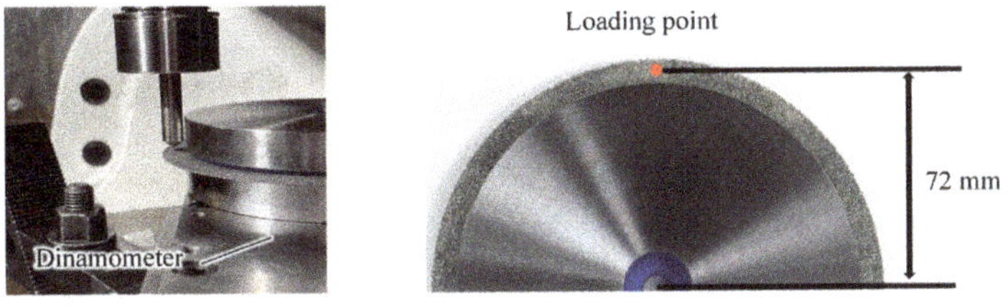

Figure 4. Experimental setup.

Figure 5. Grinding wheels.

Figure 6 shows the relationship between load and displacement for each grinding wheel. The load varies linearly with displacement for all grinding wheels. As the slope of the graph is the stiffness of each wheel, the stiffness values of the resinoid wheel, the solid wheel, and the wire mesh wheel were 63.7 N/mm, 283 N/mm, and 30.4 N/mm, respectively. The stiffness of the wire mesh wheel was about 1/2 that of the resinoid wheel and 1/10 that of the solid wheel.

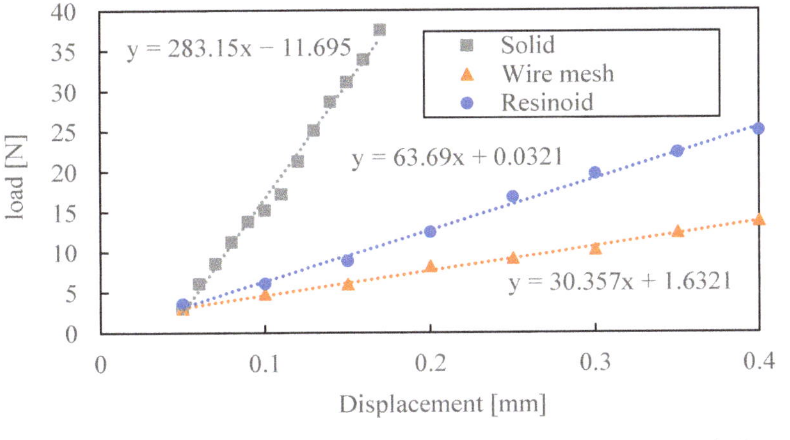

Figure 6. Relationship between bending load and displacement of each grinding wheel.

2.3. In-Plane Stiffness of Thin Grinding Wheel

The in-plane stiffnesses of thin grinding wheels were also tested as follows. Figure 7 shows a photograph of the experimental setup. In-plane compression tests were conducted using a universal material testing machine in accordance with JIS standard JISK7018. Strain and stress were measured by fixing the bottom and side of each of the three grinding wheel strips shown in Figure 5 and pressing them into the test specimens from above.

Figure 8 shows the relationship between stress and strain for each wheel. After a relatively large elastic deformation of the resinoid wheel, it experienced a brittle fracture, and the stress dropped rapidly at that instant. On the other hand, the solid and wire mesh grinding wheels deformed elastically and then plastically, and after reaching a certain stress, the stress decreased smoothly. The elastic moduli of the wheels against the in-plane load were 6.34 N/mm^2, 13.8 N/mm^2, and 5.65 N/mm^2 for the resinoid, solid, and wire mesh wheels, respectively. The stiffness of the wire mesh electroplated wheel was equivalent to that of the resinoid wheel and less rigid than that of the solid wheel. While low stiffness is a disadvantage in reproducing motion trajectory, it does contribute to extending tool life, as described below. Although not discussed in this paper, it is possible to control the stiffness of the wire mesh by changing the thickness and density of the mesh wires by making the wire mesh multilayered or by using some other method. We think that these issues should be considered in the future.

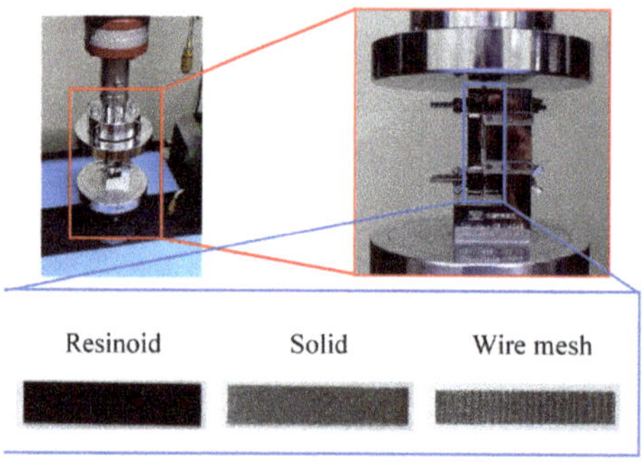

Figure 7. Experimental setup.

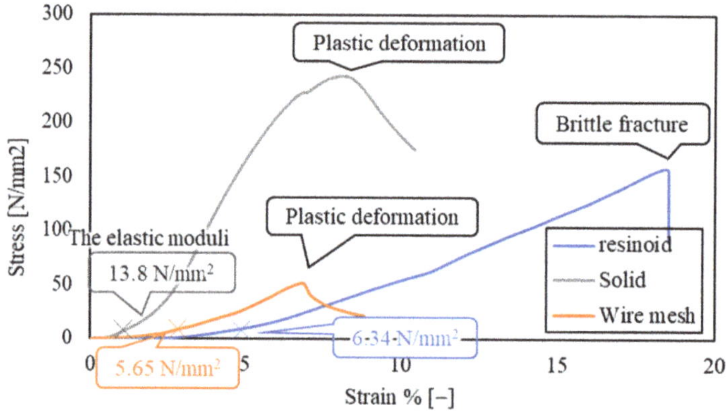

Figure 8. Relationship between compressive stress and strain in thin grinding wheel plane.

3. Experimental Setup and Method for Cutoff and Grooving

3.1. Workpiece

Table 1 shows the mechanical properties of the CFRP used as the workpiece. The CFRP plate was laminated with unidirectional prepreg (QU135-197A manufactured by Toho Tenax, Tokyo, Japan) and hardened to a thickness of 7.6 mm. The specimens were quasi-isotropic, with stacking configurations of 0°, −45°, 45°, and 90°.

Table 1. Mechanical properties of CFRP.

Carbon Fiber		Toho Tenax QU 135-197A
Resin		Epoxy resin #135
Fabric Weight	g/m^2	190
Curing Temperature	°C	180
Thickness of Prepreg	mm	0.187
Number of Layers		40
Thickness of Workpiece	mm	7.6
Length of Workpiece	mm	100

3.2. Machining Characteristics during Cutting and Grooving

Figure 9 shows the experimental setup. A vise was attached to a piezoelectric dynamometer (Kistler 9257B, Winterthur, Switzerland) set on a table of the machining center, and the CFRP plate was fixed vertically. This dynamometer has a stiffness and natural frequency high enough to measure the absolute value of the grinding force and its variation in this experiment. The sampling period in this experiment was 1 ms. Table 2 shows experimental condition A. Comparative experiments were conducted with the three grinding wheels shown in Figure 5 at a cutting speed of 1884 m/min, a feed rate of 100 mm/min, a down-cut wheel rotation direction, and a grinding fluid supply. Coolant was used here because the primary concern was to avoid the disadvantages of high temperatures above the glass transition point. A dynamometer was used to measure the grinding force, and an infrared radiation thermometer was used to measure the temperature at the backside of the machining point. The IR thermometer outputs the average temperature within an 8 mm diameter LED sight. This was supplemented by a thermographic camera to monitor and analyze temperature conditions during the experiment. In addition, the machined surface of the workpiece and the surface of the grinding wheel after machining were observed using an optical microscope.

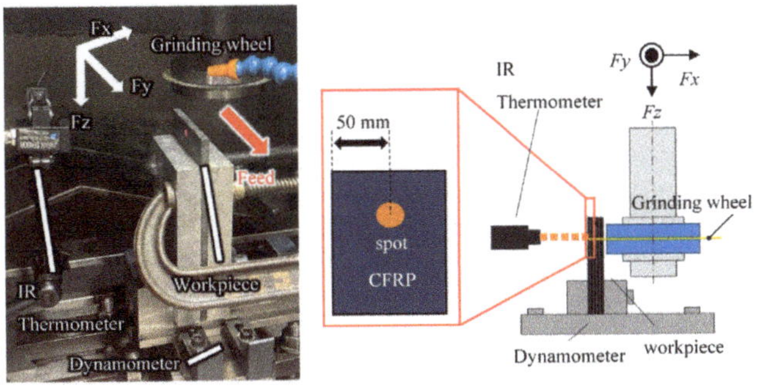

Figure 9. Experimental setup.

Table 2. Machining conditions A and B.

Length of Workpiece	mm	1000	
Grinding Speed V	m/min	1884	
Feed Rate f	mm/min	A:100	B:1000
Grinding Direction		Down cut	
Fluid Supplying Volume	L/min	A:8.0	B:Dry

The use of coolant in cutting CFRP causes cutting oil to permeate the CFRP interlayer and the interface between the fibers and matrix resin, resulting in problems such as reduced strength and dimensional defects. To avoid these problems, dry machining is recommended. For this reason, temperature comparisons were also conducted for dry processing. Table 2 shows experimental condition B. The feed rate was set at 1000 mm/min, the same level as that of the abrasive waterjet, to compare the two grinding wheels shown in Figure 5, the solid grinding wheel and the wire mesh grinding wheel. The processing temperature during the cutting process was measured with a thermographic camera. A resinoid grinding wheel was excluded from the experiment because it is quickly damaged in dry machining.

3.3. *Tool Life Test with Continuous Grooving*

Figure 10 shows the experimental setup. CFRP was fixed with bolts to a jig set on a table at the machining center. The experimental conditions followed Table 2, with a cutting speed of 2120 m/min and a feed rate of 300 mm/min. In the tool life test, the feed rate was set higher than in the force/temperature measurement, aiming to accelerate the wear rate by increasing the feed rate to the point where the grinding wheel is not damaged. Grooves with a depth of 7 mm were machined 60 times on work material 7.6 mm thick, the outer diameter of the wheel was measured, and the surface of the wheel was observed. Because the length of the workpiece was 100 mm, the machining distance for one grooving operation was 100 mm. The three grinding wheels shown in Figure 5 were used for comparison. These wheels were in unused condition at the start of the experiment.

Figure 10. Experimental setup for grooving.

4. Results and Discussion

4.1. *Machining Characteristics during Cutting and Grooving*

4.1.1. Observation of Grinding Wheel Surface

Figures 11–13 show the surfaces of the resinoid, solid, and wire mesh grinding wheels, respectively, before and after machining under experimental condition A. The figure on the right shows the grinding wheel after machining 300 mm. Chips stuck to the resinoid grinding wheel and covered the abrasive grains, resulting in clogging. In the solid grinding wheel, clogging was observed on the sides, but the tip was not covered with chips. The wire mesh grinding wheel, on the other hand, had only a few chips stuck to both the sides and the tip. These chips could be easily removed by blowing air, indicating that there was no clogging. This is because the chips were discharged through the spaces in the wire mesh. As for the solid grinding wheel, the chips could be removed by ultrasonic cleaning, but continuous machining could not be expected. On the other hand, continuous long-distance processing can be expected from a wire mesh grinding wheel if it is treated with air.

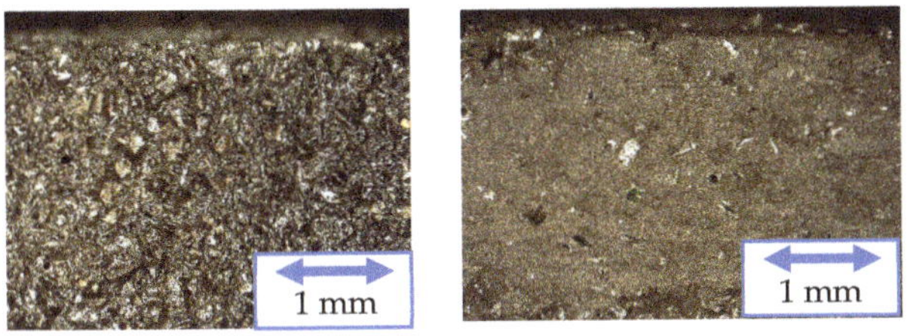

(**a**) Before cutting (**b**) After cutting

Figure 11. Grinding wheel surface (Resinoid).

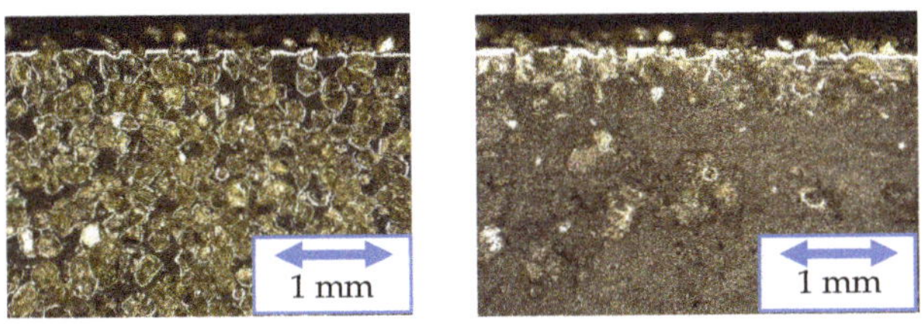

(**a**) Before cutting (**b**) After cutting

Figure 12. Grinding wheel surface (Solid).

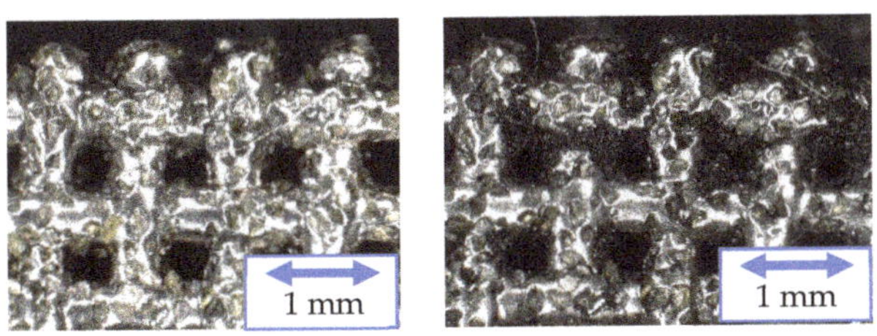

(**a**) Before cutting (**b**) After cutting

Figure 13. Grinding wheel surface (Wire mesh).

4.1.2. Grinding Force

Figures 14–16 show the vertical grinding force Fx and the tangential grinding force Fy for machining with a resinoid, solid, and wire mesh grinding wheel, respectively, under experimental condition A. Grinding forces for several spindle rotations are shown for comparison of the maximum value of the force (Time: 31.0~31.04 s). In the resinoid, the number of abrasive grains appearing on the wheel surface decreased due to clogging, causing the wheel to be pressed against the workpiece, which in turn resulted in a higher grinding force. Comparing the solid and wire mesh grinding wheels, whose tips were not clogged, the wire mesh grinding wheel had a higher grinding force. This is because the wire mesh grinding wheel has peripheral runout, and a large force is applied instantaneously because only a small portion of one revolution is involved in grinding. The sampling period for this experiment was 1 ms, and the spindle speed was 4000 rpm (0.015 s per revolution), so the grinding force was plotted 15 times per revolution. In the waveforms measured with the wire mesh grinding wheel, the force was measured once per revolution. However, this is not a problem because cutting is still possible in this state. On the other hand, the low dimensional accuracy in the z-axis direction in Figure 1 is a problem in groove machining, as discussed in Section 4.1. When grooving with a resinoid or solid grinding wheel, the groove width is about 1.1 mm, whereas a wire mesh grinding wheel produces a groove about 1.6 mm wide. Thus, care must be taken when machining.

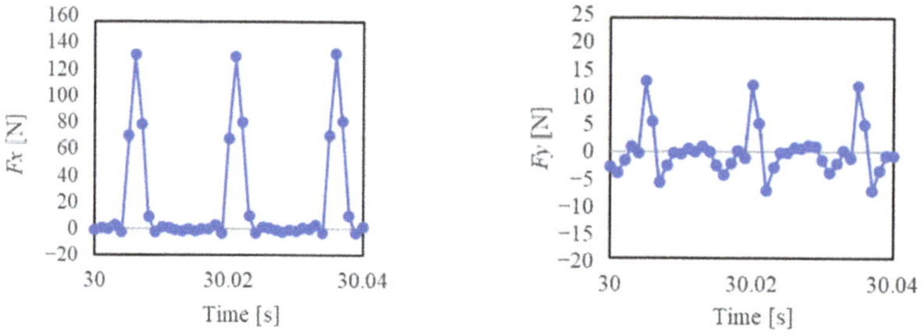

(**a**) Vertical grinding force (**b**) Tangential grinding force

Figure 14. Grinding force (Resinoid, Time: 31.0~31.04 [s]).

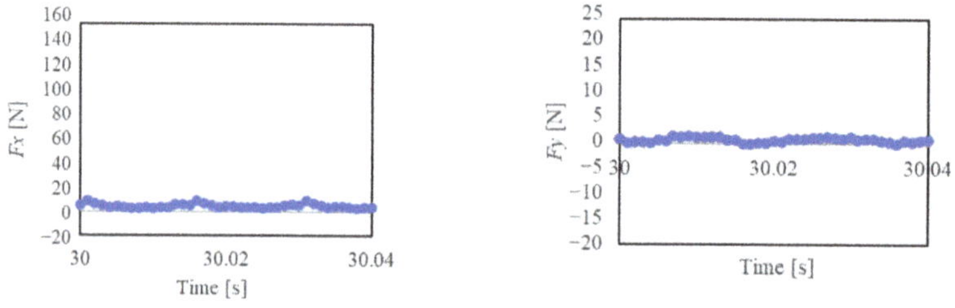

(**a**) Vertical grinding force (**b**) Tangential grinding force

Figure 15. Grinding force (Solid, Time: 31.0~31.04 [s]).

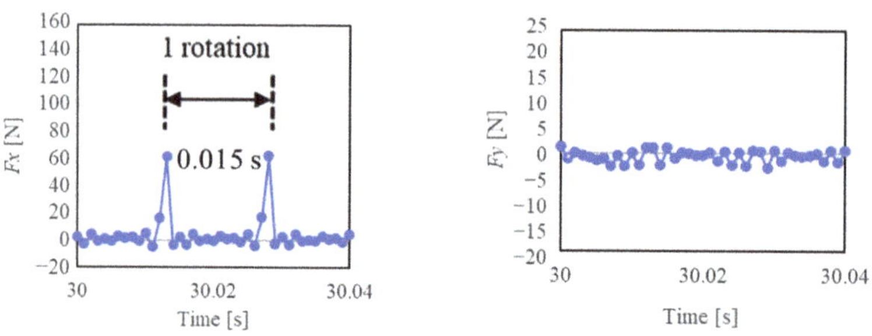

(**a**) Vertical grinding force (**b**) Tangential grinding force

Figure 16. Grinding force (Wire mesh, Time: 31.0~31.04 [s]).

4.1.3. Temperature behind the Grinding Point and Observation of the Machined Surface

The temperature behind the machining point was measured using an IR thermometer when a groove 7 mm wide was machined in 7.6 mm CFRP. A cylindrical cover was used to prevent grinding fluid from entering the optical path of the IR sensor during the measurement, and it was confirmed that the spot position was not wet after the measurement. Figure 17 shows the maximum temperature measured under experimental condition A. The temperature at the machining point is estimated to be higher because the average temperature in a spot with a diameter of 8 mm is shown for a wheel thickness of 1 mm, and the spot is placed on the backside of the machining point. The maximum temperatures of the resinoid, solid, and wire mesh grinding wheels were 53.2 °C, 36.0 °C, and 34.3 °C, respectively, under condition A. The processing temperatures of the wire mesh and resinoid grinding wheels differed by approximately 20 °C. This is because the abrasive grains in the wire mesh grinding wheel always did their work due to the suppression of clogging during machining, whereas the abrasive grains in the resinoid wheel tended to become covered with chips due to clogging during the machining process, thus reducing the effectiveness of the abrasive grains and increasing friction. Consequently, a larger temperature rise was observed in resinoid wheels compared to wire mesh grinding wheels. The processing temperature of the solid grinding wheel was higher than that of the wire mesh grinding wheel. This is presumably because it is difficult for the grinding fluid to reach the inside of the groove with a solid grinding wheel, whereas a wire mesh wheel can temporarily hold the fluid in its mesh and supply it to the machining point.

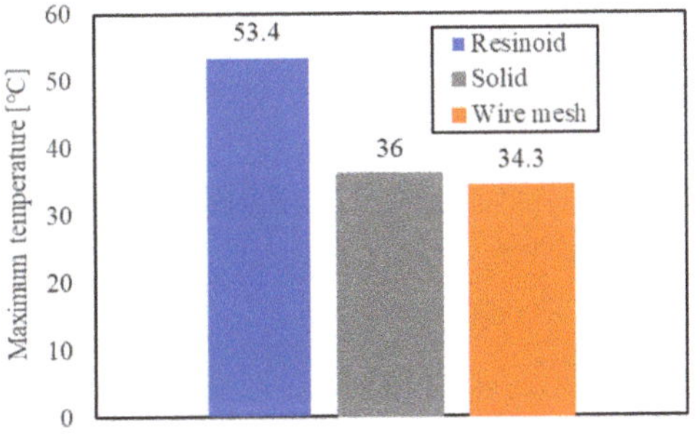

Figure 17. Maximum temperature behind the processing point.

However, the temperature difference between the solid and wire mesh grinding wheels was not significant. As described in Section 3.2, the temperature behind the machining point was measured instead of the machining point, and the measurement area of the infrared thermometer was Φ8 mm, which was wider than the grinding wheel width. These are the causes of lower temperature output than the actual temperature. Therefore, a new comparison was made in dry machining with a feed rate of 1000 mm/min (experimental condition B).

Figure 18 shows the maximum temperature measured. The machining temperature of the solid wheel was 134 °C, whereas that of the wire mesh wheel was 86.5 °C, a difference of about 50 °C. This difference is attributable to two points. The first is the presence or absence of clogging. In the case of the solid grinding wheel, the machining temperature increased due to clogging, whereas the wire mesh grinding wheel maintained its sharpness and suppressed the increase in machining temperature. The second point is thought to be that the wire mesh grinding wheel has peripheral runout, which causes intermittent machining, resulting in a high temperature at one point on the wheel but a lower average temperature over the entire circumference of the wheel.

Figure 19 shows the state of fibers and resin dissolution on the machined surface during high-speed cutting, as observed by scanning electron microscopy. When comparing the workpiece machined with a solid grinding wheel to the one machined with a wire mesh grinding wheel under a magnification of 100×, white areas can be observed on the workpiece machined with the solid grinding wheel. This can be attributed to the occurrence of charge-up due to the accumulation of electric charge on the nonconductive resin part of the workpiece. It is estimated that the resin underwent localized welding due to the temperature rise during processing. The glass transition temperature of the epoxy resin contained in the CFRP used in this experiment is approximately 180 °C. Because the material strength decreases above this temperature, it is important to control the temperature increase when processing CFRP. Although the thermographic camera measures the temperature of the grinding wheel surface, which is lower than the machining point, the actual temperature at the machining point is estimated to have been above 180 °C for the solid grinding wheel but below 180 °C for the wire mesh grinding wheel.

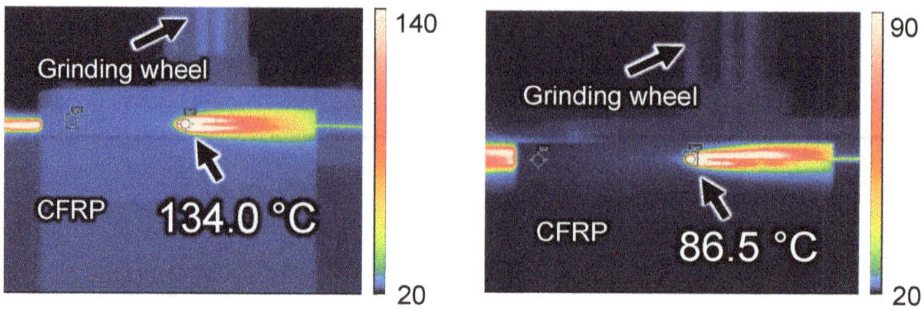

(**a**) Solid grinding wheel (**b**) Wire mesh grinding wheel

Figure 18. Maximum temperature measured by thermocamera.

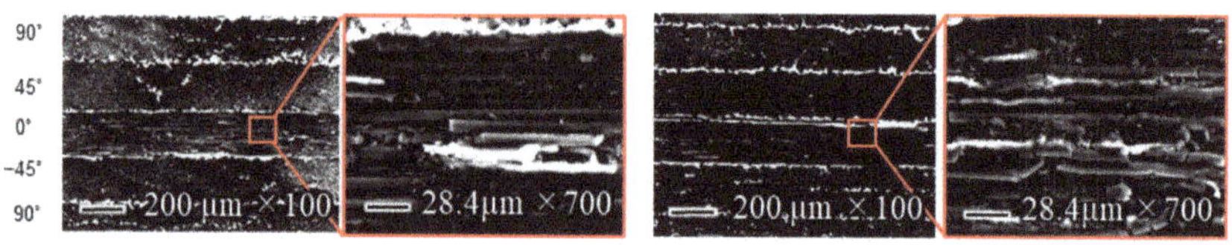

(**a**) solid (**b**) Wire mesh

Figure 19. Workpieces observed by SEM.

A comparison of the observation of the 0° fiber direction at 700× magnification shows that the workpiece machined with the wire mesh wheel had finer-cut carbon fibers than that machined with the solid wheel. This is presumably due to the low dimensional accuracy of the former, which prevents the abrasive grains that do the work from acting evenly on the surface to be machined. Grinding is a machining method in which a tool is given a certain amount of depth of cut, and the motion trajectory of the tool is transferred to the workpiece, so the accuracy of the tool's depth of cut is directly reflected in the machining accuracy. In the case of a cutoff wheel, the runout in the z direction shown in Figure 1 affects the machining accuracy. During the measurement of runout in the z-direction, the solid grinding wheel exhibited a runout of less than 0.1 mm, while the wire mesh wheel had a runout of 0.8 mm. This indicates that the height of the abrasive grains on the work surface in the *z*-axis direction is not uniform, which means that some abrasive grains are cutting less or more than the desired depth of cut. This results in a nonuniform machined surface, as shown in Figure 19.

4.1.4. Observation of Machined Surface and Surface Roughness

Figures 20–22 show the machined surfaces of the workpieces under experimental condition A for the resinoid, solid, and wire mesh grinding wheels, respectively. Photographs of the left end, right end, and side of the cut workpiece are shown. The grinding wheel was fed from the left to the right of the workpiece. Under the conditions of this experiment, although slight burrs were observed at both ends, there were no noticeable delaminations or uncut fibers. There were no burrs on the sides.

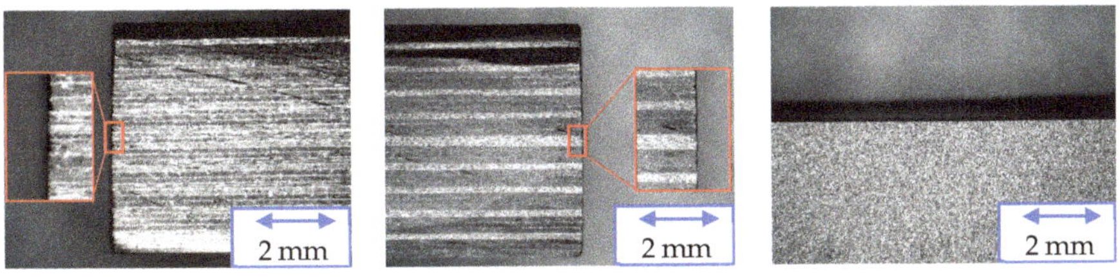

(**A**) Left end (**B**) Right end (**C**) Side

Figure 20. Machined surface (Resinoid tool).

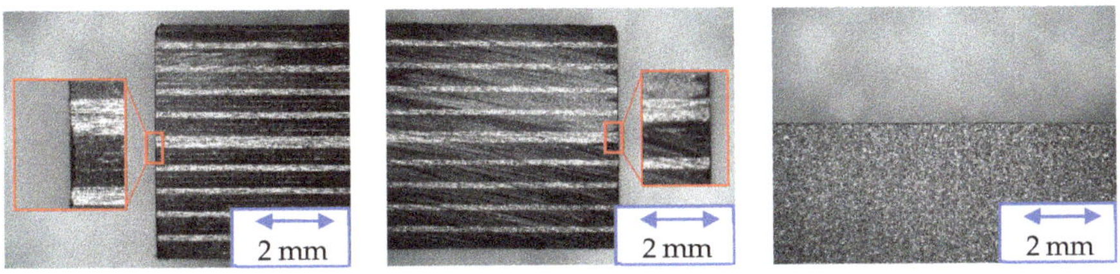

(**A**) Left end (**B**) Right end (**C**) Side

Figure 21. Machined surface (Solid tool).

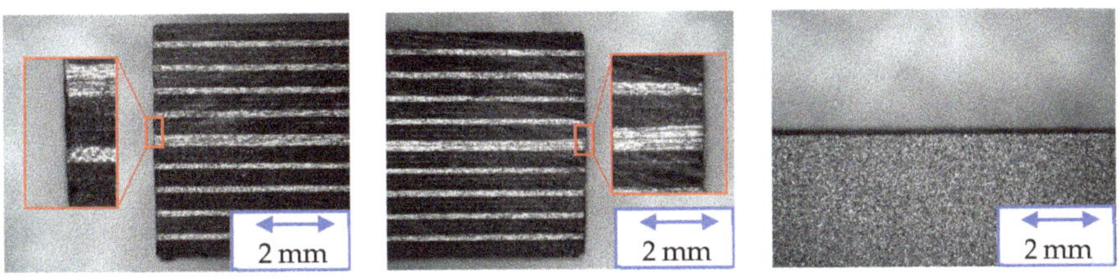

(**A**) Left end (**B**) Right end (**C**) Side

Figure 22. Machined surface (Wire mesh tool).

Figures 23–25 show the grooves machined by the resinoid, solid, and wire mesh grinding wheels, respectively, under experimental condition A. The thickness of the grinding wheel was uniformly 1 mm. Groove widths were 1.02 mm when machined with a resinoid wheel, 1.13 mm with a solid wheel, and 1.6 mm with a wire mesh wheel. The groove width from the wire mesh wheel was 1.6 times larger than the thickness of the wheel. This is thought to be due to the runout in the z direction shown in Figure 1. As mentioned above, the wire mesh grinding wheel exhibited a runout of 0.8 mm, indicating that some abrasive grains were cutting beyond the desired depth of cut. As the feed rate increased, resulting in deterioration of the groove shape accuracy and surface roughness in general. However, straight cutting and grooving were possible even when the feed rate was increased up to about 2000 mm/min. On the other hand, with a tool with a 20 mm protrusion for deeper grooving, the wheel was deformed, and straight grooving was difficult even when the feed rate was 1000 mm/min. It also shows the importance of the stiffness of the wire mesh.

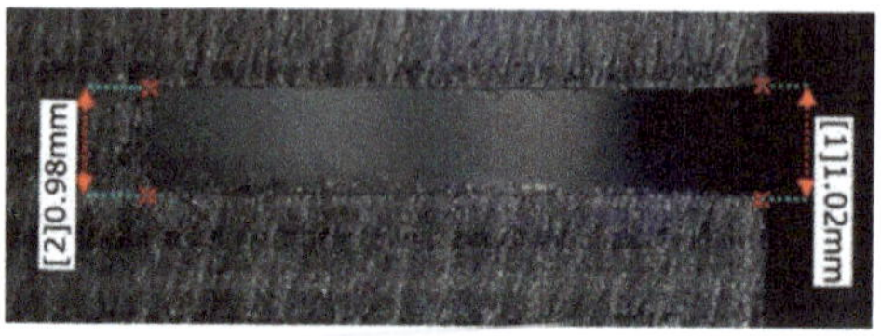

Figure 23. Machined groove (Resinoid tool).

Figure 24. Machined groove (Solid tool).

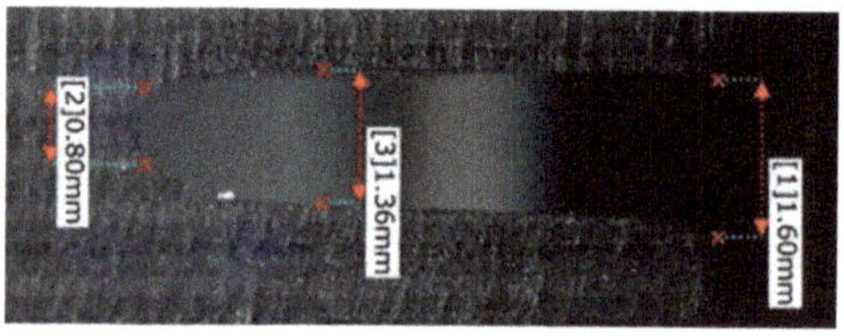

Figure 25. Machined groove (Wire mesh tool).

Table 3 shows the surface roughness values of the workpieces. While the arithmetic mean roughness of the workpiece machined with the solid grinding wheel was 1.37 µm, the wire mesh grinding wheel had a larger value, 2.76 µm. As mentioned above, this is thought to be due to the unevenness of the machined surface caused by the low dimensional accuracy of the wire mesh wheel. However, the required $Ra \leq 3.2$ µm for specific aircraft parts is met. These results indicate that CFRP machined with a wire mesh wheel can be used to fabricate aircraft parts because no burr or delamination was observed, and the surface roughness met the standard value. However, attention should be paid to dimensional accuracy when machining grooves.

Table 3. Surface roughness.

	Ra (µm)	*Rz* (µm)
Resinoid	0.502	3.18
Solid	1.37	7.25
Wire mesh	2.76	11.5

4.2. Tool Life Test with Continuous Grooving

Figure 26 shows the workpieces after machining. The solid and wire mesh grinding wheels completed 60 grooving operations on a 100 mm CFRP plate (machining distance: 6000 mm) without any problems, but the resinoid grinding wheel broke after the fourth cut (machining distance: 400 mm). Looking at the workpiece after machining, we found that the groove was curved at an angle when the resinoid grinding wheel broke. Under the axial load, the resinoid grinding wheel experienced slight elastic deformation. However, due to its limited ability to withstand the applied load, the wheel eventually fractured in a brittle manner.

(**a**) Resinoid tool (**b**) Wire mesh tool

Figure 26. Workpieces after grooving.

Table 4 shows the outer diameter of each wheel before and after machining. For the resinoid grinding wheel, the outside diameter was measured after two cuts (machining distance: 200 mm), and for the solid and wire mesh grinding wheels, the outside diameter was measured after 60 cuts. The resinoid grinding wheel showed about 0.05 mm of wear after two cuts, while the solid and wire mesh grinding wheels showed almost no wear (less than 0.02 mm), even after 60 cuts. Because no significant difference was observed in the wear of those two grinding wheels, we compared their performance in terms of grain shedding and crushing.

Table 4. Diameter of each grinding wheel (mm).

	Resinoid **Distance: 200 mm**	**Solid** **Distance: 6000 mm**	**Wire Mesh** **Distance: 6000 mm**
Before cutting	150.85	150.32	151.34
After cutting	150.80	150.30	151.33
Wear of grinding wheel	−0.05	−0.02	−0.01

Figures 27 and 28 show the outer circumference areas involved in machining with the solid and wire mesh grinding wheels, respectively. The arrows indicate abrasive grains, and the other places are Ni-plated surfaces. From left to right, the figures show the state of the wire mesh grinding wheel prior to machining, after 30 cuts (machining distance: 3000 mm), and after 60 cuts (machining distance: 6000 mm), respectively. In the solid grinding wheel, the abrasive grains were crushed in the red circles during the first 30 cuts. After the next 30 cuts, we found no abrasive grain crushing or wear. On the other hand, for the wire mesh grinding wheel, the red circles indicate that the abrasive grains fell off during the first 30 cuts. Abrasive grains with a weak fixation force, which were not securely embedded within the plating layer, became dislodged and dropped out. This phenomenon is referred to here as "initial shedding". The wire mesh grinding wheel showed only initial spilling, whereas the solid grinding wheel showed a lot of abrasive grains crushed even in the small area observed. We attribute this to the disparity in the modulus of elasticity between the two types of wheels.

As shown in Figure 8, the elastic modulus of the wire mesh wheel is 2.4 times smaller than that of the solid wheel and is easily deformed. The solid wheel has a high elastic modulus and is not easily deformed, so when a load is concentrated on the abrasive grains at the tip, the stress is high, and crushing occurs. On the other hand, the wire mesh wheel, characterized by a low modulus of elasticity, exhibits moderate deformation even under the application of impact processing forces. This deformation capability plays a crucial role in preventing excessive impact input to the diamond abrasive grains and is assumed to help prevent grain crushing.

In addition, a previous study [24] noted that wire mesh grinding wheels are self-sharpening. When abrasive grains at the tool tip become dislodged, it exposes a metal wire that is more prone to wear compared to the diamond grains. Once exposed, the wire is susceptible to immediate abrasion and quickly gets shaved off. As a result, the abrasive grains that exist on the sides of the grinding wheel appear at the tip, enabling machining by abrasive grains at all times. Although this phenomenon was not confirmed under the experimental conditions due to low wear, it is presumed that the low modulus of elasticity prevents the abrasive grains from fracturing; in addition, the self-sharpening prolongs tool life.

Figure 27. Surface of solid grinding wheel.

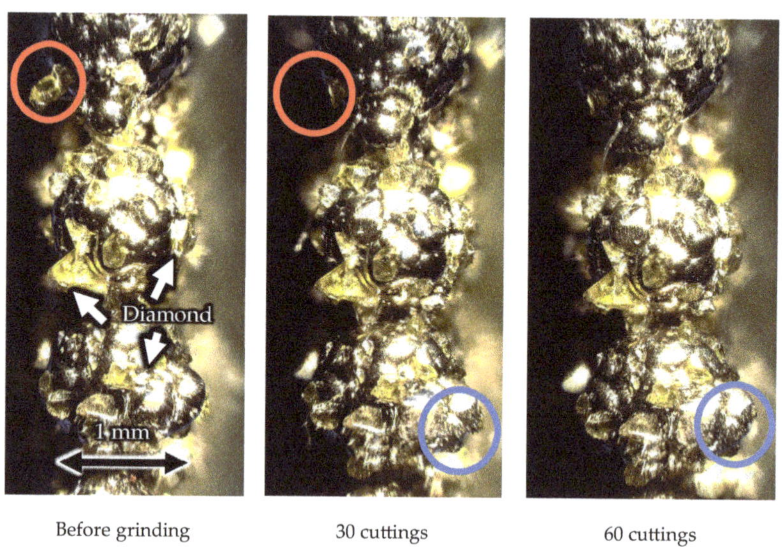

Figure 28. Surface of wire mesh grinding wheel.

5. Conclusions

In this study, we proposed a tool with diamond abrasive grains electrodeposited on a disc-shaped wire mesh. The obtained results regarding the machining characteristics and tool life when performing deep grooving and cutoff of CFRP using this tool were as follows.

1. Because chips are discharged through the spaces in the mesh, clogging is avoided, and thus the machining temperature of the wire mesh wheel is lower than that of the resinoid or solid wheel.
2. CFRP machined with a wire mesh wheel can be used to fabricate aircraft parts because no burr or delamination was observed, and the surface roughness met the standard value, although attention should be paid to dimensional accuracy when machining grooves.
3. The wire mesh grinding wheel showed almost no wear after 6000 mm of machining. It has a low modulus of elasticity, so it deforms moderately even when an impact processing force is applied. This avoids impact input to the diamond abrasive grains and is assumed to avoid grain crushing.
4. It is estimated that tool life can be prolonged due to the absence of clogging, avoidance of abrasive grain crushing, and self-sharpening.

The greatest feature of the proposed tool is its ability to machine grooves with large aspect ratios over a long life. This paper describes grooving to a depth of 7 mm, but we have confirmed that grooving to a depth of 15 mm is also possible with a larger wheel. Current issues include the effects of vibration due to the low stiffness of the tool, which increases grinding resistance, and the low axial dimensional accuracy of the grinding wheel, which results in wide grooves. In the future, we plan to develop a highly rigid and precise tool by using a fabrication method that compresses several layers of wire mesh.

Author Contributions: M.N. designed this study and manufactured the grinding wheel; S.K. performed relevant experiments and wrote the manuscript; Y.I. and Y.F. methodology and advised; H.S. supervised the experiments and data analysis and reviewed the manuscript. All authors have read and agreed to the published version of the manuscript.

Funding: This research received no external funding.

Institutional Review Board Statement: Not applicable.

Informed Consent Statement: Not applicable.

Data Availability Statement: Not applicable.

Conflicts of Interest: The authors declare no conflict of interest.

References

1. Soutis, C. Fibre reinforced composites in aircraft construction. *Prog. Aerosp. Sci.* **2005**, *41*, 143–151. [CrossRef]
2. Roberts, T. Rapid growth forecast for carbon fibre market. *Reinf. Plast.* **2017**, *51*, 10–13. [CrossRef]
3. Wang, H.; Sun, J.; Li, J.; Lu, L.; Li, N. Evaluation of cutting force and cutting temperature in milling carbon fiber-reinforced polymer composites. *Int. J. Adv. Manuf. Technol.* **2016**, *82*, 1517–1525. [CrossRef]
4. Wang, F.; Yin, J.; Ma, J.; Jia, Z.; Yang, F.; Niu, B. Effects of cutting edge radius and fiber cutting angle on the cutting-induced surface damage in machining of unidirectional CFRP composite laminates. *Int. J. Adv. Manuf. Technol.* **2017**, *91*, 3107–3120. [CrossRef]
5. Hosokawa, A.; Hirose, N.; Ueda, T.; Furumoto, T. High-quality machining of CFRP with high helix end mill. *CIRP Ann. Manuf. Technol.* **2014**, *63*, 89–92. [CrossRef]
6. Bi, G.; Wang, F.; Fu, R.; Chen, P. Wear characteristics of multi-tooth milling cutter in milling CFRP and its impact on machining performance. *J. Manuf. Process.* **2022**, *81*, 580–593. [CrossRef]
7. Voss, R.; Seeholzer, L.; Kuster, F.; Wegener, K. Influence of fibre orientation, tool geometry and process parameters on surface quality in milling of CFRP. *CIRP J. Manuf. Sci. Technol.* **2017**, *18*, 75–91. [CrossRef]
8. Ashworth, S.; Fairclough, J.; Meredith, J.; Takikawa, Y.; Kerrigan, K. Effects of tool coating and tool wear on the surface quality and flexural strength of slotted CFRP. *Wear* **2022**, *498–499*, 204340. [CrossRef]
9. Nor Khairusshima, M.; Che Hassan, C.; Jaharah, A.; Amin, A.; Md Idriss, A. Effect of chilled air on tool wear and workpiece quality during milling of carbon fibre-reinforced plastic. *Wear* **2013**, *302*, 1113–1123. [CrossRef]
10. Nguyen, D.; Abdullah, M.S.B.; Khawarizmi, R.; Kim, D.; Kwon, P. The effect of fiber orientation on tool wear in edge-trimming of carbon fiber reinforced plastics (CFRP) laminates. *Wear* **2020**, *450–451*, 203213. [CrossRef]

11. Kim, M.; Lee, M.; Cho, G.; Lee, K. Effect of the Fiber Orientation and the Radial Depth of Cut on the Flank Wear in End Milling of CFRP. *Int. J. Precis. Eng. Manuf.* **2020**, *21*, 1187–1199. [CrossRef]
12. Khairusshima, M.K.N.; Sharifah, I.S.S. Study on Tool Wear during Milling CFRP under Dry and Chilled Air Machining. *Procedia Eng.* **2017**, *184*, 78–89. [CrossRef]
13. Chenrayan, V.; Manivannan, C.; Shahapurkar, K.; Zewdu, G.; Maniselvam, N.; Alarifi, I.; Alblalaihid, K.; Tirth, V.; Algahtani, A. An experimental and empirical assessment of machining damage of hybrid glass-carbon FRP composite during abrasive water jet machining. *J. Mater. Res. Technol.* **2022**, *19*, 1148–1161. [CrossRef]
14. Demiral, M.; Abbassi, F.; Saracyakupoglu, T.; Habibi, M. Damage analysis of a CFRP cross-ply laminate subjected to abrasive water jet cutting. *Alex. Eng. J.* **2022**, *61*, 7669–7684. [CrossRef]
15. Wu, Y.; Zhang, G.; Wang, J.; Chao, Y.; Zhang, W. The cutting process and damage mechanism of large thickness CFRP based on water jet guided laser processing. *Opt. Laser Technol.* **2021**, *141*, 107140. [CrossRef]
16. El-Hofy, M.; Helmy, M.; Escobar-Palafox, G.; Kerrigan, K.; Scaife, R.; El-Hofy, H. Abrasive Water Jet Machining of Multidirectional CFRP Laminates. *Procedia CIRP* **2018**, *68*, 535–540. [CrossRef]
17. Hu, N.; Zhang, L. A study on the grindability of multidirectional carbon fibre-reinforced plastics. *J. Mater. Process. Technol.* **2003**, *140*, 152–156. [CrossRef]
18. Hu, N.; Zhang, L. Some observations in grinding unidirectional carbon fibre-reinforced plastics. *J. Mater. Process. Technol.* **2004**, *152*, 333–338. [CrossRef]
19. Gao, T.; Li, C.; Yang, M.; Zhang, Y.; Jia, D.; Ding, W.; Debnath, S.; Yu, T.; Said, Z.; Wang, J. Mechanics analysis and predictive force models for the single-diamond grain grinding of carbon fiber reinforced polymers using CNT nano-lubricant. *J. Mater. Process. Technol.* **2021**, *290*, 116976. [CrossRef]
20. Wang, H.; Ning, F.; Hu, Y.; Cong, W. Surface grinding of CFRP composites using rotary ultrasonic machining: A comparison of workpiece machining orientations. *Int. J. Adv. Manuf. Technol.* **2018**, *95*, 2917–2930. [CrossRef]
21. Ning, F.; Cong, W.; Pei, Z.; Treadwell, C. Rotary ultrasonic machining of CFRP: A comparison with grinding. *Ultrasonics* **2016**, *66*, 125–132. [CrossRef] [PubMed]
22. Yuan, Z.; Hu, J.; Wen, Q.; Cheng, K.; Zheng, P. Investigation on an innovative method for high-speed low-damage micro-cutting of CFRP composites with diamond dicing blades. *Materials* **2018**, *11*, 1974. [CrossRef]
23. Soo, S.; Shyha, I.; Barnett, T.; Aspinwall, D.; Sim, W. Grinding performance and workpiece integrity when superabrasive edge routing carbon fibre reinforced plastic (CFRP) composites. *CIRP Ann. Manuf. Technol.* **2012**, *61*, 295–298. [CrossRef]
24. Suzuki, K.; Koyasu, R.; Takeda, Y.; Sasahara, H. Application of a novel woven metal wire tool with electrodeposited diamond grains for carbon fiber reinforced plastics core drilling. *Precis. Eng.* **2019**, *56*, 386–394. [CrossRef]

materials

MDPI

Article

Statistical Study of the Process Parameters for Achieving Continuous Consolidation of a Thermoplastic Composite

Daniel Campos [1,2,*], Pere Maimí [1] and Alberto Martín [2]

1 AMADE-UdG Research Group, University of Girona, 17003 Girona, Spain; pere.maimi@udg.edu
2 Applus+ Laboratories, 08193 Bellaterra, Spain; alberto.martin@applus.com
* Correspondence: daniel.campos@applus.com

Abstract: Manufacturing components using thermoplastic composite materials necessitates a judicious balance among fabrication parameters, cost considerations and the ultimate quality of the elements produced. Continuous manufacturing technologies, exemplified by methods such as continuous compressing molding and glide forming, seek to revolutionize production through their continuous processing approach. This study aimed to investigate the effects different process parameters have on the final quality of the manufactured parts when a continuous manufacturing technology, such as glide forming, is applied to thermoplastic composite materials. An experimental rig was designed, and 19 samples were prepared using a unidirectional-carbon-fiber-reinforced LM-PAEK (low-melting polyaryletherketone) composite. The process parameters studied were temperature, pressure and forming speed. The quality of the final parts was evaluated based on their thickness and consolidation levels. The findings underscore the feasibility of leveraging continuous manufacturing technologies for producing components using thermoplastic composite materials, but the process parameters must be carefully controlled to ensure the quality of the final part. The models obtained could be used as a post-processing tool to predict thickness and consolidation levels when simulating the manufacture of a component on macroscale levels. Further research is needed to optimize the process parameters and study their effects on other thermoplastic composite materials.

Keywords: thermoplastic materials; LM-PAEK; PEEK; consolidation; manufacturing process

1. Introduction

Thermoplastic composites are the combination of fibers (such as glass, carbon or aramid) and a thermoplastic polymer matrix. The resultant is a lightweight material that combines the strength and resistance of the fibers with the malleability and chemical resistance of the polymer. Unlike thermoset polymers, thermoplastics consist of long chains of molecules that can be repeatedly softened and hardened through heating and cooling cycles without undergoing any chemical changes [1]. Thus, thermoplastics composites can be reshaped multiple times when heated making them more versatile than thermoset composites, enabling easier manufacturing and repair processes, as well as the possibility of recycling [2]. Thermoplastic composite materials have gained significant attention in the field of advanced materials due to their unique combination of high strength, low weight and the ability to be molded and reshaped under heat [3]. One important quality asset in the manufacturing of thermoplastic composite components is consolidation, which refers to the removal of excess resin and the creation of a uniform, dense structure within the composite layout. This phenomenon has a significant impact on the microstructure and porosity of the composite material, which in turn plays a crucial role in determining the final mechanical and physical properties of the composite material [4].

In recent years, there has been a significant amount of research into the optimization of thermoplastic composite consolidation processes. For example, Aquier et al. [5] reviewed the consolidation of unidirectional-carbon-fiber-reinforced LM-PAEK (low-melting polyaryletherketone) composites. They studied different fabrication techniques and the effects their parameters had on consolidation and mechanical properties. Colton et al. [6] investigated the processing parameters for consolidating PEEK (polyetheretherketone)/carbon fiber (APC-2) composites on resin flow measured by the Kozeny constant and the bulk modulus. They found that these variables depend on the consolidation pressure, temperature, the number of layers and their interactions at a 99% confidence level. Patou et al. [7] examined the influence of the consolidation process on the voids and mechanical properties of powdered and commingle glass/PPS (polyphenylene sulfide) laminates and found that increasing consolidation pressure resulted in a more homogeneous microstructure and a reduction in porosity, which improved the mechanical properties. Feng et al. [8] studied the effects of injection molding parameters on the microstructure and mechanical properties of a PE (polyethylene) composite reinforced with long glass fibers. They found that increasing injection pressure and melt temperature resulted in a more uniform distribution of fibers and improved mechanical properties. Overall, these studies showed that consolidation is highly dependent on the process parameters applied during the manufacturing of the composite part.

The emergence of new CFRTP (carbon fiber reinforced thermoplastic polymers) manufacturing technologies makes it necessary to continue investigating consolidation parameters, especially when these technologies differ from the standard due to their continuous application—such as glide forming [9–12] and Continuous Compression Molding [13–15]).

GF (glide forming) is an advanced automated manufacturing technology currently under development and is specifically tailored for crafting slender structural components with intricate double curvature profiles using composite materials. This innovative technique addresses a critical industry need—the efficient production of complex geometries such as stringers, wing reinforcements and stabilizer components, all of which pose significant manufacturing challenges due to their lengthy and curved geometries. Traditional methods often fall short when it comes to achieving high production rates for these parts. What sets the GF process apart are its key advantages, not the least being its remarkable flexibility. By employing a flexible nozzle concept and localized heating, it can seamlessly adapt to create components with varying cross-sectional shapes, thus eliminating the need for extensive tooling changes and resulting in substantial time and cost savings. Furthermore, the process boasts high levels of automation, reducing labor demands, and its environmentally friendly approach, which eschews the necessity for heated tools, positions it as a cost-effective and sustainable manufacturing solution compared to alternatives.

Likewise, CCM (continuous compression molding) is also a manufacturing technology currently undergoing development to produce CFRTP components. This process involves feeding a continuous laminate into a compression molding machine that contains a series of heated molds or dies. The material is heated above its melting point and then compressed consecutively in different stages as it passes through the machine until the final geometry is achieved. CCM is a highly efficient process for producing large quantities of constant section components with consistent quality and dimensional accuracy. This process has many advantages over other manufacturing processes for CFRTP, such as stamp forming or hot forming. CCM can produce large components, and its production rate is relatively high. Moreover, the process can be automated, making it cost-effective for high-volume production. Despite these advantages, CCM has some limitations, such as high initial tooling costs and limited flexibility for small batch production.

To address the lack of studies around these emerging technologies, a statistical study was proposed aimed at identifying the optimal process parameters—temperature, pressure and time (forming speed)—for achieving optimal consolidation levels using continuous manufacturing technologies in LM-PAEK reinforced with carbon fibers.

2. Materials and Methods

The experiment was based on the most commonly used manufacturing processes such as stamp forming [16–18], hot press forming [19] or sheet forming processes [20]. The parameters were adjusted to meet the conditions of continuous production processes, especially for the glide forming.

The statistical analysis was conducted following an experimental sequence based on a 3-factor Box–Behnken design experiment [21–23] that led to two RSMs (response surface models): one for each control variable. The RSMs were assessed using an ANOVA (analysis of variance) analysis with a confidence level of 90%. These models could be used to implement numerical post-processing models to predict forming results at macroscale levels. Similarly, they could be used directly as a guide to fine-tune a specific manufacturing process.

2.1. Material and Samples

The experiment was conducted using unidirectional-carbon-fiber-reinforced LM-PAEK composite from Toray (Toray Cetex TC1225, T700, 145 gsm FAW, 34 wt% RC). The key properties of the material are presented in Table 1.

Table 1. Material properties.

Property	TC1225/T700
Type of composite	Unidirectional Prepreg
Type of reinforcement	T700
Fiber areal weight (FAW)	145 g/m^2
Resin content by weight (RC)	34%
Consolidated ply thickness (CPT)	0.185 ± 0.016 mm

The samples utilized in the experiment consisted of 25 mm × 25 mm laminates with a symmetric ply sequence $[(0,90)_5]_s$, point-welded in the center ply by ply (see Figure 1). Each laminate was equipped with a J-type thermocouple, which was positioned on the middle ply to monitor and control the temperature.

Figure 1. Ply-by-ply point-welding process.

2.2. Experiment Setup Description

When manufacturing thermoplastic composite material components, three phases (pre-heating, impregnation and consolidation) are distinguished during the forming as described in reference [12]. They proposed to use this dynamic process in a static experimental setup which separates the whole process into two stages:

1. Pre-heating and impregnation/forming phases;
2. Consolidation phase;

Based on this concept, a special testing rig involving two presses adapted to each defined stage was designed (see Figure 2). Stage 1 press consisted of two 50 mm × 50 mm heating plates that were pressed using a 25 mm diameter pneumatic cylinder. This press simulated the application of pressure during the impregnation phase, where the molten material is shaped, and the intimate contact between the different plies is increased. The pressure application in this phase is less than that in the cooling phase. Stage 2 press consisted of a lower heating plate that simulated the heated tool and a nonheated upper plate that simulated the pressing skid (counter mold). Both plates had dimensions of 50 mm × 50 mm. The pressure was exerted by a 50 mm diameter pneumatic cylinder. The samples were wrapped in Kapton film to avoid the matrix sticking to the plates and a thermocouple was embedded in the middle ply of the sample to monitor the temperature during the test.

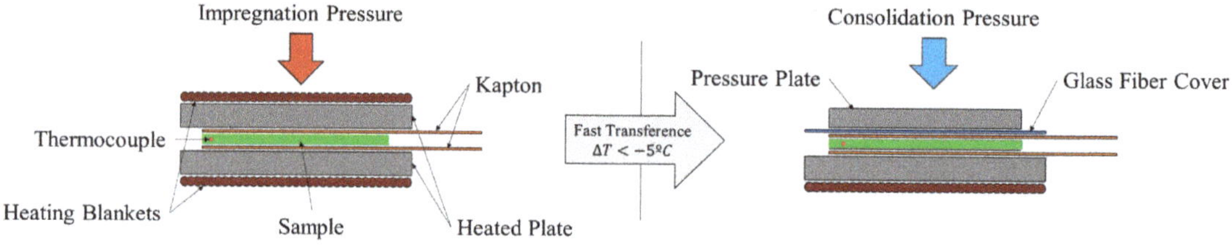

Figure 2. Experimental rig scheme. (**Left**): Stage 1 equivalent to the heating and impregnation phases of the glide forming process. (**Right**): Stage 2 equivalent to the consolidation phase of the glide forming process.

The experiment started by introducing the laminate into stage 1 press and heating the material. The heating time is equivalent to the actual heating time of the forming process. Once the temperature on the thermocouple reached the target value, the impregnation pressure was then applied for the corresponding duration of the stage in the real process. After this period, the samples were rapidly transferred to stage 2 press, where they were subjected to consolidation pressure for the designated time in the experimental sequence. To deem the sample acceptable, the temperature drops during the transfer had to be less than 5 °C.

2.2.1. Typical Process Parameters

Previous studies, such as those conducted by Khan M.A. et al. [24] and Brooks R.A. et al. [17], established that, among all the parameters that influence the consolidation of a laminate, there were three separately controllable parameters, which were key for forming processes.

- Forming temperature: The material supplier typically provided the recommended range for forming temperatures. For TC1225/T700 [25], this range is normally between 320 °C and 400 °C.
- Consolidation pressure: The consolidation pressure range could vary depending on the manufacturing process. From previous experiences with high-rate applications, good consolidation values were achieved in a range of pressures between 0.2 and 2 MPa [12].
- Time: Time is an indirect measure of the forming speed. It was feasible to approximate the forming speed by considering a standard length for the forming skids (counter molds) used in glide forming. Thus, for the impregnation/forming phase-assuming a skid length of 200 mm and forming speeds between 1 mm/s and 10 mm/s—forming equivalent time varied from 20 s to 200 s. For the consolidation phase, the equivalent time considering a 100 mm skid varied from 10 s to 100 s.

Other parameters, such as the number of plies, impregnation pressure and tool temperature, may also impact the forming process but were kept constant in this study to reduce the number of variables. The number of plies was set to 20, the impregnation pressure to 4 kPa, and the tool temperature to 250 °C. These values were selected based on previous experiences with the glide forming process and material supplier indications.

- Number of plies: The number of plies in the laminate, a key parameter in the design of the component, is typically within a range of 5 to 20 for secondary structures. However, in certain cases, local reinforcements may increase the number of plies up to 50. For the purposes of this study, a laminate with 20 plies was chosen.
- Impregnation pressure: Impregnation pressure, applied during the forming phase, is utilized to secure the cohesion of the plies while the material is at the appropriate forming temperature. This results in a greater degree of intimate contact among the plies and ensures that each fiber is fully impregnated by the thermoplastic matrix. Additionally, impregnation pressure is employed in shaping the component. Based on previous research, an appropriate value for this pressure was determined to be 4 kPa.
- Tool temperature: Tool temperature, which is typically below melting point and above the crystallization temperature, was set to 250 °C for this study.

2.2.2. Experimental Variables and Characterization

The consolidation of the component, measured by the porosity level and thickness, was studied to evaluate the results and identify the optimal process settings.

- Porosity: Porosity was quantified using the A-scan method via MUPE (manual ultrasound portable equipment). The transducer used was a mono-element straight beam transducer with a delay of 8.178 and a speed setting of 2713 m/s. The analysis involved comparing the sample attenuation to a reference standard. Micrographics were also utilized to provide a detailed examination of select samples. The methodology employed was in accordance with the Airbus AITM (Airbus Industries Test Methods) 6-4005 standard [26].
- Thickness: Thickness was measured using a micrometer and compared to the nominal thickness of a consolidated ply. The measured nominal thickness of the raw material was 0.170 mm.

2.3. *Experiment Setup Description*

A total of 19 samples were tested in accordance with the randomized experimental design. The study utilized the 3-parameter Box–Behnken design for 12 of the samples, which were set with varying parameters. Three samples were designated as the central point (0, 0, 0) for the purpose of evaluating the normal dispersion and repeatability of the experimental data. Additionally, four extra points were randomly selected to validate the models.

Randomization of the experiment design was crucial in ensuring the independence of sample conditions, as it prevented any potential influence on subsequent samples. This randomness provided a basis for drawing reliable conclusions from the test results, thus avoiding any ambiguities. The experimental factor ranges and the experimental sequence are presented in Tables 2 and 3, respectively.

The results of the experiment were processed to determine the multiple response surface for each evaluation parameter. The multiple response surface provided a mathematical representation of the relationship between the various parameters and the evaluation criteria. While this approach alone may not provide highly accurate predictions, it serves as a useful and efficient means of obtaining a preliminary approximation. Further experimentation is required to validate the results and to enhance the level of accuracy of the predictions.

Table 2. Values for the three factors.

Factor	Description	Low (−1)	Midpoint (0)	High (+1)
A	Temperature	320 °C	350 °C	400 °C
B	Consolidation Pressure	0.2 MPa	1 MPa	2 MPa
C	Consolidation Time	10 s	50 s	100 s

Table 3. Randomize experimental sequence and results.

Sample	Order	Factor A	B	C	Normalized Thickness	Backwall Echo Amplitude (%)
T01	19	−1	−1	0	1.21 ± 0.014	41 ± 1.132
T02	9	1	−1	0	0.93 ± 0.005	77 ± 1.729
T03	1	−1	1	0	1.09 ± 0.017	62 ± 1.729
T04	18	1	1	0	0.75 ± 0.006	85 ± 0.653
T05	13	−1	0	−1	1.15 ± 0.006	58 ± 2.356
T06	10	1	0	−1	0.94 ± 0.013	83 ± 2.848
T07	15	−1	0	1	1.08 ± 0.013	69 ± 2.848
T08	5	1	0	1	0.88 ± 0.006	88 ± 2.848
T09	7	0	−1	−1	1.06 ± 0.006	37 ± 1.960
T10	8	0	1	−1	1.01 ± 0.006	82 ± 2.356
T11	14	0	−1	1	1.03 ± 0.006	56 ± 4.573
T12	2	0	1	1	0.85 ± 0.006	89 ± 1.729
T13	16	0	0	0	1.00 ± 0.003	82 ± 1.802
T14	3	0	0	0	1.01 ± 0.003	79 ± 1.802
T15	6	0	0	0	1.00 ± 0.006	81 ± 1.820
T16	12	−1	−1	0	1.22 ± 0.006	42 ± 3.638
T17	17	−1	0	−1	1.13 ± 0.003	53 ± 5.696
T18	4	0	1	1	0.90 ± 0.006	84 ± 2.356
T19	11	0	0	0	1.01 ± 0.006	84 ± 1.820

2.4. Evaluation of the Models

The statistical significance of the models obtained from the experimental results were determined using Analysis of Variance (ANOVA) with a confidence level of 90%. ANOVA is a powerful tool for evaluating the relationship between control variables and the observable response.

As per references [20–22], the change in response due to variations in control variables is known as the effect of these variables, and the ANOVA results provide adequate information to determine the validity of a model in explaining this relationship. To be considered statistically significant, the p-value of the coefficient must be less than the set confidence level, which was 0.1 in this case.

3. Results

The following section presents the results of the 19 runs, along with the regression models, their ANOVAs and micrographics. A summary of the results is provided in Table 3. The thickness results were obtained by taking three measurements per sample, which were then normalized and averaged. Normalization was performed to facilitate the applicability of the model to laminates with different thicknesses and sequences. A thickness of 3.40 mm, equivalent to 20 layers of 0.17 mm each, was considered.

The porosity results were obtained through the measurement of the amplitude of the backwall echo, using the protocol specified in the AITM-6-4005 [26] regulations. Three measurements were taken for each sample and the average was calculated. The amplitude of the backwall echo is indicative of the level of porosity of the material. The greater the amplitude, the lower the presence of pores and voids in the component.

3.1. Thickness Model

The thickness response surface is represented by Equation (1), with parameters listed in Table 4. The model comprises a quadratic term (PQ) of temperature, the interaction terms (TWI) between pressure and time, the linear terms (FO) of temperature, pressure and time and a constant.

$$TRS = \beta_0 + \beta_1 \cdot T + \beta_2 \cdot P + \beta_3 \cdot t + \beta_4 \cdot P \cdot t + \beta_5 \cdot T^2 \quad (1)$$

where TRS is the thickness response surface, β_i are the model coefficients, T is the forming temperature, P is the consolidation pressure, and t is the equivalent time.

Table 4. Thickness model. Regression analysis coefficients and ANOVA. Analysis performed on RStudio. FO: first order terms; TWI: interaction terms; PQ: pure quadratic terms; Df: degrees of freedom; Sum Sq: sum of squares; Mean Sq: mean squares.

	Type and Affected Parameters		**Coefficients**		**p-Value**
β_0	Constant		5.7084×10^{00}		0.0024
β_1	Linear term of T		-2.2711×10^{-02}		0.0153
β_2	Linear term of P		-3.2726×10^{-02}		(0.1819)
β_3	Linear term of t		-5.3302×10^{-05}		(0.9088)
β_4	Interaction term of P·t		-7.8929×10^{-04}		0.0607
β_5	Squared term of T		2.7073×10^{-05}		0.0298
R^2	0.9582				
adjR^2	0.9349				
***p*-Value (fstats)**	0.0000				
	Df	**Sum Sq**	**Mean Sq**	**F Value**	**Pr (>F)**
FO (T, P, t)	3	0.1754	0.0587	64.9335	0.0000
TWI (P, t)	1	0.0042	0.0042	4.6602	0.0592
PQ (T)	1	0.0059	0.0059	6.6430	0.0298
Residuals	9	0.0081	0.0009		

3.1.1. Analysis of Variance

Table 4 displays the *p*-value of each coefficient in the thickness model. The *p*-value of the overall model (0.0000) is below the confidence level (0.1), thereby enabling the rejection of the null hypothesis. The null hypothesis states that there is no relationship between the control variables (temperature, consolidation pressure and equivalent time) and the dependent variable (thickness). This result confirms that the proposed quadratic model with interaction and linear terms accurately explains the relationship between thickness, temperature, pressure and time.

The effect of each coefficient in the equation was analyzed, and it was found that the squared term of temperature, linear term of temperature and the interaction term of pressure and time had a significant impact on the thickness results. On the other hand, the linear terms of pressure and time were found to be insignificant for the model but were retained because of the significant impact their interaction had on the final thickness of the component.

3.1.2. Response Surface Results

The study of the influence of three parameters on component manufacturing was conducted by setting the equivalent time as a constant variable.

The response surface results, shown in Figure 3, display three sections corresponding to equivalent times of 10, 50 and 100 s. The trend observed from the plots is that the maximum thickness values are achieved at low temperature and pressure, while the minimum thickness values are observed at high temperature and pressure. Moreover, the green iso-curve representing the nominal design thickness shifts towards lower temperature and pressure requirements as the equivalent time increases.

The relationship between temperature, pressure and the viscosity of the matrix (as described in references [23,24,27]) is the basis for explaining the observed behavior. As temperature increases, the viscosity of the matrix decreases, increasing matrix mobility and reducing the amount of pressure and time required. However, reducing viscosity could boost the matrix flowing outwards from the component if the applied pressure or the application time is excessive. These effects could lead to a thinner component as the matrix creeps more easily.

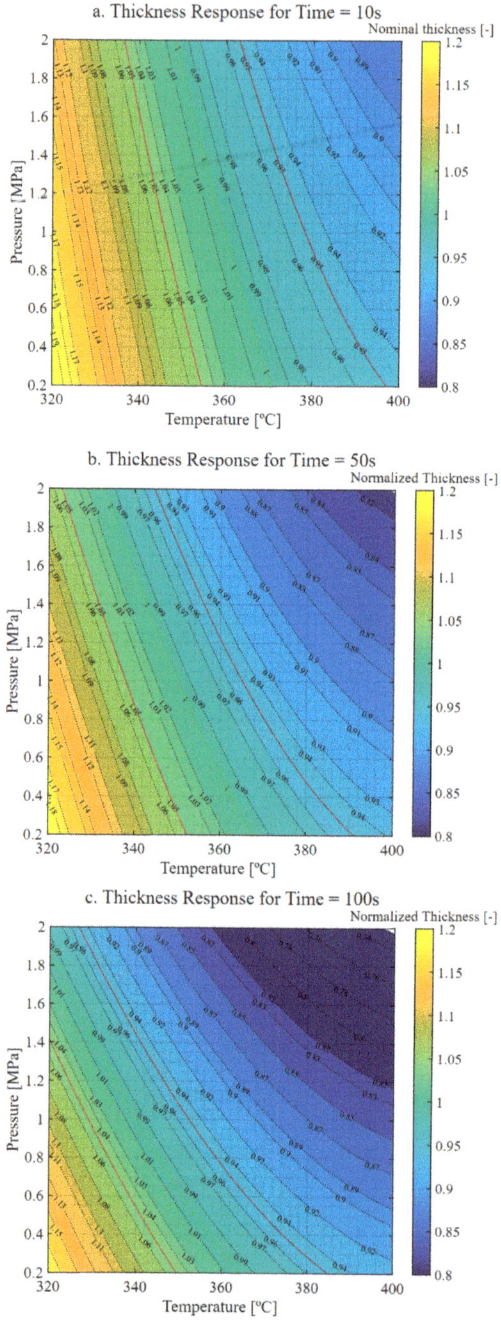

Figure 3. Thickness response. (**a**) Response for t = 10 s. (**b**) Response for t = 50 s. (**c**) Response for t = 100 s. Green indicates nominal thickness. Red indicates thickness tolerances (±5%).

3.1.3. Diagnostic Model

Figure 4a presents a comparison between the experimental values and the predictions generated by the statistical model. The results clearly demonstrate that the equation derived from the model provides accurate predictions of the part's thickness as a function of the control variables under the testing conditions. Table 4 further supports this conclusion, as the high R^2 (95.82%) and adjR^2 (93.49%) values indicate that the model is able to predict the response with over 90% accuracy.

A residual analysis, depicted in Figure 4b, reveals no discernible pattern, suggesting a homogeneous distribution of residuals and lending credibility to the model. To further evaluate the statistical significance of the model, the residuals were analyzed against the number of runs. As depicted in Figure 4c, there is no significant correlation between the residuals and the number of runs, which eliminates any potential relationship between the number of tests and the results obtained.

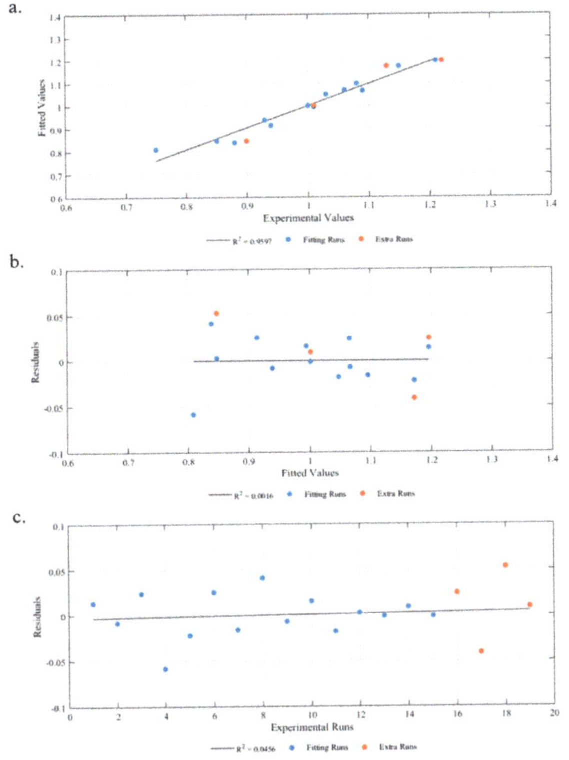

Figure 4. Thickness model diagnostic. (**a**) Fitted values vs. experimental values. (**b**) Residuals vs. fitted values. (**c**) Residuals vs. experimental runs.

3.2. Backwall Echo Model

The backwall echo response surface revealed a more complex relationship between the independent variables and the consolidation level of the final part, thus suggesting logarithmic dependencies. The model indicated that the interaction of pressure and time, as well as the linear terms of temperature and pressure, were the most significant parameters within the model. Equation (2) represents the model in its logarithmic and exponential forms.

$$BRS = P^{\beta 2} \cdot T^{\beta 3} \cdot \exp[\beta_0 + \beta_1 \cdot T + \beta_4 \cdot \log(P) \cdot \log(t)] \quad (2)$$

where BRS is the backwall echo response surface, β_i are the model coefficients, T is the forming temperature, P is the consolidation pressure, and t is the equivalent time.

3.2.1. Analysis of Variance

The variance of the model was analyzed using a 90% confidence interval, following the methodology described in a previous section. The results of the ANOVA, summarized in Table 5, reveal that the global *p*-value (0.0003) is significantly lower than the confidence interval. This result supports the rejection of the null hypothesis, indicating the feasibility of a relationship between the amplitude backwall echo and the control parameters.

Table 5. Backwall echo model. Regression analysis coefficients and ANOVA. Analysis performed on RStudio. FO: first order terms; TWI: interaction terms; PQ: pure quadratic terms; Df: degrees of freedom; Sum Sq: sum of squares; Mean Sq: mean squares.

	Type and Affected Parameters	**Coefficients**	***p*-Value**
β_0	Constant	2.4528	0.0001
β_1	Linear term of T	0.0046	0.0017
β_2	Linear term of log(P)	0.5175	0.0102
β_3	Linear term of log(t)	0.0607	(0.1423)
β_4	Interaction term of log(P)·log(t)	−0.0838	0.0823
R^2	0.8548		
adjR^2	0.7968		
***p*-Value (fstats)**	0.0003		

	Df	**Sum Sq**	**Mean Sq**	**F Value**	**Pr (>F)**
FO (T, log(P), log(t))	3	0.8534	0.2844	18.3873	0.0002
TWI (log(P), log(t))	1	0.0576	0.0576	3.7284	0.0823
Residuals	10	0.1547	0.1547		

Upon analyzing the effect of each coefficient, the amplitude backwall echo was found to be highly sensitive to variations in temperature and pressure. In contrast, the interaction between the logarithm of pressure and temperature exhibited lower significance. The logarithm of pressure did not exhibit a *p*-value lower than the confidence interval, thus rendering it statistically insignificant.

3.2.2. Response Surface Results

Figure 5 presents the porosity response surface against the control parameters. The two iso-curves highlighted in the figure represent the minimum acceptable amplitude of backwall echo according to reference [26]. The solid red line indicates the limit for thicknesses less than 5 mm, while the dashed line marks the minimum measurement for thicknesses greater than 5 mm.

The results show that the highest degree of consolidation occurs at the upper right corner of the surface, corresponding to the maximum temperature and pressure. Conversely, the minimum amplitude of backwall echo is observed at the lower left corner, where temperature and pressure are at their lowest values. As the equivalent time increases, the backwall amplitude becomes higher for lower temperatures and pressures.

These findings suggest a correlation between the control variables of thickness and porosity. The highest amplitude of the backwall echo is recorded in the case of minimum thickness. This correlation may be attributed to the creep behavior of the material, which helps in evacuating the air trapped between the plies and molecules, leading to a more compact structure with reduced porosity. This relationship agrees with Darcy's law that describes the flow of a fluid through a porous medium (applicable to laminar fluid flow through a porous media at low flow rates).

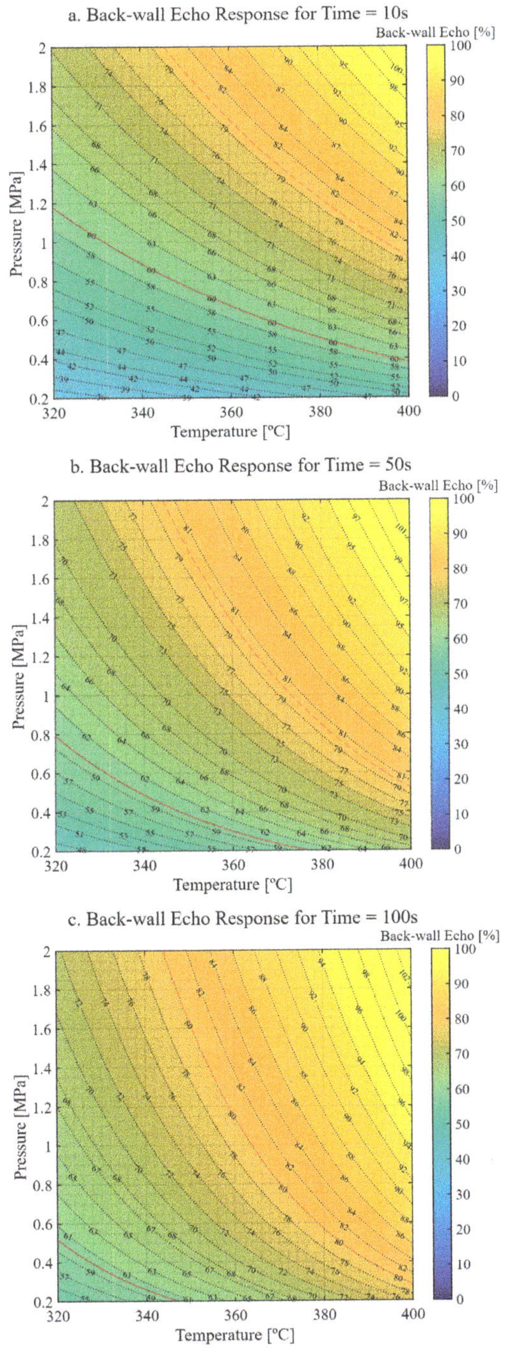

Figure 5. Backwall echo response. (**a**) Response for t = 10 s. (**b**) Response for t = 50 s. (**c**) Response for t = 100 s. Dashed red line is the minimum amplitude of backwall echo (BWE) allowed for parts with a nominal thickness greater than 5 mm. Solid red line is the minimum BWE allowed for parts with a nominal thickness less than 5 mm.

3.2.3. Diagnostic Model

The results of the model diagnosis (Figure 6a) indicate a low correlation between the predicted values and the experimental ones, as evidenced by the R^2 (0.8548) and adjR^2 (0.7968) values, which are lower than those of the thickness model. Despite this, the model still explains approximately 85% of the observed variation in the data. Further analysis of the residuals and the fitted values (Figure 6b), as well as the residuals and the experimental runs (Figure 6c), failed to reveal any significant relationships, suggesting the absence of any association between these variables.

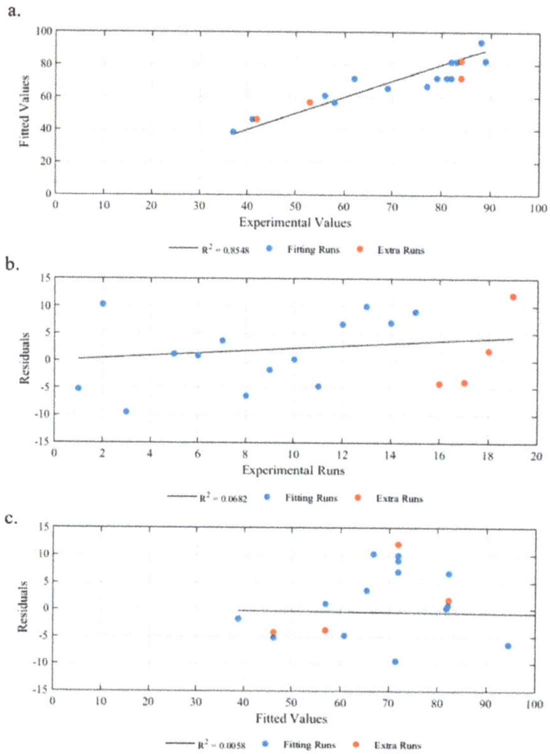

Figure 6. Backwall echo model diagnostic. (**a**) Fitted values vs. experimental values. (**b**) Residuals vs. fitted values. (**c**) Residuals vs. experimental runs.

3.3. Micrographics

Figure 7 summarizes the microstructural examination of the samples generated from the Box–Behnken experimental design. The results of the models are illustrated through these micrographics. For example, sample T04 was found to be the thinnest among the samples and displayed a homogeneous distribution of the layers without any visible pores or defects. The forming parameters of T = 400 °C, P = 2 MPa and t = 50 s resulted in a predicted thickness of 0.81 and 85% according to the model, while the experimental results were 0.75 and 85%. These values corresponded to the maximum level of amplitude of the backwall echo and the minimum level of thickness, positioning the test point in the upper right corner of both graphs. Sample T13 achieved a thickness closest to the nominal value. Under the conditions of T = 350 °C, P = 1 MPa and t = 50 s, the model predicted a thickness of 1.00 and 81%. In contrast, sample T09 showed the poorest results in the model, as evidenced by the micrograph, which reveals significant air bubbles trapped between the layers and delamination.

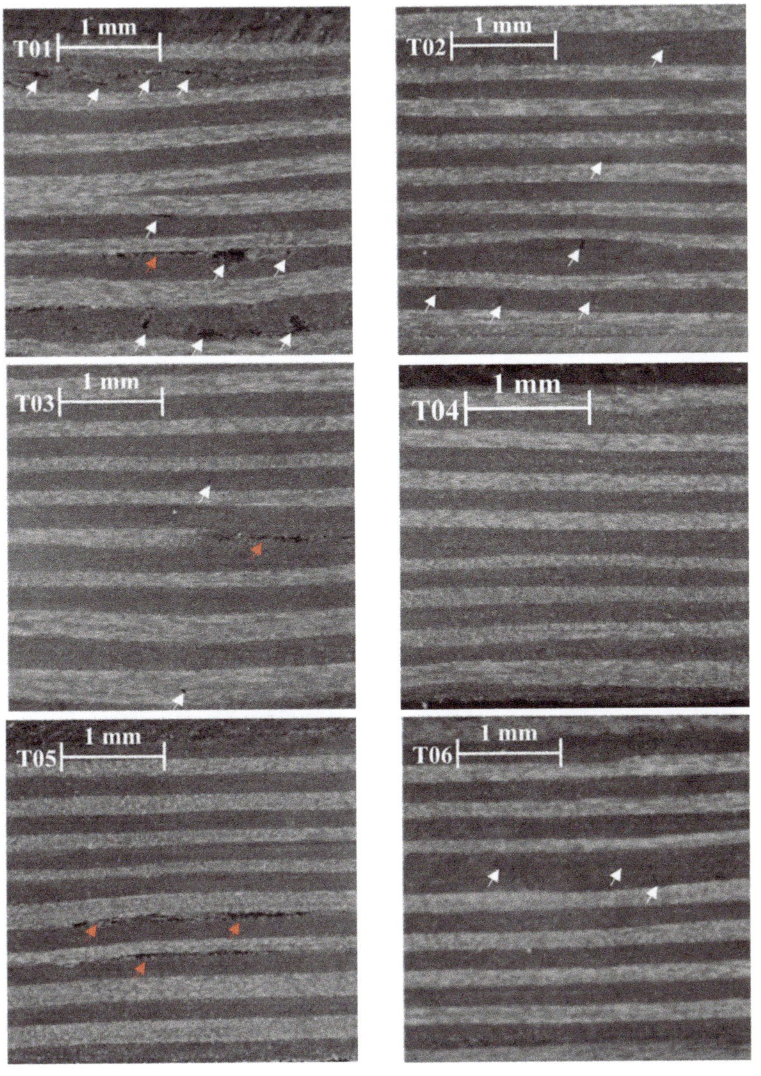

Figure 7. *Cont.*

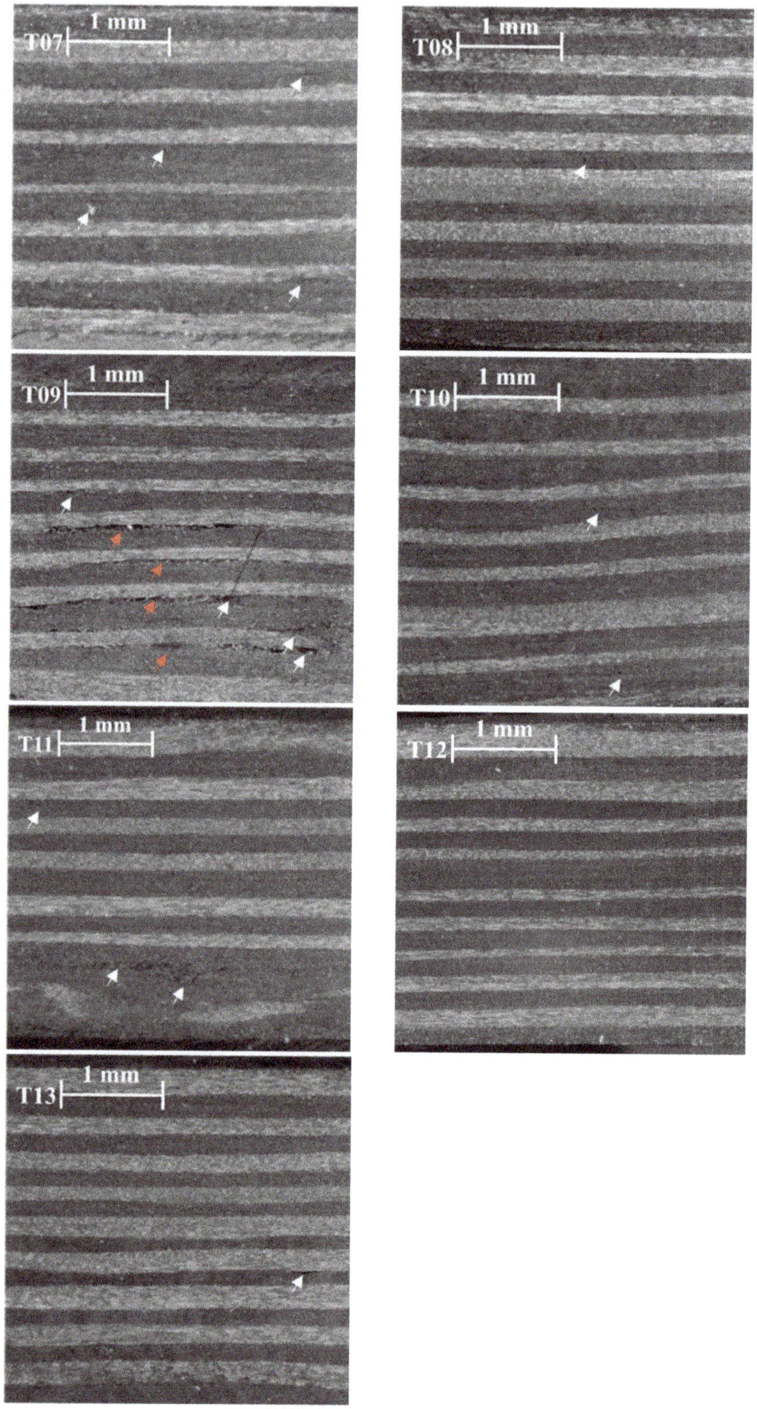

Figure 7. Micrographics of the samples used for the model fitting. White indicates voids. Red indicates delamination.

4. Discussion

4.1. Relationship between Thickness and Porosity

As previously reported, the thickness and porosity level of a manufactured part are strongly correlated. Figure 8 clearly illustrates this relationship, with a decrease in the amplitude of the backwall echo observed as the thickness increases, in accordance with Darcy's law.

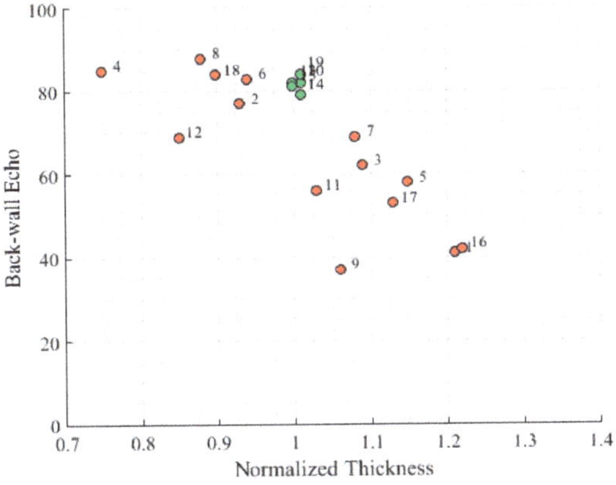

Figure 8. Amplitude of the backwall echo vs. normalized thickness. Green indicates good quality of the manufactured samples with values withing tolerances. Red indicates values outside tolerances.

The observed effects are in line with expectations, as a reduced thickness leads to increased material flow, enabling the filling of all existing pores within the matrix, thereby reaching higher levels of porosity. This effect underscores the significance of controlling the parameters, as conditions favoring the nominal thickness are found to decrease the amplitude of the backwall echo.

The superimposed graphs of the thickness and amplitude of the backwall echo are presented in Figure 9. In this figure, the manufacturing conditions that yield components meeting both criteria (thickness and amplitude of the backwall echo) are highlighted in green. The first graph depicts a consolidation time of 10 s, corresponding to a velocity of 10 mm/s. This velocity condition is the only one capable of achieving sufficient consolidation levels for components exceeding 5 mm in thickness. However, it is worth noting that such high velocity necessitates elevated temperature and pressure values to achieve favorable thickness outcomes.

Conversely, the other end of the spectrum is represented by the figure corresponding to 100 s, i.e., equivalent to a velocity of 1 mm/s. As indicated by the outcomes, the slower material flow due to the reduced velocity leads to lower temperature and pressure requirements as the material has more time to flow. Surprisingly, this prolonged flow time demonstrates a counterproductive impact, resulting in higher porosity compared to the other two time frames tested.

Another noteworthy trend that is discernible from the figures is that lower pressure expands the range of temperatures that yield thickness values within tolerance. On the other hand, as pressure increases, temperature control becomes increasingly critical, given the reduced thermal range.

In light of these findings, it is evident that the consolidation, thickness and forming conditions wield a substantial influence on the resulting composite material properties. The interaction between parameters such as consolidation time, temperature and pressure plays a decisive role in determining the porosity and thickness outcomes of the manufactured parts. The delicate balance between these parameters is crucial to striking the right balance between sufficient material flow for pore filling and avoiding excessive porosity.

Furthermore, the relationship between temperature, pressure and time underscores the trade-offs inherent in the forming process. While higher velocities facilitated quicker production and reduced cycle times, they necessitated higher temperatures and pressures to compensate for the limited material flow duration. On the other hand, lower velocities offer extended flow periods, allowing for decreased temperature and pressure requirements, but at the cost of potentially reduced material quality due to increased porosity.

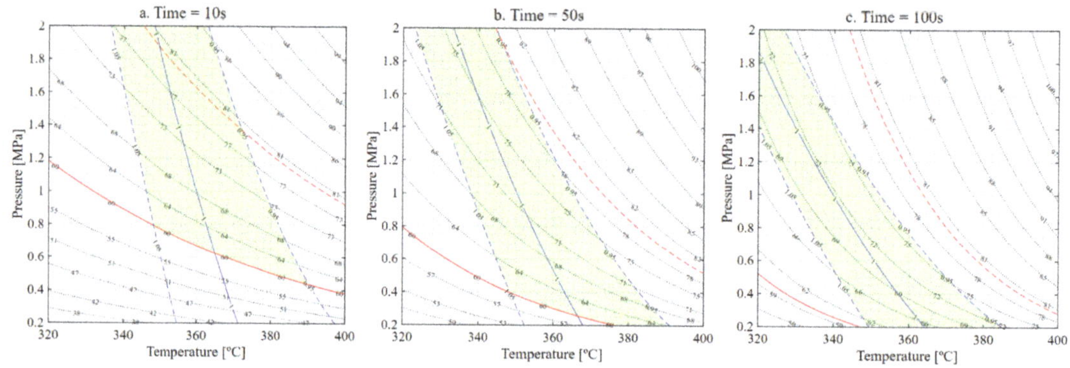

Figure 9. Amplitude of the backwall echo vs. normalized thickness. (**a**) Time slice of 10 s. (**b**) Time slice of 50 s. (**c**) Time slice of 100 s. Green indicates good quality of the manufactured samples. Dashed blue lines represent nominal thickness ± 5%. Solid blue lines indicate nominal thickness. Dashed red lines are the minimum amplitude of the backwall echo (BWE) allowed for parts with a nominal thickness greater than 5 mm. Solid red lines are the minimum BWE allowed for parts with a nominal thickness less than 5 mm.

4.2. Process Optimization

One of the objectives of the present study was to establish the parameters for the glide forming process adapted to thermoplastic composite materials. Similar to other manufacturing processes, lower values of the controllable parameters are typically preferred due to ease and cost considerations. High-temperature and high-pressure processes pose challenges because of the lack of auxiliary materials and complex heating systems, respectively. Thus, the aim was to find the minimum pressure and temperature that can meet the quality requirements of the parts being manufactured while also maintaining a high rate of production to remain commercially competitive.

The presented models can assist this determination by considering the limitations of tooling and auxiliary materials. An example of the use of these models is illustrated in Figure 10. In this scenario, a part with 15 plies (2.5 mm) must be manufactured, with strict control over its thickness and a requirement for an amplitude of the backwall echo greater than 60%. The machine has a maximum pressure limitation of 1 MPa, and the required production rate is 10 mm/s.

With this information, the models were used to set the production parameters focused on production rate and thickness. The 10 s section plots were selected to meet the manufacturing rate. This rate requires a forming temperature of 359 °C and a pressure of 1 MPa, resulting in a normalized thickness of 1.0 and an amplitude of backwall echo of 68%.

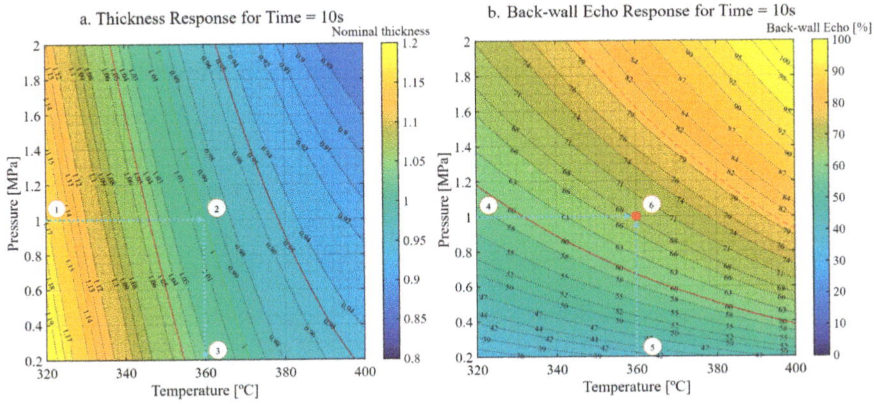

Figure 10. Process optimization sequence. 1. Maximum pressure available is selected. 2. Nominal thickness is pointed. 3. Temperature required is obtained. 4. Maximum pressure available is selected. 5. Temperature required is selected. 6. The consolidation level predicted is pointed where pressure and temperature intersect. In the left image, green solid line represents nominal thickness. Red solid lines indicated nominal thickness ± 5%. In the right image, dashed red line indicates the minimum amplitude of the back-wall echo (BWE) allowed for parts with a nominal thickness greater than 5 mm. Solid red line is the minimum BWE allowed for parts with a nominal thickness less than 5 mm. Red dot marks the predicted consolidation level.

5. Conclusions

In conclusion, the study suggested that continuous consolidation manufacturing technologies, such as glide forming, could be adapted to thermoplastic composite materials, but the process parameters must be carefully controlled to ensure the quality of the final part. The results of this study provide important information for the industrial application of the glide forming technology to thermoplastic composite materials. The ANOVA analysis allowed us to establish the significant manufacturing parameters affecting the quality of the laminate, leading to two response surface models. These models could be used as a post-processing tool to predict the thickness and consolidation level when simulating the manufacture of a component on macroscale levels. Further research is needed to optimize the process parameters and to study the effects for other thermoplastic composite materials.

Author Contributions: Conceptualization, D.C. and A.M.; methodology, D.C. and A.M.; software, D.C.; validation, D.C., A.M. and P.M.; formal analysis, D.C.; investigation, D.C.; writing—original draft preparation, D.C.; writing—review and editing, D.C., A.M. and P.M.; visualization, D.C.; supervision, P.M.; project administration, D.C. All authors have read and agreed to the published version of the manuscript.

Funding: This research was supported by the Industrial Doctorate Program of the Generalitat de Catalunya, the AMADE Research Center and Applus+ Laboratories. Grant: 2019 DI 058.

Data Availability Statement: Data are only available upon reasonable request.

Conflicts of Interest: The authors declare no conflict of interest.

References

1. Biron, M. *Thermoplastic Composites, in Thermoplastic and Thermoplastic Composites*, 3rd ed.; William Andrew Publishing: Norwich, NY, USA, 2018; pp. 821–882.
2. Steenkamer, A.; Sullivan, J.L. On the recyclability of a cyclic thermoplastic composite material. *Compos. Part B Eng.* **1998**, *29*, 745–752. [CrossRef]
3. Pegoretti, A. Towards sustainable structural composites: A review on the recycling of continuous-fiber-reinforced thermoplastics. *Adv. Ind. Eng. Polym. Res.* **2021**, *4*, 105–115. [CrossRef]

4. Martin, I.; Saenz del Castillo, D.; Fernandez, A.; Güemes, A. Advanced thermoplastic composite manufacturing by in-situ consolidation: A review. *J. Compos. Sci.* **2020**, *4*, 149–185. [CrossRef]
5. Arquier, R.; Iliopoulos, I.; Réginier, G.; Miquelard-Garnier, G. Consolidation of continuous-carbon-fiber-reinforced PAEK composites: A review. *Mater. Today Commun.* **2022**, *32*, 104036. [CrossRef]
6. Colton, J.M.; Muzzy, J.; Birger, S.; Yang, H.; Norpoth, L. Processing parameters for consolidating PEEK/carbon fibre (APC-2) composites. *Polym. Compos.* **1992**, *13*, 421–426. [CrossRef]
7. Patou, J.; Bonnaire, R.; De Luycker, E.; Bernhart, G. Influence of consolidation process on voids and mechanical properties of powdered and commingled carbon/PPS laminates. *Compos. Part. A Appl. Sci. Manuf.* **2019**, *17*, 260–275. [CrossRef]
8. Liu, F.; Li, T.; Xu, F.; Li, J.; Jiang, S. Microstructure, Tensile Property, and Surface Quality of Glass Fiber-Reinforced Polypropylene Parts by rapid heat cycle molding. *Adv. Polym. Technol.* **2020**, *2020*, 3161068. [CrossRef]
9. Brufau, J.; Lorente, A.S. System for Forming Stacks of Composite Materials. U.S. Patent Application No. US2014/0290866A1, 17 October 2017.
10. Brufau, J. System for Forming Stacks of Composite Materials. U.S. Patent Application No. US2016/0082675A1, 15 May 2018.
11. Robins, B.G.; Berrios, I.; Redondo, J.B.; Rueda, M.C.C. Heat Blankets Assembly for Forming a Composite Charge. U.S. Patent Application No. US2019/0016040A1, 23 June 2020.
12. Campos, D.; Brufau, J.; Biurrun, M.; Martín, A. A Review of the A+ Glide Forming Manufacturing Technology and its Adaptation to CFRP Thermoplastic Composite Materials. In Proceedings of the 5th International Conference & Exhibition on Thermoplastic Composites, Bremen, Germany, 13–14 October 2020.
13. Enoki, S.; Kojima, K.; Mizuno, S.; Katayama, K.; Tanaka, K. High-speed compression molding of continuous carbon fiber reinforced polypropylene. *WIT Trans. Built Environ.* **2014**, *137*, 311–315.
14. Piott, F.; Krämer, A.; Lück, A.; Hoffmann, L.; Mitschang, P.; Drummer, D. Increasing the performance of continuous compression moulding by local pressure adaption. *Adv. Manuf. Polym. Comppos. Sci.* **2021**, *7*, 1–14. [CrossRef]
15. Christmann, M.; Medina, L.; Mitschang, P. Effects of inhomogeneous temperature distribution on the impregnation process of the continuous compression molding technology. *J. Thermoplast.* **2016**, *30*, 1285–1302. [CrossRef]
16. Chih-Min, M.; Cheng-Tao, Y.; Bor-Wen, C. Optimization of stamp forming process for thermoplastic composites. *Res. J. Appl. Sci. Eng. Technol.* **2014**, *7*, 1568–1572.
17. Brooks, R.A.; Wang, H.; Zerong, D.; Jie, X.; Song, Q.; Liu, H.; Dear, J.P.; Li, N. A review on stamp forming of continuous fibre-reinforced thermoplastics. *Int. J. Lightweight Mater. Manuf.* **2022**, *5*, 411–430. [CrossRef]
18. McCool, R.; Murphy, A.; Wilson, R.; Jiang, Z.; Price, M.; Butterfield, J.; Hornsby, P. Thermoforming carbon fiber-reinforced thermoplastic composites. *Proc. Inst. Mech. Eng. Part L J. Mater. Des. Appl.* **2012**, *226*, 91–102.
19. Tatsuno, D.; Yoneyyama, T.; Okamoto, M. Hot press forming of thermoplastic CFRP sheets. In Proceedings of the 17th International Conference on Metal Forming, Toyohashi, Japan, 16–19 September 2018.
20. Yanagimoto, J.; Ikeuchi, K. Sheet forming process of carbon fiber reinforced plastics for lightweight parts. *CIRP Annals Manuf. Tech.* **2012**, *61*, 247–250. [CrossRef]
21. Tukey, J.W. *Exploratory Data Analysis*, 1st ed.; Pearson: Readind, MA, USA, 1977.
22. Kutner, M.; Nachtsheim, C.; Neter, J.; Li, W. *Applied Linear Statistical Models*, 4th ed.; McGraw-Hill Education: Homewood, IL, USA, 2004.
23. Box, G.E.P.; Cox, D.R. An analysis of transformations. *J. R. Statist. Soc. B* **1964**, *26*, 211–252. [CrossRef]
24. Khan, M.A.; Mitschang, P.; Schledjewski, R. Identification of some optimal parameters to achieve higher laminate quality through tape placement process. *Adv. Polym. Technol.* **2010**, *29*, 98–111. [CrossRef]
25. Toray Advanced Composites. Toray Cetex TC1225 Product Data Sheet. 2019. Available online: https://www.toraytac.com/media/3bd72fac-0406-48e4-bfc4-2ffd2398ac0c/zipxIA/TAC/Documents/Data_sheets/Thermoplastic/UDtapes%2Cprepregsandlaminates/Toray-Cetex-TC1225_PAEK_PDS.pdf (accessed on 5 August 2023).
26. *AITM-6-4005*; Ultrasonic Pulse-Echo Inspection of Carbon Fibre Plastic. Airbus Industries: Blagnac, France, 2018.
27. Deignan, A.; Figiel, L.; McCarthy, M.A. Insights into complex rheological behaviour of carbon fibre/PEEK from a novel numerical methodology incorporating fibre friction and melt viscosity. *Compos. Strct.* **2018**, *189*, 614–626. [CrossRef]

materials MDPI

Review

Fused Deposition Modelling (FDM) of Thermoplastic-Based Filaments: Process and Rheological Properties—An Overview

Domenico Acierno [1] and Antonella Patti [2,*]

[1] Regional Center of Competence New Technologies for Productive Activities Scarl, Via Nuova Agnano 11, 80125 Naples, Italy; acierno@crdctecnologie.it
[2] Department of Civil Engineering and Architecture (DICAr), University of Catania, Viale Andrea Doria 6, 95125 Catania, Italy
* Correspondence: antonella.patti@unict.it

Abstract: The fused deposition modeling (FDM) process, an extrusion-based 3D printing technology, enables the manufacture of complex geometrical elements. This technology employs diverse materials, including thermoplastic polymers and composites as well as recycled resins to encourage sustainable growth. FDM is used in a variety of industrial fields, including automotive, biomedical, and textiles, as a rapid prototyping method to reduce costs and shorten production time, or to develop items with detailed designs and high precision. The main phases of this technology include the feeding of solid filament into a molten chamber, capillary flow of a non-Newtonian fluid through a nozzle, layer deposition on the support base, and layer-to-layer adhesion. The viscoelastic properties of processed materials are essential in each of the FDM steps: (i) predicting the printability of the melted material during FDM extrusion and ensuring a continuous flow across the nozzle; (ii) controlling the deposition process of the molten filament on the print bed and avoiding fast material leakage and loss of precision in the molded part; and (iii) ensuring layer adhesion in the subsequent consolidation phase. Regarding this framework, this work aimed to collect knowledge on FDM extrusion and on different types of rheological properties in order to forecast the performance of thermoplastics.

Keywords: fused deposition modelling (FDM); thermoplastics; rheological properties; buckling; pressure-driven extrusion; flow instability; welding

1. Introduction

Additive manufacturing (AM) is an emerging technology for processing material and creating an object from a three-dimensional (3D) model, usually via layer-to-layer deposition. The AM process can be constituted by two phases: the first is a virtual phase during which a computer-aided design (CAD) model is prepared using CAD software packages, the second is a physical phase to develop the physical object [1].

According to the standard ISO/ASTM 52900 [2], the additive manufacturing (AM) is defined: "process of joining materials to make parts from 3D model data, usually layer upon layer, as opposed to subtractive manufacturing and formative manufacturing methodologies" [3]. Historical terms are "additive fabrication, additive processes, additive techniques, additive layer manufacturing, layer manufacturing, solid freeform fabrication and freeform fabrication".

Additive manufacturing (AM) is changing the aviation industry, space applications, and missile defense systems [4]. It opens up new opportunities for designing fine jewelry models or developing new artistic works [1], for the creation of adaptive and sophisticated buildings, or for smooth ready textures to paint surfaces [5]. The most common AM uses in biomedical applications involve the fabrication of body parts, anatomical models, implants for orthopedics, scaffolding, and drug delivery systems [6]. AM technologies are also becoming increasingly common in the automotive sector, where they are used to develop newer models and to modify existing ones in response to new design trends and technological breakthroughs [7].

The earliest patents on additive manufacturing processes were granted in 1984 to three French engineers from Cilas Alcatel (Orléans, France) ("French Patent FR2567668(B1)") [8] and to C. Hull from 3D Systems, Inc. (Rock Hill, SC, USA) ("US Patent 4,575,330") [9]. In the years since then, the AM sector has expanded to include a wide range of fast prototyping technologies, leading to the so-called "3rd Industrial Revolution" [10].

Controlling the rheological properties of molten/suspended polymers is critical to the efficiency of polymer processing. Controlling shear flow behavior requires the modification of many polymer properties (molecular weight, chain branching), as well as the application of modifiers (fillers, plasticizers, polymers) and adjusting processing factors (temperature, shear, pressure) [11].

Accordingly, the proper processing of polymeric materials requires a comprehensive understanding of their rheological properties; the characterization of polymer melt/suspended properties using relevant rheometers is very important for polymer processing management.

This review focused mainly on research articles written in English. Scopus, ScienceDirect, and Google Scholar have been considered as three common databases. More than 1000 recent articles in the last 5 years (~50% research articles; 20% reviews; ~10% encyclopedia contributions; 20% book chapters) were identified during database searching by using the following keywords and Boolean operators: "Fused deposition modelling" OR "3D printing" OR "Fused Filament Fabrication" OR "Additive Manufacturing" AND "rheological properties" AND "thermoplastics". Approximately 80% of the total works were eliminated: some research articles were excluded based on the title and abstract; patents, book chapters, and encyclopedia articles (often not available), theses and proceedings (except for specific cases) were not considered,. Based exclusively on the articles' contents, 50% of those remaining (200 articles) were removed. A table of contents was developed with the following key focuses: (i) description of the fused deposition modelling (FDM) technology; (ii) the importance of rheological properties in the processing of polymer-based materials; (iii) the important role of rheological analysis in predicting material behavior in the FDM process to encourage smooth and easy operation inside the capillary extrusion while avoiding buckling and promoting dimensional stability; (iv) a summary of recent studies from the literature involving rheological testing and FDM technology; and (v) challenges and future perspectives. The review incorporated more than 100 contributions.

2. Fused Deposition Modelling (FDM) Technology of Thermoplastic-Based Filaments

Fused deposition modeling (FDM) is an AM technique that is commonly used to create models, prototypes, and products. FDM is a fast and simple method to make personalized items at a low cost. S. Scott Crump, cofounder of Stratasys, invented this technology in the late 1980s that became commercially available in 1990. Fast forward to today, FDM, also known as extrusion-based additive manufacturing or FFF (fused filament fabrication), is the most widely used 3D printing technology.

An STL (STereo Lithography interface format or acronyms of "Standard Triangle Language" or "Standard Tessellation Language") file is processed by software, which mathematically slices and orients the model for the build process. Unwound from a coil, a solid filament (usually thermoplastic in nature) is heated past the glass transition/melting point and delivers material to an extrusion nozzle, which controls the flow. A worm-drive directs the filament into the nozzle at a predetermined rate. At the exit of the nozzle, the melted material is selectively deposited on a heated support platform to produce 3D parts directly from a CAD model in a layer-by-layer manner. This implies that after printing the first layer, the platform lowers and another layer is printed on top of the first one, and the procedure is repeated. The mechanism is frequently an X-Y-Z rectilinear design. A numerically controlled system can move the nozzle in both horizontal and vertical directions (along the X and Y axes), whereas the platform operates in the Z axis [12].

A schematic diagram of the FDM extrusion process is shown in Figure 1.

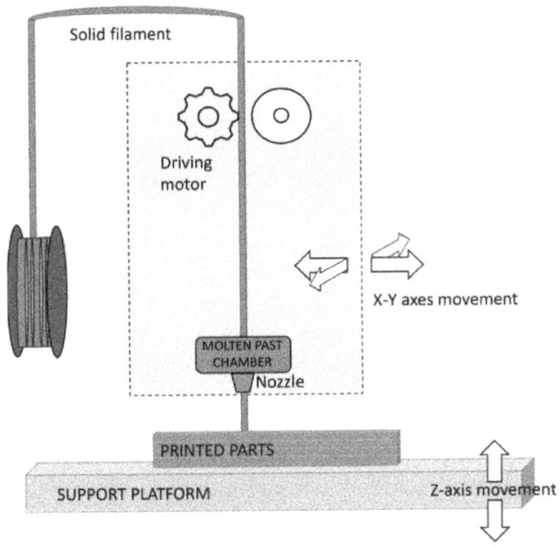

Figure 1. Schematic of FDM process.

Slicing parameters are layer thickness, flow rate, infill percentage, raster angle, raster pattern, air gap, nozzle diameter, and top and bottom width. Build orientation refers to the position of the part within the build platform in relation to the X, Y, and Z axes (horizontal, flat, or vertical alignment) [13]. Slicing parameters, building orientation, and temperature conditions are all considered critical parameters in the printing process [14].

The 3D printing process has some disadvantages, such as the low strength of the parts, rough surface finish, and slow production time [15]. For example, the layers may be too thick, leading to poor surface quality. An additional support may be needed to stabilize the 3D structure. The main cause of mechanical deficiency is deformation between layers. There is a weakness in the vertical strength of the parts due to warping and temperature fluctuations; these cause structure delamination and poor mechanical strength [14].

The maximum operating temperature of a commercially available FDM machine is commonly around 300 °C. This means that materials with exceptionally high melting points are usually inappropriate for use with this machine, and only certain types of plastics and materials that melt at moderate temperatures are suitable. ABS and PLA are the most often used thermoplastic polymers in FDM. Alternative polymers include polyamide or nylon (PA), polycarbonate (PC), polymethyl methylacrylate (PMMA), polyethylene (PE), and polypropylene (PP) [16].

Polymer-based composites and nanocomposites made with metal particles [17] (such as aluminium and iron powder), ceramics [18] (such as titanium dioxide, zirconium oxide, and aluminium oxide), nanomaterials [19] (such as carbon nanotubes, graphene, and clay), glass [20] and carbon fibres [21] (either continuous rovings or chopped fibres), and natural fibres [22] (such as wood, bamboo, flax, coir, jute, sisal, vegetable fibres, and oil palm) are commonly used in the FDM process [23].

One of the notable benefits of this growing technology is the reuse of discarded thermoplastic materials to produce quality products [24].

On one hand, convenient materials for extrusion-based printing are amorphous thermoplastic polymers. For amorphous polymers, internal tensions that occur after cooling (e.g., warpage) are limited by their low thermal expansion coefficients and extrusion temperatures [25]. On the other hand, semi-crystalline thermoplastics such as polylactic acid (PLA) [26] and thermoplastic polyurethanes [27] and polyamides (PA) [28] are employed to increase performance (increased toughness and wear resistance, as well as stiffness and strength) [29]. For example, semicrystalline printed parts are expected to exhibit a higher degree of anisotropy than amorphous ones. Semicrystalline polymers are also more challenging to work with when investigating interdiffusion across the contact, because incipient crystallization can be one of the limiting factors of the layer adhesion [25]. A high crystallization rate prevents interfacial adhesion, since the crystallized segment cannot weld to the prior layer; furthermore, high crystallinity causes internal stress and volume shrinkage [30].

The durability of FDM parts made from waste thermoplastics is lower than that of printed products made from virgin plastics [31]. Accordingly, it is important to understand the significant changes in recycled thermoplastic materials under repeated extrusions, such as chain fragmentation and changes in viscosity and strength at break [32]. It was discovered that the printing process has a strong impact on the pristine properties of extruded materials. This technique may induce a polymer decomposition reaction, resulting in a decrease in rheological properties over time. If the systems are kept at high temperatures, the destructive activity of the printing process can continue over time, further degrading the macromolecules. In composites, the damage caused by printing extrusion can be limited to the process duration [33].

Recently, new-type FDM filaments with excellent thermophysical properties have been investigated in order to improve the mechanical properties of printed parts by utilizing special engineering plastics such as polyetheretherketone (PEEK) [34], polyetherimide (PEI) [35], polyaryletherketone (PAEK) [36], and polyetherketoneketone (PEKK) [37] with excellent thermophysical properties and high melting points. This has been made possible by pushing the filament extrusion process' technological limitations and boosting the maximum operating temperatures to 400–500 °C.

3. Rheology in the Processing of Polymer-Based Materials

Rheology is the science of material deformation and flow, and viscosity is an important material characteristic that indicates flow resistance.

Consider a pair of large parallel plates, each one with area A, separated by a distance Y. One plate is set in motion with a velocity V. The space between them is filled by an incompressible Newtonian fluid. The flow is laminar. In steady-state conditions, a constant force F is required to maintain the motion of the plate. The viscosity μ is defined as constant of proportionality (for a given temperature, pressure, and compositions) between tangential force F divided by the area A (i.e., applied shear stress τ_{yx}) and the velocity V divided by the distance between plates Y (i.e., velocity gradient $-\frac{dv_x}{dy}$):

$$\frac{F}{A} = \mu\frac{V}{Y} \tag{1}$$

also rewritten as follows:

$$\tau_{yx} = -\mu\frac{dv_x}{dy} \tag{2}$$

Equation (2), known as Newton's law of viscosity, states that the shearing force per unit area is proportional to the negative of the velocity gradient [38].

This is the potential scenario of a rotational rheometer, in which a viscometric flow and a constant shear rate can be assumed across the rheometer gap.

For incompressible generalized Newtonian fluid and for extending the principle to any arbitrary flow [39], Equation (2) is expressed as follows:

$$\tau = -\eta\dot{\gamma} \tag{3}$$

where η can be a function of $\dot{\gamma}$(= the shear rate).

The decrease in viscosity with increasing the shear rate is referred as "shear-thinning" behaviour and the fluid is called "shear-thinning" or "pseudoplastic". When the viscosity increases as function of the shear rate, the fluid is called "shear thickening" or "dilatant".

Water has a viscosity of 10^{-3} Pa*s, whereas the viscosity of most polymer melts under extrusion can range from 10^2 to 10^5 Pa*s. Shear-rate dependency or non-Newtonian viscosity is an important characteristic of polymeric fluids during polymer processing: increasing the rate of shearing, i.e., extruding quicker through a die, reduces viscosity. This has been linked to molecular alignments and polymer chain disentanglements [40].

Various types of polymer processes often include different shear rate ranges. Compression moulding is an example of low shear rate process, whereas extrusion and injection moulding are an example of high shear rate techniques. Defining viscosity values in predetermined shear rate ranges is crucial for the design of plastic products, to ensure an easy process and to avoid clogging and loss of fluidity and productivity [41].

Polymeric fluids are referred to as viscoelastic fluids exhibiting both viscous and elastic characteristics when deformed under stress. Elastic property refers to a material's capacity to restore its original shape after being deformed by the action of a force. Viscous property refers to a material's irreversible deformation process, in which the basic form is no longer recoverable. The viscoelasticity of a material can be determined by applying an input and measuring the response of the material. Under linear conditions, the stress $(\tau(t))$ and the strain $(\gamma(t))$ are sinusoidal function of time (t) according to the expressions:

$$\tau(t) = \tau_0 \sin(\omega t + \delta) \tag{4}$$

$$\gamma(t) = \gamma_0 \sin(\omega t) \tag{5}$$

where τ_0 and γ_0 are the stress and strain amplitude, respectively; ω is the frequency and δ is the phase angle. $\delta = 0°$ corresponds to a perfect elastic material ("Hookenian"); $\delta = 90°$ corresponds to a perfect viscous material ("Newtonian").

Defining the complex modulus (G*) (Equation (6)), it can be divided into two components, i.e., the storage modulus-G′ (Equation (7)), and loss modulus-G″ (Equation (8)):

$$G^* = \frac{\tau(t)}{\gamma(t)} \tag{6}$$

$$G' = \frac{\tau_0}{\gamma_0} \cos\delta \tag{7}$$

$$G'' = \frac{\tau_0}{\gamma_0} \mathrm{sen}\delta \tag{8}$$

G′ represents the elastic response of the material and is related to the energy stored and recovered per cycle; G″ represents the viscous response of the material and refers to the loss, or dissipated, energy per cycle.

Modelling of Viscosity Data

The power law (also called the Ostwald–de Waele model) is the most used model for expressing the shear-thinning behaviour of polymers:

$$\eta = k\dot{\gamma}^{n-1} \tag{9}$$

where k is the consistency index and n is the power law exponent. For n = 1 the power law model reduces to Newton's law (constant viscosity). As n decreases, the polymer becomes more shear thinning. The power law exponent (n) for the most common polymers ranges from 0.25 for poly(methyl methacrylate) and acrylonitrile butadiene styrene to 0.35 for polypropylene and polyethylene, and 0.75 for nylon and polycarbonate [42].

In addition to the power law model, two other models, i.e., the Carreau–Yasuda model (Equation (10)) and the Cross model (Equation (11)), are typically used to improve data fitting over the whole range of shear rates and to include the Newtonian plateau at low shear rates:

$$\eta = \eta_\infty + \frac{\eta_0 - \eta_\infty}{\left[1 + (\lambda\dot{\gamma})^\alpha\right]^{\frac{1-n}{\alpha}}} \quad (10)$$

$$\eta = \eta_\infty + \frac{\eta_0 - \eta_\infty}{1 + (\lambda\dot{\gamma})^{1-n}} \quad (11)$$

η_0 is the zero-shear rate viscosity, $\eta\infty$ is an infinite-shear rate viscosity, and λ, α, and n are fitted parameters.

The zero-shear viscosity is a function of the weight average molecular weight (Mw):

$$\eta_0 = AM_w^\alpha \quad (12)$$

In Equation (12), the constant (A) and exponent (α) depend on polymer-based systems. Most often α is equal to 3.4.

Viscosity dependence on temperature is expressed by Equation (13):

$$\eta = \eta_{ref}\exp(-b(T - T_{ref})) \quad (13)$$

η_{ref} is the reference viscosity at a reference temperature (Tref) and b is the temperature sensitivity coefficient. Most common polymers, such as polyethylene, polypropylene, polyvinylchloride, and poly(methyl methacrylate), have b values of roughly 0.01%/°C (PE), with polystyrene reaching 0.05 [42].

The temperature dependency of viscoelastic properties can also be described using the Arrhenius law (Equation (14)). For temperatures between the glass transition temperature (T_g) and T_g + 100 °C, the Williams–Landel–Ferry (WLF) law (Equation (15)) was also discovered to be applicable to a broad range of polymers. The models are based on the time–temperature superposition (TTS) principle, entailing gradually translating the isotherms calculated at different temperatures through a shift factor (a_T) in relation to a curve taken at a reference temperature until all the curves overlap significantly. A master curve at the temperature of reference is obtained as a result of several shifts.

$$a_T = \exp\left(\frac{E_a}{R}\left(\frac{1}{T} - \frac{1}{T_{ref}}\right)\right) \quad (14)$$

$$\log a_T = \frac{-C_1(T - T_{ref})}{C_2 + (T - T_{ref})} \quad (15)$$

In Equation (14), Ea is the activation energy for flow, R is the gas constant. In the Equation (15), a_T is a shift factor, C_1 and C_2 are material-dependent coefficients.

4. Rheology in the FDM Process

The viscosity can be measured both by rotational and capillary viscometers. The ideal shear rate range during FDM extrusion is covered by a capillary rheometer, which also represents a device similar to a nozzle (i.e., Hagen–Poiseuille flow) [43]. However, during the printing process, the polymer melts undergo a wide range of shear rates, lower than 0.1 s^{-1} on the printed bed, up to 10^2–10^4 s^{-1} inside the nozzle [28], and a rotational rheometer can always be used. Curve fitting models (the power law, Cross, and Carreau–Yasuda models) or time–temperature superposition (TTS) can be considered useful approaches to extend the frequency scale from 0.01 to 1000 rad/s, i.e., beyond that achievable with the rotational viscometers, so as to accomplish the typical shear rate encountered by the melted polymer during the printing process [43,44].

However, similarities existing between the steady-state shear flow material and viscoelastic properties determined by dynamic measurements should be confirmed [28] using Equation (16):

$$\left|\eta^*(\omega)\right| = \eta(\dot{\gamma})|_{\omega=\dot{\gamma}} \quad (16)$$

This empiricism (Equation (16)), known as the Cox–Merz rule, predicts that, for different polymers, linear and branched macromolecules, high and low molecular weights, and solutions and melts, the magnitude of the complex dynamic viscosity (η^*) at frequency (ω) is comparable to the magnitude of the shear viscosity (η) at shear rate ($\dot{\gamma}$) [45].

Knowing the optimal viscosity range can help to predict whether a new melt formulation can be extruded [46–48]. Using a rheometer, comparing the viscosity of a new formulation with a successful extruded system is a useful method for testing the printability of a new/unknown resin. Different viscosity profiles do not always mean that the new molten formulation cannot be extruded, as long as it has a similar viscosity at the operating shear rate [43,47,49].

Capillary viscometers are usually used for the shear rate range from about 1 s^{-1} to 10^4 s^{-1}. Rotational viscometers are usually used for the range 10^{-2} to about 10^2 rad/s. Then, specific viscosity values should be required to extrude melted polymer through the print nozzle (shear rate in the range of 30–500 s^{-1}) and during consolidation process upon deposition on the print bed (shear rate in the range of 0.01–0.1 s^{-1}) (Figure 2a). The appropriate material characteristics, in terms of rheological features, are considered essential to the development of specific performance and products in the extrusion-based 3D printing process [50].

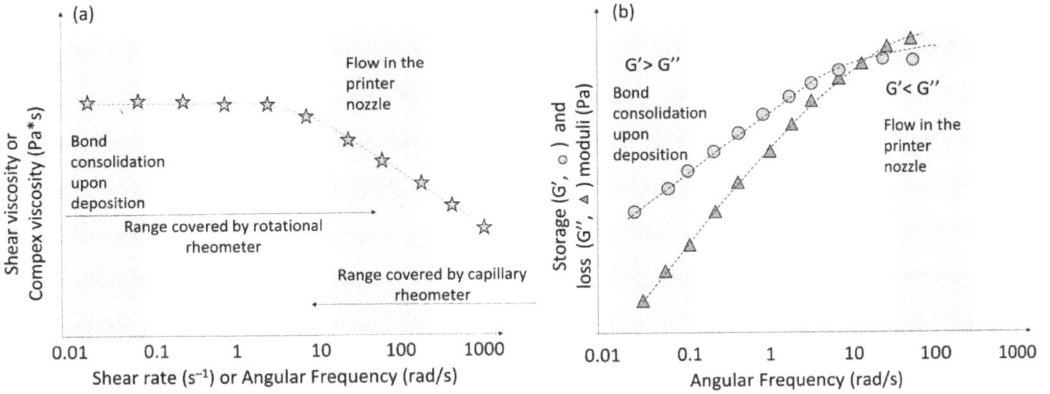

Figure 2. Typical viscosity (shear or complex) (**a**) and moduli (**b**) conditions for bond consolidation during deposition and flow in the printer nozzle.

To retain the three-dimensional design after deposition, the material should have a sufficiently high zero-shear viscosity in the terminal region (low shear rate). Furthermore, a material with good printability should have shear-thinning behaviour and low melt viscosity. The shear-thinning property of the polymer promotes smooth extrusion through small nozzles in extrusion-based 3D printing [28,51].

Warping was found in 3D objects when the material exhibited typical polymer chain relaxation time of 2×10^{-1} s, in combination with elevated η_0 values of 10^4 Pa*s. The material's response to thermal stress was improved, and no evidence of warpage was detected at an equal nozzle temperature (190 °C) as long as the polymer chains had a reduced relaxation time (about 10^{-2} s) and η_0 values in the order of magnitude of 10^3 Pa*s [44]. However, while increasing the extrusion temperature (210 °C) solved the warpage issue in the first systems (higher η_0 and relaxation time), it decreased printing accuracy due to considerable material deburring on the edges in the second systems (lower η_0 and relaxation time) [44].

Liquid-like characteristics (loss modulus (G'') higher than storage modulus (G')) are also necessary to ensure extrusion through the printing nozzle. Solid-like characteristics (storage modulus (G') higher than loss modulus (G'')) are then required to create an interlayer bond and to retain the shape in the post extrusion (Figure 2b). These conditions were validated in several investigations to demonstrate the suitability of 3D printing for various systems such as: (i) polymers (acrylonitrile butadiene styrene [52]); (ii) blends (polycaprolactone (PCL)/starch [53], poly(vinyl chloride) (PVC)/diisononyl phthalate [54]); and (iii) composites (polyamide 6/carbon fibre [28], polylactide acid (PLA)/poly(butylene adipate-co-terephthalate)/nano-talc [55], PLA/calcium peroxide [56], PLA/alumina [57], thermoplastic bio-polyurethane/carbon nanotubes [25]).

Table 1 displays the shear-thinning index and other parameters based on model fit for different polymer-based systems used in 3D printing applications.

Table 1. Parameters by fitting models for various polymer-based systems used in FDM.

Materials	Shear-Thinning Index (n)	Other Parameters	Fitting Models	References
Acrylonitrile butadiene styrene (ABS)	0.27	/	Carreau–Yasuda	[58]
	0.42–0.49	k = 6678–12189 Pa*s	Power law	[59]
ABS/carbon fibers	0.50	k = 8213 Pa*s	Power law	[59]
ABS/oil palm fiber (up to 7 wt.% in content)	0.53–0.71	k = 0.0024 − 0.0008	Power law	[60]
Ethylene-vinyl acetate (EVA)	0.40–0.47	$\eta_0 = 5.65 \times 10^2 - 2.95 \times 10^4$ Pa*s	Cross model	[61]
Polycaprolactone (PCL)	0.18	$\eta_0 = 4.55 \times 10^3$ Pa*s	Cross model	[61]
	0.47–0.49	/	/	[62]
PCL/ hydrolyzed collagen (30 wt.%)	0.6–0.7	/	/	[62]
Polyethylene oxide (PEO)	0.47	$\eta_0 = 4.18 \times 10^4$ Pa*s	Cross model	[61]
Polyethylene/wood flour (up to 50 wt%)	0.4	k = 10^4–4×10^4 Pa*s^n	Power law	[63]
Poly(D,L-lactic-acid) (PDLLA)	0.33	η_∞ = 0.1 Pa*s; η_0 = 2260 Pa*s, λ = 0.0294 s	Carreau–Yasuda	[64]
Polylactide acid (PLA)	0.32	/	Power Law	[58]
	0.69	η_∞ = 0 Pa*s; η_0 = 1945 Pa*s, λ = 0.08 s	Carreau–Yasuda	[65]
	0.29–0.37	η_∞ = 0 Pa*s; η_0 = 1480–9460 Pa*s, λ = 0.013–0.083 s	Cross	[44]
PLA/carbon nanotubes (CNT) (1.5 to 12 wt.%)	0.51–0.07	/	Power law	[66]
PLA/ graphene nanoplates (GNP) (1.5 to 12 wt.%)	1–0.15	/	Power law	[66]
PLA/copper (66 wt.%)	0.52	k = 215 Pa*s	Power law	[67]
PLA/carbon fibers (14 wt.%)	0.76	k = 222 Pa*s	Power law	[67]
Poly(methyl methacrylate) (PMMA)/ CNT(1 wt.%)/ nano hydroxyapatite up to 10 wt.%)	0.53–0.56	/	/	[68]
Polypropylene/CNT (0.3–1 wt.%)	0.27–0.28	η_0 = 440–509 Pa*s	Carreau–Winter	[69]

Several critical conditions must be met for a material to be used properly in the FDM process [70]. Most of them can be summarized as follows: pressure driven extrusion at certain flow rate through an assigned diameter nozzle, ability to preserve the shape once deposited on the support throughout the characteristic processing time, dimensional stability of the extruded structure throughout cooling, and transition to the final state.

4.1. Buckling

The printed layers must adhere well to the base support, and the printed structures must exhibit minimum warpage owing to consolidation and shrinkage. The solid filament serves as the push rod for the extrusion process. It is pulled forward by a wheel and is responsible for transmitting force to the soft material in the nozzle. Therefore, the filament must be strong enough to be handled in the printer. While it must be flexible enough to be collected as a coil, it must also be strong and rigid enough to act as a push rod during the extrusion process without breakage and deformation (buckling effect (Figure 3)) [71].

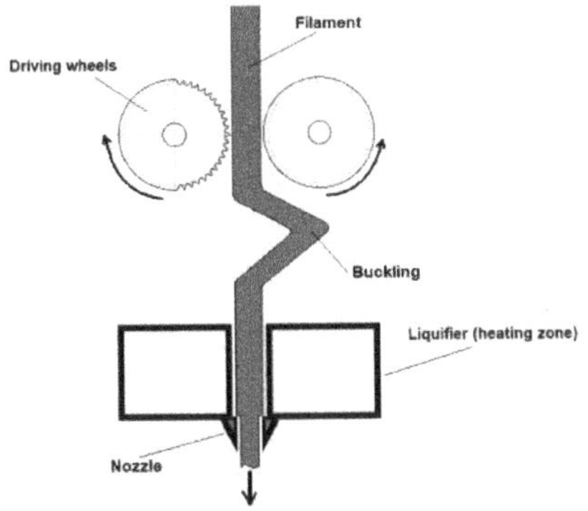

Figure 3. Schematic of buckling phenomena during printing. Reproduced from [72].

Two primary mechanical properties are required for the solid: enough strength to prevent buckling and enough surface hardness to prevent filament grinding induced by the extruder gears [73].

By considering L as a distance between the rollers and the printing head, a filament of radius R, K the young modulus of materials, and $\dot{\gamma}_w$ the wall shear rate, there is a critical stress limit (σ_c) that the filament can withstand before it buckles and becomes ineffective. To avoid filament buckling, σ_c must be higher than the required pressure to push the filament through the nozzle (Equation (17)):

$$\Delta P < \sigma_c \tag{17}$$

where (r and l are the radius and length of nozzle, respectively):

$$\Delta P = \frac{2l\eta\dot{\gamma}_w}{r} \tag{18}$$

$$\sigma_c = \frac{K\pi^2}{\left(\frac{L}{R}\right)^2} \tag{19}$$

Thus:

$$\frac{2l\eta\dot{\gamma}_w}{r} < \frac{K\pi^2}{\left(\frac{L}{R}\right)^2} \tag{20}$$

$$\frac{K}{\eta} > \frac{2l\dot{\gamma}_w}{r\pi^2}\left(\frac{L}{R}\right)^2 \tag{21}$$

Adjusting the viscosity and shear rate could contribute to reduce the buckling [43].

Novel low-density polyethylene (LDPE) composites at 15 and 30 wt.% of waste glass were proposed to FDM for use in low-duty frictional applications. Despite having a higher viscosity during printing than the matrix, composites have a higher elastic modulus than LDPE, allowing for faster printing. The higher viscosity of composites was compensated by a higher stiffness permitting to faster filament feeding [72]. According to [49], the viscosity should not surpass specified values 10^5 Pa*s in the shear rate period involved in printing to provide an easy flowing material and avoid buckling of the solid component of the filament.

Rotational rheology was utilized in [54] to explore plasticized poly(vinyl chloride) (PVC) formulations, including up to 40 wt. % of diisononyl phthalate (DINP), to produce flexible and ductile filaments for use in the 3D printing process. The problem of filament buckling has been addressed by taking into account the constraint defined by the ratio of the filament's compressibility in the solid state (K) to the viscosity in the molten state (η_{app}) (Figure 4).

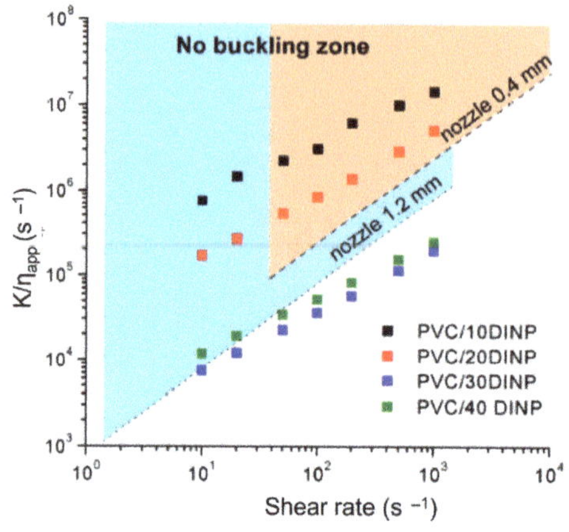

Figure 4. K/η_{app} vs the shear rate for formulations of plasticized PVC up to 40 wt. % of DINP content. The dotes lines highlighted critical conditions depending on nozzle diameter (0.4 and 1.2 mm). Coloured zone (blue/orange depending on nozzle diameter) indicates "no buckling" conditions. Reproduced from [54].

4.2. Pressure-Driven Extrusion

Materials with a low elasticity modulus are susceptible to buckling. As a result, if the pressure drops over the nozzle is reduced by increasing the nozzle diameter, the critical pressure for buckling increases. Thus, the processing window for a material with low elasticity modulus in 3D printer can be increased by adopting greater nozzle diameter. A variety of polymers were tested to determine the printability window [61]: thermoplastic polyurethanes (EG), ethylene–vinyl-acetates (EVA), polycaprolactone, polyethylene-oxide, hydroxypropylcellulose (HPC), polyvinylcaprolactam–polyvinyl acetate–polyethylene glycol graft copolymer, and copovidone. The filaments that buckled were discovered to be two grades of EVA and TPU with modulus values ranging from 14–70 MPa and 14–25 MPa, respectively. Optimization of process conditions, taking into account extruder temperatures (acting on viscosity) printing speed and nozzle diameter, may allow the printing of these elastic materials. Even by changing the process conditions, printing with an EVA grade (EVA2825A) failed at each nozzle diameter (0.4, 0.6 and 0.8 mm). Although EG and HPC polymers have a high elasticity modulus (442.2 MPa and 251.9 MPa, respectively), the print window expanded as nozzle diameter increased. For these polymers, the effect of nozzle diameter was correlated with another mechanism than Eulers' buckling. This failure mechanism was linked to the processing of a highly viscous polymer. The driving force to push the melt from the nozzle in FDM is exclusively determined by the pressure drop across the system. If the pressure drop to turn the screws is excessively high due to the material's high viscosity, the motor in the 3D printer may be unable to create the requisite torque.

The pressure driven flow of a Newtonian fluid in a capillary conduct of radius r and length l is described through Equation (22) (Hagen–Poiseuille equation):

$$Q = \frac{\pi \Delta P r^4}{8 \eta l} \tag{22}$$

where Q is the volume flow rate and ΔP is the pressure drop.

For Newtonian fluids in a capillary conduct the velocity profile is parabolic, the shear rate (apparent shear rate $\dot{\gamma}_{app}$) at the wall is given as follows:

$$\dot{\gamma}_{app} = \frac{4Q}{\pi r^3} \tag{23}$$

A correction is necessary for non-Newtonian fluids (Rabinowitsch correction), since the velocity profile is affected by the shear-thinning behaviour of the polymer.

$\dot{\gamma}$ is the true shear rate expressed by Equation (24) whereas the pressure drop ΔP is given by Equation (25):

$$\dot{\gamma} = \dot{\gamma}_{app}\left(\frac{3n+1}{4n}\right) \tag{24}$$

$$\Delta P = \frac{8Q\eta l}{\pi r^4}\left(\frac{3n+1}{4n}\right) \tag{25}$$

The material can be efficiently extruded if the pressure drop (ΔP) to achieve a given volume flow rate (Q) is less than the maximum pressure (ΔP_{max}) allowable by the printer apparatus:

$$v_{print}\eta \leq \frac{\Delta P^{max} R^2}{2l}\left(\frac{n}{3n+1}\right) \tag{26}$$

Given the printing machine (i.e., diameter and length of nozzle, and maximum allowable pressure drop), a specific combination of material viscosity (affected by extrusion temperature) and printing speed was discovered to cause the overcoming of ΔP_{max}, resulting in a discontinuous flow of molten polymer from the duct and in a printing failure [44].

4.3. Flow Instability

Larger flow rates should be set in the typical operations of polymer processing and flow to increase productivity and reduce energy consumption. These working circumstances may be accompanied with a larger shear stress acting on the materials, causing flow instability with visible evidence of irregularity and roughness on the product surface ("sharkskin", "stick-slip", and "melt fracture). Causes producing extrusion instabilities are still being debated. Wall slip, melt compressibility, melt viscoelasticity, viscous heating, viscoelastic rupture, and local stick slip are all rheological characteristic of polymer-based systems that can be responsible of flow instabilities [74]. The extrudate is initially smooth and defect free at low flow rates. Defects appear on the surface as the flow rate increases ("shark-skin effect"). As the rate increases further, the flaws turn into discontinuities on the extrudate surface ("stick-slip instability"). The extrudate returns to being fault free as the flow rate is increased further in the regime known as "superflow". Gross melt fracture occurs at very high shear rates with significant irregular and chaotic surface distortions [75].

The influence of the shear rate on the shape of the PLA extrudates was shown in [76]. The extrudate had a cylindrical shape with a smooth surface at an inlet velocity of 88 mm/s (shear rate of 1560 s^{-1}) and became irregular as the velocity increased up to 156 mm/s and appeared distorted at a velocity of 302 mm/s (shear rate of 5375 s^{-1}).

Die swell is a typical event during the polymer extrusion due to macroscopic effects of polymer viscoelasticity such as normal stress, elastic energy, entropy enlargement, orientation effect, and memory effect. It consists of the difference in diameter between the extrudate diameter and die diameter. As a result of the action of extension, shear, and compression during polymer extrusion through the conduit, the polymer macromolecules become disentangled, uncoiled, or oriented (sheared and stretched). During die flow, the resulting stress and strain cannot be completely released. When molecules leave the die, they are loosened by elastic deformation caused by entanglement and recoiling. Extrudate tends to compress in the direction of flow and expand in the opposite direction, resulting in swelling [77].

Die swell affects final product qualities and the printing precision, and necessitates process adjustment to correct the geometric distortion [78]. According to the literature [79], swelling can be reduced by lowering the extrusion rate, raising the melt temperature, increasing the die length, decreasing the die entrance angle, increasing the draw ratio, and adding lubricants or particles to formulations.

4.4. Welding

Dimensional stability of the extruded structure throughout cooling and transition to the final state was one of the investigated aspects that affected the printing quality and 3D parts properties. FDM often employs amorphous materials, which slowly change from a rubbery to a glassy state. As the material cools during this transition, the characteristics are gradually altered, and stress caused by volume variation gradually changes [80]. During the cooling stage, two neighbouring filaments should form a bond between them (intralayer bonding and interlayer bonding) throughout the viscous sintering mechanisms.

Sintering is the phenomena of particle coalescence that is caused by two temperature-dependent properties: surface tension (driving force) and viscosity (limiting force) [81]. Particle coalescence in polymers is frequently accomplished at temperatures higher than the melting point of semi-crystalline materials or higher than the glass transition temperature of amorphous materials. The sintering process has typically been investigated for ceramics and metals, but its applicability in polymer processing has attracted the curiosity of some researchers. Traditional sintering models successfully describe polymer sintering, revealing two major elements that play a role in polymer sintering: surface tension and viscosity. Rheological properties and surface energy measurements were conducted on glycerol plasticized zeins to attest the fusion bonding behaviour in FDM process. Extrudates demonstrated a viscoelastic behaviour with a low viscosity that, for a longer time, increased due to thermal protein aggregation. The surface tension of zein-based materials was comparable to that of standard polymers used in FDM such as PLA (43 mN.m^{-1}) and ABS (42 mN.m^{-1}) [82].

However, further research revealed that melt elasticity is an additional influencing parameter in polymer sintering [83].

The quality of the sintering mechanism influences the bonding of the filaments once they are deposited on the heated platform, as well as the strength of the 3D parts [84]. Tensile tests were carried out on samples prepared in two different build directions. The agreement between theoretical and experimental values of the ultimate tensile load, as confirmed by microscopical examination on the fracture surface, indicated that the strength of the FDM part was mostly related to intralayer bonding, interlayer bonding, and filament neck growth [84].

At processing temperatures above the glass transition, the bond formation between melted filaments is controlled by surface contact and intermolecular diffusion of polymer chain segments across the wetted interface [85].

Successful bonding between two adjacent layers consists of three stages (Figure 5): surface contact, neck growth due to surface tension, and molecular diffusion and entanglement on the interface [86].

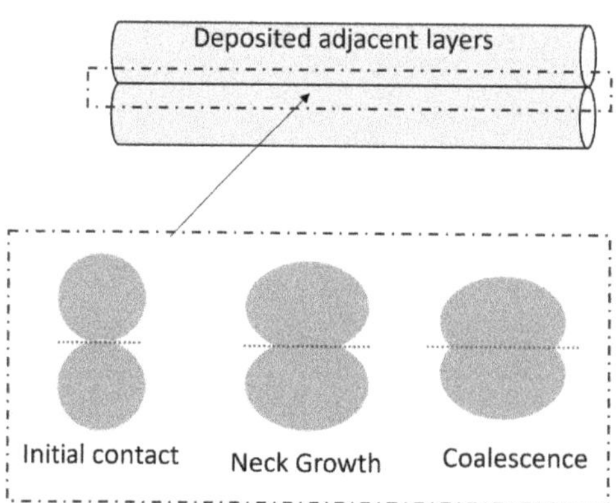

Figure 5. Schematic of interlayer bonding between two adjacent layers deposited on the support.

The successful interdiffusion and re-entanglement of the polymer melt across the layer–layer interfaces is critical to ensuring the strength of the final printed object [87]. Diffusion slows as the printed layer cools near the glass transition temperature and in response to high shear rates. High shear rates in the nozzle dramatically distort and detangle the polymer microstructure prior to welding. Because of the extrusion process' deformation and chain alignment, macromolecular structures have a lesser capacity to migrate across the interfacial contact between two adjacent beads, to form neck growth and to coalesce [26,88]. Limits in molecular diffusion are thought to hinder interlayer contact, resulting in defects and voids. Thus, diffusion is considered to be one of the primary causes of FDM products' poor mechanical performances [89].

Although interlayer bond strength had previously been calculated as a function of wetting process (due to surface forces) hindered by the viscous resistance, wetting was shown to be low during the rapid cooling of the deposited layer. More recently, pressure-driven intimate contact model was adopted to successfully predict interlayer contact. In other words, the bond width (W_{bond}) between layers was predicted through Equation (27) [90]:

$$W_{bond} = W * R_c \left(\frac{P_{contact} * t_p}{\eta(T, \dot{\gamma})} \right)^{1/5} \quad (27)$$

where W_{bond}: bond width; W: road width; Rc: roughness parameter; t_p: contact time of applied pressure; $P_{contact}$: contact pressure applied by the nozzle onto the freshly deposited layer (melt pressure attributed to the confined space created between the nozzle and the previous layer [91]); $\eta(T, \dot{\gamma})$: viscosity of material. It basically asserts that greater interlayer contact can be accomplished by applying higher pressures over longer periods of time to drive the layers into contact as hindered by melt viscosity.

The relaxation time, which may be calculated from rotational rheological data, is a useful material screening tool. Longer relaxation times imply that the material will diffuse slowly, which may indicate diffusive strength issues [90].

The time necessary for polymer relaxation, entanglement recovery, and diffusion to form full interfacial welding increases due to fast cooling and disentanglements. The relaxation time can be quantified by the crossover of the dynamic moduli in the low-frequency region or by dividing the zero-shear viscosity and the plateau modulus [88]. The relaxation time for high-density polyethylene (HDPE) was determined around 1–3 ms while a higher value (40 ms) was calculated for PETG. Thus, HDPE polymer chain could diffuse and weld at the interface faster than the PETG polymer chain, providing robust interlayer welding between deposited layers [88]. Both PLA and ABS have very short relaxation times (0.02 and 0.04 s, respectively). This has no substantial effect on the processing and final product characteristics because it could be assumed that the macromolecules quickly recover their equilibrium state after leaving the printing nozzle [58].

The coalescence phenomena during the FDM process of PLA and PEEK were investigated in [92]. PLA melted entirely at 160 °C and PEEK at 355 °C, and the complex viscosity of PLA in molten state was substantially lower than that of PEEK. The length of the bonding between two adjacent filaments has been registered with time and temperature (Figure 6). According to the predictive model, the experimental data demonstrated that the coalescence in PLA appeared faster than PEEK (due to the lower viscosity of PLA than PEEK). The higher the temperature, the lower the viscosity, and the stronger the coalescence. When the polymer was entirely melted, the coalescence process began.

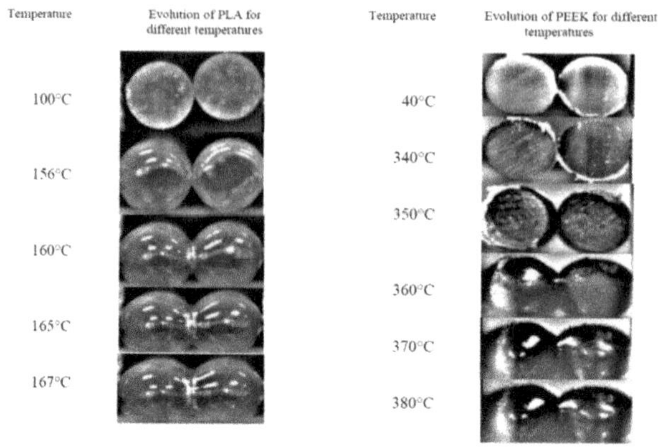

Figure 6. Coalescence as function of temperature for PLA and PEEK. Reprinted from [92] with the permission of AIP Publishing.

High printing temperatures may increase the diffusion rate, but they can also promote polymer degradation [46,93], the release of more organic volatiles into the atmosphere [94], and too low molten polymer viscosity, which can induce bubble formation and flow instability [95]. Heat-induced stress is believed to cause defects and deformation in printed parts such as shrinkage and warping, which have a negative impact on printing quality [70,96]. Increasing the nozzle speed and temperatures of surrounding atmosphere resulted in less residual stress and less warpage. This is due to an improvement in heat transfer ability between the deposited layer, and to a decrease in cooling rate and thermal gradient [97].

The dimensional accuracy of the 3D printed object can be negatively affected by high printing temperatures (too low viscosities). Larger differences from the typical nominal size are represented by the higher fluidity of the polymer [98]. Using a prismatic specimen with a theoretical volume of 9000 mm^3 dashed line, volume changes with printing temperature and filament colour (natural or black) were observed in [98] (Figure 7). The extrusion temperature has a significant impact on the dimensional accuracy of FDM-printed specimens. The dimensional errors of both kinds of PLA increased as the extrusion temperature increased. This could be explained by the higher temperature fluidity of the extruded materials, which allows the filaments to expand freely and complicates dimensional control.

Dahlquist criterion, which was defined in the context of adhesive materials, refers to the ability to form a bond and resist debonding. Dahlquist discovered that only materials with sufficiently high compliance at the testing temperature were sticky by examining the rheological properties as a function of time and temperature. He attested the presence of a minimum value of compliance (3×10^{-6} Pa^{-1}) or, equivalently, a maximum value of modulus (3×10^{5} Pa) to provide good tack or instantaneous adhesion. This criterion has been applied in the case of FDM technology to determine adhesion conditions [49]. Stress relaxation studies were carried out to determine the temperature at which the Dahlquist criterion, and hence good adhesion between layers, was satisfied. Figure 8a depicts the shear stress (G(t)) as a function of time for a recycled PLA-based filament from packaging applications. G(t) typically decreased over time, with the effect becoming more pronounced as the testing temperature rose [99]. After only 0.01 s, the shear stress measurement was lower 3×10^{5} Pa. This limit was met in correspondence of 120 °C by displaying G(t) value at 0.01 s as a function of testing temperature (Figure 8b). 1 sec was chosen as a typical cooling time for polymers when welding occurred [54]. Thus, once deposited on the platform, the extruded material should be at temperatures higher than 120 °C for at least 1 s to ensure good layer bonding.

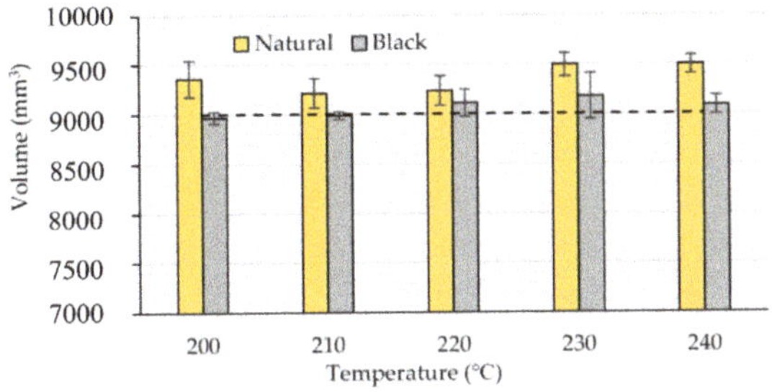

Figure 7. Volume of prismatic specimens as function of extruder temperature for two different types of PLA filaments (natural and black) [98].

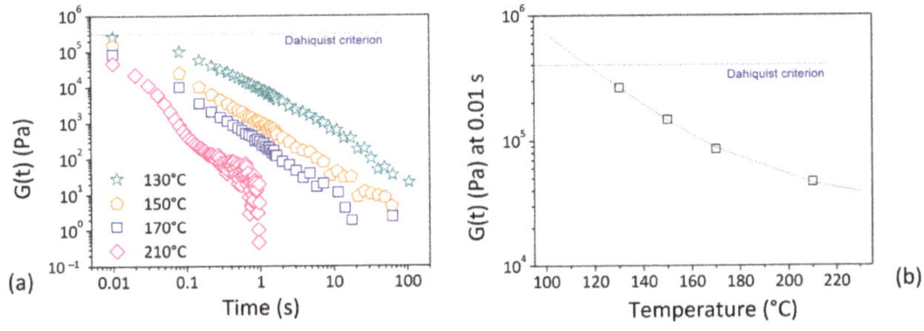

Figure 8. (**a**) Stress relaxation data in terms of shear stress vs time; (**b**) shear stress measured at 0.01 s as function of testing temperature. The critical Dalquist value was meet in correspondence of 120 °C. Reproduced from [99] with the permission of Chemical Engineering Transaction.

5. Rheological Measurements to Attest the Printability of Polymer-Based Materials

Relevant rheological characterization, to attest the extrudability of materials in 3D printing technologies, consists of [50]: (i) frequency sweep test to ensure melt flow into the extruding die and potential clogging of the conducts; due to the sudden rise in viscosity and/or particles agglomeration (in the case of composite-based filaments), the arrangement of polymer chains can be hindered causing melt flow restriction and nozzle obstruction [66,100]; (ii) rheological measurements in extensional field [101] to gain information on the melt elasticity, die swell, and shape retention, (iii) transient shear stress to chain diffusion and bond healing [25,54,99], and (iv) time sweep test to verify the thermal stability of materials at the processing temperatures [102,103].

Recent research activities on new thermoplastic-based systems (blends and composites) and main results regarding the rheological properties to optimize the FDM process and meet specific requirements are detailed in Table 2.

Table 2. Current studies from the literature on FDM thermoplastic materials, rheological characterization at specific temperatures, main results to develop specific requirements.

Thermoplastic-Based Systems	Scope	Testing	Results	Reference
Blends				
Polycaprolactone (PCL)/hydrolysed collagen (HC)	To produce biodegradable items for applications in agriculture and plant nurseries.	Capillary rheometer and flow curves at 130, 140, 150 °C	A decrease in the melt viscosity was observed with the addition of HC due to its plasticizing effect	Seggiani et al., 2018 [62]
Polybutylene succinate (PBS)/ Poly (butylene succinateran-adipate) (PBSA)	To develop semi-crystalline biodegradable filaments	Frequency sweep from 150 °C to 230 °C (TTS principle). Continuous flow measurements (Cox–Merz rule).	The viscosity values of samples lie below 10^4 Pa*s, which allows forecasting a suitable flow in the nozzle	Candal et al., 2020 [49]

Table 2. *Cont.*

Thermoplastic-Based Systems	Scope	Testing	Results	Reference
Poly(lactide) blends containing low molecular weight polymers of chemically identical but enantiomerically different nature (poly(L-lactide) (PLLA) and poly(D-Lactide) (PDLA).)	To promote interfacial weld and stiffness	Oscillatory frequency sweep tests at 180 °C	Any composition possessing η_0 > 5000 Pa*s is too viscous for extrusion. Any composition characterized by η_0 < 500 Pa*s, could not be printed due to uncontrolled fluctuations in volume flow rate and consequential loss in print resolution	Srinivas et al., 2020 [26]
Brominated butyl-rubber (BIIR)/polypropylene (PP) thermoplastic vulcanizate (BIIR/PP-TPV)	To flexible FDM	Frequency sweep at 180 °C	The low viscosity of blends prepared by masterbatch procedure and interfacial compatibilization effectively improves the bonding strength between the adjacent layers of the 3D printed product.	Hou et al., 2020 [104]
Polypropylene (PP)/ elastomeric ethylene-octene copolymer (EOC)	To address the deficiencies of PP in melt extrusion processing (warpage and poor layer adhesion)	Frequency sweep at 190 and 210 °C	The EOC addition did not change substantially the complex viscosity of the blends	Ho and Kontopoulou 2022 [30]
Thermoplastic starch (TPS)/polylactic acid (PLA) /poly(butyleneadipate-co-terephthalate) (PBAT) and chain extender (CE)	Highly renewable filaments for 3D printing	Frequency sweep at 180 °C.	Complex viscosity and modulus increment of blends with the addition of chain extender	Ju et al., 2022 [105]
Composites				
Wood flour (WF)/thermoplastic polyurethane (TPU) and modification with ethylene-propylene-diene-monomer grafted maleic anhydride (EPDM-g-MAH)	Adding low-cost natural fibre to make a low cost, biodegradable, and ecofriendly material	Frequency sweep at 200 °C.	Moduli and complex viscosity increase in composites after the modification with EPDM-g-MAH	Bi et al., 2018 [106]
Polylactide/hemp hurd	To valuable reinforcement of PLA-based blend	Frequency sweep at 190 °C	The biocomposites showed shear-thinning behaviour. Composites at the highest filler loading (40 wt.%) displayed lower melt flow and lower ease of processability than other biocomposite blends.	Xiao et al., 2019 [107]
Inorganic Fullerene Tungsten Sulphide (IF-WS2) nanoparticles/ poly-ether-ketone-ketone (PEEK)	Nanoparticles addition to enhance the flowability of PEEK	Dynamic shear tests at 400 °C	At low shear rate, the shear viscosity of PEEK was reduced with the addition of 2 wt% IF-WS2. The difference in viscosity of samples becomes less pronounced in the higher-shear rate range.	Golbang et al., 2020 [108]
Carbon fibre (CF)/ polyetherimide (PEI)/oligophenylene sulfone (OPSU)/polycarbonate (PC)	Plasticizing high-performance polymers	Capillary rheometer at a temperature of 380 °C	OPSU and PC decrease the melt viscosity of carbon-filled composite preserving the mechanical properties and heat resistance at a sufficiently high level.	Slonov et al., 2020 [109]
Wood flour (WF)/polyhydroxyalkanoates (PHA)	To cost reduction in PHA uses	Frequency sweep at 190 °C	The fluidity of the composites decreased with the increase in WF content	Tian et al., 2021 [110]
Polylactic acid (PLA)/ maraging steel particles	To develop composite functional filaments with additional properties (magnetic, electrical, optical)	Frequency and flow tests at 160 °C (Cox–Merz rule)	The viscous component predominates over the elastic component for all the samples. Remarkable decrease in the viscoelastic moduli as the particle content increases	Díaz-García et al., 2022 [111]

Table 2. *Cont.*

Thermoplastic-Based Systems	Scope	Testing	Results	Reference
Boron nitride nanosheets/thermoplastic polyurethane (TPU)	To high-power integrated electronic devices for 5 G system	Rotational rheometer in the small-amplitude shear oscillation mode at 225 °C	G′ of the composites diverged from the behaviour of the pure TPU. However, liquid-to-solid transition not significantly affected the viscosity within the shear rate range of the 3D printing process	Gao et al., 2022 [27]
Polylactic acid (PLA)/ thermoplastic polyurethane (TPU) blend with enzymatically modified lignin (EL)	To enhance mechanical and thermal properties of bio-based polymers	Frequency sweep at 170–200 °C depending on materials	A gradual decrease in complex viscosity at higher EL concentrations	Murillo-Morales et al., 2023 [112]
Polyetherketoneketone (PEKK)/mica platelets	To provide added mechanical strength to the PEKK	Frequency sweep at 360 °C and temperature ramp	Mica doping does not significantly alter the viscoelastic properties inherent to unfilled PEKK	Kennedy et al., 2022 [37]
Poly (L-lactic acid)/ cellulose nanocrystals	To produce fully green, high-performance consumables	Capillary rheology measurements and small amplitude oscillatory shear experiments	Composites exhibited shear-thinning behaviour favourable for the stable extrusion at nozzle, and G″ > G′ beneficial to the interfuse adhesion during welding	Wu et al., 2022 [113]

6. Challenges and Future Perspectives

The challenges and future perspectives for the next generation of 3D printing technology are primarily related to increasing process and product sustainability; reducing cost and processing times, increasing the mechanical performance and durability of 3D printed parts compared to products made with traditional processes, improving the resolution and printing quality.

Directing 3D printing towards a more sustainable development can be achieved by lowering energy consumption, using renewable natural polymers, consuming biomass sources and/ or usage of recyclable materials, reducing polluting emissions in the surrounding environment, and enabling effective waste utilization to obtain consumable products.

Because of their low mechanical strength and thermal stability, the direct use of biomass and recycled polymers in the production of 3D printing filaments needs to be supplemented with additives, particularly macromolecules with linear structures [114]. Further study on renewable natural resources for 3D printing is needed to customize tuneable properties, such as desirable processability, printability, mechanics, bioactivity, and biodegradability [115]. New recycling procedures are required to provide recycled fibres [116] and polymers [117] with outstanding characteristics while keeping costs and carbon dioxide (CO_2) emissions low.

The utilization of low processing temperatures and low-emitting materials as well as establishing control measures, such as employing an enclosure surrounding the printer in conjunction with an appropriate filter, are ways to reduce polluting emissions during 3D printing [118].

Energy efficiency can be correlated to lower support and extruder temperatures, as well as shorter processing times [119]. This latter can be performed by increasing printing speed or reducing the melt viscosity without sacrificing the mechanical performance.

The first point (increased printing speed) can be attained by selecting a more powerful printer and/or concentrating on machine development to help raise build speed [120]. Extrusion-based AM systems capable of producing considerably more parts than the traditional FDM technique are also currently being manufactured [121]. One example is Big Area Additive Manufacturing (BAAM). BAAM is built on a single-screw extruder. It has a large build volume, uses pelletized feedstock, and can process thermoplastics at higher rates than a filament-based system. The BAAM process allows several of the limitations of the FDM process to be overcome, such as the use of more expensive filaments in comparison to polymers in pellets form and the buckling effect during the feeding. However, this method has significant drawbacks, such as the elimination of support structures with adequate post processing, reduced printable resolution due to larger bead size, poor surface finish, and slow cooling due to quick deposition [122].

Increasing printing speed by lowering viscosity can be accomplished by using low-molecular-weight polymers and integrating flow enhancers or even plasticizers; the subsequent drop in mechanical characteristics can be addressed via reinforcement or fillers [120]. The incorporation of advanced materials as fillers into polymer matrices for FDM filaments improved mechanical properties, stiffness and toughness, thermal conductivity, electrical conductivity, and flame retardancy, broadening the range of potential applications for 3D printed components [123]. Improved thermal conductivity and lower thermal expansion coefficient, in particular, can be two critical features for promoting excellent bonding of the deposited filaments. [122]. However, one of the most significant disadvantages of filled 3D printing filaments is the potential increase in melt viscosity, which complicates the extrusion phase and raises the risk of nozzle clogging, stringing, and warping [123].

7. Conclusions

Among additive manufacturing technologies, fused deposition modelling (FDM) is the most widely utilized. This process consists of several stages: (i) the pushing and melting of solid filament, (ii) the material extrusion along the conduit, (iii) the deposition of melted material on a support base (at room temperature or heated) following the layer-by-layer method, and (iv) the consolidation of neighbouring deposited layers to form a three-dimensional structure. The rheological properties of molten polymers and composites used in FDM are crucial for the efficient development of the 3D printing process. A thorough understanding of the viscoelastic properties of the materials to be printed enables prediction of their behaviour throughout each step of the FDM technique.

The efficacy of materials in 3D printing technology has been studied using both capillary and rotational rheometers.

A good material for printing possesses a high zero-shear viscosity at low shear rates and low relaxation time and shows shear-thinning behaviour at high shear rates. The first characteristics influence the consolidation process, whilst the second is relevant to ensuring smooth extrusion through the nozzle. Given the printing apparatus, an accurate combination of viscosity (and therefore implicitly of extruder temperature) and printing speed was discovered to be determinant in favouring a continuous flow of molten material from the nozzle and avoiding a printing failure. However, too high extruder temperatures should be avoided in order to prevent polymer decomposition within the extruder chamber and poor printing quality with loss of precision in details and edges. Adjusting viscosity and shear rate could also be a valuable attempt to minimize solid filament breaking and distortion (the "Buckling" effect) when pushed and fed in the heating chamber.

Experimental evidence indicated that to favour extrusion, storage modulus (G') should be lower than the loss modulus (G''). To maintain the shape, G' should be higher than G''. The lower the relaxation time of materials, the higher the dimensional stability and the welding between filaments once deposited on the support plate. Finally, the Dalquist criterion was used to assess the adhesion properties between adjacent layers and the quality of layer welding. In this case, a useful condition was that the shear stress ($G(t)$) should be lower than Dalquist limit (3×10^5 Pa).

Author Contributions: Conceptualization, A.P. and D.A.; writing—original draft preparation, A.P.; and writing—review and editing, D.A. All authors have read and agreed to the published version of the manuscript.

Funding: This research received no external funding.

Acknowledgments: A.P. wishes to thank the Italian Ministry of Education, Universities and Research (MIUR) under the framework of Action 1.2 "Researcher Mobility" of the Axis I of PON R&I 2014–2020 under the call "AIM- Attrazione e Mobilità Internazionale".

Conflicts of Interest: The authors declare no conflict of interest.

References

1. Negi, S.; Kumar Sharma, R.; Dhiman, S. Basics, applications and future of additive manufacturing technologies: A review. *J. Manuf. Technol. Res.* **2013**, *5*, 75–96.
2. *ISO/ASTM 52900*; Additive Manufacturing—General Principles—Terminology. ASTM International: West Conshohocken, PA, USA, 2022.
3. Bikas, H.; Koutsoukos, S.; Stavropoulos, P. A decision support method for evaluation and process selection of Additive Manufacturing. *Procedia CIRP* **2019**, *81*, 1107–1112. [CrossRef]
4. Najmon, J.C.; Raeisi, S.; Tovar, A. Review of additive manufacturing technologies and applications in the aerospace industry. In *Additive Manufacturing for the Aerospace Industry*; Froes, F.H., Boyer, R., Eds.; Elsevier Inc.: Amsterdam, The Netherlands, 2019; pp. 7–31. ISBN 9780128140635.
5. Al Rashid, A.; Khan, S.A.; Al-Ghamdi, S.G.; Koç, M. Additive manufacturing: Technology, applications, markets, and opportunities for the built environment. *Autom. Constr.* **2020**, *118*, 103268. [CrossRef]
6. Kumar, R.; Kumar, M.; Chohan, J.S. The role of additive manufacturing for biomedical applications: A critical review. *J. Manuf. Process.* **2021**, *64*, 828–850. [CrossRef]
7. Leal, R.; Barreiros, F.M.; Alves, L.; Romeiro, F.; Vasco, J.C.; Santos, M.; Marto, C. Additive manufacturing tooling for the automotive industry. *Int. J. Adv. Manuf. Technol.* **2017**, *92*, 1671–1676. [CrossRef]
8. Andre, J.-C.; Le Mehaute, A.; Witte, O. De Device for Producing a Model of an Industrial Part. FR2567668A1, 17 January 1986. Available online: https://patents.google.com/patent/FR2567668A1/en (accessed on 10 December 2023).
9. Charles, W. Hull Apparatus for Production of Three-Dimensional Objects by Stereolithography. U.S. Patent US4575330A, 19 December 1989.
10. Ryder, M.A.; Lados, D.A.; Iannacchione, G.S.; Peterson, A.M. Fabrication and properties of novel polymer-metal composites using fused deposition modeling. *Compos. Sci. Technol.* **2018**, *158*, 43–50. [CrossRef]
11. Zhang, W.; Chen, J.; Zeng, H. Polymer processing and rheology. In *Polymer Science and Nanotechnology: Fundamentals and Applications*; Narain, R., Ed.; Elsevier: Amsterdam, The Netherlands, 2020; pp. 149–178. ISBN 9780128168066.
12. Mohd Pu'ad, N.A.S.; Abdul Haq, R.H.; Mohd Noh, H.; Abdullah, H.Z.; Idris, M.I.; Lee, T.C. Review on the fabrication of fused deposition modelling (FDM) composite filament for biomedical applications. *Mater. Today Proc.* **2020**, *29*, 228–232. [CrossRef]
13. Solomon, I.J.; Sevvel, P.; Gunasekaran, J. A review on the various processing parameters in FDM. *Mater. Today Proc.* **2021**, *37*, 509–514. [CrossRef]
14. Rajan, K.; Samykano, M.; Kadirgama, K.; Harun, W.S.W.; Rahman, M.M. Fused deposition modeling: Process, materials, parameters, properties, and applications. *Int. J. Adv. Manuf. Technol.* **2022**, *120*, 1531–1570. [CrossRef]
15. Kumar, R.; Kumar, M.; Chohan, J.S. Material-specific properties and applications of additive manufacturing techniques: A comprehensive review. *Bull. Mater. Sci.* **2021**, *44*, 1–19. [CrossRef]
16. Mohan, N.; Senthil, P.; Vinodh, S.; Jayanth, N. A review on composite materials and process parameters optimisation for the fused deposition modelling process. *Virtual Phys. Prototyp.* **2017**, *12*, 47–59. [CrossRef]
17. Nurhudan, A.I.; Supriadi, S.; Whulanza, Y.; Saragih, A.S. Additive manufacturing of metallic based on extrusion process: A review. *J. Manuf. Process.* **2021**, *66*, 228–237. [CrossRef]
18. Dadkhah, M.; Tulliani, J.M.; Saboori, A.; Iuliano, L. Additive manufacturing of ceramics: Advances, challenges, and outlook. *J. Eur. Ceram. Soc.* **2023**, *43*, 6635–6664. [CrossRef]
19. Wu, H.; Fahy, W.P.; Kim, S.; Kim, H.; Zhao, N.; Pilato, L.; Kafi, A.; Bateman, S.; Koo, J.H. Recent developments in polymers/polymer nanocomposites for additive manufacturing. *Prog. Mater. Sci.* **2020**, *111*, 100638. [CrossRef]
20. Sodeifian, G.; Ghaseminejad, S.; Yousefi, A.A. Preparation of polypropylene/short glass fiber composite as Fused Deposition Modeling (FDM) filament. *Results Phys.* **2019**, *12*, 205–222. [CrossRef]
21. Le, A.D.; Akué Asséko, A.C.; Cosson, B.; Krawczak, P. Investigating the Effect of Interface Temperature on Molecular Interdiffusion during Laser Transmission Welding of 3D-Printed Composite Parts. *Materials* **2023**, *16*, 6121. [CrossRef] [PubMed]
22. Bi, X.; Huang, R. 3D printing of natural fiber and composites: A state-of-the-art review. *Mater. Des.* **2022**, *222*, 111065. [CrossRef]
23. Angelopoulos, P.M.; Samouhos, M.; Taxiarchou, M. Functional fillers in composite filaments for fused filament fabrication; a review. *Mater. Today Proc.* **2021**, *37*, 4031–4043. [CrossRef]
24. Patti, A.; Acierno, S.; Cicala, G.; Zarrelli, M.; Acierno, D. Recovery of Waste Material from Biobags: 3D Printing Process and Thermo-Mechanical Characteristics in Comparison to Virgin and Composite Matrices. *Polymers* **2022**, *14*, 1943. [CrossRef]
25. Candal, M.V.; Calafel, I.; Fernández, M.; Aranburu, N.; Aguirresarobe, R.H.; Gerrica-Echevarria, G.; Santamaría, A.; Müller, A.J. Study of the interlayer adhesion and warping during material extrusion-based additive manufacturing of a carbon nanotube/biobased thermoplastic polyurethane nanocomposite. *Polymer* **2021**, *224*, 123734. [CrossRef]
26. Srinivas, V.; van Hooy-Corstjens, C.S.J.; Rastogi, S.; Harings, J.A.W. Promotion of molecular diffusion and/or crystallization in fused deposition modeled poly(lactide) welds. *Polymer* **2020**, *202*, 122637. [CrossRef]
27. Gao, J.; Hao, M.; Wang, Y.; Kong, X.; Yang, B.; Wang, R.; Lu, Y.; Zhang, L.; Gong, M.; Zhang, L.; et al. 3D printing boron nitride nanosheets filled thermoplastic polyurethane composites with enhanced mechanical and thermal conductive properties. *Addit. Manuf.* **2022**, *56*, 102897. [CrossRef]

28. Das, A.; Etemadi, M.; Davis, B.A.; McKnight, S.H.; Williams, C.B.; Case, S.W.; Bortner, M.J. Rheological investigation of nylon-carbon fiber composites fabricated using material extrusion-based additive manufacturing. *Polym. Compos.* **2021**, *42*, 6010–6024. [CrossRef]
29. Vaes, D.; Van Puyvelde, P. Semi-crystalline feedstock for filament-based 3D printing of polymers. *Prog. Polym. Sci.* **2021**, *118*, 101411. [CrossRef]
30. Ho, Q.B.; Kontopoulou, M. Improving the adhesion and properties in the material extrusion of polypropylene by blending with a polyolefin elastomer. *Addit. Manuf.* **2022**, *55*, 102818. [CrossRef]
31. Gomes, T.E.; Cadete, M.S.; Dias-de-Oliveira, J.; Neto, V. Controlling the properties of parts 3D printed from recycled thermoplastics: A review of current practices. *Polym. Degrad. Stab.* **2022**, *196*, 109850. [CrossRef]
32. Mishra, V.; Negi, S.; Kar, S. FDM-based additive manufacturing of recycled thermoplastics and associated composites. *J. Mater. Cycles Waste Manag.* **2023**, *25*, 758–784. [CrossRef] [PubMed]
33. Patti, A.; Acierno, S.; Cicala, G.; Acierno, D. Changes in Chemical and Physical Properties of Bio-based Thermoplastics During Fused Deposition Modelling (FDM) Process. *Chem. Eng. Trans.* **2023**, *100*, 439–444. [CrossRef]
34. Wang, P.; Zou, B.; Ding, S.; Li, L.; Huang, C. Effects of FDM-3D printing parameters on mechanical properties and microstructure of CF/PEEK and GF/PEEK. *Chin. J. Aeronaut.* **2021**, *34*, 236–246. [CrossRef]
35. Wang, P.; Zou, B.; Xiao, H.; Ding, S.; Huang, C. Effects of printing parameters of fused deposition modeling on mechanical properties, surface quality, and microstructure of PEEK. *J. Mater. Process. Technol.* **2019**, *271*, 62–74. [CrossRef]
36. Garcia-Leiner, M.; Streifel, B.; Başgül, C.; MacDonald, D.W.; Kurtz, S.M. Characterization of polyaryletherketone (PAEK) filaments and printed parts produced by extrusion-based additive manufacturing. *Polym. Int.* **2021**, *70*, 1128–1136. [CrossRef]
37. Kennedy, Z.C.; Christ, J.F.; Fenn, M.D.; Zhong, L.; Chouyyok, W.; Arnold, A.M.; Denny, A.C.; Albrecht, A.M.; Silverstein, J.A.; Erikson, R.L.; et al. Mica filled polyetherketoneketones for material extrusion 3D printing. *Addit. Manuf.* **2022**, *49*, 102492. [CrossRef]
38. Bird, R.B.; Stewart, W.E.; Lightfoot, E.N. *Transport Phenomena*; John Wiley & Sons, Inc.: New York City, NY, USA, 2002.
39. Bird, R.B.; Armstrong, R.C.; Hassager, O. *Dynamics of Polymeric Liquids*; John Wiley & Sons, Inc.: New York City, NY, USA, 1987.
40. Polychronopoulos, N.D.; Vlachopoulos, J. Polymer Processing and Rheology. In *Functional Polymers*; Mazumder, M.A.J., Sheardown, H., Al-Ahmed, A., Eds.; Springer: New York City, NY, USA, 2019; pp. 1–47.
41. Carnicer, V.; Alcázar, C.; Orts, M.J.; Sánchez, E.; Moreno, R. Microfluidic rheology: A new approach to measure viscosity of ceramic suspensions at extremely high shear rates. *Open Ceram.* **2021**, *5*, 100052. [CrossRef]
42. Vlachopoulos, J.; Polychronopoulos, N.D. *Understanding Rheology and Technology of Polymer Extrusion*; Polydynamics Inc.: Dundas, ON, Canada, 2019.
43. Elbadawi, M. Polymeric Additive Manufacturing: The Necessity and Utility of Rheology. In *Polymer Rheology*; Rivera-Armenta, O.L., Cruz, B.A.S., Eds.; IntechOpen: London, UK, 2018; pp. 43–63. ISBN 978-1-78984-002-5.
44. Patti, A.; Acierno, S.; Cicala, G.; Acierno, D. Predicting the Printability of Poly(Lactide) Acid Filaments in Fused Deposition Modeling (FDM) Technology: Rheological Measurements and Experimental Evidence. *ChemEngineering* **2022**, *7*, 1. [CrossRef]
45. Kulicke, W.-M.; Porter, R.S. Relation between steady shear flow and dynamic rheology. *Rheol. Acta* **1980**, *19*, 601–605. [CrossRef]
46. Ritzen, L.; Montano, V.; Garcia, S.J. 3D Printing of a Self-Healing Thermoplastic Polyurethane through FDM: From Polymer Slab to Mechanical Assessment. *Polymers* **2021**, *13*, 305. [CrossRef] [PubMed]
47. Esposito Corcione, C.; Gervaso, F.; Scalera, F.; Montagna, F.; Sannino, A.; Maffezzoli, A. The feasibility of printing polylactic acid–nanohydroxyapatite composites using a low-cost fused deposition modeling 3D printer. *J. Appl. Polym. Sci.* **2017**, *134*, 44656. [CrossRef]
48. Bermudez, D.; Quiñonez, P.A.; Vasquez, E.J.; Carrete, I.A.; Word, T.J.; Roberson, D.A. A Comparison of the physical properties of two commercial 3D printing PLA grades. *Virtual Phys. Prototyp.* **2021**, *16*, 178–195. [CrossRef]
49. Candal, M.V.; Calafel, I.; Aranburu, N.; Fernández, M.; Gerrica-Echevarria, G.; Santamaría, A.; Müller, A.J. Thermo-rheological effects on successful 3D printing of biodegradable polyesters. *Addit. Manuf.* **2020**, *36*, 101408. [CrossRef]
50. Das, A.; Gilmer, E.L.; Biria, S.; Bortner, M.J. Importance of Polymer Rheology on Material Extrusion Additive Manufacturing: Correlating Process Physics to Print Properties. *ACS Appl. Polym. Mater.* **2021**, *3*, 1218–1249. [CrossRef]
51. Nguyen, N.A.; Bowland, C.C.; Naskar, A.K. A general method to improve 3D-printability and inter-layer adhesion in lignin-based composites. *Appl. Mater. Today* **2018**, *12*, 138–152. [CrossRef]
52. Khaliq, M.H.; Gomes, R.; Fernandes, C.; Nóbrega, J.; Carneiro, O.S.; Ferrás, L.L. On the use of high viscosity polymers in the fused filament fabrication process. *Rapid Prototyp. J.* **2017**, *23*, 727–735. [CrossRef]
53. Zhao, Y.Q.; Yang, J.H.; Ding, X.; Ding, X.; Duan, S.; Xu, F.J. Polycaprolactone/polysaccharide functional composites for low-temperature fused deposition modelling. *Bioact. Mater.* **2020**, *5*, 185–191. [CrossRef] [PubMed]
54. Calafel, I.; Aguirresarobe, R.H.; Peñas, M.I.; Santamaria, A.; Tierno, M.; Conde, J.I.; Pascual, B. Searching for Rheological Conditions for FFF 3D Printing with PVC Based Flexible Compounds. *Materials* **2020**, *13*, 178. [CrossRef] [PubMed]
55. Prasong, W.; Muanchan, P.; Ishigami, A.; Thumsorn, S.; Kurose, T.; Ito, H. Properties of 3D Printable Poly(lactic acid)/Poly(butylene adipate-co-terephthalate) Blends and Nano Talc Composites. *J. Nanomater.* **2020**, *2020*, 8040517. [CrossRef]
56. Mohammed, A.H.; Kovacev, N.; Elshaer, A.; Melaibari, A.A.; Iqbal, J.; Hassanin, H.; Essa, K.; Memić, A. Preparation of Polylactic Acid/Calcium Peroxide Composite Filaments for Fused Deposition Modelling. *Polymers* **2023**, *15*, 2229. [CrossRef] [PubMed]

57. Smirnov, A.; Seleznev, A.; Peretyagin, P.; Bentseva, E.; Pristinskiy, Y.; Kuznetsova, E.; Grigoriev, S. Rheological Characterization and Printability of Polylactide (PLA)-Alumina (Al_2O_3) Filaments for Fused Deposition Modeling (FDM). *Materials* **2022**, *15*, 8399. [CrossRef]
58. Sanchez, L.C.; Beatrice, C.A.G.; Lotti, C.; Marini, J.; Bettini, S.H.P.; Costa, L.C. Rheological approach for an additive manufacturing printer based on material extrusion. *Int. J. Adv. Manuf. Technol.* **2019**, *105*, 2403–2414. [CrossRef]
59. Chen, J.; Smith, D.E. Filament rheological characterization for fused filament fabrication additive manufacturing: A low-cost approach. *Addit. Manuf.* **2021**, *47*, 102208. [CrossRef]
60. Ahmad, M.N.; Ishak, M.R.; Taha, M.M.; Mustapha, F.; Leman, Z. Rheological and Morphological Properties of Oil Palm Fiber-Reinforced Thermoplastic Composites for Fused Deposition Modeling (FDM). *Polymers* **2021**, *13*, 3739. [CrossRef]
61. Henry, S.; Samaro, A.; Marchesini, F.H.; Shaqour, B.; Macedo, J.; Vanhoorne, V.; Vervaet, C. Extrusion-based 3D printing of oral solid dosage forms: Material requirements and equipment dependencies. *Int. J. Pharm.* **2021**, *598*, 120361. [CrossRef] [PubMed]
62. Seggiani, M.; Altieri, R.; Puccini, M.; Stefanelli, E.; Esposito, A.; Castellani, F.; Stanzione, V.; Vitolo, S. Polycaprolactone-collagen hydrolysate thermoplastic blends: Processability and biodegradability/compostability. *Polym. Degrad. Stab.* **2018**, *150*, 13–24. [CrossRef]
63. Patti, A.; Cicala, G.; Acierno, S. Rotational Rheology of Wood Flour Composites Based on Recycled Polyethylene. *Polymers* **2021**, *13*, 2226. [CrossRef] [PubMed]
64. Lacambra-Andreu, X.; Morelle, X.P.; Maazouz, A.; Chenal, J.M.; Lamnawar, K. Rheological investigation and modeling of healing properties during extrusion-based 3D printing of poly(lactic-acid). *Rheol. Acta* **2023**, *62*, 31–44. [CrossRef]
65. Balani, S.B.; Mokhtarian, H.; Salmi, T.; Coatanéa, E. An Investigation of the Influence of Viscosity and Printing Parameters on the Extrudate Geometry in the Material Extrusion Process. *Polymers* **2023**, *15*, 2202. [CrossRef] [PubMed]
66. Ivanova, R.; Kotsilkova, R. Rheological study of poly(lactic) acid nanocomposites with carbon nanotubes and graphene additives as a tool for materials characterization for 3D printing application. *Appl. Rheol.* **2018**, *28*, 201854014. [CrossRef]
67. Thumsorn, S.; Prasong, W.; Kurose, T.; Ishigami, A.; Kobayashi, Y.; Ito, H. Rheological Behavior and Dynamic Mechanical Properties for Interpretation of Layer Adhesion in FDM 3D Printing. *Polymers* **2022**, *14*, 2721. [CrossRef]
68. Esmi, A.; Jahani, Y.; Yousefi, A.A.; Zandi, M. PMMA-CNT-HAp nanocomposites optimized for 3D-printing applications. *Mater. Res. Express* **2019**, *6*, 085405. [CrossRef]
69. Stanciu, N.V.; Stan, F.; Fetecau, C.; Susac, F. On the Feasibility of Printing 3D Composite Objects Based on Polypropylene/Multi-walled Carbon Nanotubes. *MATEC Web Conf.* **2019**, *290*, 03017. [CrossRef]
70. Duty, C.; Ajinjeru, C.; Kishore, V.; Compton, B.; Hmeidat, N.; Chen, X.; Liu, P.; Hassen, A.A.; Lindahl, J.; Kunc, V. What makes a material printable? A viscoelastic model for extrusion-based 3D printing of polymers. *J. Manuf. Process.* **2018**, *35*, 526–537. [CrossRef]
71. Riecker, S.; Hein, S.; Studnitzky, T.; Andersen, O.; Kieback, B. 3D printing of metal parts by means of fused filament fabrication—A non beam-based approach. In Proceedings of the Euro PM 2017 Congress & Exhibition, Milan, Italy, 1–5 October 2017.
72. Olesik, P.; Godzierz, M.; Kozioł, M. Preliminary Characterization of Novel LDPE-Based Wear-Resistant Composite Suitable for FDM 3D Printing. *Materials* **2019**, *12*, 2520. [CrossRef] [PubMed]
73. Shaqour, B.; Samaro, A.; Verleije, B.; Beyers, K.; Vervaet, C.; Cos, P. Production of Drug Delivery Systems Using Fused Filament Fabrication: A Systematic Review. *Pharmaceutics* **2020**, *12*, 517. [CrossRef] [PubMed]
74. Achilleos, E.C.; Georgiou, G.; Hatzikiriakos, S.G. Role of processing aids in the extrusion of molten polymers. *J. Vinyl Addit. Technol.* **2002**, *8*, 7–24. [CrossRef]
75. Agassant, J.F.; Arda, D.R.; Combeaud, C.; Merten, A.; Münstedt, H.; Mackley, M.R.; Robert, L.; Vergnes, B. Polymer processing extrusion instabilities and methods for their elimination or minimisation. *Int. Polym. Process.* **2006**, *21*, 239–255. [CrossRef]
76. Bakrani Balani, S.; Chabert, F.; Nassiet, V.; Cantarel, A. Influence of printing parameters on the stability of deposited beads in fused filament fabrication of poly(lactic) acid. *Addit. Manuf.* **2019**, *25*, 112–121. [CrossRef]
77. Wang, K. Die Swell of Complex Polymeric Systems. In *Viscoelasticity from Theory to Biological Applications*; de Vicente, J., Ed.; IntechOpen: London, UK, 2012; pp. 77–96.
78. Colon, A.R.; Kazmer, D.O.; Peterson, A.M.; Seppala, J.E. Characterization of die-swell in thermoplastic material extrusion. *Addit. Manuf.* **2023**, *73*, 103700. [CrossRef]
79. Nisticò, R. Polyethylene terephthalate (PET) in the packaging industry. *Polym. Test.* **2020**, *90*, 106707. [CrossRef]
80. Zawaski, C.E.; Wilts, E.M.; Chatham, C.A.; Stevenson, A.T.; Pekkanen, A.M.; Li, C.; Tian, Z.; Whittington, A.R.; Long, T.E.; Williams, C.B. Tuning the material properties of a water-soluble ionic polymer using different counterions for material extrusion additive manufacturing. *Polymer* **2019**, *176*, 283–292. [CrossRef]
81. Garzon-Hernandez, S.; Garcia-Gonzalez, D.; Jérusalem, A.; Arias, A. Design of FDM 3D printed polymers: An experimental-modelling methodology for the prediction of mechanical properties. *Mater. Des.* **2020**, *188*, 108414. [CrossRef]
82. Chaunier, L.; Duchemin, B.; Couallier, E.; Valle, G.D.; Lourdin, D.; Reguerre, A.L.; Leroy, E. Fusion-bonding behavior of plasticized corn proteins in fused deposition modeling process. *AIP Conf. Proc.* **2019**, *2065*, 030055. [CrossRef]
83. Bellehumeur, C.T.; Kontopoulou, M.; Vlachopoulos, J. The role of viscoelasticity in polymer sintering. *Rheol. Acta* **1998**, *37*, 270–278. [CrossRef]
84. Gurrala, P.K.; Regalla, S.P. Part strength evolution with bonding between filaments in fused deposition modelling. *Virtual Phys. Prototyp.* **2014**, *9*, 141–149. [CrossRef]

85. Yin, J.; Lu, C.; Fu, J.; Huang, Y.; Zheng, Y. Interfacial bonding during multi-material fused deposition modeling (FDM) process due to inter-molecular diffusion. *Mater. Des.* **2018**, *150*, 104–112. [CrossRef]
86. Gao, X.; Qi, S.; Kuang, X.; Su, Y.; Li, J.; Wang, D. Fused filament fabrication of polymer materials: A review of interlayer bond. *Addit. Manuf.* **2021**, *37*, 101658. [CrossRef]
87. McIlroy, C.; Olmsted, P.D. Disentanglement effects on welding behaviour of polymer melts during the fused-filament-fabrication method for additive manufacturing. *Polymer* **2017**, *123*, 376–391. [CrossRef]
88. Naqi, A.; Swain, Z.; Mackay, M.E. Dual Material Fused Filament Fabrication via Core-Shell Die Design. *ACS Appl. Polym. Mater.* **2023**, *5*, 2481–2489. [CrossRef]
89. Srinivas, V.; van Hooy-Corstjens, C.S.J.; Harings, J.A.W. Correlating molecular and crystallization dynamics to macroscopic fusion and thermodynamic stability in fused deposition modeling; a model study on polylactides. *Polymer* **2018**, *142*, 348–355. [CrossRef]
90. Coogan, T.J.; Kazmer, D.O. Prediction of interlayer strength in material extrusion additive manufacturing. *Addit. Manuf.* **2020**, *35*, 101368. [CrossRef]
91. Coogan, T.J.; Kazmer, D.O. Modeling of interlayer contact and contact pressure during fused filament fabrication. *J. Rheol.* **2019**, *63*, 655–672. [CrossRef]
92. Shahriar, B.B.; France, C.; Valerie, N.; Arthur, C.; Christian, G. Toward improvement of the properties of parts manufactured by FFF (fused filament fabrication) through understanding the influence of temperature and rheological behaviour on the coalescence phenomenon. *AIP Conf. Proc.* **2017**, *1896*, 040008. [CrossRef]
93. Patti, A.; Acierno, S.; Cicala, G.; Acierno, D. Evaluating Thermal Ageing of Recycled Bio-based Filament for 3D Printing as Function of Extruder Temperature. *Chem. Eng. Trans.* **2023**, *98*, 57–62. [CrossRef]
94. Patti, A.; Acierno, S.; Cicala, G.; Tuccitto, N.; Acierno, D. Refining the 3D Printer Set-up to Reduce the Environmental Impact of the Fused Deposition Modelling (fdm) Technology. *Chem. Eng. Trans.* **2022**, *91*, 415–420. [CrossRef]
95. Ferreira, R.T.L.; Amatte, I.C.; Dutra, T.A.; Bürger, D. Experimental characterization and micrography of 3D printed PLA and PLA reinforced with short carbon fibers. *Compos. Part B Eng.* **2017**, *124*, 88–100. [CrossRef]
96. Mustapha, K.B.; Metwalli, K.M. A review of fused deposition modelling for 3D printing of smart polymeric materials and composites. *Eur. Polym. J.* **2021**, *156*, 110591. [CrossRef]
97. Samy, A.A.; Golbang, A.; Harkin-Jones, E.; Archer, E.; Tormey, D.; McIlhagger, A. Finite element analysis of residual stress and warpage in a 3D printed semi-crystalline polymer: Effect of ambient temperature and nozzle speed. *J. Manuf. Process.* **2021**, *70*, 389–399. [CrossRef]
98. Frunzaverde, D.; Cojocaru, V.; Ciubotariu, C.-R.; Miclosina, C.-O.; Ardeljan, D.D.; Ignat, E.F.; Marginean, G. The Influence of the Printing Temperature and the Filament Color on the Dimensional Accuracy, Tensile Strength, and Friction Performance of FFF-Printed PLA Specimens. *Polymers* **2022**, *14*, 1978. [CrossRef] [PubMed]
99. Patti, A.; Acierno, S.; Cicala, G.; Acierno, D. Recycling Waste from Film Packaging to 3D Printing Applications: A Prospective Study to Identify the Processing Temperature. *Chem. Eng. Trans.* **2022**, *96*, 55–60. [CrossRef]
100. Guo, Z.; Peng, X.; Sang, L.; Guo, Z.; Peng, X.; Sang, L. Enhancement of Mechanical Properties of PA6 Blending with Talcum for Fused Deposition Modeling. *Macromol. Mater. Eng.* **2020**, *305*, 2000355. [CrossRef]
101. Torabi, H.; McGreal, H.; Zarrin, H.; Behzadfar, E. Effects of Rheological Properties on 3D Printing of Poly(lactic acid) (PLA) and Poly(hydroxy alkenoate) (PHA) Hybrid Materials. *ACS Appl. Polym. Mater.* **2023**, *5*, 4034–4044. [CrossRef]
102. Kovalcik, A.; Smilek, J.; Machovsky, M.; Kalina, M.; Enev, V.; Dugova, H.; Cernekova, N.; Kovacova, M.; Spitalsky, Z. Properties and structure of poly(3-hydroxybutyrate-co-4-hydroxybutyrate) filaments for fused deposition modelling. *Int. J. Biol. Macromol.* **2021**, *183*, 880–889. [CrossRef]
103. Kovalcik, A.; Sangroniz, L.; Kalina, M.; Skopalova, K.; Humpolíček, P.; Omastova, M.; Mundigler, N.; Müller, A.J. Properties of scaffolds prepared by fused deposition modeling of poly(hydroxyalkanoates). *Int. J. Biol. Macromol.* **2020**, *161*, 364–376. [CrossRef]
104. Hou, J.; Zhong, M.; Pan, X.; Chen, L.; Wu, X.; Kong, Z.; Yuan, Y.; Yan, S.; Zhang, J.; Duan, Y. Fabricating 3D printable BIIR/PP TPV via masterbatch and interfacial compatibilization. *Compos. Part B Eng.* **2020**, *199*, 108220. [CrossRef]
105. Ju, Q.; Tang, Z.; Shi, H.; Zhu, Y.; Shen, Y.; Wang, T. Thermoplastic starch based blends as a highly renewable filament for fused deposition modeling 3D printing. *Int. J. Biol. Macromol.* **2022**, *219*, 175–184. [CrossRef] [PubMed]
106. Bi, H.; Ren, Z.; Guo, R.; Xu, M.; Song, Y. Fabrication of flexible wood flour/thermoplastic polyurethane elastomer composites using fused deposition molding. *Ind. Crops Prod.* **2018**, *122*, 76–84. [CrossRef]
107. Xiao, X.; Chevali, V.S.; Song, P.; He, D.; Wang, H. Polylactide/hemp hurd biocomposites as sustainable 3D printing feedstock. *Compos. Sci. Technol.* **2019**, *184*, 107887. [CrossRef]
108. Golbang, A.; Harkin-Jones, E.; Wegrzyn, M.; Campbell, G.; Archer, E.; McIlhagger, A. Production and characterization of PEEK/IF-WS2 nanocomposites for additive manufacturing: Simultaneous improvement in processing characteristics and material properties. *Addit. Manuf.* **2020**, *31*, 100920. [CrossRef]
109. Slonov, A.; Musov, I.; Zhansitov, A.; Rzhevskaya, E.; Khakulova, D.; Khashirova, S. The Effect of Modification on the Properties of Polyetherimide and Its Carbon-Filled Composite. *Polymers* **2020**, *12*, 1056. [CrossRef] [PubMed]
110. Tian, J.; Zhang, R.; Wu, Y.; Xue, P. Additive manufacturing of wood flour/polyhydroxyalkanoates (PHA) fully bio-based composites based on micro-screw extrusion system. *Mater. Des.* **2021**, *199*, 109418. [CrossRef]

111. Díaz-García, Á.; Law, J.Y.; Felix, M.; Guerrero, A.; Franco, V. Functional, thermal and rheological properties of polymer-based magnetic composite filaments for additive manufacturing. *Mater. Des.* **2022**, *219*, 110806. [CrossRef]
112. Murillo-Morales, G.; Sethupathy, S.; Zhang, M.; Xu, L.; Ghaznavi, A.; Xu, J.; Yang, B.; Sun, J.; Zhu, D. Characterization and 3D printing of a biodegradable polylactic acid/thermoplastic polyurethane blend with laccase-modified lignin as a nucleating agent. *Int. J. Biol. Macromol.* **2023**, *236*, 123881. [CrossRef]
113. Wu, X.; Liu, Y.; Wu, H.; Duan, Y.; Zhang, J. Melt-processed poly (L-lactic acid)/cellulose nanocrystals biocomposites for 3D printing: Improved melt processibility and inter-fuse adhesion. *Compos. Sci. Technol.* **2022**, *218*, 109135. [CrossRef]
114. Liu, J.; Sun, L.; Xu, W.; Wang, Q.; Yu, S.; Sun, J. Current advances and future perspectives of 3D printing natural-derived biopolymers. *Carbohydr. Polym.* **2019**, *207*, 297–316. [CrossRef] [PubMed]
115. Shah, S.W.A.; Xu, Q.; Ullah, M.W.; Zahoor; Sethupathy, S.; Morales, G.M.; Sun, J.; Zhu, D. Lignin-based additive materials: A review of current status, challenges, and future perspectives. *Addit. Manuf.* **2023**, *74*, 103711. [CrossRef]
116. Ateeq, M.; Shafique, M.; Azam, A.; Rafiq, M. A review of 3D printing of the recycled carbon fiber reinforced polymer composites: Processing, potential, and perspectives. *J. Mater. Res. Technol.* **2023**, *26*, 2291–2309. [CrossRef]
117. Mikula, K.; Skrzypczak, D.; Izydorczyk, G.; Warchoł, J.; Moustakas, K.; Chojnacka, K.; Witek-Krowiak, A. 3D printing filament as a second life of waste plastics—A review. *Environ. Sci. Pollut. Res.* **2021**, *28*, 12321–12333. [CrossRef] [PubMed]
118. Kwon, O.; Yoon, C.; Ham, S.; Park, J.; Lee, J.; Yoo, D.; Kim, Y. Characterization and Control of Nanoparticle Emission during 3D Printing. *Environ. Sci. Technol.* **2017**, *51*, 10357–10368. [CrossRef] [PubMed]
119. Quanjin, M.; Rejab, M.R.M.; Idris, M.S.; Kumar, N.M.; Abdullah, M.H.; Reddy, G.R. Recent 3D and 4D intelligent printing technologies: A comparative review and future perspective. *Procedia Comput. Sci.* **2020**, *167*, 1210–1219. [CrossRef]
120. Thompson, M.S. Current status and future roles of additives in 3D printing—A perspective. *J. Vinyl Addit. Technol.* **2022**, *28*, 3–16. [CrossRef]
121. Ajinjeru, C.; Kishore, V.; Lindahl, J.; Sudbury, Z.; Hassen, A.A.; Post, B.; Love, L.; Kunc, V.; Duty, C. The influence of dynamic rheological properties on carbon fiber-reinforced polyetherimide for large-scale extrusion-based additive manufacturing. *Int. J. Adv. Manuf. Technol.* **2018**, *99*, 411–418. [CrossRef]
122. Pricci, A.; de Tullio, M.D.; Percoco, G. Analytical and Numerical Models of Thermoplastics: A Review Aimed to Pellet Extrusion-Based Additive Manufacturing. *Polymers* **2021**, *13*, 3160. [CrossRef]
123. Kantaros, A.; Soulis, E.; Petrescu, F.I.T.; Ganetsos, T. Advanced Composite Materials Utilized in FDM/FFF 3D Printing Manufacturing Processes: The Case of Filled Filaments. *Materials* **2023**, *16*, 6210. [CrossRef]

materials

MDPI

Review

Bridging Nanomanufacturing and Artificial Intelligence—A Comprehensive Review

Mutha Nandipati [1], Olukayode Fatoki [1] and Salil Desai [1,2,*]

[1] Department of Industrial and Systems Engineering, North Carolina Agricultural and Technical State University, Greensboro, NC 27411, USA; mrnandipati@aggies.ncat.edu (M.N.); omfatoki@aggies.ncat.edu (O.F.)
[2] Center of Excellence in Product Design and Advanced Manufacturing, North Carolina Agricultural and Technical State University, Greensboro, NC 27411, USA
* Correspondence: sdesai@ncat.edu; Tel.: +1-336-285-3725

Abstract: Nanomanufacturing and digital manufacturing (DM) are defining the forefront of the fourth industrial revolution—Industry 4.0—as enabling technologies for the processing of materials spanning several length scales. This review delineates the evolution of nanomaterials and nanomanufacturing in the digital age for applications in medicine, robotics, sensory technology, semiconductors, and consumer electronics. The incorporation of artificial intelligence (AI) tools to explore nanomaterial synthesis, optimize nanomanufacturing processes, and aid high-fidelity nanoscale characterization is discussed. This paper elaborates on different machine-learning and deep-learning algorithms for analyzing nanoscale images, designing nanomaterials, and nano quality assurance. The challenges associated with the application of machine- and deep-learning models to achieve robust and accurate predictions are outlined. The prospects of incorporating sophisticated AI algorithms such as reinforced learning, explainable artificial intelligence (XAI), big data analytics for material synthesis, manufacturing process innovation, and nanosystem integration are discussed.

Keywords: artificial intelligence; digital manufacturing; nanocharacterization; nanomaterials; nanomanufacturing

1. Nanoscale Science and Technology

Nanotechnology refers to technologies that are applied at the nanometer level and have applications in real life [1]. These technologies involve controlling and manipulating matter and operating in dimensions from 1 to 100 nm. It can be further streamlined as the design, characterization, manufacture, and shape- and size-controlled application of matter whose smallest functional organization is on the scale of one billionth of a meter to make matter, systems, and devices with fundamentally novel properties and functions [1–6]. Nanotechnology has resulted in breakthrough solutions in many fields and real-life applications, such as smart materials, nanomanufacturing, electronics, water, drug delivery, and national security. The pivotal components of nanotechnology include nanomaterials, nanomanufacturing, and the application of various systems, such as physical, chemical, and biological, to create new materials and systems at nanoscale with enhanced properties [3–5]. The subsequent sections provide an in-depth understanding of nanomaterials, nanomanufacturing, digital manufacturing, and artificial intelligence. They set the groundwork on each topic individually, before delving into exploring the intricate relationship between artificial intelligence and nanomanufacturing.

1.1. Nanomaterials

Nanomaterials are the key elements of nanomanufacturing, with dimensions of one billionth of a meter [7]. According to author Vollath, nanomaterials possess physicochemical properties that differ from the parent material based on their size and shape. They also exhibit a variety of geometric configurations classified into zero-dimensional (nanoparticles), one-dimensional (nanorods), two-dimensional (films, layers, and sheets), and three-dimensional (nanocomposites, core shells, multi nanolayers, bundles of nanowires and tubes) [8]. Nanomaterials are categorized according to their composition, with distinctions based on form, size, characteristics, and constituents, resulting in polymeric, carbon-based, metal oxides, semiconductor nanomaterials, and lipid-based variants. Among the widely employed nanomaterials, silver, carbon-based materials, and tungsten oxide are prevalent choices [9]. Nano-silver is used for antimicrobial applications due to its physicochemical and biological properties [10]. Carbon-based nanomaterials, with their extraordinary mechanical and electrical properties, are widely used in battery technologies [5–7]. Nano ceramics are used in wear-resistant coatings due to their higher hardness, toughness, and wear-resistance properties [11].

1.2. Nanomanufacturing

Nanomanufacturing denotes the production processes of fabricating nanomaterials, structures, devices, and systems at one billionth of a meter scale, utilizing either a top-down or a bottom-up method [12,13], as shown in Figure 1. Top-down approaches start with a macroscopic material and incorporate their way down to smaller-scale details. Some of the most widely used top-down processes are milling, machining, lithography, and micromachining. In milling, a suitable powder charge is placed in a high-energy mill, along with a suitable milling medium. This is to reduce the particle size and blending of particles in new phases [1]. Machining includes material removal at the nanometric level, conventional cutting, grinding, and polishing [2]. Top-down lithography is suitable to fabricate fine, mass-producible patterns. Nanoimprint lithography is a method of transferring a pattern shape of a mold on the target materials by contacting a mold. In dip-pen lithography, high-resolution nanopatterns are obtained by delivering materials via capillary transport using scanning probe microscope tips [3]. Bulk micromachining is a prevalent technique utilized in the creation of a vast range of microstructures, varying in complexity, within silicon wafers. This procedure predominantly incorporates wet bulk micromachining, which in turn employs wet etching. Wet anisotropic etching serves as the central process in the fabrication of a myriad of components, all uniquely shaped, being used extensively within microelectromechanical system (MEMS) applications [4]. Electron beam lithography (E-Beam) uses an electron beam at a voltage of approximately 15 kV and is a fabrication of nanopatterns of conductive thin films down to 100 nm [5]. Bottom-up techniques, on the other hand, begin by creating and synthesizing the desired molecules using various nanomanufacturing processes, such as self-assembly, vapor phase deposition, liquid-phase methods, pulse-laser deposition, and molecular beam epitaxy [1,12–15]. Gas-phase synthesis and liquid-phase formation are the two basic methods in bottom-up processes. Some of these widely employed methods are vapor phase deposition, molecular beam epitaxy, colloidal methods, sol–gel, pulse-laser deposition, and magnetron sputtering [6]. In vapor phase deposition, a substrate is exposed to one or more volatile precursors, which then decompose the substrate and react with it to produce the desired deposit. Molecular beam epitaxy works on vacuum evaporation, where direct impingement of thermal molecular and atomic beams takes place under ultra-high vacuum conditions. In the sol–gel process, metal oxides are fabricated from small solid molecules [16]. The deposition in pulse-laser technology involves focusing a pulsed laser beam on the source material so that it deposits on a substrate inside a high vacuum chamber. Another thin-film deposition method in use, known as magnetron sputtering, is a process where atoms are ejected from a target by energetic ions [17].

Different nanomanufacturing processes are in practice in the manufacturing industry, and the top five nanoscale manufacturing processes [18], according to the American Society of Mechanical Engineers (ASME), are printing of 3D solid materials, photolithography, microscale assembly, 3D printed memory polymers, and nanomanufacturing of polymer 2D materials. The principle behind the process of printing integrated 3D solid materials is via multiple layering of melted polymers or metals [19,20]. Photolithography is a process where the optimal mask is subjected to an ultraviolet (UV) light to facilitate the transfer of geometric pattern from the mask to a substrate [21]. Assembling components across multiple-size scales is the process behind the microscale assembly of nanostructured metamaterials. Modulated surface energy is used in the adhesion of flexible tools to accommodate assembly [22]. In the process of 3D-printed customized shape memory polymers, small and lightweight actuator systems are 3D printed with the material's response to its environment in contrast to the physical integration of mechanical components [23–25]. In the scalable nanomanufacturing of 2D polymer materials, it employs the principle of photolithography to generate high-density small patterns on polymers [26,27].

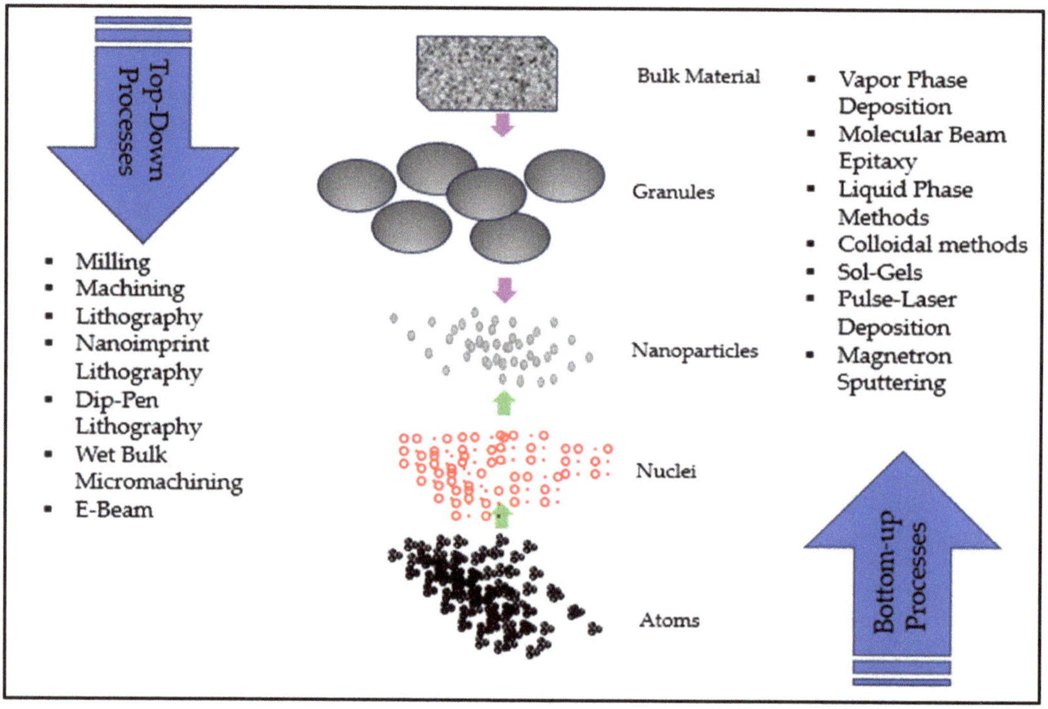

Figure 1. Production of nanomaterials (top-down and bottom-up processes).

1.3. *Digital Manufacturing*

According to author Li, conventional nanomanufacturing processes aim towards enhancing the quality and purity of particles, which has enormous potential for optimization [28]. Manufacturing firms across sectors encounter tremendous problems in effectively designing and managing modern products [29]. To address the intricacies in the development of modern products, digital product development is utilized, along with the fusion of technologies to build cyber-physical systems (CPSs) and smart factories [30].

As the era of industry 4.0 continues to broaden, organizations are driven to diversify their products into various nanoscales, resulting in the adoption of more efficient and scalable production methods [31]. CPSs are crucial in various domains, like robotics, defense systems, and electric infrastructure, due to security threats [32]. Industries are transitioning to digital manufacturing from the age-old traditional types, as stipulated by Industry 4.0. The relationship between nanomanufacturing and digital manufacturing is depicted in Table 1.

Table 1. Nanomanufacturing and digital manufacturing relationship.

Context	Relation between Nanomanufacturing (NM) and Digital Manufacturing (DM)
Industry 4.0	Industry 4.0 is facilitated by digitalization and integration technologies to build cyber-physical systems and smart factories [31]. Almakaeel et al., discussed extensively the combination of digital concepts with several artificial neural networks to predict the optimal nano/micro-manufacturing process [33].
Computer systems	NM and DM both use computer systems to design, simulate, analyze, and produce a product.
Requirement	NM and DM require a high level of research and development and require highly advanced equipment and technology. In addition, a highly skilled workforce to operate and manage the technology is required.
Control	In most cases, nanomanufacturing processes are controlled using digital manufacturing techniques.
End goal	Both NM and DM offer opportunities for greater efficiency, cost savings, and customization in production.
Future	The integration of NM and DM could lead to the development of newer materials and processes, thereby enhancing the production capabilities.

1.4. Artificial Intelligence

Artificial intelligence (AI) is a pervasive subdivision of computer science that chiefly emphasizes the development of intelligent machines competent in executing tasks that conventionally necessitate human intelligence, as illustrated by Wang et al. [34]. It is an inherently data-dependent technology that focuses on automatic learning as well as predictive capabilities. AI shows excellence in carrying out tasks in an unsupervised manner and avoids human intervention in decision-making. It also has greater fitness ability with functions and fast computing speed. AI has proved its superiority in various fields, such as autonomous driving [35], natural language processing [36], medical diagnosis [37], computer vision [38], intelligent manufacturing [39], computer (imaging) vision [40], spectral informatics [41], ultrafast photonics [42], fiber lasers [43], and drug delivery [44]. AI is increasingly being integrated into various nanomanufacturing processes, such as nano assembly, the creation of nanotubes, nanomedicine, nanorobotics, and nanosensors, denoting its essential role in the advancement of nanotechnology [45–51]. Table 2 shows an exhaustive list of such nanomanufacturing processes that incorporate the use of AI.

Table 2. Nanomanufacturing with the role of AI.

Nanomanufacturing Process	Role of AI	Ref.
Nano assembly: use of dielectrophoresis (DEP), the utilization of electrokinetic force employed during the assembly at the nanoscale.	Automated ascertainment of frequency range for positive and negative dielectrophoresis.	[52]
Carbon nanotubes and graphene were used in the preparation of grease as additives to improve the grease properties. Microwave-assisted ball milling was the process.	The performance of the grease was continuously monitored using an artificial neural network (ANN)	[53]
Nanomedicine where nanometric structures provide a highly sophisticated, targeted approach to diagnosing health conditions.	AI is used to optimize the properties of materials relative to anticipated interactions with the designated medication to prepare the cancer medicine.	[54]
Nanorobotics, characterized by their nanoscale sizes, are employed in the field of biomedical applications.	Artificial intelligence and biomedicine are used in creating nanorobots.	[55,56]
Nanocomposite multifunctional mechanical wearable sensors.	AI is used to predict the outcomes for parameters such as breathing, heart rate, cardiovascular system signals, and pressure in joint surgeries.	[56–60]
Luminescence sensors manufacturing with nanomaterials.	Machine-learning (ML) models have significantly contributed to the improvement of efficiency, sensitivity, and selectivity within luminescence platforms.	[61]

Authors Varahramyan and Lvov discussed how nanomanufacturing methods are saving human lives through nanoengineered capsules for sustained release drug delivery [62]. These nanoengineered capsules can provide cancer treatment through the integration of AI, as proposed by Adir et al. [54]. The mechanical, electrochemical, and spectral properties of nanosensors enhance the diagnostic profile for cancer patients, allowing them to have early detection, a treatment plan, and monitoring of the disease. As shown in Figure 2, during the diagnostic stage, nanosensors can detect extremely low concentrations of cancer biomarkers [63,64]. Integration of AI algorithms in the data analysis can allow recognition of combinatory treatments. Prescriptions are prepared with nanomedicine (nano-capsules), which control drug dosing by adjusting the drug release rate based on the patient's diagnosis and the treatments identified. Nanoparticle-based image analysis using convolutional neural networks has been widely used during the monitoring stage, and follow-up is conducted based on the AI model's outcome.

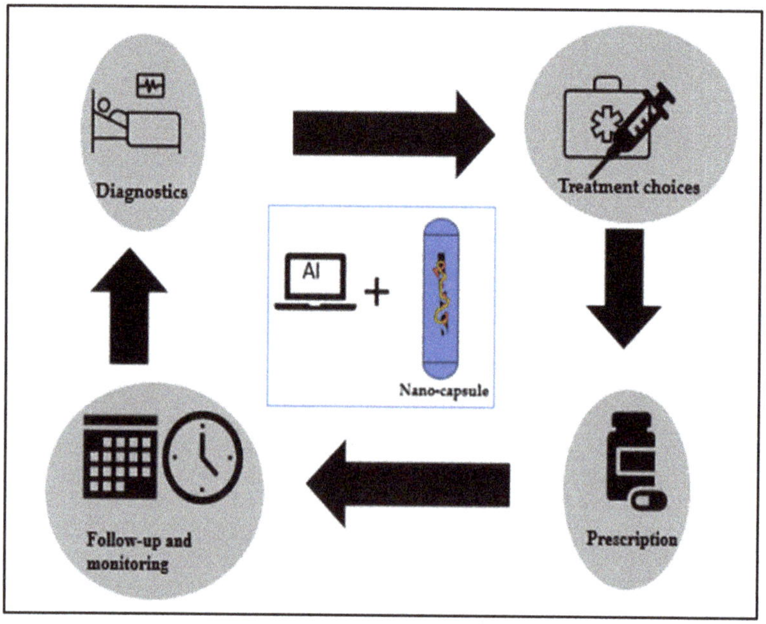

Figure 2. Nanofabricated drug delivery capsule application with AI.

1.5. *Interdependency of Nanomaterials, Nanomanufacturing, Digital Manufacturing, and Artificial Intelligence*

Nanomaterials, nanomanufacturing, digital manufacturing, and artificial intelligence are pivotal elements in industry 4.0. The application of predictive algorithms has resulted in novel paths in material discovery and optimization. Digital manufacturing, which centers on digital design and control of the manufacturing processes, emerges as a promising approach to addressing challenges in complex geometries, material optimization, and nanoscale material design [65]. According to Dopico et al. [66], this approach integrates information, computation, sensors, software, networking, and automation to carry out assigned tasks. The combination of AI and DM drives nanomanufacturing processes, streamlining the flow from inputs known as data. These data can be extracted from nano/micro-scale designs, CAD inputs, manuals, or published technical journals and give desired outputs known as outcomes [67–73]. Artificial intelligence proves instrumental in enhancing production processes, predicting novel nanostructured materials, improving efficiency and ensuring product quality (predictive maintenance) and supply chain optimization [74]. Figure 3 exemplifies a manufacturing process where input materials such as nanotubes, quantum dots, and dendrimers are utilized to produce outputs such as nanofilms, nanosensors, and nanorobots.

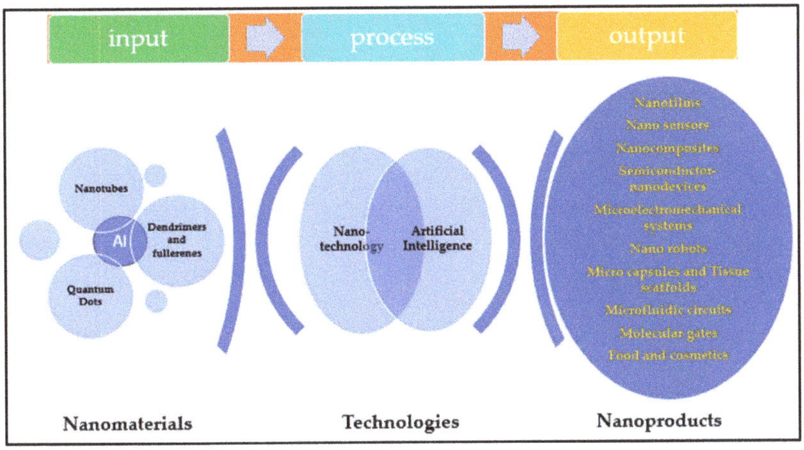

Figure 3. Alignment of artificial intelligence in the nanomanufacturing process.

Optimally integrating artificial intelligence (AI) within nanomanufacturing, in line with the benchmarks set by Industry 4.0 and digital manufacturing, equips manufacturing companies with a competitive edge and the agility to meet the rapidly changing demands of the global market. Such coordination also reveals opportunities for comprehensive optimization of production processes, real-time monitoring capabilities, and the freedom to undertake research and development. The ensuing segment of this manuscript delves deeper into the specifics of utilizing AI in the realm of nanomanufacturing.

2. Artificial Intelligence in Nanomanufacturing

In the field of exploring nanomanufacturing, AI is effective in predicting the properties and behavior of nanomaterials. Additionally, AI algorithms have been employed to optimize various aspects of nanomanufacturing processes, such as improving efficiency and reducing defects. the conversion of nanomaterials into final products strategically leverages artificial intelligence to achieve precision, minimize wastage, and optimize efficiency [74]. Support vector machines (SVMs) and artificial neural networks (ANNs) serve as pivotal tools to find the relationships between raw materials and the nano products produced. The application of AI extends beyond relationship exploration; it underscores its instrumental role in material synthesis and nanocharacterization.

2.1. Material Synthesis

Nanoparticles, owing to their distinctive properties, find extensive applications in fields such as medicine, material science, chemistry, and optoelectronic sensing. The synthesis of nanoparticles involves navigating an extensive array of synthetic paths, resulting in challenges in achieving precise control over particle sizes and the desired properties of the synthesized nanoparticles. As expressed by Chen et al., optimizing this synthesis becomes complex, requiring consideration of multiple parameters [75]. In addressing these challenges, Mijwel et al. [76] explored the application of AI, specifically employing feedforward neural network (FFNN) and recurrent neural network (RNN) models to replicate the adsorption capability of functionalized carbon nanotubes (CNTs) for methylene blue removal. It was further revealed that increasing the number of neurons in a neural network is a critical parameter that influences the model's capacity to learn and represent complex relationships in the context of material synthesis. The results revealed exemplary performance, with correlation coefficients of 0.9471 and 0.9658. Water plays an integral role in material synthesis, particularly in water treatment modeling, as emphasized by Safeer et al. [77]. Methylene blue (MB), a synthetic dye with a deep blue color, has an array of applications in the synthesis of nanoparticles because it is utilized for the fabrication, imaging, and modification of nanoparticles. For safe disposal and improving water quality assessment, neural networks are employed for water treatment modeling [77]. Several neural networks have also been applied in the synthesis, characterization, parameter optimization of nanocrystals, detection, and measurement of nanocantilever positions, fabrication, and post-fabrication imaging of structures via focused ion beam milling (FIB) [78–80].

These structures are extensively applied in the nanophotonic, seeded growth of nanostructures, and many other applications [81–85]. In addition, several machine-learning algorithms have been employed to assist in the synthesis of nanoparticles, as shown in Table 3.

Table 3. Machine-learning algorithms used in the synthesis of nanoparticles.

Nanoparticles	Description	Machine-Learning Models	Ref.
Gold nanoparticles	ML models were utilized to analyze the correlation between nanoparticle size, quantity of reagents, and bacteria strain effects on nanoparticles.	LASSO regression, RR, ENR, SVM, RF	[86]
NiO and TiO	Employed ANN to demonstrate effective techniques for utilizing ultra-low power inert gas condensation (ULPING) to create nanoscale structures.	ANN	[87]
Gold nanoparticles	Utilized ANN to understand the correlation between the ratio of sodium citrate to gold salt for gold nanoparticle generation.	ANN	[88]
Silver nanoparticles	Two ML (BO and DNN) were utilized to improve the Accuracy of model predictions for optimizing the synthesis of silver nanoparticles.	BO, DNN	[89]
CsPbBr3 type perovskite	Several models, such as SVM regression, linear regression, and quadratic regression, were developed to control the synthesis of perovskite halide nanoplatelets.	SVM	[90]
Colloidal quantum dots	ML was used to speed up the synthesis of colloidal quantum dots.	ANN	[91]
Perovskite oxides	ML analysis of perovskite oxides grown by molecular beam epitaxy.	K means	[92]
CNT forest	Using ML to predict carbon nanotube forest attributes	DL, RFF	[93]
$SrRuO_3$	Utilized BO in molecular beam epitaxy	BO	[94]
CsPbBr3	Utilized active learning algorithm to determine the optimal synthesis route	Active learning algorithm	[95]

Nanocomposites

Nanocomposites, heterogeneous materials with structures, components, or phases ranging from 1 to 100 nm, find applications in diverse fields, including food, biomedical [96–99], and electroanalysis [26,100–102]. They offer a unique approach for enhancing properties through the integration of nanoscale reinforcements developed through material synthesis. In the work by Mazahery et al. [103], a machine-learning algorithm was proposed as an alternative to costly experimental investigations in various manufacturing areas. The study employed artificial neural network (ANN) models for the numerical modeling of metal composites reinforced with nano-sized ceramic particulates. This approach, combining the finite element method (FEM) and ANN algorithms, accurately predicted wear behavior. The Levenberg–Newton method algorithm trained an ANN with a 1-hidden-layer, 8-neuron architecture. In the research by Harsha et al. [104], ANN models were developed to predict the tensile strength of Al-Al_2O_3 nanocomposites fabricated through powder metallurgy. Achieving a 98% accuracy with a 5-5 hidden layer architecture, the study highlighted the significant impact of the number of neurons and hidden layers on ANN model efficacy. Unlike Mazahery et al. [103], it was evident that ANN models vary with problem-specific parameters. Leveraging machine-learning models enables researchers to gain insights into the integral role of material synthesis and nanomaterials in the fabrication and customization of nanocomposite materials. Table 4 succinctly outlines studies utilizing ML models for predicting the properties of polymeric nanocomposites.

Table 4. Studies use ML models to predict the properties of polymeric nanocomposites.

Nanocomposite	Description	ML Model	Ref.
Polymer/SiO_2	ANN and ANFIS were developed to predict the fracture energy of polymer nanocomposites using data from various experimental datasets. Results indicated it was proportional to the experimental data.	ANN + ANFIS	[105]
Epoxy/SiO_2	ANN was utilized to predict the fracture toughness of particulate polymer composites. The inputs found important, in this order, were time > aspect ratio > elastic modulus > volume fraction. Results indicated that rod-shaped fillers exhibited enhanced fracture toughness.	ANN	[106]
PTFE/CF/TiO_2	ANN was utilized to predict the tribology properties of short-fiber composite.	ANN	[107]
PLA/GNP	Development of a back-propagation ANN to predict the mechanical property of SPS-processed GNP/PLA polymer nanocomposites. Predictions were proportional to experimental test results.	ANN	[108]
Polymer/nanofiller	A deep multi-task convolutional network was utilized to analyze polymer nanocomposite structure–property relationships and to predict its mechanical properties.	CNN	[109]
PPy/MWCNT	Novel MWCNTs-PPy-embedded RO nanocomposite membranes were synthesized then ANN was utilized for the prediction of model flux measurements.	ANN	[110]
PA-6/Nanoclay	ANN was utilized as a predictive model for optimizing surface roughness in Polyamide-6 nanocomposites based on milling parameters and nanoclay content. GA was utilized to obtain optimized machining conditions.	ANN + GA	[111]
Polymer/CNT	ANN was developed to predict the behavior pattern of the Young's modulus of polymer/CNTs composites.	ANN	[112]
Polymer/GNP	Proposed neural networks to describe the nonlinear electric conduction in random graphene-polymer nanocomposites.	ANN	[113]
Starch/Clay/AgNps	ANN with LM feed forward was utilized for optimization, evaluation and prediction analysis, and modeling of chemical reactions (AGNP size in the bio-nanocomposite matrix).	ANN + LM	[114]
CNT/Epoxy	SVR with PSO were employed to develop a model to predict the mechanical properties of CNT/epoxy composites. It successfully achieved high accuracy and consistency with experimental results.	SVR + PSO	[115]
Clay/Silica	ML models were used to predict the mechanical properties of a polymer clay nanocomposite before its synthesis. Characterization was also performed by XRD and SEM.	ANN	[116]

2.2. Nanomanufacturing Process

Nanomanufacturing forms the foundation for almost all areas of nanomaterial research and development, specifically focusing on the utilization of these materials for sophisticated, versatile devices across diverse technological sectors. These sectors include electronics, energy systems, medicine, and health care [117–121]. Techniques for creating nanostructured materials and nanoscale structures are categorized into top-down and bottom-up methods. The top-down approach applies lithography techniques to pattern nanoscale structures, often used in electronics industries, whereas the bottom-up method assembles discrete nanoscales structures through molecular or colloidal interactions in two and three dimensions [117,122]. The integration of AI has proven its superiority across diverse fields, one of which is the semiconductor industry.

Lithography–AI Case Study

The significance of lithography in the semiconductor industry and the increasing use of machine-learning algorithms, particularly deep learning, is applied in lithography simulation tasks. Lithography involves transferring a designed mask pattern onto a wafer, and it is composed of two stages: optical and resist simulation [123]. Due to high cost and time-consuming experimental verification, lithography simulation for process development and performance verification has been relied on. Various machine-learning algorithms are applied in lithography simulation for tasks like performance modeling, hotspot detection, and mask optimization. In performance modeling, Lin et al. [124] investigated performance-learning modeling under uncertainty and lithography modeling with active learning. Watanabe et al. [125] proposed a compact resist model using CNN for real-time chip simulation and achieved a 70% reduction in prediction errors compared to conventional models. Lin et al. [126] presented a novel approach to lithography simulation using transfer learning, in which residual neural networks (ResNet) were utilized for high-performance resist modeling that reduces the amount of data needed for accurate modeling. Other machine-learning-based techniques have been proposed to improve simulation quality and to predict height of resist after exposure in constructing accurate and efficient resist models [127]. Also, owing to the advancement in deep learning, other attempts have been made to solve the mask 3D effect problems using deep neural networks [128–130]. Tenable et al. [131] proposed an AI algorithm (CNN) to reduce the calculation time for diffracted waves from EUV masks. This CNN prediction is 5000 times faster than rigorous electromagnetic simulations. The authors further stated that the eigenvalue decomposition method can be used to accelerate the calculation.

Nanoimprint lithography (NIL) is characterized as a high-resolution, throughput, and cost-effective technology for developing nanoscale features in a relatively short amount of time [132–134]. Achieving defect-free imprints requires meticulous control and optimization of process parameters. Akter and Desai [135] predicted defect-free imprints by utilizing ANN algorithms. These ANN models (GRNN, BPNN, PNN) were trained using several input factors and of these three models GRNN predicted an accuracy of 55% and 90% for stage 1 and stage 2, respectively.

2.3. Nanocharacterization

Material science has seen a gradual transition over the past decades from creating innovative materials using raw computational approaches to creating linked methods that increase the reliability of the results through experimental validation and computational predictions. Products manufactured at the nanoscale necessitate the use of extremely sensitive and highly precise instruments for appropriate measurement. This process, known as nanocharacterization, involves the detailed analysis and measurement of nanoscale materials and structures [136]. Additionally, machine learning (ML) has found applications in polymeric nanocomposites for material property prediction, microstructural characterization, process optimization, and quantification of uncertainties arising from complex production processes [137,138].

2.3.1. AI in Nanocharacterization

Salah et al. [139] applied ML models to predict the absorption of CNT composites. In another study, Yusoff et al. [140] used ML and a multilayer perceptron network technique to improve process parameters and predict rheological properties such as the absorption index of nanosilica. Khanam et al. [141] utilized ANNs to optimize and predict the mechanical and chemical properties of graphene nanoplatelets nanocomposites. Other studies employed various computational techniques such as ANN, Bayesian inference, and other optimization algorithms to predict and analyze the material properties of hybrid nanocomposites [142], electrical conductivity of composites [137], and potential applications of CNT nanocomposites in damage detecting sensors [143].

Fatigue and creep are examples of some mechanical properties that are suitable for ANN analysis and ANFISs (adaptive neuro fuzzy inference systems) [144], while ANNs are suitable for predicting impact strength and yield strength. In addition, CNN has been applied to analyze images and make quantitative predictions about the mechanical behavior of composites by using the various microstructure grid sizes of the composites that are obtained from SEM analysis [105]. Several other studies have applied AI methods in the characterization of nanomaterials [145,146]. In other studies [147–149], the Levenberg–Marquardt gradient descent, Gauss–Newton algorithm, and genetic algorithms are commonly used for thin-film nanocharacterization.

2.3.2. Hotspot Detection

Hotspot detection refers to the identification and analysis of areas on a semiconductor wafer or mask that are prone to manufacturing defects or variations. It is crucial for ensuring the reliability and performance of semiconductor products. Various data-mining and machine-learning techniques, including ANN, SVM, deep CNN, and GRNN, have been integrated into recent works to enable quick and precise hotspot detection. Nagase et al. [150] trained an ANN kernel directly using 2D hotspot image patterns, achieving approximately 55% hotspot detection accuracy. Ma et al. [151] developed data-mining algorithms for clustering 2D hotspot patterns, resulting in a library to automatically categorize hotspots. Ding et al. [152] proposed hotspot detection with critical feature extraction and classification using ANNs, achieving a small detection false alarm rate and an average detection accuracy of over 90%. Furthermore, Duo Ding et al. [153] used machine-learning algorithms to predict hotspots, employing deep CNN to precisely identify hotspots, outperforming other neural networks and fuzzy-matching techniques. Yang et al. [154] utilized deep CNN to identify false negative results, demonstrating superior performance compared to existing hotspot detectors in [155,156]. In addition, Shin and Lee [157] described a deep convolutional neural network approach for precisely identifying lithography hotspots in semiconductor layout designs, outperforming previous machine-learning and fuzzy-matching techniques with high detection rate and false alarm reduction. These studies have contributed to advancements in hotspot detection using machine-learning techniques, demonstrating improved accuracy and performance in semiconductor manufacturing.

2.3.3. Spectroscopy

Machine learning provides unprecedented opportunities in the analytical sciences for extracting data from complex datasets in fields such as spectroscopy, mass spectrometry, and NMR, among other fields. Particularly in techniques like Raman and surface-enhanced Raman scattering (SERS), extensive datasets of complex chemical mixtures' vibrational spectra are collected for the analysis or imaging of chemical systems [158]. In addition, by leveraging the valuable information they provide about the chemical composition, structure, and surface characteristics of nanomaterials, researchers and manufacturers can gain a better understanding of the features and optimize their synthesis and production processes [159]. In nanomanufacturing, fabrication of SERS nanostructures can be achieved through either a top-down or a bottom-up approach. Among the different types of nanofabrication methods, lithographic techniques with electron beams [159–161] and focused ion beams and UV light have been greatly explored [162–165]. Bottom-up approaches have also been used to fabricate SER nanostructures via self-assembly, nanosphere lithography, and solution-based synthesis methods [166–169].

AI algorithms have highlighted the significant potential of SERS-based platforms for bio-analytical applications in the analysis of complex Raman spectra [159]. By utilizing spectrum data and AI-based algorithms, the accuracy of determining the findings of Raman spectra can be significantly improved. Important parameters in spectra preprocessing tasks are noise reduction, baseline correction, training, validation, normalization, and testing of the obtained raw datasets [170,171]. A kernel function is proposed for the optimal separation of the Raman spectrum to achieve more complex Raman spectral analysis in areas such as bacteria isolation [172]. Furthermore, the application of ANNs, deep learning, and machine learning has been seen in various tasks such as image recognition, photoinduced DNA damage [173], biomedical applications in cancer diagnosis, and detection of exosome proteins [174–179]. In NMR, 2D NMR reveals more information about the spectrum; however, the measuring times of 2D NMR increases astronomically with sampling numbers. In Kong et al. [180], a deep CNN was applied to speed up 2D nanoscale NMR spectroscopy, demonstrating enhanced sensitivity and accuracy in molecular structure analysis. The researchers demonstrated the effectiveness of the approach in pattern identification and noise cancellation, leading to enhanced sensitivity and accuracy in molecular structure analysis using NMR. Several authors have discussed how machine learning can be utilized with spectroscopic ellipsometry to also replace complex optical modeling [181,182]. Raman spectroscopy can also be used to identify known compounds and unknown materials and determine their number of layers [183–186].

In conclusion, the integration of machine learning (ML) with advanced spectroscopic techniques has revolutionized the field of nanomanufacturing. The ability to extract valuable insights from complex chemical mixtures' vibrational spectra, particularly in the context of Raman and SERS techniques, has significantly advanced the characterization and optimization of nanomaterials and nanomanufacturing processes.

2.3.4. Multilayer Device Measurement Methods

Electron microscopy, including critical-dimension scanning electron microscopy (SEM) and transmission electron microscopy (TEM), plays a crucial role in the analysis of multilayer thickness, defect analysis, and analyzing nanoscale structures [187–192]. In Kondo et al. [189], the authors revealed that ML can be applied to improve the imaging speed of SEM, allowing for the detection of faults or defects at high resolution using rapidly obtained low-quality images as inputs and slowly acquired high-resolution images as outputs. This allows researchers to visualize the distinct layers within a multilayer device and examine their composition, structure, and performance characterization. According to Choi et al. [193], either a multi-sample or a multi-angle can be used to increase the amount of information with regards to the measurement, and having samples of fixed optical constants as added information can result in improvements in the accuracy and uniqueness of thickness characterization when using these measurement devices. ML can save time during characterization, which results in substituting optical modeling [194,195].

AI plays an integral role in the nanomanufacturing process, particularly in material synthesis, lithography, hotspot detection, spectroscopy, and characterization methods. The use of AI in these areas has significantly transformed the way nanoparticles are synthesized and semiconductors are manufactured. AI models continue to evolve, becoming more sophisticated and efficient, enhancing the capabilities of nanomanufacturing with the applications of various machine-learning and deep-learning models, as detailed in the following section.

3. Machine Learning and Deep Learning in Nanomanufacturing

As an emerging technology that primarily focuses on data-driven analysis and processing, AI has significantly contributed to the evolution of various manufacturing processes.

In particular, the application of AI has brought about groundbreaking insights and advancements in the field of nanomanufacturing. One of the challenges in adopting AI into the nanomanufacturing process is the identification of appropriate machine-learning algorithms in the material's design. Costly experiments and the complex calculations needed to conduct research on nanomaterials (NMs) and nanotechnology opened gates to advanced technologies like artificial intelligence. The various structures and properties of NMs are quickly revealed by machine-learning algorithms. It was highlighted that the huge data generated from thousands of papers published per year related to the structures, properties, adsorption properties, and catalytic behavior of NMs is not sufficiently used for NMs [196]. This research developed the quantitative structure–property relationship (QSPR) models and identified promising materials for specific applications with machine-learning algorithms. As mentioned by Jia. Y et.al [197], machine-learning algorithms are widely used in the exploration of new materials to study the defects and properties of nanomaterials. With various machine-learning algorithms in practice, researchers are often ambiguous about which algorithm to use. After reviewing various algorithms used by researchers, a consolidated table is made for easy reference, as shown in Table 5. All these algorithms are used in machine-learning and deep-learning algorithms, which are part of artificial intelligence. AI systems are designed to mimic human thinking and work autonomously. The Venn diagram given by Goodfellow et al. [198] shows the relation between AI, ML, and DL, and the same is used to show the relation between them with tasks performed by ML and DL, as shown in Figure 4. AI systems consist of machine-learning and deep-learning algorithms to predict the outcomes.

Table 5. AI algorithms in the exploration of nanomaterials.

Area of Application	AI Algorithm(s)	Purpose	Reference(s)
Discovery of new nanomaterials	Recurrent neural networks (RNNs) and Monte Carlo tree search.	Utilized in predicting the growth of clusters into nanoparticles and recovery of structures in thin films.	[199–201]
Nanoscale structure descriptors	Multichannel convolutional neural network (CNN).	Utilized in predicting the cytotoxicity of virtual carbon nanoparticles, predicting disease-prone protein binding sites, and nanoparticle material properties.	[202–205]
Defect analysis (surface and internal structural)	Linear regression (LR), random forest (RF), and deep learning.	To detect the defects in nanofibrous materials, microencapsulated materials, lattice defects, and polymer composites.	[206,207]
Exploring phase-changing materials (PCMs) and thermoelectric effect materials	Digital tree, elastic net (EN) quadratic polynomial LASSO (QP-LASSO), and neural networks.	To explore highly active catalysts for energy conservation.	[208]
Optical properties of the materials	Deep learning with bi-directional neural networks.	To predict the refractive index of organic polymers.	[209,210]
Thermal properties of the materials	Linear regression, polynomial regression, decision trees, random forest, and artificial neural networks (ANN).	To predict interfacial thermal resistance thermal interface materials.	[211]
Electronic properties of materials	Multiple linear regression (MLR), DT, KNN, ANN, and SVM models.	Utilized in predicting electronic properties and energy differences in graphene nanoflakes.	[212,213]
Exploring nanomaterials with high adsorption capacity	Deep neural networks	Utilized in predicting the adsorption of organic pollutants.	[214]
To study the interaction between nanomaterials and biology	Random forest (RF)	To predict the formation and composition of protein coronas.	[68,215]

The AI algorithms used in nanomanufacturing are from four different categories, as shown in Table 6. These categories comprise supervised, unsupervised, reinforcement, deep, and hybrid learnings. A detailed understanding of each method is fundamental for effective application. In essence, supervised learning entails the training of models based on known input and output data, which consequently facilitates the prediction of future outputs. In unsupervised learning, algorithms aim to find patterns and structures in input data. The algorithms allow models to learn based on feedback from their actions in an environment in reinforcement learning. The final and most important one is deep learning, a sub-field of machine learning, which uses neural networks with multiple layers for complex pattern recognition.

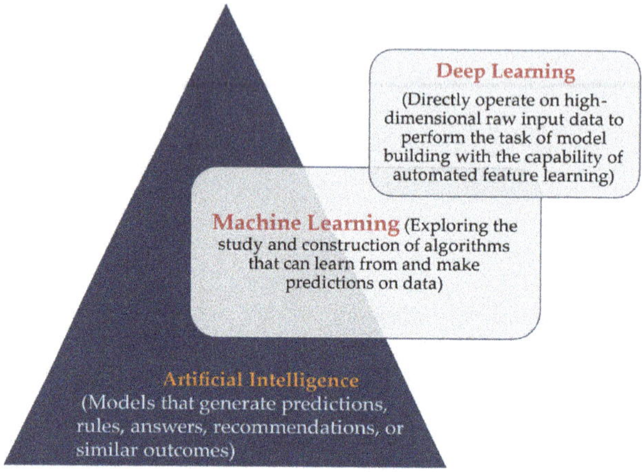

Figure 4. Machine learning and deep learning as AI tools.

Table 6. Commonly used AI algorithms.

Type of Algorithm	Algorithm Name	Description
Supervised Learning [216]	Linear Regression	Linear regression models are constructed with input variables relevant to the output variable and they are not highly correlated, used for regression problems.
	Logistic Regression	Logistic regression is for the numeric input variables to model binary classification problems, used for classification problems.
	Random Forest	In a random forest model, instances from the training dataset are selected, utilizing a replacement methodology. Further, the development of trees is strategically performed to mitigate the correlation observed among individual classifiers, thus enhancing the efficacy and reliability of the model.
	Gradient Boosting	Gradient boosting is constructed for classification using the gradient boosting classifier class.
	Support Vector Machines (SVM)	SVM is prepared with a line that separates the two classes. The data closest to the line are called support vectors, used for multiple classes.
	Decision Trees	Decision trees construct a binary tree from the data and split the data to minimize a cost function.
	Naïve Bayes	Naïve Bayes calculates the probability of each class based on each input value.
	Neural Networks	Neural networks are predominantly composed of intricately interconnected nodes, also known as neurons, which are systematically organized into various layers. Each corresponding connection within the network is affiliated with a specific weight. This said weight is duly adjusted over the course of the training process to decrease the error rate to the lowest possible extent.

Table 6. *Cont.*

Type of Algorithm	Algorithm Name	Description
Unsupervised Learning [217]	Principal Component Analysis (PCA)	PCA uses linear algebra to convert the dataset into a compressed form to reduce the data, called feature extraction.
	K-Means Clustering	In K-means clustering, data are paired with values as clusters.
	Hierarchical Clustering	Establishing a hierarchy of clusters is the main idea behind the hierarchical clustering, to check the data anomalies that go beyond the clusters.
	Apriori	Apriori is for data mining, to extract data from a large pool of data with various transactions.
	ECLAT	Equivalence class clustering and bottom-up lattice traversal (ECLAT) is also for data mining, to locate frequent items.
Reinforcement Learning [218]	Q-Learning	Q-learning finds the optimal solution in a Markov decision process by maintaining a Q-table with the combination of states and actions.
	State-Action-Reward-State-Action (SARSA)	In the SARSA algorithm, the current state action is updated based on the next action.
	Deep Q Network	The deep Q network works with discrete actions. It combines convolutional and pooling layers, followed by fully connected layers.
	Monte Carlo Methods	The Monte-Carlo algorithm works with episodic cases and considers complete episodes to obtain gradient values.
Deep Learning [219]	Conventional Neural Networks (CNN)	CNNs are applied for object recognition, detecting digits, faces, and objects with different orientations. Features are learned and used across the whole image.
	Recurrent Neural Networks (RNN)	RNNs are designed for sequence problems. Additional loops are used in the network in addition to the feedforward multilayer perceptron network.
	Long Short-Term Memory Networks (LSTM)	LSTM is an RNN, trained using backpropagation. These networks use memory blocks that are connected into layers.
	Autoencoders (AE)	Encoders work by encoding the strings to integers, then convert the vector of integers to a one-hot encoding.

Implementing AI in nanomanufacturing is a complex process that presents various challenges and prospects for various reasons, such as precision requirements, quick and accurate detection of defects, data security [220,221], technological limitations, and sensitive predictive analysis, and this has made people think about the challenges and prospects of implementing AI, which are emphasized in the following section.

4. Challenges and Prospects in Implementing AI in Nanomanufacturing

With AI adoption still at the initial stages in nanomanufacturing, it faces challenges in implementation, such as data availability, understandability, security, meeting regulatory requirements, etc. To mitigate all the challenges, researchers have proposed AI algorithms such as convolutional and deep neural networks, which are more powerful than traditional algorithms such as classifiers and regression models.

4.1. Challenges and Potential Solutions

Data availability: Despite the vast potential of nanotechnology across diverse domains, it is imperative to address the prevailing challenges concerning data assembly and alignment to facilitate the evolution of AI models [222]. For an AI model to function efficiently, it needs large quantities of quality data, which could be challenging to collect. The authors in papers [223,224] shed light on data-cleaning issues, specifically mentioning the deliberate or inadvertent inclusion of wrong data in training sets, which drastically affect the accuracy of the model and predictions. The authors also mentioned that the management of different versions of the datasets is a difficult task and errors that occur during this process are often hard to detect and correct. Data sparsity and quality are also considered as challenges due to missing data in the high-dimensional data. Noisy or incomplete data may pose a challenge in achieving the accuracy of machine-learning models [225].

Overcoming data availability challenges necessitates adopting a systematic approach towards data collection and cleansing, as endorsed by researchers. Additionally, they suggest employing data imputation strategies in managing absent data. This methodology enhances data input quality into AI models, thereby refining their precision and predictability [223,225]. For addressing the complexity and multi-scaled dimensions of nanomanufacturing data, the application of data fusion techniques has been proposed [224]. Furthermore, the establishment of public–private collaboration is advocated to stimulate data accessibility and dissemination [222].

Understandability: Results produced by AI and ML are often not clear (lack of transparency) and are sometimes like a black box. This could lead to a challenge while performing sensitive analysis during nano-characterization. A large concern in the implementation of AI is addressed in the paper by Von Eschenbach [226]: the lack of transparency and understandability in how AI algorithms make predictions. Wischmeyer's paper [227] further emphasizes the black box issue. The author argues that the opacity of AI systems needs to be analyzed so that their predictions can be interpreted and understood. Carabantes' paper [228] also stresses the importance of understandability in AI systems. It highlighted the issue of AI systems in decision-making, which can have a significant impact on the decision-making process. All these authors highlight the serious challenge of data understandability in AI. The authors emphasize the necessity for transparency to enable the effective regulation of AI.

To address the challenge of comprehending artificial intelligence (AI) models, the implementation of explainable AI (XAI) models has been suggested [226]. These models deliver an in-depth explanation for each decision they make, providing enhanced transparency and interpretability. This advantage is particularly useful in sensitive operations during nano-characterization. Additionally, establishing a systematic process for AI model documentation can aid in decoding the intricate internal functions of such models, thereby boosting their credibility [228].

Complexity: For a complex process with dynamic and stochastic effects, multi-scale effects, and many influencing variables, implementing AI will be a challenge. Multi-scale modeling to integrate models at different scales to develop a holistic understanding of manufacturing systems for their predictive capabilities comes with huge complexity. Peng et al. [225] explained several challenges regarding the integration of multiscale modeling and machine learning. Exponential increases in complexity can arise due to the complex methodologies used in machine-learning and multiscale modeling. Computationally intensive multiscale modeling and machine-learning models are a challenge to manage in terms of computational resources efficiently. The authors also mentioned that linking multiple scales in a dynamic rather than in a sequential manner is a challenge due to the requirement of consistent communication between different scales.

Strategies that bridge various scales, meticulously adapted to handle dynamic inter-scale interactions, can tackle the intricacies pertaining to multiscale modeling and machine learning, as identified in reference [225]. It is further suggested that a synthesis of data-driven and physics-oriented models can effectively mitigate the complexities inherent in systems associated with nanomanufacturing, as recommended by the study in reference [229].

Security and privacy: All the AI models, like any other computer models, are vulnerable to attacks that generate adversarial predictions [33]. Nano-security is a national priority program for nano-electronics, funded by the German Research Council (DFG), and shows the challenge of vulnerability [230]. Oseni et al. [231] highlighted the importance of security and privacy in AI models and clearly explained the issues of integrity, confidentiality, authenticity, data poisoning, privacy, trojan attacks, and fairness and bias, referencing examples. Some of the challenges mentioned include the following: inputs created with intentional but subtle changes to cause incorrect output, susceptibility to spoofing attacks, minor adversarial instances inserted into the training set causing a model to learn wrong features that erode the performance of the AI models, ML models can sometimes infer sensitive attributes that might compromise individual privacy, trojan attacks by inserting hidden malicious function into the model during the training phase and replicate and amplify the social biases in the training data sets, and inadequate laws and regulations for new technologies. Authors Solanas and Martinez also mentioned similar security and privacy challenges [232]. In their paper, the following were mentioned: improperly anonymized or protected information could be maliciously exploited, a bias in the data may lead to discriminatory practices, cyber-attacks in the manipulation of AI systems, and misuse of AI for nefarious purposes such as deepfake creation. Curzon et al. [233] also agreed to the existence of all the above-mentioned security and privacy issues.

Effective utilization of comprehensive cybersecurity protocols, encryption, and data anonymization techniques can significantly minimize potential security risks. Initiatives to raise awareness and enact appropriate legal regulations are paramount in ensuring security and privacy within AI models. The implementation of deep-learning models has been shown to be a potent tool in the detection and mitigation of adversarial data poisoning in several studies [232]. Additionally, continual monitoring and model updates for the purpose of detecting anomalies or attacks have proven to be helpful in preventing security breaches [233].

Training and implementation: AI systems require extensive training to function efficiently. In nanomanufacturing, where the processes are highly intricate, extensive training would be required. Moreover, changes, evolutions, or alterations to manufacturing processes would require additional system training. Many AI technologies, notably neural networks, are frequently perceived as 'black boxes.' This metaphor conveys the complexity and opaque nature of their decision-making processes, making comprehension challenging for many. Implementation of AI involves understanding AI models, their interactions, capabilities, and limitations. The lack of understanding and technical knowledge among professionals could result in resistance to adoption [234]. It was also mentioned that the higher costs to the administrators associated with training AI systems became a challenge in implementing AI.

Addressing the challenge of training and implementation necessitates increased investment in AI education across all levels to enhance the comprehension and acceptance of AI technologies [234]. Herein, regulatory bodies could have a pivotal role by providing incentives to stimulate relevant training programs. Additionally, cultivating a corporate culture that respects and understands AI, and nurturing collaborative relationships between academic institutions and industries, could prove to be mutually beneficial.

Technical issues: AI's abilities are limited to doing what it has been trained for, and it may be challenging to implement it in settings like nanomanufacturing, where several tasks may need real-time decisions and unexpected problem-solving. Managing and allocating computational resources to process information and make decisions within a given time frame is a key challenge in real-time, according to Musliner et al. [235]. It was mentioned that problems also arise related to task scheduling. In real-time, is difficult to determine the precedence, order, and priority of tasks. Environmental changes or unforeseen circumstances may pose uncertainty and make AI predictions a significant challenge. Apart from the above-mentioned challenges, author Seguin et al. [229] highlighted some more challenges, as follows: the computational time to provide a solution within the required timeframe, computationally expensive solutions under changing conditions, a trade-off between the quality of the solution and the response time, and the requirement of substantial memory space.

The application of real-time machine-learning algorithms, supported by potent computational resources, can provide effective solutions to the challenges of real-time decision-making [235]. Furthermore, the construction of sturdy scheduling algorithms can enhance task scheduling within the nanomanufacturing industry.

Regulations: There are strict regulations in the nanomanufacturing industry to ensure safety and quality standards, especially in the food industry [222]. Fitting AI into these regulations can be a challenge, especially given the dynamic nature of AI and its evolving interpretation of rules [236]. Author Hoffmann raises several challenges regarding law and regulation. One of the major challenges mentioned is the ambiguity and complexity associated with AI, making it difficult to attribute liability due to harm or damage. With AI working autonomously, accountability becomes complex and unidentified. The diversity of AI applications causes regulatory challenges, making it difficult to create comprehensive rules with wider areas of application. With rapid advancements in AI technologies, it is difficult for the law to keep up. AI can have reasonable impacts on society that may go beyond legal considerations, including implications for ethics and human rights, for which careful thought is required regarding law and policy.

Addressing the complexities of regulatory hurdles necessitates active participation and collaboration from all interested parties, notably developers and regulatory bodies. It is crucial for policymakers to be not only proactive but also adaptable in devising regulatory frameworks. They are expected to adequately predict potential developments, regulate with foresight, and modify policies as required [236].

Equipment compatibility: Implementing AI might require new equipment or might not be compatible with existing machinery or systems, which means companies need to incur a high cost to update their existing systems. Sharma et al. [237] mentioned challenges due to infrastructure. They include a lack of infrastructure, such as high-speed internet, digital devices, or software, and a lack of understanding in identifying the equipment needed to perform the task. Emerging economies might lack the equipment to support AI implementation, and maintenance will be difficult and costly, even with equipment in place.

Addressing equipment compatibility issues necessitates the development of cost-effective tools and methodologies for seamless integration of AI with pre-existing technologies [237]. It is imperative to conduct research focused on the creation of economical and trustworthy AI-compatible machinery. Such advancements can significantly diminish the entry barrier for nano-manufacturers, most notably within developing economies.

Flexibility in material properties: The inherent variability in the properties of nanoscale materials complicates the task of developing a reliable and generalized AI model for nanomanufacturing [238]. At a tiny nanoscale, nanomaterials can exhibit significantly different properties compared with their bulk counterparts. Nanoscale materials and devices can be sensitive to their environment, and their properties can change over time. Training an AI model with dynamics poses a challenge. Even with the same manufacturing conditions, the properties of nanoscale materials can be variable due to the inherent uncertainties in the properties of nanoscale materials.

Researchers have suggested the creation of comprehensive artificial intelligence models, fundamentally embedded with the intrinsic uncertainties associated with nanoscale properties. They posit that these AI models can be equipped to conduct robust analyses by accommodating the imprecisions and inconsistencies inherently present in the properties of nanoscale materials [238].

Higher costs: Developing and implementing AI algorithms for nanomanufacturing is expensive. It requires investment in extensive data collection, processing, and advanced computing systems, not to mention the expense of hiring experts in both AI and nanomanufacturing to manage the process [239]. The paper by Davenport et al. [239] discusses the economic implications of integrating AI into operations, which include cost challenges. One of the major challenges is the initial investment required for implementation because it involves big data infrastructure, software/hardware, and expert AI practitioners. Even after successful implementation, funds are required for ongoing maintenance and upgrades.

Recognizing regions of strategic investment and prospective cost savings over an extended period is crucial in counterbalancing implementation expenses. The development of machine-learning algorithms, intended to enhance operational efficiencies, can play a significant role in curtailing operational costs. Moreover, engaging in public–private partnerships could proffer an effective method of distributing the risks and expenses affiliated with the integration of AI technologies. This approach can potentially expedite adoption and contrive a cost-effective strategy for implementing these tech-driven solutions.

4.2. Discussion and Future Prospects

Bridging nanomanufacturing and artificial intelligence (AI) involves integrating AI technologies into the processes of designing, fabricating, and characterizing nanoscale materials and devices. This review paper highlights the transformative potential of integrating AI technologies into nanoscale manufacturing processes. This integration has the potential to revolutionize the field of nanomanufacturing by enabling advanced capabilities in material discovery, process optimization, quality control, and predictive modeling. In this paper, some key aspects we considered are:

Nanomanufacturing and Industry 4.0: This paper introduces nanomanufacturing as a critical component of Industry 4.0, emphasizing its role in producing materials and devices at the nanoscale. It discusses how nanomanufacturing enables the creation of innovative products with unique properties and functionalities, setting the stage for the integration of advanced technologies like AI.

Role of AI in Nanomanufacturing: This paper explores the diverse applications of AI in nanomanufacturing, highlighting its significance in predicting material properties, optimizing manufacturing processes, data-driven decision support, and enhancing operational efficiency. The prospects of AI in nanomanufacturing encompass accelerated material discovery, customized nanomaterial design, advanced nanodevices, and efficient resource utilization, paving the way for transformative advancements in nanotechnology and advanced manufacturing. AI algorithms are portrayed as powerful tools for real-time monitoring, quality control, and predictive maintenance in nanomanufacturing operations [220,221].

Material Discovery and Exploration: AI's role in material discovery and exploration within nanomanufacturing is discussed in detail. By analyzing vast datasets and identifying patterns, AI algorithms facilitate the discovery of new materials with unique characteristics, potentially leading to breakthroughs in diverse fields such as nanoelectronics, photonics, and biomedicine. This capability drives innovation in product development and design, enabling companies to stay at the forefront of technological advancements. The prospects include the rapid development of materials for emerging technologies, the creation of novel nanomaterial libraries, and the integration of diverse expertise for breakthrough innovations. These prospects contribute to the continual advancement of nanotechnology and its diverse applications across various industries and scientific domains [199–201].

Process Optimization and Efficiency: This paper delves into how AI-driven process optimization enhances efficiency and productivity in nanomanufacturing. AI algorithms analyze data in real-time, recommend adjustments, and minimize wastage, leading to improved operational efficiency and reduced defects. The automation and streamlining of processes through AI technologies contribute to overall process optimization. These prospects contribute to the continual advancement and widespread adoption of nanotechnology in various industrial sectors, driving economic and technological progress.

Quality Control and Predictive Maintenance: AI's critical role in quality control and predictive maintenance is emphasized. AI-powered quality control mechanisms ensure consistent product quality and minimize defects, while predictive maintenance strategies optimize maintenance schedules and prevent equipment failures. This involves the utilization of advanced characterization techniques such as electron microscopy, spectroscopy, and atomic force microscopy to assess the structural, chemical, and morphological properties of nanomaterials and products. This ensures adherence to stringent quality standards and specifications. This proactive approach to maintenance enhances operational efficiency and ensures continuous operation of manufacturing systems [222,223].

Innovation, Research Advancements, and Competitive Edge: AI fosters innovation, drives research advancements, and enhances competitiveness in nanomanufacturing. By applying AI technologies, companies gain a competitive edge by improving efficiency, product quality, and market relevance. AI-driven innovation enables companies to differentiate themselves, optimize operations, and meet evolving customer demands. Future research directions may focus on further refining AI algorithms for nanomanufacturing applications, addressing challenges such as data availability, security, and cost to unlock the full potential of AI in driving innovation and competitiveness.

Market Differentiation and Long-Term Sustainability: The integration of AI into nanomanufacturing not only provides immediate benefits in terms of efficiency and quality but also ensures long-term sustainability and competitiveness. Companies that invest in AI technologies for nanomanufacturing are better positioned to innovate, adapt to market changes, and sustain their competitive edge over time. By offering customized nanoproducts, embracing sustainable practices, and addressing ethical considerations, nanomanufacturers can position themselves for long-term success, growth, and positive industry impact.

Production systems are becoming more sophisticated, dynamic, and interconnected, causing nonlinear and stochastic activity issues. AI, particularly machine learning, has the potential to transform manufacturing by processing large volumes of industrial data, known as big data [240]. However, gaps must be addressed before AI can be integrated into nanomanufacturing operations.

Optimum performance necessitates a substantive understanding of the intricacies involved in the relationships between material processing and property. Development and implementation of conventional modeling and control mechanisms can be integrated into the process to address the change in material properties during the process and to solve the operational issues on a real-time basis [241]. For a different challenge, Malaca et al. [242] proposed computer vision-based AI models be developed to match the variable process operating conditions. The same concept is also proposed by Guven et al. [243] to address ever-changing laboratory operations. The intricate and unpredictable dynamics of manufacturing systems necessitate a comprehensive understanding of the multifaceted processes involved. The vast quantities of generated data, diverse in their characteristics and dependent on numerous variables, require the application of suitable algorithms to aid in the effective analysis and interpretation of these complex systems [196,240]. Machine learning with image recognition, such as microscopy, can generate more data, which is essential to getting high accuracy levels for an ML algorithm. Since manufacturing operations are based on physics, not on probability theories as used in AI, more transparent, physics-guided process models such as explainable AI (XAI), which supports mental support models [240], can be studied further. With very limited research conducted on the biological effects of nanoplastics, researchers can use ML algorithms to explore the same as a part of artificial intelligence [197].

The implementation of AI in nanomanufacturing presents several challenges, along with emerging solutions and prospects. Some of the main challenges include issues with data availability and quality, the understandability of AI and ML results, complexity in multiscale modeling, issues with security and privacy, training hurdles, technical issues in real-time decision-making, stringent regulations, equipment compatibility, and costs, as well as variability in material properties. However, recent developments in AI show the potential to transform the manufacturing domain with the use of advanced analytic tools. Potential solutions proposed by researchers include the development of conventional modeling and control mechanisms, computer vision-based AI models, reinforced learning algorithms, and physics-guided models like explainable AI. The exploration of these new solutions could mitigate the challenges and pave the way for seamless integration.

5. Conclusions

This paper fills a significant gap in the literature by providing a comprehensive review of the integration of artificial intelligence in nanomanufacturing processes. It focuses on how AI can optimize material synthesis, process control, and quality assurance in nanomanufacturing. Furthermore, it reveals the potential of AI in enhancing competitiveness and market relevance by improving product quality, reduce costs, and accelerate time-to-market, hence offering strategic insights for organizations seeking to expand and enhance their manufacturing capabilities. Another key justification for this review is that it discusses how AI resonates with the principles of Industry 4.0, emphasizing digitalization, automation, and data-driven decision-making in nanomanufacturing. By fostering cross-disciplinary collaboration and knowledge exchange between experts in materials science, engineering, and AI, these efforts can lead to breakthroughs in nanomanufacturing technology. In addition, by identifying key challenges, opportunities, and emerging trends, this paper serves as a guide for further exploration and innovation in this rapidly evolving area.

The scientific paradigm underlying materials at the nanoscale, developments in nanomanufacturing, digital fabrication, and the advent of emerging artificial intelligence frameworks represent the frontier of the fourth industrial revolution. In this review, we discuss the integration of artificial intelligence in nanomaterial exploration, manufacturing, and metrology, offering exciting opportunities for optimization in synthesis, processing, and characterization. Different AI algorithms that include supervised, unsupervised, reinforcement, and deep learning are discussed in the context of nanomaterial discovery, property enhancement, and nanomanufacturing predictive analytics. Moreover, significant challenges need to be overcome for the implementation of AI, such as data availability, security, high costs, lack of expertise, and legal and ethical issues. Hybrid models combining AI with other heuristic algorithms can pave the way for real-time control and feedback of nanomanufacturing processes. A notable development is the application of reinforced learning concepts that could significantly increase productivity levels. Furthermore, the advent of explainable AI (XAI) is a step towards creating a more reliable and user-friendly digital platform as a physics-guided process model to make the AI's decision-making process understandable for its users. It ensures a transparent interplay between data analysis and physical constraints, to provide judicious guidance for material designers and manufacturers. Future research is needed to fully integrate and exploit AI algorithms to further revolutionize material synthesis and the manufacturing industry, paving the way for next-generation solutions.

Author Contributions: M.N., O.F. and S.D. contributed to conceptualization, methodology, background literature, and formatting of this review article. S.D. contributed to reviewing and editing, supervision, and funding acquisition. All authors have read and agreed to the published version of the manuscript.

Funding: The authors would like to express their gratitude for funding support from the National Science Foundation Grant (NSF CMMI Award #2100739, #2100850) and the Center of Excellence in Product Design and Advanced Manufacturing at North Carolina A&T State University.

Data Availability Statement: The data presented in this study are available on request from the correspondence author.

Conflicts of Interest: The authors declare no conflicts of interest. The funders had no role in the design of the study; in the collection, analyses, or interpretation of data; in the writing of the manuscript; or in the decision to publish the results.

References

1. Siegel, R.W.; Hu, E.; Cox, D.M.; Goronkin, H.; Jelinski, L.; Koch, C.C.; Mendel, J.; Roco, M.C.; Shaw, D.T. *Nanostructure Science and Technology: R&D Status and Trends in Nanoparticles, Nanostructured Materials, and Nanodevices*; Springer: Dordrecht, The Netherlands, 1999; ISBN 978-90-481-5277-3.
2. Nasrollahzadeh, M.; Sajadi, S.M.; Sajjadi, M.; Issaabadi, Z. An Introduction to Nanotechnology. In *Interface Science and Technology*; Elsevier: Amsterdam, The Netherlands, 2019; Volume 28, pp. 1–27, ISBN 978-0-12-813586-0.
3. Silva, G.A. Introduction to Nanotechnology and Its Applications to Medicine. *Surg. Neurol.* **2004**, *61*, 216–220. [CrossRef] [PubMed]
4. Bhushan, B. (Ed.) *Encyclopedia of Nanotechnology*; Springer: Dordrecht, The Netherlands, 2012; ISBN 978-90-481-9750-7.
5. Bhushan, B.; Luo, D.; Schricker, S.R.; Sigmund, W.; Zauscher, S. (Eds.) *Handbook of Nanomaterials Properties*; Springer: Berlin/Heidelberg, Germany, 2017; ISBN 978-3-662-51048-3.
6. Bhushan, B. *Springer Handbook of Nanotechnology*, 4th ed.; Springer handbooks; Springer: Berlin/Heidelberg, Germany; New York, NY, USA, 2017; ISBN 978-3-662-54355-9.
7. Kolahalam, L.A.; Kasi Viswanath, I.V.; Diwakar, B.S.; Govindh, B.; Reddy, V.; Murthy, Y.L.N. Review on Nanomaterials: Synthesis and Applications. *Mater. Today Proc.* **2019**, *18*, 2182–2190. [CrossRef]
8. Vollath, D. *Nanomaterials: An Introduction to Synthesis, Properties and Applications*, 2nd ed.; Wiley-VCH Verlag GmbH & Co., KGaA: Weinheim, Germany, 2013; ISBN 978-3-527-33379-0.
9. McIntyre, R.A. Common Nano-Materials and Their Use in Real World Applications. *Sci. Prog.* **2012**, *95*, 1–22. [CrossRef] [PubMed]
10. Vigneshwaran, N.; Kathe, A.A.; Varadarajan, P.V.; Nachane, R.P.; Balasubramanya, R.H. Functional Finishing of Cotton Fabrics Using Silver Nanoparticles. *J. Nanosci. Nanotechnol.* **2007**, *7*, 1893–1897. [CrossRef] [PubMed]
11. Ma, J.; Wong, H.; Kong, L.B.; Peng, K.W. Biomimetic Processing of Nanocrystallite Bioactive Apatite Coating on Titanium. *Nanotechnology* **2003**, *14*, 619–623. [CrossRef]
12. Khanna, P.; Kaur, A.; Goyal, D. Algae-Based Metallic Nanoparticles: Synthesis, Characterization and Applications. *J. Microbiol. Methods* **2019**, *163*, 105656. [CrossRef]
13. Baig, N.; Kammakakam, I.; Falath, W. Nanomaterials: A Review of Synthesis Methods, Properties, Recent Progress, and Challenges. *Mater. Adv.* **2021**, *2*, 1821–1871. [CrossRef]
14. McGehee, M.D.; Gruner, S.M.; Yao, N.; Chun, C.M.; Navrotsky, A.; Aksay, I.A. Synthesis of Mesoscopic Structures by Co-Assembly. *Proc. Annu. Meet. Electron. Microsc. Soc. Am.* **1994**, *52*, 448–449. [CrossRef]
15. Stupp, S.I.; Braun, P.V. Molecular Manipulation of Microstructures: Biomaterials, Ceramics, and Semiconductors. *Science* **1997**, *277*, 1242–1248. [CrossRef]
16. Kessler, V.G.; Seisenbaeva, G.A. Molecular Mechanisms of the Metal Oxide Sol-Gel Process and Their Application in Approaches to Thermodynamically Challenging Complex Oxide Materials. *J. Sol.-Gel Sci. Technol.* **2023**, *107*, 190–200. [CrossRef]
17. Bhattacharyya, A.S. Modeling of Sputter-Based Atomic Layer Deposition with Altered Parameters. *Mater. Today Commun.* **2024**, *38*, 107697. [CrossRef]
18. Giges, N.S. Top 5 Nanoscale Manufacturing Processes. Available online: https://www.asme.org/topics-resources/content/top-5-nanoscale-manufacturing-processes (accessed on 2 November 2023).
19. Daryadel, S. *Additive Manufacturing and In Situ Mechanical and Material Characterization of Metallic Structures at Micro/Nanoscale*; University of Texas at Dallas: Dallas, DT, USA, 2018.
20. Leong, Y.J. *3D Printing Functional Nanocomposites*; University of California: San Diego, CA, USA, 2016.
21. Li, C.-W.; Wang, G.-J. MEMS Manufacturing Techniques for Tissue Scaffolding Devices. In *MEMS for Biomedical Applications*; Elsevier: Amsterdam, The Netherlands, 2012; pp. 192–217, ISBN 978-0-85709-129-1.
22. Zheng, X.; Smith, W.; Jackson, J.; Moran, B.; Cui, H.; Chen, D.; Ye, J.; Fang, N.; Rodriguez, N.; Weisgraber, T.; et al. Multiscale Metallic Metamaterials. *Nat. Mater.* **2016**, *15*, 1100–1106. [CrossRef] [PubMed]
23. Ehrmann, G.; Ehrmann, A. 3D Printing of Shape Memory Polymers. *J. Appl. Polym. Sci.* **2021**, *138*, 50847. [CrossRef]
24. Li, Y.; Zhang, F.; Liu, Y.; Leng, J. 4D Printed Shape Memory Polymers and Their Structures for Biomedical Applications. *Sci. China Technol. Sci.* **2020**, *63*, 545–560. [CrossRef]
25. Zhang, J.; Yin, Z.; Ren, L.; Liu, Q.; Ren, L.; Yang, X.; Zhou, X. Advances in 4D Printed Shape Memory Polymers: From 3D Printing, Smart Excitation, and Response to Applications. *Adv. Mater. Technol.* **2022**, *7*, 2101568. [CrossRef]
26. Desai, S.; Shankar, M.R. Emerging Trends in Polymers, Composites, and Nano Biomaterial Applications. In *Bio-Materials and Prototyping Applications in Medicine*; Bártolo, P.J., Bidanda, B., Eds.; Springer International Publishing: Cham, Switzerland, 2021; pp. 19–34, ISBN 978-3-030-35875-4.
27. Rahman, S.; Lu, Y. Nano-Engineering and Nano-Manufacturing in 2D Materials: Marvels of Nanotechnology. *Nanoscale Horiz.* **2022**, *7*, 849–872. [CrossRef] [PubMed]
28. Li, L.; Hong, M.; Schmidt, M.; Zhong, M.; Malshe, A.; Huis in'tVeld, B.; Kovalenko, V. Laser Nano-Manufacturing—State of the Art and Challenges. *CIRP Ann.* **2011**, *60*, 735–755. [CrossRef]
29. Promyoo, R.; Alai, S.; El-Mounayri, H. Innovative Digital Manufacturing Curriculum for Industry 4.0. *Procedia Manuf.* **2019**, *34*, 1043–1050. [CrossRef]

30. Singh, C.D.; Kaur, H. (Eds.) *Factories of the Future Technological Advancements in the Manufacturing Industry*, 1st ed.; Wiley-Blackwell: Hoboken, NJ, USA, 2023; ISBN 978-1-119-86519-3.
31. Almakayeel, N.; Desai, S.; Alghamdi, S.; Qureshi, M.R.N.M. Smart Agent System for Cyber Nano-Manufacturing in Industry 4.0. *Appl. Sci.* **2022**, *12*, 6143. [CrossRef]
32. Vettiger, P.; Cross, G.; Despont, M.; Drechsler, U.; Durig, U.; Gotsmann, B.; Haberle, W.; Lantz, M.A.; Rothuizen, H.E.; Stutz, R.; et al. The "Millipede"—Nanotechnology Entering Data Storage. *IEEE Trans. Nanotechnol.* **2002**, *1*, 39–55. [CrossRef]
33. Almakaeel, H.; Albalawi, A.; Desai, S. Artificial Neural Network Based Framework for Cyber Nano Manufacturing. *Manuf. Lett.* **2018**, *15*, 151–154. [CrossRef]
34. Wang, H.; Fu, T.; Du, Y.; Gao, W.; Huang, K.; Liu, Z.; Chandak, P.; Liu, S.; Van Katwyk, P.; Deac, A.; et al. Scientific Discovery in the Age of Artificial Intelligence. *Nature* **2023**, *620*, 47–60. [CrossRef] [PubMed]
35. Hu, Y.; Yang, J.; Chen, L.; Li, K.; Sima, C.; Zhu, X.; Chai, S.; Du, S.; Lin, T.; Wang, W.; et al. Planning-Oriented Autonomous Driving. In Proceedings of the 2023 IEEE/CVF Conference on Computer Vision and Pattern Recognition (CVPR), Vancouver, BC, Canada, 17–24 June 2023; IEEE: Vancouver, BC, Canada, 2023; pp. 17853–17862.
36. Chowdhary, K.R. Natural Language Processing. In *Fundamentals of Artificial Intelligence*; Springer: New Delhi, India, 2020; pp. 603–649, ISBN 978-81-322-3970-3.
37. Hamet, P.; Tremblay, J. Artificial Intelligence in Medicine. *Metabolism* **2017**, *69*, S36–S40. [CrossRef] [PubMed]
38. Voulodimos, A.; Doulamis, N.; Doulamis, A.; Protopapadakis, E. Deep Learning for Computer Vision: A Brief Review. *Comput. Intell. Neurosci.* **2018**, *2018*, 7068349. [CrossRef] [PubMed]
39. Peres, R.S.; Jia, X.; Lee, J.; Sun, K.; Colombo, A.W.; Barata, J. Industrial Artificial Intelligence in Industry 4.0—Systematic Review, Challenges and Outlook. *IEEE Access* **2020**, *8*, 220121–220139. [CrossRef]
40. Barbastathis, G.; Ozcan, A.; Situ, G. On the Use of Deep Learning for Computational Imaging. *Optica* **2019**, *6*, 921. [CrossRef]
41. Mishra, P.; Passos, D.; Marini, F.; Xu, J.; Amigo, J.M.; Gowen, A.A.; Jansen, J.J.; Biancolillo, A.; Roger, J.M.; Rutledge, D.N.; et al. Deep Learning for Near-Infrared Spectral Data Modelling: Hypes and Benefits. *TrAC Trends Anal. Chem.* **2022**, *157*, 116804. [CrossRef]
42. Genty, G.; Salmela, L.; Dudley, J.M.; Brunner, D.; Kokhanovskiy, A.; Kobtsev, S.; Turitsyn, S.K. Machine Learning and Applications in Ultrafast Photonics. *Nat. Photonics* **2021**, *15*, 91–101. [CrossRef]
43. Jiang, M.; Wu, H.; An, Y.; Hou, T.; Chang, Q.; Huang, L.; Li, J.; Su, R.; Zhou, P. Fiber Laser Development Enabled by Machine Learning: Review and Prospect. *PhotoniX* **2022**, *3*, 16. [CrossRef]
44. Jitendra, G.; Reena, G.; Abhishek, T. Nanobot: Artificial Intelligence, Drug Delivery and Diagnostic Approach. *JPRI* **2021**, *33*, 189–206. [CrossRef]
45. Ogunsanya, M.; Isichei, J.; Desai, S. Grid Search Hyperparameter Tuning in Additive Manufacturing Processes. *Manuf. Lett.* **2023**, *35*, 1031–1042. [CrossRef]
46. Ogunsanya, M.; Desai, S. Physics-Based and Data-Driven Modeling for Biomanufacturing 4.0. *Manuf. Lett.* **2023**, *36*, 91–95. [CrossRef]
47. Altubaishe, B.; Desai, S. Multicriteria Decision Making in Supply Chain Management Using FMEA and Hybrid AHP-PROMETHEE Algorithms. *Sensors* **2023**, *23*, 4041. [CrossRef] [PubMed]
48. Ogunsanya, M.; Isichei, J.; Parupelli, S.K.; Desai, S.; Cai, Y. In-Situ Droplet Monitoring of Inkjet 3D Printing Process Using Image Analysis and Machine Learning Models. *Procedia Manuf.* **2021**, *53*, 427–434. [CrossRef]
49. Elhoone, H.; Zhang, T.; Anwar, M.; Desai, S. Cyber-Based Design for Additive Manufacturing Using Artificial Neural Networks for Industry 4.0. *Int. J. Prod. Res.* **2020**, *58*, 2841–2861. [CrossRef]
50. Desai, S.; Dean, C.; Desai, Y. Cyber-Enabled Concurrent Material and Process Selection in a Flexible Design for Manufacture Paradigm. *Int. J. Adv. Manuf. Technol.* **2018**, *97*, 1719–1731. [CrossRef]
51. Ogunsanya, M.; Desai, S. Predictive Modeling of Additive Manufacturing Process Using Deep Learning Algorithm. In Proceedings of the IISE Annual Conference & Expo 2022, Seattle, WA, USA, 21–24 May 2022; IISE: Norcross, GA, USA, 2022; pp. 1–6.
52. Michaels, M.; Yu, S.-Y.; Zhou, T.; Du, F.; Al Faruque, M.A.; Kulinsky, L. Artificial Intelligence Algorithms Enable Automated Characterization of the Positive and Negative Dielectrophoretic Ranges of Applied Frequency. *Micromachines* **2022**, *13*, 399. [CrossRef]
53. Wang, S.; Liang, Z.; Liu, L.; Wan, P.; Qian, Q.; Chen, Y.; Jia, S.; Chen, D. Artificial Intelligence-Based Rapid Design of Grease with Chemically Functionalized Graphene and Carbon Nanotubes as Lubrication Additives. *Langmuir* **2023**, *39*, 647–658. [CrossRef] [PubMed]
54. Adir, O.; Poley, M.; Chen, G.; Froim, S.; Krinsky, N.; Shklover, J.; Shainsky-Roitman, J.; Lammers, T.; Schroeder, A. Integrating Artificial Intelligence and Nanotechnology for Precision Cancer Medicine. *Adv. Mater.* **2020**, *32*, 1901989. [CrossRef] [PubMed]
55. Li, M.; Xi, N.; Wang, Y.; Liu, L. Progress in Nanorobotics for Advancing Biomedicine. *IEEE Trans. Biomed. Eng.* **2021**, *68*, 130–147. [CrossRef]
56. Wang, Z.; Li, D.; Zhang, J.; Ji, Z.; Qiu, R. Robotic Nanoassembly: Current Developments and Challenges. *IJCAT* **2011**, *41*, 185. [CrossRef]
57. Mahanfar, A.; Nikfarjam, A.; Mohammadi, A.S. *Nanocomposite Multifunctional Mechanical Wearable Sensors and Investigating Their Different Modes Using Machine Learning*; SSRN: Amsterdam, The Netherlands, 2023.

58. Lu, S.; Ma, J.; Ma, K.; Wang, X.; Wang, S.; Yang, X.; Tang, H. Highly Sensitive Graphene Platelets and Multi-Walled Carbon Nanotube-Based Flexible Strain Sensor for Monitoring Human Joint Bending. *Appl. Phys. A* **2019**, *125*, 471. [CrossRef]
59. Ren, B.; Liu, J. Design of a Plantar Pressure Insole Measuring System Based on Modular Photoelectric Pressure Sensor Unit. *Sensors* **2021**, *21*, 3780. [CrossRef] [PubMed]
60. Jung, M.; Vishwanath, S.K.; Kim, J.; Ko, D.-K.; Park, M.-J.; Lim, S.-C.; Jeon, S. Transparent and Flexible Mayan-Pyramid-Based Pressure Sensor Using Facile-Transferred Indium Tin Oxide for Bimodal Sensor Applications. *Sci. Rep.* **2019**, *9*, 14040. [CrossRef] [PubMed]
61. Mousavizadegan, M.; Firoozbakhtian, A.; Hosseini, M.; Ju, H. Machine Learning in Analytical Chemistry: From Synthesis of Nanostructures to Their Applications in Luminescence Sensing. *TrAC Trends Anal. Chem.* **2023**, *167*, 117216. [CrossRef]
62. Varahramyan, K.; Lvov, Y. Nanomanufacturing by Layer-by-Layer Assembly—from Nanoscale Coating to Device Applications. *Proc. Inst. Mech. Eng. Part N J. Nanoeng. Nanosyst.* **2006**, *220*, 29–37. [CrossRef]
63. Parupelli, S.K.; Desai, S. The 3D Printing of Nanocomposites for Wearable Biosensors: Recent Advances, Challenges, and Prospects. *Bioengineering* **2023**, *11*, 32. [CrossRef] [PubMed]
64. Alonzo, S.M.M.; Bentley, J.; Desai, S.; Bastakoti, B.P. Hydrothermal Synthesis of Hierarchical Microstructure Tungsten Oxide/Carbon Nanocomposite for Supercapacitor Application. *Sci. Rep.* **2023**, *13*, 21732. [CrossRef] [PubMed]
65. Lin, L.; Kollipara, P.S.; Zheng, Y. Digital Manufacturing of Advanced Materials: Challenges and Perspective. *Mater. Today* **2019**, *28*, 49–62. [CrossRef]
66. Dopico, M.; De La Fuente, D.; Garcia, N.; Rosillo, R. A Vision of Industry 4.0 from an Artificial Intelligence Point of View. In *Proceedings of the International Conference on Artificial Intelligence (ICAI)*; CSREA Press: Athens, Greece, 2016; p. 407.
67. Adarkwa, E.; Desai, S. Scalable Droplet Based Manufacturing Using In-Flight Laser Evaporation. *J. Nanoeng. Nanomanuf.* **2016**, *6*, 87–92. [CrossRef]
68. Yang, M.; Xu, Z.; Desai, S.; Kumar, D.; Sankar, J. Fabrication of Micro Single Chamber Solid Oxide Fuel Cell Using Photolithography and Pulsed Laser Deposition. *J. Fuel Cell Sci. Technol.* **2015**, *12*, 021004. [CrossRef]
69. Desai, S.; de Fatima Gomes, P. Design for Nano/Micro Manufacturing: A Holistic Approach towards Achieving Manufacturing Excellence. *J. Udyog Pragati* **2015**, *39*, 18–25J.
70. Desai, S. Methods and Apparatus for Manufacturing Micro-and/or Nano-Scale Features. U.S. Patent Application No. 13/959,849, 28 November 2013. pp. 11–21.
71. Desai, S.; Craps, M.; Esho, T. Direct Writing of Nanomaterials for Flexible Thin-Film Transistors (fTFTs). *Int. J. Adv. Manuf. Technol.* **2013**, *64*, 537–543. [CrossRef]
72. Desai, S.; Bidanda, B.; Lovell, M.R. Material and Process Selection in Product Design Using Decision-Making Technique (AHP). *EJIE* **2012**, *6*, 322. [CrossRef]
73. Gaikwad, A.; Olowe, M.; Desai, S. Deformation Mechanism of Aluminum, Copper, and Gold in Nanoimprint Lithography Using Molecular Dynamics Simulation. *Nanomaterials* **2023**, *13*, 3104. [CrossRef] [PubMed]
74. Zhavoronkov, A.; Ivanenkov, Y.A.; Aliper, A.; Veselov, M.S.; Aladinskiy, V.A.; Aladinskaya, A.V.; Terentiev, V.A.; Polykovskiy, D.A.; Kuznetsov, M.D.; Asadulaev, A.; et al. Deep Learning Enables Rapid Identification of Potent DDR1 Kinase Inhibitors. *Nat. Biotechnol.* **2019**, *37*, 1038–1040. [CrossRef] [PubMed]
75. Chen, X.; Lv, H. Intelligent Control of Nanoparticle Synthesis on Microfluidic Chips with Machine Learning. *NPG Asia Mater.* **2022**, *14*, 69. [CrossRef]
76. Mijwel, A.-A.S.; Ahmed, A.N.; Afan, H.A.; Alayan, H.M.; Sherif, M.; Elshafie, A. Artificial Intelligence Models for Methylene Blue Removal Using Functionalized Carbon Nanotubes. *Sci. Rep.* **2023**, *13*, 18260. [CrossRef] [PubMed]
77. Safeer, S.; Pandey, R.P.; Rehman, B.; Safdar, T.; Ahmad, I.; Hasan, S.W.; Ullah, A. A Review of Artificial Intelligence in Water Purification and Wastewater Treatment: Recent Advancements. *J. Water Process Eng.* **2022**, *49*, 102974. [CrossRef]
78. Li, J.; Li, J.; Liu, R.; Tu, Y.; Li, Y.; Cheng, J.; He, T.; Zhu, X. Autonomous Discovery of Optically Active Chiral Inorganic Perovskite Nanocrystals through an Intelligent Cloud Lab. *Nat. Commun.* **2020**, *11*, 2046. [CrossRef]
79. Tadokoro, Y.; Funayama, K.; Kawano, K.; Miura, A.; Hirotani, J.; Ohno, Y.; Tanaka, H. Artificial-Intelligence-Assisted Mass Fabrication of Nanocantilevers from Randomly Positioned Single Carbon Nanotubes. *Microsyst. Nanoeng.* **2023**, *9*, 32. [CrossRef] [PubMed]
80. Buchnev, O.; Grant-Jacob, J.A.; Eason, R.W.; Zheludev, N.I.; Mills, B.; MacDonald, K.F. Deep-Learning-Assisted Focused Ion Beam Nanofabrication. *Nano Lett.* **2022**, *22*, 2734–2739. [CrossRef] [PubMed]
81. Stockman, M.I.; Kneipp, K.; Bozhevolnyi, S.I.; Saha, S.; Dutta, A.; Ndukaife, J.; Kinsey, N.; Reddy, H.; Guler, U.; Shalaev, V.M.; et al. Roadmap on Plasmonics. *J. Opt.* **2018**, *20*, 043001. [CrossRef]
82. Luk'yanchuk, B.; Zheludev, N.I.; Maier, S.A.; Halas, N.J.; Nordlander, P.; Giessen, H.; Chong, C.T. The Fano Resonance in Plasmonic Nanostructures and Metamaterials. *Nat. Mater.* **2010**, *9*, 707–715. [CrossRef]
83. Qiu, C.-W.; Zhang, T.; Hu, G.; Kivshar, Y. Quo Vadis, Metasurfaces? *Nano Lett.* **2021**, *21*, 5461–5474. [CrossRef] [PubMed]
84. Fan, M.; Andrade, G.F.S.; Brolo, A.G. A Review on the Fabrication of Substrates for Surface Enhanced Raman Spectroscopy and Their Applications in Analytical Chemistry. *Anal. Chim. Acta* **2011**, *693*, 7–25. [CrossRef] [PubMed]
85. Hugall, J.T.; Singh, A.; Van Hulst, N.F. Plasmonic Cavity Coupling. *ACS Photonics* **2018**, *5*, 43–53. [CrossRef]
86. Mirzaei, M.; Furxhi, I.; Murphy, F.; Mullins, M. A Machine Learning Tool to Predict the Antibacterial Capacity of Nanoparticles. *Nanomaterials* **2021**, *11*, 1774. [CrossRef] [PubMed]

87. Khosravinia, K.; Kiani, A. Optimizing the Ultrashort Laser Pulses for In Situ Nanostructure Generation Technique for High-Performance Supercapacitor Electrodes Using Artificial Neural Networks and Simulated Annealing Algorithms. *ACS Omega* **2023**, *8*, 17220–17233. [CrossRef] [PubMed]
88. Gherman, A.M.M.; Tosa, N.; Cristea, M.V.; Tosa, V.; Porav, S.; Agachi, P.S. Artificial Neural Networks Modeling of the Parameterized Gold Nanoparticles Generation through Photo-Induced Process. *Mater. Res. Express* **2018**, *5*, 085011. [CrossRef]
89. Mekki-Berrada, F.; Ren, Z.; Huang, T.; Wong, W.K.; Zheng, F.; Xie, J.; Tian, I.P.S.; Jayavelu, S.; Mahfoud, Z.; Bash, D.; et al. Two-Step Machine Learning Enables Optimized Nanoparticle Synthesis. *NPJ Comput. Mater.* **2021**, *7*, 55. [CrossRef]
90. Braham, E.J.; Cho, J.; Forlano, K.M.; Watson, D.F.; Arròyave, R.; Banerjee, S. Machine Learning-Directed Navigation of Synthetic Design Space: A Statistical Learning Approach to Controlling the Synthesis of Perovskite Halide Nanoplatelets in the Quantum-Confined Regime. *Chem. Mater.* **2019**, *31*, 3281–3292. [CrossRef]
91. Voznyy, O.; Levina, L.; Fan, J.Z.; Askerka, M.; Jain, A.; Choi, M.-J.; Ouellette, O.; Todorović, P.; Sagar, L.K.; Sargent, E.H. Machine Learning Accelerates Discovery of Optimal Colloidal Quantum Dot Synthesis. *ACS Nano* **2019**, *13*, 11122–11128. [CrossRef] [PubMed]
92. Provence, S.R.; Thapa, S.; Paudel, R.; Truttmann, T.K.; Prakash, A.; Jalan, B.; Comes, R.B. Machine Learning Analysis of Perovskite Oxides Grown by Molecular Beam Epitaxy. *Phys. Rev. Mater.* **2020**, *4*, 083807. [CrossRef]
93. Hajilounezhad, T.; Bao, R.; Palaniappan, K.; Bunyak, F.; Calyam, P.; Maschmann, M.R. Predicting Carbon Nanotube Forest Attributes and Mechanical Properties Using Simulated Images and Deep Learning. *NPJ Comput. Mater.* **2021**, *7*, 134. [CrossRef]
94. Wakabayashi, Y.K.; Otsuka, T.; Krockenberger, Y.; Sawada, H.; Taniyasu, Y.; Yamamoto, H. Machine-Learning-Assisted Thin-Film Growth: Bayesian Optimization in Molecular Beam Epitaxy of SrRuO3 Thin Films. *APL Mater.* **2019**, *7*, 101114. [CrossRef]
95. Abdel-Latif, K.; Epps, R.W.; Bateni, F.; Han, S.; Reyes, K.G.; Abolhasani, M. Self-Driven Multistep Quantum Dot Synthesis Enabled by Autonomous Robotic Experimentation in Flow. *Adv. Intell. Syst.* **2021**, *3*, 2000245. [CrossRef]
96. Gaikwad, A.; Desai, S. Molecular Dynamics Investigation of the Deformation Mechanism of Gold with Variations in Mold Profiles during Nanoimprinting. *Materials* **2021**, *14*, 2548. [CrossRef] [PubMed]
97. Adarkwa, E.; Roy, A.; Ohodnicki, J.; Lee, B.; Kumta, P.N.; Desai, S. 3D Printing of Drug-Eluting Bioactive Multifunctional Coatings for Orthopedic Applications. *Int. J. Bioprint* **2023**, *9*, 661. [CrossRef]
98. Marquetti, I.; Desai, S. An Atomistic Investigation of Adsorption of Bone Morphogenetic Protein-2 on Gold with Nanoscale Topographies. *Surfaces* **2022**, *5*, 176–185. [CrossRef]
99. Adarkwa, E.; Kotoka, R.; Desai, S. 3D Printing of Polymeric Coatings on AZ31 Mg Alloy Substrate for Corrosion Protection of Biomedical Implants. *Med. Devices Sens.* **2021**, *4*, e10167. [CrossRef]
100. Kumar Parupelli, S.; Saudi, S.; Bhattarai, N.; Desai, S. 3D Printing of PCL-Ceramic Composite Scaffolds for Bone Tissue Engineering Applications. *IJB* **2023**, *9*, 0196. [CrossRef]
101. Marquetti, I.; Desai, S. Nanoscale Topographical Effects on the Adsorption Behavior of Bone Morphogenetic Protein-2 on Graphite. *Int. J. Mol. Sci.* **2022**, *23*, 2432. [CrossRef] [PubMed]
102. Bentley, J.; Desai, S.; Bastakoti, B.P. Porous Tungsten Oxide: Recent Advances in Design, Synthesis, and Applications. *Chem. A Eur. J.* **2021**, *27*, 9241–9252. [CrossRef] [PubMed]
103. Mazahery, A.; Shabani, M.O. Assistance of Novel Artificial Intelligence in Optimization of Aluminum Matrix Nanocomposite by Genetic Algorithm. *Metall. Mater. Trans. A* **2012**, *43*, 5279–5285. [CrossRef]
104. Harsha, N.; Sita Rama Raju, K.; Venkata Ramana, V.S.N.; Reddy, V. Fabrication and Prediction of Tensile Strength of Al-Al2O3 Nano Composites. *Mater. Today Proc.* **2019**, *18*, 2197–2200. [CrossRef]
105. Hamdia, K.M.; Lahmer, T.; Nguyen-Thoi, T.; Rabczuk, T. Predicting the Fracture Toughness of PNCs: A Stochastic Approach Based on ANN and ANFIS. *Comput. Mater. Sci.* **2015**, *102*, 304–313. [CrossRef]
106. Sharma, A.; Anand Kumar, S.; Kushvaha, V. Effect of Aspect Ratio on Dynamic Fracture Toughness of Particulate Polymer Composite Using Artificial Neural Network. *Eng. Fract. Mech.* **2020**, *228*, 106907. [CrossRef]
107. Zhang, Z.; Friedrich, K.; Velten, K. Prediction on Tribological Properties of Short Fibre Composites Using Artificial Neural Networks. *Wear* **2002**, *252*, 668–675. [CrossRef]
108. Adesina, O.T.; Jamiru, T.; Daniyan, I.A.; Sadiku, E.R.; Ogunbiyi, O.F.; Adesina, O.S.; Beneke, L.W. Mechanical Property Prediction of SPS Processed GNP/PLA Polymer Nanocomposite Using Artificial Neural Network. *Cogent Eng.* **2020**, *7*, 1720894. [CrossRef]
109. Wang, Y.; Zhang, M.; Lin, A.; Iyer, A.; Prasad, A.S.; Li, X.; Zhang, Y.; Schadler, L.S.; Chen, W.; Brinson, L.C. Mining Structure–Property Relationships in Polymer Nanocomposites Using Data Driven Finite Element Analysis and Multi-Task Convolutional Neural Networks. *Mol. Syst. Des. Eng.* **2020**, *5*, 962–975. [CrossRef]
110. Farahbakhsh, J.; Delnavaz, M.; Vatanpour, V. Simulation and Characterization of Novel Reverse Osmosis Membrane Prepared by Blending Polypyrrole Coated Multiwalled Carbon Nanotubes for Brackish Water Desalination and Antifouling Properties Using Artificial Neural Networks. *J. Membr. Sci.* **2019**, *581*, 123–138. [CrossRef]
111. Moghri, M.; Madic, M.; Omidi, M.; Farahnakian, M. Surface Roughness Optimization of Polyamide-6/Nanoclay Nanocomposites Using Artificial Neural Network: Genetic Algorithm Approach. *Sci. World J.* **2014**, *2014*, 485205. [CrossRef] [PubMed]
112. Ho, N.X.; Le, T.-T.; Le, M.V. Development of Artificial Intelligence Based Model for the Prediction of Young's Modulus of Polymer/Carbon-Nanotubes Composites. *Mech. Adv. Mater. Struct.* **2022**, *29*, 5965–5978. [CrossRef]

113. Lu, X.; Giovanis, D.G.; Yvonnet, J.; Papadopoulos, V.; Detrez, F.; Bai, J. A Data-Driven Computational Homogenization Method Based on Neural Networks for the Nonlinear Anisotropic Electrical Response of Graphene/Polymer Nanocomposites. *Comput. Mech.* **2019**, *64*, 307–321. [CrossRef]
114. Thapliyal, A.; Khar, R.K.; Chandra, A. Artificial Neural Network Modelling of Green Synthesised Silver Nanoparticles in Bentonite/Starch Bio-Nanocomposite. *CNANO* **2018**, *14*, 239–251. [CrossRef]
115. Cheng, W.D.; Cai, C.Z.; Luo, Y.; Li, Y.H.; Zhao, C.J. Mechanical Properties Prediction for Carbon Nanotubes/Epoxy Composites by Using Support Vector Regression. *Mod. Phys. Lett. B* **2015**, *29*, 1550016. [CrossRef]
116. Lingaraju, D.; Ramji, K.; Rao, N.B.R.M.; Lakshmi, U.R. Characterization and Prediction of Some Engineering Properties of Polymer—Clay/Silica Hybrid Nanocomposites through ANN and Regression Models. *Procedia Eng.* **2011**, *10*, 9–18. [CrossRef]
117. Gates, B.D.; Xu, Q.; Stewart, M.; Ryan, D.; Willson, C.G.; Whitesides, G.M. New Approaches to Nanofabrication: Molding, Printing, and Other Techniques. *Chem. Rev.* **2005**, *105*, 1171–1196. [CrossRef] [PubMed]
118. Ariga, K.; Hill, J.P.; Lee, M.V.; Vinu, A.; Charvet, R.; Acharya, S. Challenges and Breakthroughs in Recent Research on Self-Assembly. *Sci. Technol. Adv. Mater.* **2008**, *9*, 014109. [CrossRef]
119. Li, L.; Gattass, R.R.; Gershgoren, E.; Hwang, H.; Fourkas, J.T. Achieving $\lambda/20$ Resolution by One-Color Initiation and Deactivation of Polymerization. *Science* **2009**, *324*, 910–913. [CrossRef] [PubMed]
120. Krämer, S.; Fuierer, R.R.; Gorman, C.B. Scanning Probe Lithography Using Self-Assembled Monolayers. *Chem. Rev.* **2003**, *103*, 4367–4418. [CrossRef] [PubMed]
121. Smith, J.C.; Lee, K.-B.; Wang, Q.; Finn, M.G.; Johnson, J.E.; Mrksich, M.; Mirkin, C.A. Nanopatterning the Chemospecific Immobilization of Cowpea Mosaic Virus Capsid. *Nano Lett.* **2003**, *3*, 883–886. [CrossRef]
122. Biswas, A.; Bayer, I.S.; Biris, A.S.; Wang, T.; Dervishi, E.; Faupel, F. Advances in Top–down and Bottom–up Surface Nanofabrication: Techniques, Applications & Future Prospects. *Adv. Colloid Interface Sci.* **2012**, *170*, 2–27. [CrossRef] [PubMed]
123. Mack, C. *Fundamental Principles of Optical Lithography: The Science of Microfabrication*, 1st ed.; Wiley: Hoboken, NJ, USA, 2007; ISBN 978-0-470-01893-4.
124. Lin, Y.; Li, M.; Watanabe, Y.; Kimura, T.; Matsunawa, T.; Nojima, S.; Pan, D.Z. Data Efficient Lithography Modeling With Transfer Learning and Active Data Selection. *IEEE Trans. Comput. Aided Des. Integr. Circuits Syst.* **2019**, *38*, 1900–1913. [CrossRef]
125. Watanabe, Y.; Kimura, T.; Matsunawa, T.; Nojima, S. *Accurate Lithography Simulation Model Based on Convolutional Neural Networks*; Erdmann, A., Kye, J., Eds.; SPIE: San Jose, CA, USA, 2017; p. 101470K.
126. Lin, Y.; Alawieh, M.B.; Ye, W.; Pan, D.Z. Machine Learning for Yield Learning and Optimization. In Proceedings of the 2018 IEEE International Test Conference (ITC), Phoenix, AZ, USA, 29 October–1 November 2018; IEEE: Phoenix, AZ, USA, 2018; pp. 1–10.
127. Shim, S.; Choi, S.; Shin, Y. *Machine Learning-Based 3D Resist Model*; Erdmann, A., Kye, J., Eds.; SPIE: San Jose, CA, USA, 2017; p. 101471D.
128. Lan, S.; Li, J.; Liu, J.; Wang, Y.; Zhao, K. Deep Learning Assisted Fast Mask Optimization. In Proceedings of the Optical Microlithography XXXI, San Jose, CA, USA, 27 February–March 2018; Kye, J., Owa, S., Eds.; SPIE: San Jose, CA, USA, 2018; p. 17.
129. Liu, P. Mask Synthesis Using Machine Learning Software and Hardware Platforms. In Proceedings of the Optical Microlithography XXXIII, San Jose, CA, USA, 25–26 February 2020; Owa, S., Phillips, M.C., Eds.; SPIE: San Jose, San Jose, CA, USA, 2020; p. 5.
130. Jiang, B.; Zhang, H.; Yang, J.; Young, E.F.Y. A Fast Machine Learning-Based Mask Printability Predictor for OPC Acceleration. In Proceedings of the 24th Asia and South Pacific Design Automation Conference, Tokyo Japan, 21–24 January 2019; ACM: New York, NY, USA, 2019; pp. 412–419.
131. Tanabe, H.; Sato, S.; Takahashi, A. Fast 3D Lithography Simulation by Convolutional Neural Network. In Proceedings of the Design-Process-Technology Co-Optimization XV, Online, 22–26 February 2021; Yuan, C.-M., Kim, R.-H., Eds.; SPIE: San Jose, CA, USA, 2021; p. 21.
132. Odujole, J.; Desai, S. Atomistic Investigation of Material Deformation Behavior of Polystyrene in Nanoimprint Lithography. *Surfaces* **2020**, *3*, 649–663. [CrossRef]
133. Odujole, J.I.; Desai, S. Molecular Dynamics Investigation of Material Deformation Behavior of PMMA in Nanoimprint Lithography. *AIP Adv.* **2020**, *10*, 095102. [CrossRef]
134. Gaikwad, A.; Odujole, J.; Desai, S. Atomistic Investigation of Process Parameter Variations on Material Deformation Behavior in Nanoimprint Lithography of Gold. *Precis. Eng.* **2020**, *64*, 7–19. [CrossRef]
135. Akter, T.; Desai, S. Developing a Predictive Model for Nanoimprint Lithography Using Artificial Neural Networks. *Mater. Des.* **2018**, *160*, 836–848. [CrossRef]
136. Xie, L.; Zhang, Z.; Wu, Q.; Gao, Z.; Mi, G.; Wang, R.; Sun, H.; Zhao, Y.; Du, Y. Intelligent Wearable Devices Based on Nanomaterials and Nanostructures for Healthcare. *Nanoscale* **2023**, *15*, 405–433. [CrossRef] [PubMed]
137. Doh, J.; Park, S.-I.; Yang, Q.; Raghavan, N. Uncertainty Quantification of Percolating Electrical Conductance for Wavy Carbon Nanotube-Filled Polymer Nanocomposites Using Bayesian Inference. *Carbon* **2021**, *172*, 308–323. [CrossRef]
138. Tettey, F.; Saudi, S.; Davies, D.; Shrestha, S.; Johnson, K.; Fialkova, S.; Subedi, K.; Bastakoti, B.P.; Sankar, J.; Desai, S.; et al. Fabrication and Characterization of Zn Particle Incorporated Fibrous Scaffolds for Potential Application in Tissue Healing and Regeneration. *ACS Appl. Mater. Interfaces* **2023**, *15*, 48913–48929. [CrossRef] [PubMed]
139. Sidi Salah, L.; Chouai, M.; Danlée, Y.; Huynen, I.; Ouslimani, N. Simulation and Optimization of Electromagnetic Absorption of Polycarbonate/CNT Composites Using Machine Learning. *Micromachines* **2020**, *11*, 778. [CrossRef] [PubMed]

140. Yusoff, N.I.M.d.; Ibrahim Alhamali, D.; Ibrahim, A.N.H.; Rosyidi, S.A.P.; Abdul Hassan, N. Engineering Characteristics of Nanosilica/Polymer-Modified Bitumen and Predicting Their Rheological Properties Using Multilayer Perceptron Neural Network Model. *Constr. Build. Mater.* **2019**, *204*, 781–799. [CrossRef]
141. Khanam, P.N.; AlMaadeed, M.; AlMaadeed, S.; Kunhoth, S.; Ouederni, M.; Sun, D.; Hamilton, A.; Jones, E.H.; Mayoral, B. Optimization and Prediction of Mechanical and Thermal Properties of Graphene/LLDPE Nanocomposites by Using Artificial Neural Networks. *Int. J. Polym. Sci.* **2016**, *2016*, 5340252. [CrossRef]
142. Zakaulla, M.; Pasha, Y.; Siddalingappa, S.K. Prediction of Mechanical Properties for Polyetheretherketone Composite Reinforced with Graphene and Titanium Powder Using Artificial Neural Network. *Mater. Today Proc.* **2022**, *49*, 1268–1274. [CrossRef]
143. Matos, M.A.S.; Pinho, S.T.; Tagarielli, V.L. Application of Machine Learning to Predict the Multiaxial Strain-Sensing Response of CNT-Polymer Composites. *Carbon* **2019**, *146*, 265–275. [CrossRef]
144. Lynch, M.; Patel, H.; Abrahamse, A.; Rupa Rajendran, A.; Medsker, L. Neural Network Applications in Physics. In *Proceedings of the IJCNN'01. International Joint Conference on Neural Networks. Proceedings (Cat. No.01CH37222)*; IEEE: Washington, DC, USA, 2001; Volume 3, pp. 2054–2058.
145. Leitherer, A.; Ziletti, A.; Ghiringhelli, L.M. Robust Recognition and Exploratory Analysis of Crystal Structures via Bayesian Deep Learning. *Nat. Commun.* **2021**, *12*, 6234. [CrossRef]
146. Varol, T.; Canakci, A.; Ozsahin, S. Prediction of Effect of Reinforcement Content, Flake Size and Flake Time on the Density and Hardness of Flake AA2024-SiC Nanocomposites Using Neural Networks. *J. Alloys Compd.* **2018**, *739*, 1005–1014. [CrossRef]
147. Fedorenko, Y.S. Statistical Testing Technique for Comparison Machine Learning Models Performance. *Vkit* **2019**, *16*, 10–17. [CrossRef]
148. Ruhe, A. Accelerated Gauss-Newton Algorithms for Nonlinear Least Squares Problems. *BIT* **1979**, *19*, 356–367. [CrossRef]
149. Moré, J.J. The Levenberg-Marquardt Algorithm: Implementation and Theory. In *Numerical Analysis*; Watson, G.A., Ed.; Lecture Notes in Mathematics; Springer: Berlin/Heidelberg, Germany, 1978; Volume 630, pp. 105–116, ISBN 978-3-540-08538-6.
150. Nagase, N.; Suzuki, K.; Takahashi, K.; Minemura, M.; Yamauchi, S.; Okada, T. *Study of Hot Spot Detection Using Neural Network Judgment*; Watanabe, H., Ed.; SPIE: Yokohama, Japan, 2007; p. 66071B.
151. Ma, N.; Ghan, J.; Mishra, S.; Spanos, C.; Poolla, K.; Rodriguez, N.; Capodieci, L. *Automatic Hotspot Classification Using Pattern-Based Clustering*; Singh, V.K., Rieger, M.L., Eds.; SPIE: San Jose, CA, USA, 2008; p. 692505.
152. Ding, D.; Wu, X.; Ghosh, J.; Pan, D.Z. Machine Learning Based Lithographic Hotspot Detection with Critical-Feature Extraction and Classification. In *Proceedings of the 2009 IEEE International Conference on IC Design and Technology*; IEEE: Austin, TX, USA, 2009; pp. 219–222.
153. Ding, D.; Torres, J.A.; Pan, D.Z. High Performance Lithography Hotspot Detection With Successively Refined Pattern Identifications and Machine Learning. *IEEE Trans. Comput. Aided Des. Integr. Circuits Syst.* **2011**, *30*, 1621–1634. [CrossRef]
154. Yang, H.; Lin, Y.; Yu, B.; Young, E.F.Y. Lithography Hotspot Detection: From Shallow to Deep Learning. In *Proceedings of the 2017 30th IEEE International System-on-Chip Conference (SOCC)*; IEEE: Munich, Germany, 2017; pp. 233–238.
155. Yu, Y.-T.; Lin, G.-H.; Jiang, I.H.-R.; Chiang, C. Machine-Learning-Based Hotspot Detection Using Topological Classification and Critical Feature Extraction. *IEEE Trans. Comput. Aided Des. Integr. Circuits Syst.* **2015**, *34*, 460–470. [CrossRef]
156. Yu, Y.-T.; Lin, G.-H.; Jiang, I.H.-R.; Chiang, C. Machine-Learning-Based Hotspot Detection Using Topological Classification and Critical Feature Extraction. In *Proceedings of the 50th Annual Design Automation Conference*; ACM: Austin, TX, USA, 2013; pp. 1–6.
157. Shin, M.; Lee, J.-H. Accurate Lithography Hotspot Detection Using Deep Convolutional Neural Networks. *J. Micro/Nanolith. MEMS MOEMS* **2016**, *15*, 043507. [CrossRef]
158. Lussier, F.; Thibault, V.; Charron, B.; Wallace, G.Q.; Masson, J.-F. Deep Learning and Artificial Intelligence Methods for Raman and Surface-Enhanced Raman Scattering. *TrAC Trends Anal. Chem.* **2020**, *124*, 115796. [CrossRef]
159. Kumar, P.P.P.; Kaushal, S.; Lim, D.-K. Recent Advances in Nano/Microfabricated Substrate Platforms and Artificial Intelligence for Practical Surface-Enhanced Raman Scattering-Based Bioanalysis. *TrAC Trends Anal. Chem.* **2023**, *168*, 117341. [CrossRef]
160. Duan, H.; Hu, H.; Kumar, K.; Shen, Z.; Yang, J.K.W. Direct and Reliable Patterning of Plasmonic Nanostructures with Sub-10-Nm Gaps. *ACS Nano* **2011**, *5*, 7593–7600. [CrossRef] [PubMed]
161. Luo, S.; Hoff, B.H.; Maier, S.A.; De Mello, J.C. Scalable Fabrication of Metallic Nanogaps at the Sub-10 Nm Level. *Adv. Sci.* **2021**, *8*, 2102756. [CrossRef]
162. Chen, Y.; Bi, K.; Wang, Q.; Zheng, M.; Liu, Q.; Han, Y.; Yang, J.; Chang, S.; Zhang, G.; Duan, H. Rapid Focused Ion Beam Milling Based Fabrication of Plasmonic Nanoparticles and Assemblies via "Sketch and Peel" Strategy. *ACS Nano* **2016**, *10*, 11228–11236. [CrossRef]
163. Punj, D.; Mivelle, M.; Moparthi, S.B.; Van Zanten, T.S.; Rigneault, H.; Van Hulst, N.F.; García-Parajó, M.F.; Wenger, J. A Plasmonic 'Antenna-in-Box' Platform for Enhanced Single-Molecule Analysis at Micromolar Concentrations. *Nat. Nanotechnol.* **2013**, *8*, 512–516. [CrossRef] [PubMed]
164. Jha, S.K.; Ahmed, Z.; Agio, M.; Ekinci, Y.; Löffler, J.F. Deep-UV Surface-Enhanced Resonance Raman Scattering of Adenine on Aluminum Nanoparticle Arrays. *J. Am. Chem. Soc.* **2012**, *134*, 1966–1969. [CrossRef] [PubMed]
165. Qin, L.; Huang, Y.; Xia, F.; Wang, L.; Ning, J.; Chen, H.; Wang, X.; Zhang, W.; Peng, Y.; Liu, Q.; et al. 5 Nm Nanogap Electrodes and Arrays by Super-Resolution Laser Lithography. *Nano Lett.* **2020**, *20*, 4916–4923. [CrossRef] [PubMed]
166. Lim, D.-K.; Jeon, K.-S.; Kim, H.M.; Nam, J.-M.; Suh, Y.D. Nanogap-Engineerable Raman-Active Nanodumbbells for Single-Molecule Detection. *Nat. Mater.* **2010**, *9*, 60–67. [CrossRef] [PubMed]

167. Lee, S.Y.; Hung, L.; Lang, G.S.; Cornett, J.E.; Mayergoyz, I.D.; Rabin, O. Dispersion in the SERS Enhancement with Silver Nanocube Dimers. *ACS Nano* **2010**, *4*, 5763–5772. [CrossRef] [PubMed]
168. Luo, S.; Mancini, A.; Berté, R.; Hoff, B.H.; Maier, S.A.; De Mello, J.C. Massively Parallel Arrays of Size-Controlled Metallic Nanogaps with Gap-Widths Down to the Sub-3-nm Level. *Adv. Mater.* **2021**, *33*, 2100491. [CrossRef] [PubMed]
169. Khlebtsov, B.N.; Khlebtsov, N.G. Surface Morphology of a Gold Core Controls the Formation of Hollow or Bridged Nanogaps in Plasmonic Nanomatryoshkas and Their SERS Responses. *J. Phys. Chem. C* **2016**, *120*, 15385–15394. [CrossRef]
170. Houhou, R.; Bocklitz, T. Trends in Artificial Intelligence, Machine Learning, and Chemometrics Applied to Chemical Data. *Anal. Sci. Adv.* **2021**, *2*, 128–141. [CrossRef]
171. Cui, F.; Yue, Y.; Zhang, Y.; Zhang, Z.; Zhou, H.S. Advancing Biosensors with Machine Learning. *ACS Sens.* **2020**, *5*, 3346–3364. [CrossRef]
172. Rahman, A.; Kang, S.; Wang, W.; Huang, Q.; Kim, I.; Vikesland, P.J. Lectin-Modified Bacterial Cellulose Nanocrystals Decorated with Au Nanoparticles for Selective Detection of Bacteria Using Surface-Enhanced Raman Scattering Coupled with Machine Learning. *ACS Appl. Nano Mater.* **2022**, *5*, 259–268. [CrossRef]
173. Guselnikova, O.; Trelin, A.; Skvortsova, A.; Ulbrich, P.; Postnikov, P.; Pershina, A.; Sykora, D.; Svorcik, V.; Lyutakov, O. Label-Free Surface-Enhanced Raman Spectroscopy with Artificial Neural Network Technique for Recognition Photoinduced DNA Damage. *Biosens. Bioelectron.* **2019**, *145*, 111718. [CrossRef] [PubMed]
174. Qin, Y.-F.; Lu, X.-Y.; Shi, Z.; Huang, Q.-S.; Wang, X.; Ren, B.; Cui, L. Deep Learning-Enabled Raman Spectroscopic Identification of Pathogen-Derived Extracellular Vesicles and the Biogenesis Process. *Anal. Chem.* **2022**, *94*, 12416–12426. [CrossRef] [PubMed]
175. Peng, M.; Wang, Z.; Sun, X.; Guo, X.; Wang, H.; Li, R.; Liu, Q.; Chen, M.; Chen, X. Deep Learning-Based Label-Free Surface-Enhanced Raman Scattering Screening and Recognition of Small-Molecule Binding Sites in Proteins. *Anal. Chem.* **2022**, *94*, 11483–11491. [CrossRef] [PubMed]
176. Ding, L.; Liu, L.; He, L.; Effah, C.Y.; Yang, R.; Ouyang, D.; Jian, N.; Liu, X.; Wu, Y.; Qu, L. Magnetic-Nanowaxberry-Based Simultaneous Detection of Exosome and Exosomal Proteins for the Intelligent Diagnosis of Cancer. *Anal. Chem.* **2021**, *93*, 15200–15208. [CrossRef] [PubMed]
177. Huang, J.; Wen, J.; Zhou, M.; Ni, S.; Le, W.; Chen, G.; Wei, L.; Zeng, Y.; Qi, D.; Pan, M.; et al. On-Site Detection of SARS-CoV-2 Antigen by Deep Learning-Based Surface-Enhanced Raman Spectroscopy and Its Biochemical Foundations. *Anal. Chem.* **2021**, *93*, 9174–9182. [CrossRef] [PubMed]
178. He, H.; Yan, S.; Lyu, D.; Xu, M.; Ye, R.; Zheng, P.; Lu, X.; Wang, L.; Ren, B. Deep Learning for Biospectroscopy and Biospectral Imaging: State-of-the-Art and Perspectives. *Anal. Chem.* **2021**, *93*, 3653–3665. [CrossRef] [PubMed]
179. Huang, L.; Sun, H.; Sun, L.; Shi, K.; Chen, Y.; Ren, X.; Ge, Y.; Jiang, D.; Liu, X.; Knoll, W.; et al. Rapid, Label-Free Histopathological Diagnosis of Liver Cancer Based on Raman Spectroscopy and Deep Learning. *Nat. Commun.* **2023**, *14*, 48. [CrossRef] [PubMed]
180. Kong, X.; Zhou, L.; Li, Z.; Yang, Z.; Qiu, B.; Wu, X.; Shi, F.; Du, J. Artificial Intelligence Enhanced Two-Dimensional Nanoscale Nuclear Magnetic Resonance Spectroscopy. *NPJ Quantum Inf.* **2020**, *6*, 79. [CrossRef]
181. Rédei, L.; Fried, M.; Bársony, I.; Wallinga, H. A Modified Learning Strategy for Neural Networks to Support Spectroscopic Ellipsometric Data Evaluation. *Thin Solid Film.* **1998**, *313–314*, 149–155. [CrossRef]
182. Liu, J.; Zhang, D.; Yu, D.; Ren, M.; Xu, J. Machine Learning Powered Ellipsometry. *Light. Sci. Appl.* **2021**, *10*, 55. [CrossRef] [PubMed]
183. Kumar, N.; Nezhdanov, A.V.; Smertin, R.M.; Polkovnikov, V.N.; Yunin, P.A.; Garakhin, S.A.; Chkhalo, N.I.; Mashin, A.I.; Kudryashov, M.A.; Usanov, D.A. Phase-Microstructure of Mo/Si Nanoscale Multilayer and Intermetallic Compound Formation in Interfaces. *Intermetallics* **2020**, *125*, 106872. [CrossRef]
184. No, Y.-S.; Choi, H.K.; Kim, J.-S.; Kim, H.; Yu, Y.-J.; Choi, C.-G.; Choi, J.S. Layer Number Identification of CVD-Grown Multilayer Graphene Using Si Peak Analysis. *Sci. Rep.* **2018**, *8*, 571. [CrossRef] [PubMed]
185. Stenger, I.; Schué, L.; Boukhicha, M.; Berini, B.; Plaçais, B.; Loiseau, A.; Barjon, J. Low Frequency Raman Spectroscopy of Few-Atomic-Layer Thick hBN Crystals. *2D Mater.* **2017**, *4*, 031003. [CrossRef]
186. Hajiyev, P.; Cong, C.; Qiu, C.; Yu, T. Contrast and Raman Spectroscopy Study of Single- and Few-Layered Charge Density Wave Material: 2H-TaSe2. *Sci. Rep.* **2013**, *3*, 2593. [CrossRef] [PubMed]
187. Egerton, R.F. *Physical Principles of Electron Microscopy*; Springer International Publishing: Cham, Switzerland, 2016; ISBN 978-3-319-39876-1.
188. Muller, D.A. Structure and Bonding at the Atomic Scale by Scanning Transmission Electron Microscopy. *Nat. Mater.* **2009**, *8*, 263–270. [CrossRef] [PubMed]
189. Kondo, T.; Ban, N.; Ebizuka, Y.; Toyoda, Y.; Yamada, Y.; Kashiwa, T.; Koike, H.; Shindo, H.; Charley, A.-L.; Saib, M.; et al. Massive Metrology and Inspection Solution for EUV by Area Inspection SEM with Machine Learning Technology. In Proceedings of the Metrology, Inspection, and Process Control for Semiconductor Manufacturing XXXV, Online, 22 February 2021; Adan, O., Robinson, J.C., Eds.; SPIE: San Jose, CA, USA, 2021; p. 31.
190. Baek, K.; Park, S.; Park, J.; Kim, Y.-M.; Hwang, H.; Oh, S.H. In Situ TEM Observation on the Interface-Type Resistive Switching by Electrochemical Redox Reactions at a TiN/PCMO Interface. *Nanoscale* **2017**, *9*, 582–593. [CrossRef] [PubMed]
191. McGahan, W.A.; Johs, B.; Woollam, J.A. Techniques for Ellipsometric Measurement of the Thickness and Optical Constants of Thin Absorbing Films. *Thin Solid. Films* **1993**, *234*, 443–446. [CrossRef]

192. Polgár, O.; Fried, M.; Lohner, T.; Bársony, I. Comparison of Algorithms Used for Evaluation of Ellipsometric Measurements Random Search, Genetic Algorithms, Simulated Annealing and Hill Climbing Graph-Searches. *Surf. Sci.* **2000**, *457*, 157–177. [CrossRef]
193. Choi, G.; Kim, M.; Kim, J.; Pahk, H.J. Angle-Resolved Spectral Reflectometry with a Digital Light Processing Projector. *Opt. Express* **2020**, *28*, 26908. [CrossRef] [PubMed]
194. Choi, J.E.; Song, J.; Lee, Y.H.; Hong, S.J. Deep Neural Network Modeling of Multiple Oxide/Nitride Deposited Dielectric Films for 3D-NAND Flash. *Appl. Sci. Converg. Technol.* **2020**, *29*, 190–194. [CrossRef]
195. Tian, S.I.P.; Liu, Z.; Chellappan, V.; Lim, Y.-F.; Ren, Z.; Oviedo, F.; Teo, B.H.; Thapa, J.; Dutta, R.; MacLeod, B.P.; et al. Rapid and Accurate Thin Film Thickness Extraction via UV-Vis and Machine Learning. In Proceedings of the 2020 47th IEEE Photovoltaic Specialists Conference (PVSC), Calgary, AB, Canada, 14 June 2020; IEEE: Calgary, AB, Canada, 2020; pp. 0128–0132.
196. Cheng, Y.; Chen, K.; Sun, H.; Zhang, Y.; Tao, F. Data and Knowledge Mining with Big Data towards Smart Production. *J. Ind. Inf. Integr.* **2018**, *9*, 1–13. [CrossRef]
197. Jia, Y.; Hou, X.; Wang, Z.; Hu, X. Machine Learning Boosts the Design and Discovery of Nanomaterials. *ACS Sustain. Chem. Eng.* **2021**, *9*, 6130–6147. [CrossRef]
198. Goodfellow, I.; Bengio, Y.; Courville, A. *Deep Learning*; Adaptive computation and machine learning; The MIT press: Cambridge, MA, USA, 2016; ISBN 978-0-262-03561-3.
199. Takahashi, K.; Takahashi, L. Data Driven Determination in Growth of Silver from Clusters to Nanoparticles and Bulk. *J. Phys. Chem. Lett.* **2019**, *10*, 4063–4068. [CrossRef] [PubMed]
200. Ali, A.; Park, H.; Mall, R.; Aïssa, B.; Sanvito, S.; Bensmail, H.; Belaidi, A.; El-Mellouhi, F. Machine Learning Accelerated Recovery of the Cubic Structure in Mixed-Cation Perovskite Thin Films. *Chem. Mater.* **2020**, *32*, 2998–3006. [CrossRef]
201. Timoshenko, J.; Wrasman, C.J.; Luneau, M.; Shirman, T.; Cargnello, M.; Bare, S.R.; Aizenberg, J.; Friend, C.M.; Frenkel, A.I. Probing Atomic Distributions in Mono- and Bimetallic Nanoparticles by Supervised Machine Learning. *Nano Lett.* **2019**, *19*, 520–529. [CrossRef] [PubMed]
202. Koohi-Moghadam, M.; Wang, H.; Wang, Y.; Yang, X.; Li, H.; Wang, J.; Sun, H. Predicting Disease-Associated Mutation of Metal-Binding Sites in Proteins Using a Deep Learning Approach. *Nat. Mach. Intell.* **2019**, *1*, 561–567. [CrossRef]
203. Yan, X.; Sedykh, A.; Wang, W.; Yan, B.; Zhu, H. Construction of a Web-Based Nanomaterial Database by Big Data Curation and Modeling Friendly Nanostructure Annotations. *Nat. Commun.* **2020**, *11*, 2519. [CrossRef] [PubMed]
204. Russo, D.P.; Yan, X.; Shende, S.; Huang, H.; Yan, B.; Zhu, H. Virtual Molecular Projections and Convolutional Neural Networks for the End-to-End Modeling of Nanoparticle Activities and Properties. *Anal. Chem.* **2020**, *92*, 13971–13979. [CrossRef] [PubMed]
205. Liu, G.; Yan, X.; Sedykh, A.; Pan, X.; Zhao, X.; Yan, B.; Zhu, H. Analysis of Model PM2.5-Induced Inflammation and Cytotoxicity by the Combination of a Virtual Carbon Nanoparticle Library and Computational Modeling. *Ecotoxicol. Environ. Saf.* **2020**, *191*, 110216. [CrossRef]
206. Maksov, A.; Dyck, O.; Wang, K.; Xiao, K.; Geohegan, D.B.; Sumpter, B.G.; Vasudevan, R.K.; Jesse, S.; Kalinin, S.V.; Ziatdinov, M. Deep Learning Analysis of Defect and Phase Evolution during Electron Beam-Induced Transformations in WS2. *NPJ Comput Mater* **2019**, *5*, 12. [CrossRef]
207. Carrera, D.; Manganini, F.; Boracchi, G.; Lanzarone, E. Defect Detection in SEM Images of Nanofibrous Materials. *IEEE Trans. Ind. Inf.* **2017**, *13*, 551–561. [CrossRef]
208. Iwasaki, Y.; Takeuchi, I.; Stanev, V.; Kusne, A.G.; Ishida, M.; Kirihara, A.; Ihara, K.; Sawada, R.; Terashima, K.; Someya, H.; et al. Machine-Learning Guided Discovery of a New Thermoelectric Material. *Sci. Rep.* **2019**, *9*, 2751. [CrossRef] [PubMed]
209. Copp, S.M.; Gorovits, A.; Swasey, S.M.; Gudibandi, S.; Bogdanov, P.; Gwinn, E.G. Fluorescence Color by Data-Driven Design of Genomic Silver Clusters. *ACS Nano* **2018**, *12*, 8240–8247. [CrossRef] [PubMed]
210. Copp, S.M.; Swasey, S.M.; Gorovits, A.; Bogdanov, P.; Gwinn, E.G. General Approach for Machine Learning-Aided Design of DNA-Stabilized Silver Clusters. *Chem. Mater.* **2020**, *32*, 430–437. [CrossRef]
211. Yang, H.; Zhang, Z.; Zhang, J.; Zeng, X.C. Machine Learning and Artificial Neural Network Prediction of Interfacial Thermal Resistance between Graphene and Hexagonal Boron Nitride. *Nanoscale* **2018**, *10*, 19092–19099. [CrossRef] [PubMed]
212. Fernandez, M.; Shi, H.; Barnard, A.S. Geometrical Features Can Predict Electronic Properties of Graphene Nanoflakes. *Carbon* **2016**, *103*, 142–150. [CrossRef]
213. Fernandez, M.; Bilić, A.; Barnard, A.S. Machine Learning and Genetic Algorithm Prediction of Energy Differences between Electronic Calculations of Graphene Nanoflakes. *Nanotechnology* **2017**, *28*, 38LT03. [CrossRef] [PubMed]
214. Sigmund, G.; Gharasoo, M.; Hüffer, T.; Hofmann, T. Deep Learning Neural Network Approach for Predicting the Sorption of Ionizable and Polar Organic Pollutants to a Wide Range of Carbonaceous Materials. *Environ. Sci. Technol.* **2020**, *54*, 4583–4591. [CrossRef] [PubMed]
215. Findlay, M.R.; Freitas, D.N.; Mobed-Miremadi, M.; Wheeler, K.E. Machine Learning Provides Predictive Analysis into Silver Nanoparticle Protein Corona Formation from Physicochemical Properties. *Environ. Sci. Nano* **2018**, *5*, 64–71. [CrossRef]
216. Brownlee, J. *Machine Learning Mastery with Python: Understand Your Data, Create Accurate Models, and Work Projects End-to-End*; Machine Learning Mastery: Victoria, VT, USA, 2016.
217. Naeem, S.; Ali, A.; Anam, S.; Ahmed, M.M. An Unsupervised Machine Learning Algorithms: Comprehensive Review. *IJCDS* **2023**, *13*, 911–921. [CrossRef]

218. AlMahamid, F.; Grolinger, K. Reinforcement Learning Algorithms: An Overview and Classification. In Proceedings of the 2021 IEEE Canadian Conference on Electrical and Computer Engineering (CCECE), Virtual, 12–17 September 2021; IEEE: London, ON, Canada, 2021; pp. 1–7.
219. Brownlee, J. *Deep Learning with Python: Develop Deep Learning Models on Theano and TensorFlow Using Keras*; Machine Learning Mastery: Victoria, VT, USA, 2016.
220. Isichei, J.C.; Khorsandroo, S.; Desai, S. Cybersecurity and Privacy in Smart Bioprinting. *Bioprinting* **2023**, *36*, e00321. [CrossRef]
221. Tettey, F.; Parupelli, S.K.; Desai, S. A Review of Biomedical Devices: Classification, Regulatory Guidelines, Human Factors, Software as a Medical Device, and Cybersecurity. *Biomed. Mater. Devices* **2023**, *2*, 25. [CrossRef]
222. Zhang, P.; Guo, Z.; Ullah, S.; Melagraki, G.; Afantitis, A.; Lynch, I. Nanotechnology and Artificial Intelligence to Enable Sustainable and Precision Agriculture. *Nat. Plants* **2021**, *7*, 864–876. [CrossRef] [PubMed]
223. Adadi, A. A Survey on Data-efficient Algorithms in Big Data Era. *J. Big Data* **2021**, *8*, 24. [CrossRef]
224. Munappy, A.R.; Bosch, J.; Olsson, H.H.; Arpteg, A.; Brinne, B. Data Management for Production Quality Deep Learning Models: Challenges and Solutions. *J. Syst. Softw.* **2022**, *191*, 111359. [CrossRef]
225. Peng, G.C.Y.; Alber, M.; Buganza Tepole, A.; Cannon, W.R.; De, S.; Dura-Bernal, S.; Garikipati, K.; Karniadakis, G.; Lytton, W.W.; Perdikaris, P.; et al. Multiscale Modeling Meets Machine Learning: What Can We Learn? *Arch. Comput. Methods Eng.* **2021**, *28*, 1017–1037. [CrossRef] [PubMed]
226. Von Eschenbach, W.J. Transparency and the Black Box Problem: Why We Do Not Trust AI. *Philos. Technol.* **2021**, *34*, 1607–1622. [CrossRef]
227. Wischmeyer, T. Artificial Intelligence and Transparency: Opening the Black Box. In *Regulating Artificial Intelligence*; Wischmeyer, T., Rademacher, T., Eds.; Springer International Publishing: Cham, Switzerland, 2020; pp. 75–101, ISBN 978-3-030-32360-8.
228. Carabantes, M. Black-Box Artificial Intelligence: An Epistemological and Critical Analysis. *AI Soc.* **2020**, *35*, 309–317. [CrossRef]
229. Séguin, R.; Potvin, J.-Y.; Gendreau, M.; Crainic, T.G.; Marcotte, P. Real-Time Decision Problems: An Operational Research Perspective. *J. Oper. Res. Soc.* **1997**, *48*, 162–174. [CrossRef]
230. Polian, I.; Altmann, F.; Arul, T.; Boit, C.; Brederlow, R.; Davi, L.; Drechsler, R.; Du, N.; Eisenbarth, T.; Guneysu, T.; et al. Nano Security: From Nano-Electronics to Secure Systems. In Proceedings of the 2021 Design, Automation & Test in Europe Conference & Exhibition (DATE), Virtual, 1 February 2021; IEEE: Grenoble, France, 2021; pp. 1334–1339.
231. Oseni, A.; Moustafa, N.; Janicke, H.; Liu, P.; Tari, Z.; Vasilakos, A. Security and Privacy for Artificial Intelligence: Opportunities and Challenges. *arXiv* **2021**, arXiv:2102.04661.
232. Solanas, A.; Martínez-Ballesté, A. (Eds.) *Advances in Artificial Intelligence for Privacy Protection and Security*; Intelligent information systems; World Scientific: Hackensack, NJ, USA, 2010; ISBN 978-981-279-032-3.
233. Curzon, J.; Kosa, T.A.; Akalu, R.; El-Khatib, K. Privacy and Artificial Intelligence. *IEEE Trans. Artif. Intell.* **2021**, *2*, 96–108. [CrossRef]
234. Aung, Y.Y.M.; Wong, D.C.S.; Ting, D.S.W. The Promise of Artificial Intelligence: A Review of the Opportunities and Challenges of Artificial Intelligence in Healthcare. *Br. Med. Bull.* **2021**, *139*, 4–15. [CrossRef] [PubMed]
235. Musliner, D.J.; Hendler, J.A.; Agrawala, A.K.; Durfee, E.H.; Strosnider, J.K.; Paul, C.J. The Challenges of Real-Time AI. *Computer* **1995**, *28*, 58–66. [CrossRef]
236. Hoffmann-Riem, W. Artificial Intelligence as a Challenge for Law and Regulation. In *Regulating Artificial Intelligence*; Wischmeyer, T., Rademacher, T., Eds.; Springer International Publishing: Cham, Switzerland, 2020; pp. 1–29, ISBN 978-3-030-32360-8.
237. Sharma, M.; Luthra, S.; Joshi, S.; Kumar, A. Implementing Challenges of Artificial Intelligence: Evidence from Public Manufacturing Sector of an Emerging Economy. *Gov. Inf. Q.* **2022**, *39*, 101624. [CrossRef]
238. Abbasi Moud, A. Recent Advances in Utility of Artificial Intelligence towards Multiscale Colloidal Based Materials Design. *Colloid Interface Sci. Commun.* **2022**, *47*, 100595. [CrossRef]
239. Davenport, T.H. *The AI Advantage: How to Put the Artificial Intelligence Revolution to Work*; The MIT Press: Cambridge, MA, USA, 2018; ISBN 978-0-262-35063-1.
240. Arinez, J.F.; Chang, Q.; Gao, R.X.; Xu, C.; Zhang, J. Artificial Intelligence in Advanced Manufacturing: Current Status and Future Outlook. *J. Manuf. Sci. Eng.* **2020**, *142*, 110804. [CrossRef]
241. Regli, W.; Rossignac, J.; Shapiro, V.; Srinivasan, V. The New Frontiers in Computational Modeling of Material Structures. *Comput. Aided Des.* **2016**, *77*, 73–85. [CrossRef]
242. Malaca, P.; Rocha, L.F.; Gomes, D.; Silva, J.; Veiga, G. Online Inspection System Based on Machine Learning Techniques: Real Case Study of Fabric Textures Classification for the Automotive Industry. *J. Intell. Manuf.* **2019**, *30*, 351–361. [CrossRef]
243. Tuv, E.; Guven, M.; Ennis, P.; Lee, D.H.L.L. Faster, More Accurate Defect Classification Using Machine Vision. Available online: https://media18.connectedsocialmedia.com/intel/11/16872/Faster_More_Accurate_Defect_Classification_Using_Machine_Vision.pdf (accessed on 12 December 2023).

materials

MDPI

Article

Nanoparticle Metrology of Silicates Using Time-Resolved Multiplexed Dye Fluorescence Anisotropy, Small Angle X-ray Scattering, and Molecular Dynamics Simulations

Daniel Doveiko [1,*], Alan R. G. Martin [2], Vladislav Vyshemirsky [3], Simon Stebbing [4], Karina Kubiak-Ossowska [5], Olaf Rolinski [1], David J. S. Birch [1] and Yu Chen [1,*]

1 Photophysics Group, Department of Physics, University of Strathclyde, Glasgow G4 0NG, UK; djs.birch@strath.ac.uk (D.J.S.B.)
2 EPSRC Future Continuous Manufacturing and Advanced Crystallisation National Facility, University of Strathclyde, 99 George Street, Glasgow G1 1RD, UK; alan.r.martin@strath.ac.uk
3 School of Mathematics and Statistics, University of Glasgow, Glasgow G12 8QQ, UK; vladislav.vyshemirsky@glasgow.ac.uk
4 PQ Silicas UK Limited, Warrington WA5 1AB, UK
5 ARCHIE-WeSt, Department of Physics, University of Strathclyde, Glasgow G4 0NG, UK
* Correspondence: daniel.doveiko.2018@uni.strath.ac.uk (D.D.); y.chen@strath.ac.uk (Y.C.)

Abstract: We investigate the nanometrology of sub-nanometre particle sizes in industrially manufactured sodium silicate liquors at high pH using time-resolved fluorescence anisotropy. Rather than the previous approach of using a single dye label, we investigate and quantify the advantages and limitations of multiplexing two fluorescent dye labels. Rotational times of the non-binding rhodamine B and adsorbing rhodamine 6G dyes are used to independently determine the medium microviscosity and the silicate particle radius, respectively. The anisotropy measurements were performed on the range of samples prepared by diluting the stock solution of silicate to concentrations ranging between 0.2 M and 2 M of NaOH and on the stock solution at different temperatures. Additionally, it was shown that the particle size can also be measured using a single excitation wavelength when both dyes are present in the sample. The recovered average particle size has an upper limit of 7.0 ± 1.2 Å. The obtained results were further verified using small-angle X-ray scattering, with the recovered particle size equal to 6.50 ± 0.08 Å. To disclose the impact of the dye label on the measured complex size, we further investigated the adsorption state of rhodamine 6G on silica nanoparticles using molecular dynamics simulations, which showed that the size contribution is strongly impacted by the size of the nanoparticle of interest. In the case of the higher radius of curvature (less curved) of larger particles, the size contribution of the dye label is below 10%, while in the case of smaller and more curved particles, the contribution increases significantly, which also suggests that the particles of interest might not be perfectly spherical.

Keywords: sodium silicates; fluorescence anisotropy decay; particle metrology; rhodamine 6G; rhodamine B; SAXS; MD; adsorption; molecular dynamics

1. Introduction

Sodium silicates are very versatile inorganic chemicals, manufactured on industrial scale by combining silica sand and soda ash (sodium carbonate) under high temperatures. They are often used in coating and bonding applications when in an aqueous solution. Additionally, they exhibit various attractive characteristics, such as being odourless, non-toxic, having high strength and rigidity, being resistant to high temperatures, and having an overall low cost [1]. Furthermore, they serve as a primary source for producing colloidal silica and silica gels, and are widely employed commercially as thickeners and absorbents. Given their versatility, there is an urgent need to deepen understanding of their chemistry to enhance control over production processes, optimize them, and thereby improve overall sustainability [2].

Important characteristics of silicates are the ratio of silica to soda and the size of the dissolved species. The gold standard used by industrial manufacturers to identify the presence of different oligomer species in sodium silicates is ^{29}Si nuclear magnetic resonance (NMR) spectroscopy [3]. However, this method does not provide any quantitative data regarding the size of different species. Furthermore, there are forty-eight different silica structure species identified up to date, corresponding to only 85% of silicon in the solution [2]. Recently, there have been some promising developments in determining the size and shape of different oligomer species using small-angle X-ray scattering (SAXS) and dynamic light scattering (DLS), where it was shown that the dominating fraction of silica is present in small amorphous clusters with a nominal radius of 0.7 nm and a minor component consisting of larger colloidal species of around 15 nm in radius [4–6]. Unfortunately, both of these methods are far from ideal and have significant drawbacks: DLS becomes difficult when trying to determine the particle sizes below the 10 nm range [7], and SAXS is expensive and complex [8]. Given small particles within sodium silicates, measurements must be performed at the nanoscale with nanometre precision, commonly referred to as nanometrology [9]. Furthermore, characterizing sodium silicates presents an additional layer of complexity compared to typical nanoparticles like colloidal silica or gold nanoparticles. This complexity arises from the variety of structures and their polydispersity, which in turn limits the precision of nanometrology [2].

The alternative approach discussed here and developed originally in our laboratory is based on the measurement of time-resolved fluorescence anisotropy of fluorescent probes, which utilizes the relationship between the silica particle size and the rate of the probe's rotational diffusion when bound to the silica particle [10]. This fluorescence technique offers high sensitivity and has an appropriate timescale (ps to ns), and, due to the high variety of fluorescent probes available, the experiment can be adapted to the specification of the medium to be researched, for example, pH and temperature. Furthermore, the required fluorescence time-resolved instrumentation is relatively inexpensive and easy to use.

However, previous anisotropy studies have shown [11] that the variety of different mechanisms of depolarization, such as dye diffusion on the particle surface and wobbling, may cause additional complications and require careful data interpretation and analysis before the particle size can be determined.

Here we exploit the fact that rhodamine 6G (R6G) physically binds to silica particles [11,12], while rhodamine B (RB) does not bind due to the electrostatic repulsion from its carboxylic acid group [13]. This allows using the rotational time of the non-binding RB to independently determine the microviscosity, which can then be used with the rotational time of R6G to estimate the size of the particle, as previously applied by D. Tleugabulova et al. [14] using R6G (excitation 495 nm) and pyranine (excitation 404 nm) to study Sol-to-Gel transitions in sodium silicate. In our case both R6G and RB can also be excited with a single light source at around 500 nm if required. The effect of this multiplexed dye approach is to eliminate the decrease in abundance of the free probe with time and the concomitant increase in the error in the measured viscosity and measured particle radius, as occurred in previous studies using a single dye probe [15]. To ensure a more precise determination of the silica particle size, we used two methods of modifying microviscosity in the samples: dilution of the original sample in water, and change of temperature. Furthermore, we focus on raw, poly-dispersed sodium silicates and validate the anisotropy-measured particle size against SAXS measurements combined with Molecular Dynamics (MD) simulations to investigate the impact of the dye's size on the measured average size of the silica nanoparticles. The findings further support the interpretation of time-resolved fluorescence anisotropy in terms of silica particle size and provide insight as to its measurement limits in this application. Moreover, the methods described herein can be applied to other nanoscale systems, e.g., the G4PAMAM complex with the anti-cancer drug 5-fluorouracil, which change their conformation in different conditions. By enabling the identification of such alterations without the need for complex SAXS measurements [16], this approach presents the pharmaceutical industry with a powerful tool for precise characterization of nanoscale structures, facilitating enhanced quality control and optimization of drug formulations to bolster efficacy and safety.

2. Materials and Methods

Sample Preparation. The sodium silicate solution selected for the investigation was purchased from Sigma-Aldrich (Darmstadt, Germany) as a product of Millipore (Darmstadt, Germany). According to the product description, the solution in its undiluted form has a pH of 11–11.5 at 50 g/L at 20 °C and dynamic viscosity of 130 mPa·s. It contains 25.5–28.5% SiO_2 and 7.5–8.5% Na_2O (3.7 silica to soda ratio), and has a density of 1.296–1.396 g/mL at room temperature. The content of Fe and other heavy metals (such as Pb) is $\leq$50 ppm [17].

R6G (Bioreagent, Suitable for Fluorescence), Sodium Silicate Extra Pure (Millipore), and NaOH 1N volumetric standard solution were purchased from Sigma-Aldrich, and RB (99%, Laser Grade) was purchased from Fisher Scientific (Waltham, MA, USA). Both fluorescent probes were added directly to the stock solutions (100%—not diluted) of the sodium silicate samples, creating samples of the dye concentration of 10 μM.

The diluted sodium silicate samples were prepared by mixing them with distilled water starting from nine-parts stock silicate/NaOH and one-part water (90%) until nine-parts water and one-part stock silicate solution (10%) has been reached. The dye was added after the dilution at 10 μM concentration.

Samples in plastic cuvettes, 1 × 1 × 4 cm, with relevant solutions were prepared immediately before the measurements. The sodium silicate solution is quite stable in the sealed container, but the continuous exposure to CO_2 present in the atmosphere has a tendency of lowering the pH because of formation of carbonic acid in the silicate. When combined with sample evaporation, this leads to gelation. Therefore, the cuvettes were covered with parafilm to minimize the exposure of the samples to CO_2 to prevent gel formation.

Temperature-Controlled Measurements. The samples were prepared by taking 3 mL of stock sodium silicate and adding 10 μM of the probe directly into the sample without prior dilution. The temperature in the chamber was controlled using NesLab RTE-111 Bath circulator connected to the sample holder. The sample temperature was monitored using an Omega HH804 (Omega Engineering, Norwalk, CT, USA) probe, which was kept in the cuvette at all times during the experiment. The sample chamber was purged at low temperature with nitrogen to prevent water condensation.

Time-resolved measurements. Fluorescence intensity decay measurements to obtain both the lifetime and anisotropy data were based on the time-correlated single-photon counting (TSCPC) [18,19] technique, and performed using a HORIBA-IBH (Glasgow, UK) DeltaFlex system with both excitation and emission Seya-Namioka monochromators which incorporate holographic diffraction gratings in order to minimise the detection of scattered light. The R6G-labelled samples were excited using a HORIBA-IBH NanoLED [20] with a centre wavelength of 494 nm, pulse duration of 1.5 ns and the repetition rate of 1 MHz. The emission was collected at 548 nm. RB samples were excited using a NanoLED laser diode with peak wavelength of 509 nm, pulse duration of 200 ps, and repetition rate of 1 MHz, with the fluorescence being collected at 570 nm. Accurate R6G and RB fluorescence lifetime measurements have been previously successfully demonstrated using LED excitation [21]. Fluorescence decay data were collected using the FluoroHub-A electronics containing a time-to-amplitude (TAC) converter operated in reverse mode with the start-to-stop rate kept below 1% to minimize data pile-up [18,19]. The individual fluorescence decays were measured at a magic angle 54.7° in order to eliminate orientational effects [18,19].

Time-resolved decay data analysis. Various methods exist for analysing time-resolved decays, each offering unique advantages. For instance, the maximum entropy method excels in representing decay without imposing a specific model [22]. Alternatively, the lifetime distribution approach simplifies decay analysis by reducing required parameters [23], while the maximum likelihood model effectively handles data noise through

Poisson statistics [24]. In this study, we employ a multi-exponential least-squares model known for its simplicity and robustness, with demonstrated exceptional performance over time [25–27]. Furthermore, for fluorescent dyes exhibiting fluorescence intensity decays following a simple mono-exponential model, each recovered component corresponds to a fluorophore exposed to a different environment [28]. However, in the case of protein, the justification of employing the methods presented herein becomes more challenging due to the inherently complex photophysics involved [29]. Strictly speaking, n-exponential decay times describes n excited states, but of course, even if this is not appropriate to the kinetics, multi-exponentials can still be used to describe a decay. In fact, for the purpose of calculating anisotropy decay, the model used does not have to describe the actual kinetics, but merely provide a good fit to the data, as demonstrated by the χ^2 goodness of fit. This is a real strength of the fluorescence approach to nanometrology we are using.

Reconvolution software (v1) generally assumes n-exponential decay times [18,19], but can be modified to also include the detection of scattered excitation light [30,31] when fitting the fluorescence decay parameters to the raw decay data, according to the formula:

$$F(t) = a + bL(t+\Delta) + c\int_0^t L(t+\Delta-t')I(t')dt'. \tag{1}$$

where $L(t)$ is the instrumental response function (IRF), a is the background noise count level, b is the contribution of the scattered light, and c is the scaling parameter. The temporal difference in response between the IRF and the fluorescence decay caused by the effect of their wavelength differences on the light detector and light path can be approximated using a linear shift parameter Δ, while the $I(t)$

$$I(t) = \sum_i^n b_i(t)\exp\left[-\frac{t}{t_i}\right] \tag{2}$$

is the n-exponential model of the decay.

As time-resolved fluorescence anisotropy theory has been widely reviewed, we present only a short summary here that is of relevance to nanoparticle rotation.

The anisotropy decay function $R(t)$ can be generated by recording vertically $I_V(t)$ and horizontally $I_H(t)$ polarized fluorescence decay curves. Both decays are then combined, forming the anisotropy function [32]:

$$R(t) = \frac{I_V(t) - I_H(t)}{I_V(t) + 2I_H(t)} \tag{3}$$

The sum and the difference of vertically and horizontally polarized fluorescence decay curves have to be reconvoluted and fitted to obtain the anisotropy decay curve. First, the denominator, which describes the fluorescence decay, is fitted using the nonlinear least square method, and the resultant impulse response is iteratively convoluted with $R(t)$ to provide the best fit to the numerator using appropriate models for rotation until the minimum in χ^2 is determined. The anisotropy $R(t)$ was calculated using the HORIBA IBH DAS6 (v 6.8) software package. The polarizer dwell time was set to 60 s, and the photons were counted until the peak difference of 10,000 counts was reached. For both sets of excitation and emission wavelengths, the G-factor correction for the relative detection efficiencies of $I_V(t)$ and $I_H(t)$ was applied.

In the simplest case of spherical rigid rotor in an isotropic medium, the fluorescence depolarization due to the Brownian rotation can be described by

$$R(t) = R_0 \exp\left(-\frac{t}{\varnothing_r}\right). \tag{4}$$

For one photon excitation, the initial anisotropy R_0 has a value of 0.4, and the rotational correlation time $\varnothing_r$ can be described using the Stokes–Einstein equation

$$\varnothing_r = \frac{\eta V}{kT} \quad (5)$$

where T is the temperature, k is the Boltzmann constant, η is the microviscosity of the medium, and V is the hydrodynamic volume. Assuming spherical shape of the rotating particles, we can express the hydrodynamic radius R_p as

$$R_p = \sqrt[3]{\frac{3kT\varnothing_r}{4\pi\eta}}. \quad (6)$$

If the radius of the spherical rotor R_p is known, the rotational time can be used to determine the microviscosity of the medium using

$$\eta = \frac{3kT\varnothing_r}{4\pi R_p^3}. \quad (7)$$

In the approach here, we use the rotational correlation time of RB to calculate the microviscosity of the medium from Equation (7) and then use it in Equation (6) to estimate the radius of R6G–silicate particles.

Small-angle X-ray scattering. The SAXS data were collected with a Xenocs Xeuss 2.0 SAXS/WAXS (Grenoble, France) using Cu Kα^1 radiation Genix3D microfocus source with a wavelength of 1.5406 Å, with an operating voltage of 50 kV and operating current of 0.6 mA. Single-reflection multilayer optics with 2D collimation were used, and the scattered beam was collimated through two sets of motorised scatterless slits with variable apertures. A DECTRIS Pilatus 1M detector (Baden, Switzerland) was used to collect the SAXS data with the sample to detector distance calibrated at 2507.35 mm and a pixel size of 172 µm. Five 30 min images were recorded and accumulated by measuring the scattering intensity as a function of scattering vector $q = 2\pi\sin\theta/\lambda$, where 2θ is the scattering angle and λ is the wavelength of the scattering beam. Samples in their original concentration were prepared in stainless steel sample holders sealed with Kapton dots to limit the air exposure. The sample-detector path was *in-vacuo* to avoid any scattering or absorption from air. The scattering curves of the samples were corrected for the background scattering from the instrument, Kapton dots, and residual atmosphere [33]. The data were analysed using BioXTAS RAW (v2.2.1) [34] software.

MD Protocol. Following the MD simulations and protocol presented previously [12], the starting structure for the R6G was taken from the protein databank entry 2v3l.pdb [35], which was modified to match the experimental one. Furthermore, the partial charges were corrected to match the values obtained by Chuichay et al. [36]. The SNPs were built using the CHARMM-GUI Nanomaterial Modeller [37,38], while the rest of the system preparation was performed using VMD [39]. All simulations were performed using NAMD3 CUDA version [40,41]. Interface Force-Field (FF) [42] was used for the SNPs and CHARMM36 [43] was used for the rest of the system components. As usually done in MD simulations, the system minimization was a two-step process. Initially, a water only minimization was performed, consisting from 1000 minimization steps followed by 100 ps of equilibration in 300 K temperature. Afterwards, the whole system underwent a minimization, consisting of 10,000 minimization steps followed by 30 ps of heating to 300 K and 270 ps of thermalisation with a 1 fs step. For the production runs, the integration step was set to 1 fs to minimize errors and ensure system stability, with the total length of the trajectory being 100 ns. VdW cut-off was set to 12 Å, while the Particle Mesh Ewald (PME) was used to calculate the electrostatic interactions. The system was solvated using the TIP3P [44] water model, while the internal water molecule vibrations were constrained. The neutralization was done using NaCl, resulting in six Cl^- ions added to neutralize the cationic charge of R6G, while all the Na^+ ions present in the system came from ionizing the SNP to the appropriate pH via deprotonating Si-O-H. The anisotropic cell fluctuations ensured that the desired pressure of 1 atm at 300 K was reached.

3. Results and Discussion

3.1. Steady-State Measurements

The results of steady-state measurements of R6G samples in water at various concentrations of NaOH and sodium silicate and RB samples in sodium silicate are shown in Figure 1.

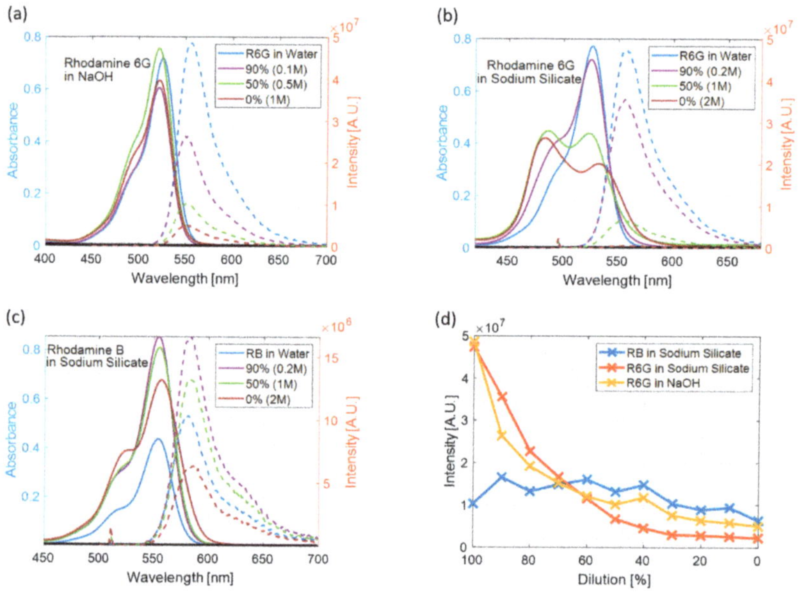

Figure 1. Steady-State results where percentages represent the level of dilution and the molarity represents the concentration of NaOH in the sample. The solid line represents the absorption spectra while the dashed line represents emission spectra. (**a**) R6G absorption and emission spectra in water at various concentrations of NaOH; (**b**) R6G absorption and emission spectra in sodium silicate; (**c**) RB absorption and emission spectra in sodium silicate; (**d**) RB emission intensity as a function of dilution. Samples in water were prepared by diluting stock solution in distilled water to 10 μM dye concentration.

In the case of R6G in sodium silicate (Figure 1b), there is the emergence of a new peak at around 470 nm in the absorption spectrum, which is growing with increasing concentration of sodium silicate and pH in the sample. This effect is widely known [45,46], as R6G tends to form non-fluorescent aggregates (most probably dimers), also known as H-type. Interactions of the dye molecules with the polar surface of the silica structures is very clearly represented in the emergence of the higher energy peak in the absorption spectra. R6G dimers have a strong absorption band at a shorter wavelength when compared with monomers. However, they do not fluoresce, and are mainly responsible for quenching the fluorescence. This process does not occur in at this concentration where there are no silica particles present (Figure 1a), hence the R6G aggregation due to high dye concentration can be excluded. Both sets of plots show the dominating absorption peak at 530 nm, typical for R6G, and a vibronic shoulder at around 500 nm, which overlaps with the H-Type dimer absorption band.

Samples were excited at 494 nm for R6G and 509 nm for RB to measure the emission spectra. Fluorescence emission spectra were recorded in 1 nm increments. The acquired spectra show that the emission intensity is heavily dependent on the sodium hydroxide content in the sample. Namely, with increasing concentration of NaOH, the intensity of the R6G emission decreases, which suggests that increasing pH with the addition of NaOH quenches the R6G fluorescence. On the other hand, when looking at the absorption and emission spectra of RB in sodium silicate, the trend of decreased intensity of both the absorption and emission spectra with pH is much less evident, as further visualised in Figure 1d. It seems to be evident that in cases when R6G is used, the emission intensity drops when the sample is less diluted, which corresponds to increasing the concentration of NaOH. The drop in R6G intensity in the basic regime is caused by the quenching of R6G excited states due to the high concentration of OH^- ions, which happens in a multi-step way. Initially, at pH between 10 and 12, T_1 triplet state is affected, and when pH rises above 12, the S_1 singlet state is additionally depopulated through a non-radiative decay channel [47].

Based on the steady-state measurements, it might be concluded that R6G forms non-fluorescent aggregates when exposed to silica particles, but the dominating factor controlling its fluorescence intensity is the concentration of sodium hydroxide in the sample.

3.2. Fluorescence Decays

Fluorescence lifetime measurements were performed and fitted to several exponential decay models. The 1-exponential model (R6G in water) and 2-exponential model (R6G in silicate and RB in silicate) were found to be the simplest functions that result in acceptable values for the best fit of $\chi^2 < 1.2$ (see Tables S1 and S2 in Supplementary Information). The changes in fluorescence decay components and their contributions when the concentration of NaOH and sodium silicate is changing are shown in Figure 2.

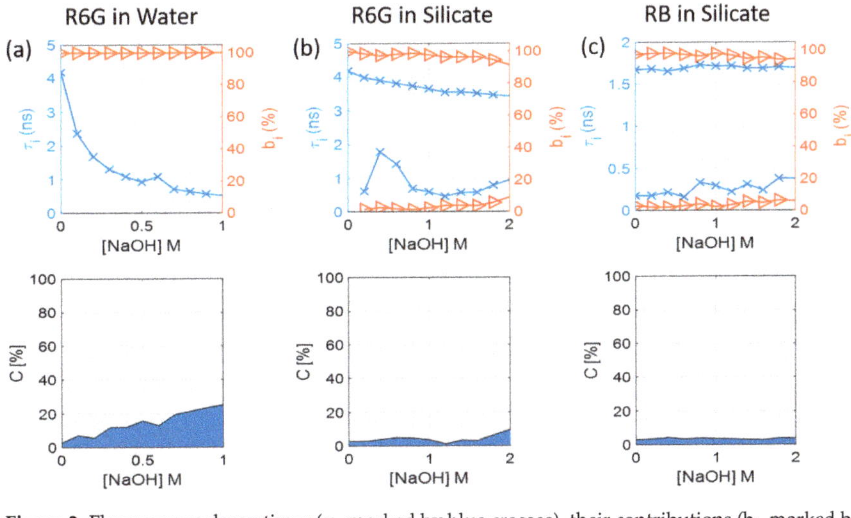

Figure 2. Fluorescence decay times (τ_i, marked by blue crosses), their contributions (b_i, marked by orange triangles), and scattered excitation light contribution (c) depending on the NaOH concentration for each sample group, where the blue area represents the scatter while the white area represents the fluorescence. (**a**) R6G in water at different concentrations of NaOH; (**b**) R6G in sodium silicate; (**c**) RB in sodium silicate.

The pure NaOH samples show the biggest decrease in R6G lifetime at higher pH (see Figure 2a). In the case of R6G in sodium silicate samples, the τ_1 component is also shortening as the sodium silicate concentration increases, which corresponds to the increasing amount of NaOH in the sample that is present in the manufactured sodium silicates (see Figure 2b). It is worth noting that when silicate particles are present in the system, the drop in τ_1 component is not as significant as for pure NaOH samples, which suggests that the presence of silica particles in the samples helps to maintain the decay time of R6G. On the other hand, RB lifetime stays constant in all the samples, and is only slightly affected by the presence of sodium hydroxide or silica particles. Additionally, the scattering increases with the growing concentration of NaOH, which suggests an increase in nanoscale particulates. It is also important to discuss the emergence of the second lifetime component in the sodium silicate samples. As is illustrated on Figure 2b, in the presence of sodium silicates, there is a second lifetime component with a fluorescence decay time shorter than 1 ns. It might be that this component corresponds to free dye in the sample, as its decay component at 1 M of NaOH closely matches the value of R6G in NaOH at 1 M concentration. The slight increase of its contribution at the 2 M concentration might also suggest that most of the dye has adsorbed to the silica particle surface; therefore, any additional dye that would be added to the sample would not be adsorbed, but will stay in the sample in its free form instead. Furthermore, RB also exhibits a biexponential behaviour, suggesting that there is an additional excited state present. The origin of this second component is not entirely clear; nevertheless, it may be that a fraction of RB could adsorb on some of the silicate oligomers. Given this short component is a small fraction of the total emission, its presence does not seem to impact our particle size measurements, as we will show in the following sections.

3.3. *Time-Resolved Fluorescence Anisotropy in Diluted Samples*

The time-resolved fluorescence anisotropy decays of R6G–silicate complexes and RB in silicate, the latter being used to determine the microviscosity of the samples, are presented on Figure 3. Both dyes are comparable in size, and the only difference comes from the presence of the negatively charged carboxylic acid group in RB, which will result in electrostatic repulsion and hence will not bind to the negatively charged silicate particle [13]. The obtained plots are consistent with this, indicating that the decay of anisotropy is notably longer in the R6G–silicate complex when compared with the RB decays. Furthermore, it is important to discuss the discrepancy of the r_0 values from the theoretical value of 0.4. Theoretically, in time-resolved anisotropy measurements, the r_0 value starts at 0.4. In our case, as indicated by Figure 3, in the case of R6G, this is correct only for the 2 M, 1.8 M, and 1.6 M samples, while for the case of RB, the initial anisotropy in all cases is lower than 0.4. Those lower values are well-known to be due to rapid intramolecular reorientation from the absorption to the emission dipole.

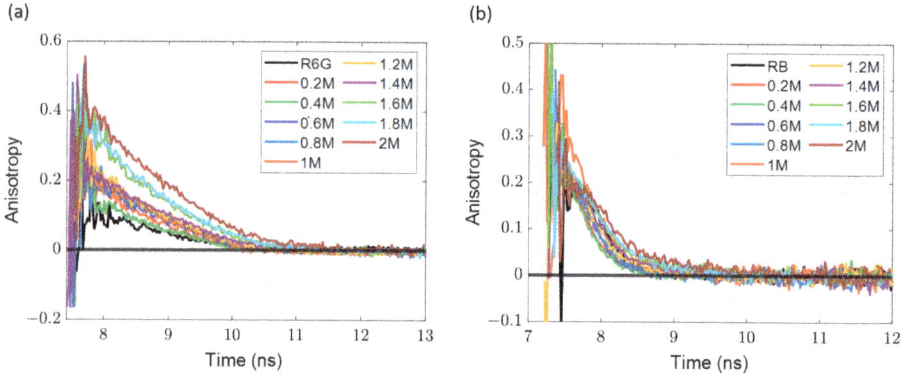

Figure 3. Time-Resolved Fluorescence Anisotropy Decays as a function of sodium hydroxide concentration. (**a**) R6G–silicate complex; (**b**) RB in sodium silicate samples.

The results of anisotropy decay fittings to (4) are shown in Tables S3 and S4. Given that the χ^2 values are acceptable in most cases, we deemed that mono-exponential model of anisotropy decay is appropriate. The measurements for the pairs of samples of the same presence of silicate were performed: one for R6G (attached to silicates), and the second for RB (freely rotating in the solution). By combining those two measurements, and assuming that the microviscosity in each pair is the same, we can determine the particle size in the sodium silicate solutions using the method described previously [13]. The results are listed in Table 1.

Table 1. Calculated microviscosities using RB rotational time, and the estimated upper size limit for the particle sizes using R6G rotational time.

SiO_2 (%) (Estimated)	[NaOH] M (Estimated)	Rotational Time of Non-Binding Probe (RB) (ns)	Microviscosity (mPa·s)	Rotational Time of Binding Probe (R6G) (ns)	Silica Particle Size (Å)
27.0	2.0	0.54 ± 0.04	2.47 ± 0.19	1.16 ± 0.03	7.7 ± 0.8
24.3	1.8	0.46 ± 0.03	1.99 ± 0.12	0.84 ± 0.02	7.4 ± 0.7
21.6	1.6	0.41 ± 0.03	1.92 ± 0.12	0.40 ± 0.01	5.9 ± 0.5
18.9	1.4	0.35 ± 0.02	1.56 ± 0.11	0.44 ± 0.02	6.5 ± 0.7
16.2	1.2	0.32 ± 0.03	1.42 ± 0.13	0.27 ± 0.01	5.7 ± 0.8
13.5	1.0	0.40 ± 0.03	1.69 ± 0.14	0.19 ± 0.02	4.8 ± 0.8
10.8	0.8	0.30 ± 0.03	1.15 ± 0.14	0.24 ± 0.01	5.8 ± 1.0
8.1	0.6	0.25 ± 0.03	1.02 ± 0.12	0.17 ± 0.02	5.4 ± 1.2
5.4	0.4	0.26 ± 0.02	1.04 ± 0.16	0.19 ± 0.003	5.6 ± 1.4
2.7	0.2	0.22 ± 0.04	1.25 ± 0.11	0.14 ± 0.02	4.7 ± 1.2

To precisely determine the hydrodynamic radii of both R6G and RB, we prepared two aqueous dye solutions of 10 μM concentration each and performed time-resolved anisotropy measurements. The data were collected until specified count difference was obtained, and the measurements were repeated in 5000 count difference steps until we acquired six decays, starting with 5000 counts and continuing all the way until 30,000 were reached. The data were fitted next, the corresponding hydrodynamic radii obtained and the average size of the dye were calculated. Resulting anisotropy measurements of both dyes in water gave the hydrodynamic radius of 5.9 ± 0.6 Å for R6G and 5.7 ± 0.2 Å for RB (assuming viscosity of water at 20 °C 1 cP and the temperature 293.15 K), which agree well with values obtained previously using the picosecond polarization grating technique and multi-colour dual-focus fluorescence correlation spectroscopy experiments [48,49]. As both dyes are comparable in size, slower decays of the R6G samples suggest that R6G is indeed attached to silica particles (Figure 3). Having the radius of the RB, we were able to determine the microviscosity of each sample. Then, using the obtained microviscosities and the rotational time of particles labelled with R6G allowed us to calculate the size of the labelled silica particles.

Based on calculated microviscosities using RB rotational times and the estimated particle sizes using the R6G rotational times, as listed in Table 1, it is evident that the measured sizes are significantly smaller in the samples diluted to 1.6 M NaOH and lower, when compared with the undiluted and 1.8 M samples. Additionally, the recovered sizes are the upper limit, as our measured size of R6G is 5.9 ± 0.6 Å and we do not know how the particle orientates on the silica surface.

The microviscosity and particle size changes upon sample dilution are illustrated in Figure 4. The microviscosity is going down upon dilution, which is expected, as is the detected particle size. Moreover, starting from 1.6 [NaOH] M, the particle size is equal to the free R6G size. It might be explained by the fact that the dilution decreases the amount of silicate particles in the sample, while the dye concentration is maintained at the value of 10 μM; therefore, the amount of free dye is growing. Hence, the anisotropy decay will be dominated by the free dye.

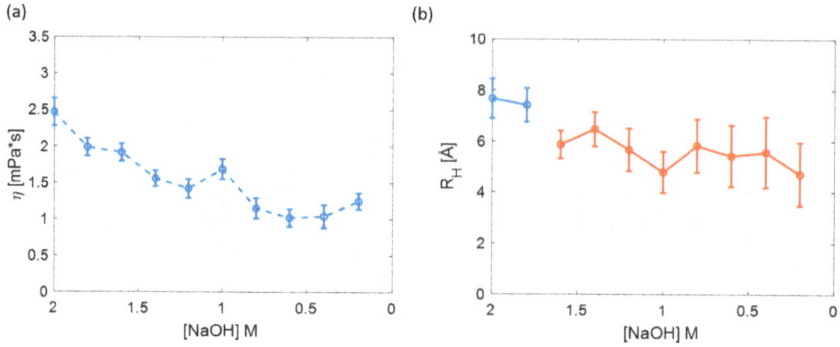

Figure 4. The microviscosity and particle size dependence on sample dilution. (**a**) Microviscosity; (**b**) particle size.

The formation of the non-fluorescent H-type dimers seems to not affect the particle size measurements in any way. When looking at the UV-Vis absorption spectra in Supplementary Figure S1, we can clearly see that the H-type dimers are present in all samples from 2 M to 0.6 M of NaOH, but neither the anisotropy decays shown in Figure 3 nor the particle sizes listed in Table 1 show any discrepancies. This proves that only R6G monomers fluoresce when measuring the particle sizes, and the dimers have no visible effect on the course of the experiment.

3.4. Fluorescence Intensity Decays at Different Temperatures

As in the case of dilution experiments fluorescence, decay measurements were performed and fitted to several models. The 2-exponential model was found to be the simplest that can generate acceptable values of the χ^2 (Tables S5 and S6). Figure 5 shows the changes in fluorescence decay components and their relative contributions when the temperature is changed. It is noted that both dyes behave in a very different way. R6G (Figure 5a) decay times stay almost the same over the whole range of temperatures, suggesting that the complexation with silica particles protects it from the effects of collisional quenching at high temperatures. Moreover, the scatter goes down at higher temperatures.

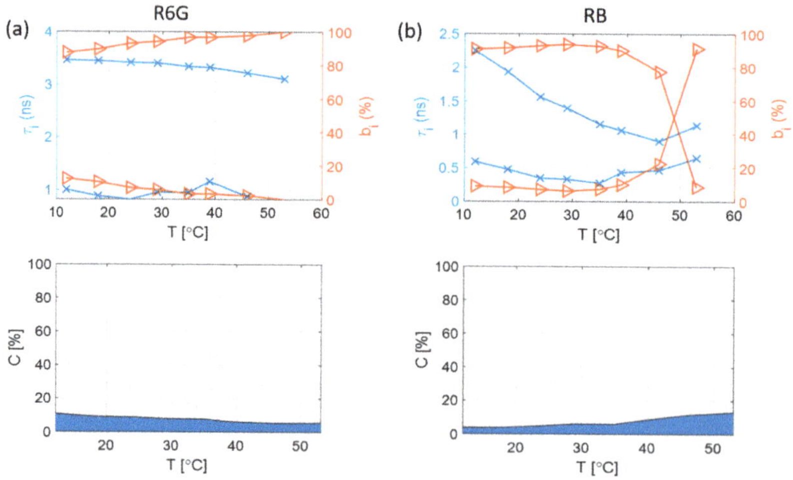

Figure 5. Fluorescence decay components (τ_i, marked by blue crosses), their contributions (b_i, marked by orange triangles), and scattered excitation light contribution (c) depending on the temperature for each sample group at pH 11, where the blue area represents the scatter and the white area represents the fluorescence. (**a**) R6G in sodium silicate; (**b**) RB in sodium silicate.

In the case of RB molecules (Figure 5b), the most abundant decay time drops sharply with the increase in temperature as a result of faster molecular movement, more frequent collisions, and hence stronger dynamic quenching. Furthermore, the amount of scattered light increases with increasing temperature, which is a result of the short decay time at high temperatures, suggesting that at higher temperatures the anisotropy decay measurement may not be so accurate.

Lastly, as it was the case in Section 3.2 with diluted sodium silicate samples, here there are also two decay components present, which suggests that at all times there is a fraction of the dye which is free (shorter component), while the adsorbed dye decay component dominates in the sample (longer component).

3.5. Time Resolved Fluorescence Anisotropy at Different Temperatures

The fluorescence anisotropy decay of R6G–silicate complexes and RB in silicate at different temperatures are shown in Figure 6. In both cases, the decay follows a very clear trend: at lower temperatures the decay is longer, while at higher temperatures the decay is faster. This is directly correlated with the microviscosity, which is higher at low temperatures and lower at high temperatures. As a result, here we have a much more controlled experiment, where the only factors that are changing are the temperature and viscosity, with sample composition staying the same during the whole duration of the experiment.

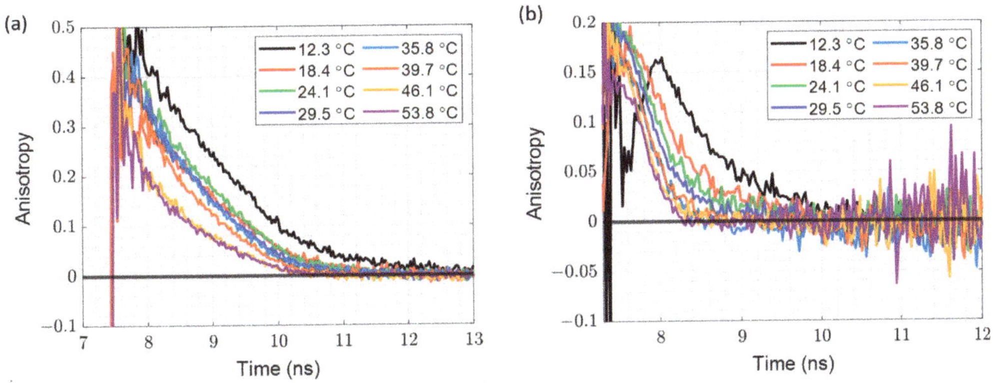

Figure 6. Time resolved fluorescence anisotropy decay at different temperatures of (**a**) R6G–Sodium Silicate Complex; (**b**) RB in Sodium Silicate solution.

Supplementary Tables S7 and S8 provide the details of the fitting results for both samples at different temperatures. As it was in the case of dilution, given that the χ^2 values are acceptable, we deemed that the mono-exponential model was again appropriate. As in the case of the dilution experiment, the rotational time of RB reported on the microviscosity at different temperatures. and the rotation time of the R6G–silicate complex combined with microviscosity, allowed the determination of the size of the complex.

Table 2 lists the calculated microviscosities using RB rotational time and the particle sizes determined using R6G rotational times. It is reassuring to note that the size is more-or-less constant over a wide range of temperatures when compared with the dilution experiment.

The temperature dependence of microviscosity and particle radius is shown in Figure 7. The microviscosity (Figure 7a) drops with increasing temperature. Comparison with Figure 4a indicates that microviscosity at 2 M of NaOH matches the microviscosities at 18 °C and 24 °C (Figure 7a).

Table 2. Calculated microviscosities using RB rotational time, and the estimated upper limit of the particle sizes using R6G rotational time.

T (°C)	Rotational Time of Non-Binding Probe (RB) (ns)	Microviscosity (mPa·s)	Rotational Time of Binding Probe (R6G) (ns)	Silica Particle Size (Å)
12 ± 1	0.72 ± 0.03	3.04 ± 0.12	1.17 ± 0.03	7.1 ± 1.1
18 ± 1	0.53 ± 0.03	2.28 ± 0.13	0.76 ± 0.03	6.8 ± 1.1
24 ± 1	0.54 ± 0.03	2.36 ± 0.13	0.82 ± 0.06	7.0 ± 1.4
29 ± 1	0.37 ± 0.03	1.66 ± 0.13	0.51 ± 0.03	6.7 ± 1.3
35 ± 1	0.23 ± 0.03	1.05 ± 0.13	0.44 ± 0.03	7.5 ± 1.5
39 ± 1	0.29 ± 0.02	1.37 ± 0.09	0.32 ± 0.03	6.2 ± 1.2
46 ± 1	0.19 ± 0.02	0.90 ± 0.09	0.18 ± 0.03	5.8 ± 1.5
53 ± 1	0.14 ± 0.02	0.69 ± 0.10	0.22 ± 0.03	7.0 ± 1.6

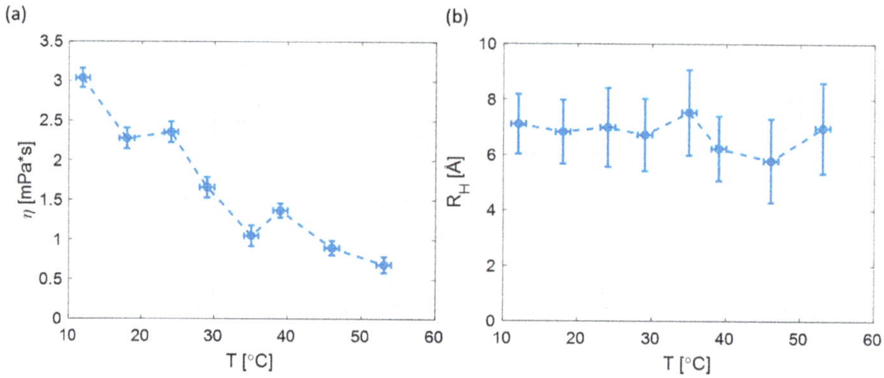

Figure 7. The microviscosity and particle size dependence on temperature (**a**) Microviscosity; (**b**) Particle Size.

Furthermore, a comparison of Figures 4b and 7b suggests that in the case of changing temperature, the detected particle size is constant and the size remains constant at the wide range of the temperature. Additionally, the measured particle size at various temperatures matches the results obtained for 2 M and 1.8 M samples in the dilution experiment. All the above suggests that changing the temperature is a better method for changing viscosity when measuring silicate particle size, as the sample composition does not change with the temperature.

The subsequent step is to determine the size of the silica particle (without the R6G label dye). The determined hydrodynamic radius of R6G is 5.9 ± 0.6 Å, and the size of the silicate-R6G complex is 7.0 ± 1.2 Å. If it is assumed that the dye attaches to the surface of the particle without any alterations to its shape, then the silica particle size would be 1.1 Å, which is less than a water molecule size (~2.8 Å [50]). It is worth mentioning that the Si-O bond length is longer than that of O-H [51]; therefore, the simplest estimation leads to impossible results. This in turn strongly suggests that the dye must adjust its orientation and shape upon binding to the silica particles.

3.6. Multiplexed Time-Resolved Measurements

We prepared a sample of pure sodium silicate containing 10 μM of both R6G and RB to simplify the method of multiplexing two dyes even further. Using λ_{EX} = 494 nm, we measured both fluorescence intensity decays and anisotropy decays, starting from λ_{Em} = 525 nm and measuring in 5 nm increments until λ_{Em} = 595 nm. The measured fluorescence intensity decays and time-resolved anisotropy decays are shown on Figures S6 and S7.

The results of fitting fluorescence intensity decays to a bi-exponential function and anisotropy decays to a mono-exponential function are shown in Figure 8.

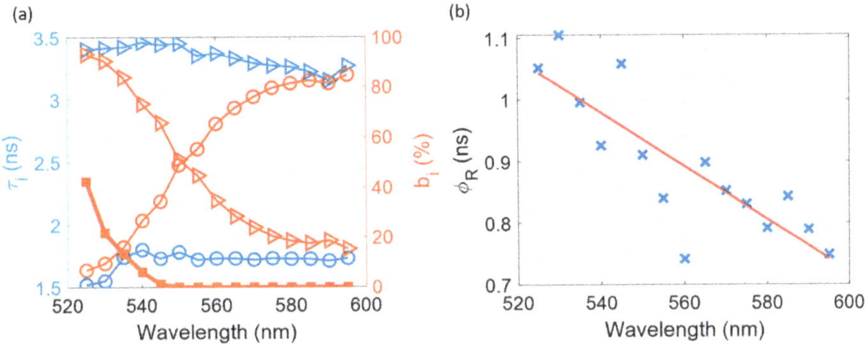

Figure 8. Fitting results for the Multiplexed Time-Resolved Measurements. (**a**) Fluorescence decay components (τ_i) and their contributions (b_i) as a function of $\lambda_{Em.}$, while the thick line with squares shows the contribution of the scattered excitation light. Triangles represent recovered R6G parameters, while circles represent the recovered RB parameters. (**b**) Recovered rotational times as a function of λ_{Em}.

The wavelength-dependent decay time component contribution is clearly seen in Figure 8a, where, at λ_{Em} = 525 nm, the R6G contribution to the overall decay is over 90% while RB is below 10%. The opposite is observed when the λ_{Em} is set to 595 nm, where the decay is dominated by the RB fluorescence with its contribution being over 80% and the R6G contribution dropping below 20%. Between those two points, the contribution of the R6G gradually decreases and RB increases as we move towards longer wavelengths, which is to be expected, as the emission wavelength at 595 nm is far from the peak emission of R6G at 548 nm, and vice versa for the shorter wavelengths.

The rotational times acquired from the anisotropy decay fitting to a mono-exponential model are shown in Figure 8b. The recovered rotational times have a clear wavelength dependence, and can be separated into three groups. The first group is for the emission between 525 and 545 nm, where the decays are dominated by the R6G, hence the recovered rotational times are dominated by the R6G–silicate complex. The second group is for the emission range of 580 nm to 595 nm, where our decay is dominated by the RB, which is freely rotating in the solution and reports on the microviscosity of the solution. Finally, the third group are the decays in between the first and second group, where we see comparable contributions from both dyes, and the recovered rotational time is the mixture of the two dyes, one adsorbed to the particle surface and the other free in the solution. This is caused by the simultaneous fluorescence of both dyes, resulting from a significant overlap between their absorption and emission spectra.

Upon closer analysis, we can see that the precision with which we can measure the rotational time of the R6G–silica complex is limited by the scattered excitation light (see Figure 8a). At shorter wavelengths, the contribution from the scattered light grows exponentially, and initial anisotropy is above the theoretical limit of 0.4 (see Figure S7) [52]. On the other hand, as the scattering drops to zero at longer wavelengths and initial anisotropy decreases below the theoretical limit of 0.4 (see Figure S7), the only factors limiting the precision of RB rotational time measurement are the excitation source power and repetition rate. As a result, it is possible to minimize the R6G contribution even further by moving to even longer wavelengths, and thereby obtaining a more precise measurement of microviscosity. Finally, using the average value from the second group of the rotational times (free RB in solution) and Equation (7), we were able to determine the microviscosity of the sample η = (3.0 ± 0.5) mPa·s. While using the average value from the first group (R6G-silica complex) and Equation (6), we recovered the hydrodynamic radius of the particle R_H = (7.1 ± 0.9) Å. Both of these results agree with the obtained values for both the η and R_H discussed in Sections 3.3 and 3.5. It is important to mention that although two measurements were needed to determine particle size successfully, due to the fact that both dyes can be excited using a single excitation wavelength, this measurement could also be performed using a T-format spectrometer were we can monitor two emission wavelengths simultaneously [53,54], or across the whole fluorescence spectrum using a linear SPAD array combined with a spectrograph [55,56]. The choice of binding and non-binding dyes with overlapping but not identical absorption and fluorescence spectra, as demonstrated here, offers a lot of flexibility in optimising precision when measuring viscosity and particle hyrodynamic radius.

3.7. Small Angle X-ray Scattering and Comparison with Time-Resolved Anisotropy Measurements

To cross-validate our anisotropy measurements and independently determine the average particle size of the sodium silicate, we performed SAXS measurements on the unlabelled stock solution. Figure 9 shows the summed and corrected SAXS data over five 30 min images. Using the reduced SAXS data, the R_g (radius of gyration) was calculated by fitting the corrected scattering profile of the undiluted silicate to the Guinier model:

$$lnI(q) \approx -\left(\frac{R_g^2}{3}\right)q^2 \quad (8)$$

where $I(q)$ is the scattering intensity and q is the scattering vector. Due to the exponential behaviour of the Guinier approximation, the R_g values can be determined by plotting ln(I) vs. q^2, which is shown in Figure 9b. The obtained value for the R_g was 6.50 ± 0.08 Å with R^2 equal to 0.95. This agrees well with results obtained by J. Nordström et al. [4] for the 3.3 silica-to-soda-ratio silicate, exhibiting very close composition to the silicate user herein. Furthermore, a characteristic turn in the residuals plot is evident (Figure 9b) and caused by the polydispersity in the silicate species present in the sample [5].

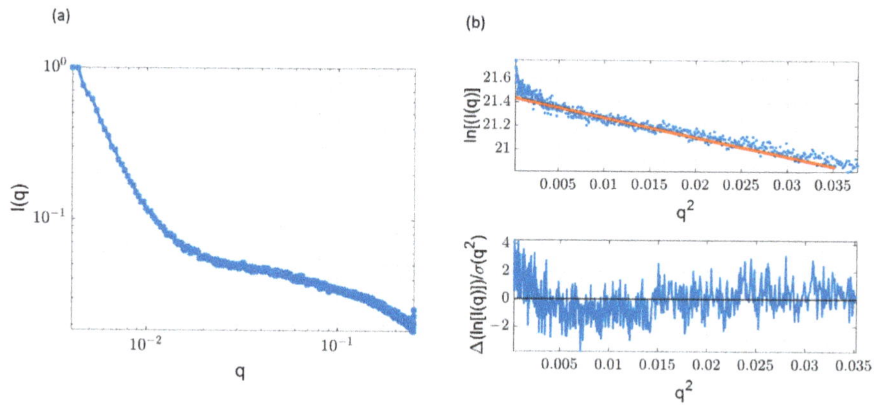

Figure 9. SAXS Results. (**a**) Corrected sample scattering profile. (**b**) Guinier fit of the sample with corresponding residuals plot. Blue dots represent the experimental data and the orange line represents the Guinier fit.

As expected, the radius of gyration is lower than the hydrodynamic radius due to the inclusion of a hydration shell in the case of R_H (6.50 ± 0.08 Å vs. 7.0 ± 1.2 Å). However, it is important to note that the ratio R_g/R_H is between 0.75–0.87, which is very close to the reference value of 0.77 obtained for the solid sphere [57]. Nevertheless, the obtained R_g/R_H ratio is not centred around the 0.77 value, but slightly displaced towards 1.0, which might suggest a slightly elongated shape of the particles. The details regarding the shape of the complex are further investigated via the MD method and are described in the next section.

Lastly, the effect of the free dye in the sample when performing the anisotropy measurements needs to be considered as well. Due to the lack of strong intrinsic fluorescence and the need to label sodium silicates, the fraction of free dye will be present in the solution, contributing to the anisotropy decays. As a result, the measured average rotational time and resulting particle size will be smaller than that it would be in the case when there is no labelling dye present.

3.8. R6G Adsorption to Silica Nanoparticles

To fully understand the influence of the R6G size on the particle size determined using fluorescence anisotropy decay, we performed a range of MD simulations and measured the size of the complex to elucidate, on an atomistic level, the impact of the dye on the measured size.

The measured diameter of the silica nanoparticle (SNP) alongside the diameter of the nanoparticle–R6G complex as a function of the simulation time is shown in Figure 10. In the case of the 40 Å diameter particle, even when the biggest possible distance between the furthest R6G and a silicate oxygen atom on the other side is taken into consideration, the dye contribution to the measured complex size is in the range of 10% (yellow line). However, by looking at the size of the R6G on its own and adding it to the SNP size, it is evident that the obtained value is significantly larger than the actual measured size of the R6G–SNP complex (purple line). This suggests that the shape of the SNP plays a crucial role when measuring the size. Moreover, the dye may not lie entirely flat on the surface, but marginally adjusts its shape to match that of the nanoparticle. The complex size is therefore smaller (yellow line) than the added sizes of separated counterparts (purple line). Similar analysis for 20 Å diameter SNPs is presented in Figure S6.

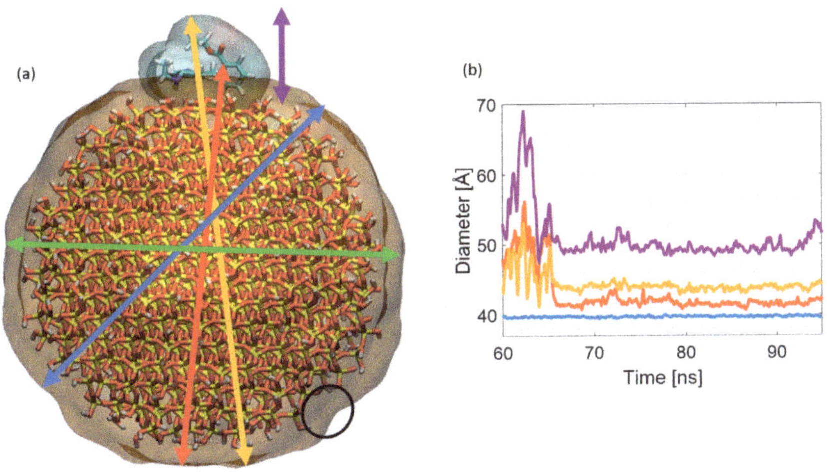

Figure 10. 40 Å diameter SNP size measurements. (**a**) SNP–R6G complex. The black circle indicates local SNP surface irregularity, while the arrows indicate directions of distance measurements. The colour code: green and blue indicate the SNP diameter measured in two different directions, red indicates the distance between R6G xanthene core atom and the SNP atom on its opposite side, yellow points the largest distance between any R6G and SNP atom, while purple shows the maximal thickness of the adsorbed R6G molecules. All atoms are represented by thick sticks coloured by atom type: oxygen is red, silica is yellow, hydrogen is white, carbon is cyan, while nitrogen is blue. To indicate the size of the complex, the dye's surface is shown with transparent representation. (**b**) Measured diameter of the SNP–R6G complex as a function of the simulation time. Colour code corresponds to the arrows code, while the purple line is simply a sum of the R6G thickness itself and the SNP diameter.

The situation is somewhat different in the case of smaller SNPs. The size contribution of R6G is significantly larger in these examples, and can be up to 50% of the measured size. The main reason is that 20 Å SNPs are significantly more curved when compared with 40 Å diameter SNPs. R6G, which is relatively rigid at its xanthene core plane, therefore cannot bend enough to match precisely 20 Å SNPs' curvature. Furthermore, our recent studies have shown [12] it is impossible to have a layer of the R6G formed on the SNP surface because R6G adsorbs on SNP only via its xanthene core and because of the requirement for antiparallel orientation of the constituent molecules dipole moments. Additionally, as proved with the experimental results presented herein and confirmed with our previous MD work [12], the R6G dimers do not have any impact on the measured size, as they cannot interact with SNPs due to the aforementioned restrictions based on geometric constraints and dipole moment orientation. Finally, it is important to mention the shape of the SNP. In all cases, we considered that the SNP was perfectly spherical. However, this may not be the case. At 20 Å–40 Å scale, the atomistic structure and discrepancies on the atomic scale of the nanoparticle may also have some impact. One of such irregularities is indicated in Figure 9a by the black circle, and a small (~4 Å) discrepancy in diameter measured in different direction is illustrated in Figure 10b (blue and green lines). Based on that, we might anticipate that if the nanoparticle would be more ellipsoidal, then the dye contribution would be strongly impacted by the adsorption location. Namely, the dye size contribution to the size of entire complex would be smaller than measured here if R6G would adsorb on the less curved part of the ellipsoid, and probably similar to the contribution measured for spherical particles if the adsorption would occur on the more curved ellipsoid end. In a recent study conducted by G. Hu et al., it was shown using SAXS that for silicates with a ratio less than 4.2, the primary particles are ellipsoidal, which confirms our findings and further justifies why the dye impact on the measured size is small, compared to the MD-obtained results [6].

The discrepancy between the MD results and SAXS results allows us to draw a few important conclusions. The aforementioned simulation results demonstrate what is perhaps obvious, i.e., that the impact of R6G on the measured size is strongly influenced by the size of the labelled particle where the dyes contribution grows significantly with decreasing SNP size. This trend, however, is not observed when comparing the anisotropy and SAXS-obtained results, where $R_g = 6.50 \pm 0.08$ Å (no R6G) while the $R_H = 7.0 \pm 1.2$ Å (with adsorbed R6G), suggesting that the adsorbed dye has only a fractional contribution to the measured average particle size. Additionally, the obtained R_g/R_H ratio is not centred around the 0.77 value, but slightly displaced towards 1.0, which is the case for elongated particles. This discrepancy in MD results comes from the fact that modelled particles used in the simulations were considered spherical, as modelling non-spherical and amorphous particles is very complex due to the lack of well-defined structures and parameter files, and therefore would require additional parametrization using quantum chemical methods. Therefore, the MD results presented here should be treated as an indicator of possible trends observed in the experiments. Due to the matching chemical composition and physical properties (charge, hydrophobicity, etc.) of crystalline SNPs used above, and the amorphous sodium silicates used in the experiments, the outcome of the simulations involving elongated SNPs made from amorphous silica would most probably result in a very comparable outcome to the one discussed above. Nonetheless, performing simulations involving elongated amorphous SNPs would most definitely help to broaden the understanding of this labelling mechanism. Considering all of the above, and the fact that R6G can only marginally adjust its shape in the case when the SNPs have smaller curvature, this suggests that the measured particles are elongated and R6G potentially adsorbs to the side of the SNP which is longer and has smaller curvature, allowing it to marginally adjust its orientation and marginally adjust its shape to match the curvature of the nanoparticle. In fluorescence anisotropy decay measurements, it is of course the hydrodynamic radius which determines the kinetics, and this effectively can take no account of a structure being non-spherical unless much higher statistical precision is obtained than that required for the metrology objective described here.

4. Conclusions

We have presented two approaches based on time-resolved fluorescence anisotropy that allow the determination of particle size of sodium silicate oligomers. The viscosity has been altered in two ways: by sample dilution, and by changing the temperature of the sample, the latter offering advantages. Finally, we showed that the particle size can be successfully measured using a single excitation wavelength when both dyes are present in the sample simultaneously. The measured particle size for Rhodamine 6G agrees with previous work [58], which leads to confidence that the measured sizes are not greater than reported here. Moreover, the data imply that the dye can marginally adjust its shape upon binding. We performed SAXS measurements on the undiluted sample to cross-validate the anisotropy results reported here. Guinier analysis of the SAXS data allowed for determination of the radius of gyration, which is equal to 6.50 ± 0.08 Å. This result agrees well with the previous results obtained for sodium silicates, and it simultaneously confirms the results and conclusions obtained from the anisotropy measurements. Finally, to get a better insight into the mechanism of sodium silicate particle labelling, we performed MD simulations of 40 Å and 20 Å diameter nanoparticles. MD trajectories showed that R6G can marginally adjust its shape to match the curvature of the nanoparticle. This implies that in the case of bigger and less curved particles the impact of the dye on the measured complex size is up to 10%, while for the more curved smaller particles, that size contribution grows substantially and can potentially be up to 50%. Combining all of the results above, we can speculate that there is a high possibility that the particles of interest are not perfectly spherical because of the agreement between anisotropy and SAXS measurements and the shape adjustment observed in MD simulations. This is consistent with a previous study [6].

In summary, provided the pH and dye are compatible, i.e., the dye is stable at highly alkaline pH and maintains a sufficiently long lifetime in that environment, the simple methods described here allow efficient determination of average silicate oligomer particle sizes. Furthermore, as already pointed out, a simple multi-exponential model can be utilized to describe the fluorescence decay kinetics, as the use of fluorescence anisotropy decay does not require an understanding of complex kinetics; the fluorescence is just providing a marker in time, related to particle rotation. A major benefit is that the samples are relatively simple to label and do not require any modifications that may alter the oligomer speciation, thus providing a simple method of monitoring silicates. Furthermore, the methods presented herein can be adapted to other systems involving labelled nanoparticles to determine the average size in a more simple and cost-effective manner. It is important to note that due to the complexity of sodium silicates and the variety of structures present, the method is limited to measuring the average size and cannot distinguish between different species within the silicates or determine size distributions.

Supplementary Materials: The supporting information can be downloaded at: https://www.mdpi.com/article/10.3390/ma17071686/s1, Figure S1: UV-Vis absorption Spectra of (a) R6G at different concentrations of NaOH; (b) R6G in Sodium silicate containing different concentrations of NaOH; Figure S2: Emission Spectra (λ_{EX} = 494 nm) of (a) R6G at different concentrations of NaOH; (b) R6G in Sodium silicate containing different concentrations of NaOH; Figure S3: Excitation Spectra (λem = 590 nm) of (a) R6G at different concentrations of NaOH; (b) R6G in Sodium silicate containing different concentrations of NaOH; Figure S4: Fluorescence intensity decays of (a) R6G in NaOH, (b) R6G-silicate complex; (c) RB in sodium silicate; Figure S5: Fluorescence intensity decays at different temperatures (a) R6G-Silicate complex; (b) RB in Sodium Silicate; Figure S6: Fluorescence intensity decays for the multiplexed measurements; Figure S7: Anisotropy decays for the multiplexed measurements; Figure S8: 20 Å diameter SNP measurement results. Colour code follows the one introduced on Figure 9 while the sum of separated sizes is ignored for clarity; Table S1: R6G Parameters recovered from Fluorescence Intensity Decay fitting; Table S2: RhB Parameters recovered from Fluorescence Intensity Decay Fitting; Table S3: Anisotropy Fitting Results for R6G-Silicate complex; Table S4: Anisotropy Fitting Results for RhB in Sodium Silicates; Table S5: R6G Parameters recovered from Fluorescence Intensity Decay fitting; Table S6: RB Parameters recovered from Fluorescence Intensity Decay fitting; Table S7: Anisotropy Fitting Results for R6G-Silicate Complex; Table S8: Anisotropy Fitting Results for RhB in Sodium Silicate.

Author Contributions: Conceptualization, D.D., K.K.-O., O.R., D.J.S.B. and Y.C.; methodology, D.D., A.R.G.M., K.K.-O., O.R., D.J.S.B. and Y.C.; software, V.V. and O.R.; validation, D.D., A.R.G.M., S.S., K.K.-O., O.R., D.J.S.B. and Y.C.; formal analysis, D.D.; writing—original draft preparation, D.D.; writing—review and editing, D.D., A.R.G.M., S.S., K.K.-O., O.R., D.J.S.B. and Y.C.; supervision, S.S., K.K.-O., O.R. and Y.C. All authors have read and agreed to the published version of the manuscript.

Funding: This research was funded by EPSRC and PQ Silicas, grant number EP/T517938/1.

Institutional Review Board Statement: Not applicable.

Informed Consent Statement: Not applicable.

Data Availability Statement: Data are contained within the article and Supplementary Materials. Any additional data needed will be shared on a reasonable request to the corresponding author.

Acknowledgments: D.D. would like to thank PQ Corporation, University of Strathclyde, and EPSRC for the PhD studentship (EP/T517938/1). MD simulation results were obtained using the ARCHIE-WeSt High-Performance Computer (www.archie-west.ac.uk, accessed on 7 March 2024) based at the University of Strathclyde. The authors would like to acknowledge that part of this work was carried out in the CMAC National Facility supported by UKRPIF (UK Research Partnership Fund) award from the Higher Education Funding Council for England (HEFCE) (grant ref.: HH13054). Authors would like to thank John Revie for his assistance in setting up the temperature-controlled measurements.

Conflicts of Interest: Author Simon Stebbing was employed by the company PQ Silicas UK Limited. The remaining authors declare that the research was conducted in the absence of any commercial or financial relationships that could be construed as a potential conflict of interest.

References

1. Yang, X.; Zhu, W.; Yang, Q. The viscosity properties of sodium silicate solutions. *J. Solut. Chem.* **2008**, *37*, 73–83. [CrossRef]
2. Matinfar, M.; Nychka, J.A. A review of sodium silicate solutions: Structure, gelation, and syneresis. *Adv. Colloid Interface Sci.* **2023**, *322*, 103036. [CrossRef] [PubMed]
3. Stebbins, J.F. Identification of multiple structural species in silicate glasses by ^{29}Si NMR. *Nature* **1987**, *330*, 465–467. [CrossRef] [PubMed]
4. Nordström, J.; Sundblom, A.; Jensen, G.V.; Pedersen, J.S.; Palmqvist, A.; Matic, A. Silica/alkali ratio dependence of the microscopic structure of sodium silicate solutions. *J. Colloid Interface Sci.* **2013**, *397*, 9–17. [CrossRef]
5. Nordström, J.; Nilsson, E.; Jarvol, P.; Nayeri, M.; Palmqvist, A.; Bergenholtz, J.; Matic, A. Concentration-and pH-dependence of highly alkaline sodium silicate solutions. *J. Colloid Interface Sci.* **2011**, *356*, 37–45. [CrossRef] [PubMed]
6. Hu, G.; Jin, S.; Liu, K. Structure-Directing Effect on Silica Nanoparticle Growth in Sodium Silicate Solutions through Small-Angle X-ray Scattering. *J. Phys. Chem. C* **2023**, *127*, 10585–10593. [CrossRef]
7. Usković, V. Dynamic light scattering based microelectrophoresis: Main prospects and limitations. *J. Dispers. Sci. Technol.* **2012**, *33*, 1762–1786. [CrossRef] [PubMed]
8. Pauw, B.R. Everything SAXS: Small-angle scattering pattern collection and correction. *J. Phys. Condens. Matter* **2013**, *25*, 383201. [CrossRef] [PubMed]
9. López-Lorente, Á.I.; Valcárcel, M. Chapter 1—Analytical Nanoscience and Nanotechnology. In *Comprehensive Analytical Chemistry*; Valcárcel, M., López-Lorente, Á.I., Eds.; Elsevier: Amsterdam, The Netherlands, 2014; Volume 66, pp. 3–35.
10. Birch, D.J.; Geddes, C.D. Sol-gel particle growth studied using fluorescence anisotropy: An alternative to scattering techniques. *Phys. Rev. E* **2000**, *62*, 2977. [CrossRef]
11. Yip, P.; Karolin, J.; Birch, D.J. Fluorescence anisotropy metrology of electrostatically and covalently labelled silica nanoparticles. *Meas. Sci. Technol.* **2012**, *23*, 084003. [CrossRef]
12. Doveiko, D.; Kubiak-Ossowska, K.; Chen, Y. Impact of the Crystal Structure of Silica Nanoparticles on Rhodamine 6G Adsorption: A Molecular Dynamics Study. *ACS Omega* **2024**, *9*, 4123–4136. [CrossRef] [PubMed]
13. Birch, D.J.; Yip, P. Nanometrology. In *Fluorescence Spectroscopy and Microscopy*; Springer: Berlin/Heidelberg, Germany, 2014; pp. 279–302.
14. Tleugabulova, D.; Zhang, Z.; Brennan, J.D. Evolution of sodium silicate sols through the sol-to-gel transition assessed by the fluorescence-based nanoparticle metrology approach. *J. Phys. Chem. B* **2003**, *107*, 10127–10133. [CrossRef]
15. Geddes, C.; Birch, D. Nanometre resolution of silica hydrogel formation using time-resolved fluorescence anisotropy. *J. Non-Cryst. Solids* **2000**, *270*, 191–204. [CrossRef]
16. Szota, M.; Wolski, P.; Carucci, C.; Marincola, F.C.; Gurgul, J.; Panczyk, T.; Salis, A.; Jachimska, B. Effect of Ionization Degree of Poly(amidoamine) Dendrimer and 5-Fluorouracil on the Efficiency of Complex Formation—A Theoretical and Experimental Approach. *Int. J. Mol. Sci.* **2023**, *24*, 819. [CrossRef]

17. Burgert, D.R. *Specification 1.05621.0000 Sodium Silicate Solution Extra Pure*; EMD Millipore Corporation—A Subsidiary of Merck KgaA: Darmstadt, Germany, 2016.
18. Birch, D.J.; Imhof, R.E. Time-domain fluorescence spectroscopy using time-correlated single-photon counting. In *Topics in Fluorescence Spectroscopy*; Springer: Berlin/Heidelberg, Germany, 2002; pp. 1–95.
19. O'Connor, D.; Phillips, D. *Time-Correlated Single Photon Counting*; Academic Press: New York, NY, USA, 1984.
20. McGuinness, C.; Sagoo, K.; McLoskey, D.; Birch, D. A new sub-nanosecond LED at 280 nm: Application to protein fluorescence. *Meas. Sci. Technol.* **2004**, *15*, L19. [CrossRef]
21. O'Hagan, W.; McKenna, M.; Sherrington, D.; Rolinski, O.; Birch, D. MHz LED source for nanosecond fluorescence sensing. *Meas. Sci. Technol.* **2001**, *13*, 84. [CrossRef]
22. Brochon, J.-C. Maximum entropy method of data analysis in time-resolved spectroscopy. In *Methods in Enzymology*; Elsevier: Amsterdam, The Netherlands, 1994; Volume 240, pp. 262–311.
23. Bharill, S.; Sarkar, P.; Ballin, J.D.; Gryczynski, I.; Wilson, G.M.; Gryczynski, Z. Fluorescence intensity decays of 2-aminopurine solutions: Lifetime distribution approach. *Anal. Biochem.* **2008**, *377*, 141–149. [CrossRef] [PubMed]
24. Bajzer, Ž.; Therneau, T.M.; Sharp, J.C.; Prendergast, F.G. Maximum likelihood method for the analysis of time-resolved fluorescence decay curves. *Eur. Biophys. J.* **1991**, *20*, 247–262. [CrossRef]
25. Li, Y.; Natakorn, S.; Chen, Y.; Safar, M.; Cunningham, M.; Tian, J.; Li, D.D.-U. Investigations on average fluorescence lifetimes for visualizing multi-exponential decays. *Front. Phys.* **2020**, *8*, 576862. [CrossRef]
26. Luchowski, R.; Gryczynski, Z.; Sarkar, P.; Borejdo, J.; Szabelski, M.; Kapusta, P.; Gryczynski, I. Instrument response standard in time-resolved fluorescence. *Rev. Sci. Instrum.* **2009**, *80*, 033109. [CrossRef]
27. Lakowicz, J.R.; Shen, B.; Gryczynski, Z.; D'Auria, S.; Gryczynski, I. Intrinsic Fluorescence from DNA Can Be Enhanced by Metallic Particles. *Biochem. Biophys. Res. Commun.* **2001**, *286*, 875–879. [CrossRef] [PubMed]
28. Selanger, K.; Falnes, J.; Sikkeland, T. Fluorescence lifetime studies of Rhodamine 6G in methanol. *J. Phys. Chem.* **1977**, *81*, 1960–1963. [CrossRef]
29. Stryer, L. Fluorescence Spectroscopy of Proteins: Fluorescent probes provide insight into the structure, interactions, and dynamics of proteins. *Science* **1968**, *162*, 526–533. [CrossRef] [PubMed]
30. Alghamdi, A.; Vyshemirsky, V.; Birch, D.J.; Rolinski, O.J. Detecting beta-amyloid aggregation from time-resolved emission spectra. *Methods Appl. Fluoresc.* **2018**, *6*, 024002. [CrossRef]
31. Rolinski, O.J.; McLaughlin, D.; Birch, D.J.; Vyshemirsky, V. Resolving environmental microheterogeneity and dielectric relaxation in fluorescence kinetics of protein. *Methods Appl. Fluoresc.* **2016**, *4*, 024001. [CrossRef] [PubMed]
32. Visser, A.; Rolinski, O.J. Basic photophysics. *Am. Soc. Photobiol.* **2010**. Available online: http://www.photobiology.info/Visser-Rolinski.html (accessed on 1 April 2024).
33. Pilz, I.; Glatter, O.; Kratky, O. [11] Small-angle X-ray scattering. In *Methods in Enzymology*; Elsevier: Amsterdam, The Netherlands, 1979; Volume 61, pp. 148–249.
34. Hopkins, J.B.; Gillilan, R.E.; Skou, S. BioXTAS RAW: Improvements to a free open-source program for small-angle X-ray scattering data reduction and analysis. *J. Appl. Crystallogr.* **2017**, *50*, 1545–1553. [CrossRef] [PubMed]
35. Neubauer, H.; Gaiko, N.; Berger, S.; Schaffer, J.; Eggeling, C.; Tuma, J.; Verdier, L.; Seidel, C.A.; Griesinger, C.; Volkmer, A. Orientational and dynamical heterogeneity of rhodamine 6G terminally attached to a DNA helix revealed by NMR and single-molecule fluorescence spectroscopy. *J. Am. Chem. Soc.* **2007**, *129*, 12746–12755. [CrossRef]
36. Chuichay, P.; Vladimirov, E.; Siriwong, K.; Hannongbua, S.; Rösch, N. Molecular-dynamics simulations of pyronine 6G and rhodamine 6G dimers in aqueous solution. *J. Mol. Model.* **2006**, *12*, 885–896. [CrossRef] [PubMed]
37. Jo, S.; Kim, T.; Iyer, V.G.; Im, W. CHARMM-GUI: A web-based graphical user interface for CHARMM. *J. Comput. Chem.* **2008**, *29*, 1859–1865. [CrossRef]
38. Choi, Y.K.; Kern, N.R.; Kim, S.; Kanhaiya, K.; Afshar, Y.; Jeon, S.H.; Jo, S.; Brooks, B.R.; Lee, J.; Tadmor, E.B. CHARMM-GUI nanomaterial modeler for modeling and simulation of nanomaterial systems. *J. Chem. Theory Comput.* **2021**, *18*, 479–493. [CrossRef]
39. Humphrey, W.; Dalke, A.; Schulten, K. VMD: Visual molecular dynamics. *J. Mol. Graph.* **1996**, *14*, 33–38. [CrossRef] [PubMed]
40. Phillips, J.C.; Braun, R.; Wang, W.; Gumbart, J.; Tajkhorshid, E.; Villa, E.; Chipot, C.; Skeel, R.D.; Kale, L.; Schulten, K. Scalable molecular dynamics with NAMD. *J. Comput. Chem.* **2005**, *26*, 1781–1802. [CrossRef]
41. Phillips, J.C.; Hardy, D.J.; Maia, J.D.; Stone, J.E.; Ribeiro, J.V.; Bernardi, R.C.; Buch, R.; Fiorin, G.; Hénin, J.; Jiang, W. Scalable molecular dynamics on CPU and GPU architectures with NAMD. *J. Chem. Phys.* **2020**, *153*, 044130. [CrossRef] [PubMed]
42. Heinz, H.; Lin, T.-J.; Kishore Mishra, R.; Emami, F.S. Thermodynamically consistent force fields for the assembly of inorganic, organic, and biological nanostructures: The INTERFACE force field. *Langmuir* **2013**, *29*, 1754–1765. [CrossRef] [PubMed]
43. Huang, J.; MacKerell, A.D., Jr. CHARMM36 all-atom additive protein force field: Validation based on comparison to NMR data. *J. Comput. Chem.* **2013**, *34*, 2135–2145. [CrossRef] [PubMed]
44. Mark, P.; Nilsson, L. Structure and dynamics of the TIP3P, SPC, and SPC/E water models at 298 K. *J. Phys. Chem. A* **2001**, *105*, 9954–9960. [CrossRef]
45. Malfatti, L.; Kidchob, T.; Aiello, D.; Aiello, R.; Testa, F.; Innocenzi, P. Aggregation states of rhodamine 6G in mesostructured silica films. *J. Phys. Chem. C* **2008**, *112*, 16225–16230. [CrossRef]

46. Anedda, A.; Carbonaro, C.M.; Corpino, R.; Ricci, P.C.; Grandi, S.; Mustarelli, P. Formation of fluorescent aggregates in Rhodamine 6G doped silica glasses. *J. Non-Cryst. Solids* **2007**, *353*, 481–485. [CrossRef]
47. Haimerl, J.M.; Ghosh, I.; König, B.; Lupton, J.M.; Vogelsang, J. Chemical photocatalysis with rhodamine 6g: Investigation of photoreduction by simultaneous fluorescence correlation spectroscopy and fluorescence lifetime measurements. *J. Phys. Chem. B* **2018**, *122*, 10728–10735. [CrossRef]
48. Müller, C.; Loman, A.; Pacheco, V.; Koberling, F.; Willbold, D.; Richtering, W.; Enderlein, J. Precise measurement of diffusion by multi-color dual-focus fluorescence correlation spectroscopy. *Europhys. Lett.* **2008**, *83*, 46001. [CrossRef]
49. Gumy, J.-C.; Vauthey, E. Picosecond polarization grating study of the effect of excess excitation energy on the rotational dynamics of rhodamine 6G in different electronic states. *J. Phys. Chem.* **1996**, *100*, 8628–8632. [CrossRef]
50. D'Arrigo, J.S. Screening of membrane surface charges by divalent cations: An atomic representation. *Am. J. Physiol.-Cell Physiol.* **1978**, *235*, C109–C117. [CrossRef] [PubMed]
51. Trease, N.M.; Clark, T.M.; Grandinetti, P.J.; Stebbins, J.F.; Sen, S. Bond length-bond angle correlation in densified silica—Results from 17O NMR spectroscopy. *J. Chem. Phys.* **2017**, *146*, 184505. [CrossRef]
52. Lakowicz, J. *Principles of Fluorescence Spectroscopy*; Springer: Boston, MA, USA, 2006.
53. Aguilar-Caballos, M.P.; Gómez-Hens, A.; Pérez-Bendito, D. Simultaneous stopped-flow determination of butylated hydroxyanisole and propyl gallate using a T-format luminescence spectrometer. *J. Agric. Food Chem.* **2000**, *48*, 312–317. [CrossRef] [PubMed]
54. Birch, D.; Holmes, A.; Gilchrist, J.; Imhof, R.; Al Alawi, S.; Nadolski, B. A multiplexed single-photon instrument for routine measurement of time-resolved fluorescence anisotropy. *J. Phys. E Sci. Instrum.* **1987**, *20*, 471. [CrossRef]
55. Sutherland, J.C. Simultaneous measurement of circular dichroism and fluorescence polarization anisotropy. In Proceedings of the Clinical Diagnostic Systems: Technologies and Instrumentation, San Jose, CA, USA, 22–24 January 2002; pp. 126–136.
56. Kufcsák, A.; Erdogan, A.; Walker, R.; Ehrlich, K.; Tanner, M.; Megia-Fernandez, A.; Scholefield, E.; Emanuel, P.; Dhaliwal, K.; Bradley, M. Time-resolved spectroscopy at 19,000 lines per second using a CMOS SPAD line array enables advanced biophotonics applications. *Opt. Express* **2017**, *25*, 11103–11123. [CrossRef] [PubMed]
57. Tande, B.M.; Wagner, N.J.; Mackay, M.E.; Hawker, C.J.; Jeong, M. Viscosimetric, hydrodynamic, and conformational properties of dendrimers and dendrons. *Macromolecules* **2001**, *34*, 8580–8585. [CrossRef]
58. Eichler, H.; Klein, U.; Langhans, D. Measurement of orientational relaxation times of rhodamine 6G with a streak camera. *Chem. Phys. Lett.* **1979**, *67*, 21–23. [CrossRef]

materials

MDPI

Article

Experimental Measurement of Emissivity of Polished Steel Strips from a Continuous Annealing Line

Šimon Staško [1,*], Gustáv Jablonský [1], Augustín Varga [1], Róbert Dzurňák [1] and Jan Kizek [2]

1 Institute of Metallurgy, Faculty of Materials, Metallurgy and Recycling, Technical University of Kosice, Letna 1/9, 042 00 Kosice, Slovakia; gustav.jablonsky@tuke.sk (G.J.); augustin.varga@tuke.sk (A.V.); robert.dzurnak@tuke.sk (R.D.)

2 Department of Process Technique, Faculty of Manufacturing Technologies with a Seat in Presov, Technical University of Kosice, Bayerova 1, 080 01 Presov, Slovakia; jan.kizek@tuke.sk

* Correspondence: simon.stasko@tuke.sk

Abstract: The long-term use of steel strip in various industries makes it an important semi-finished product, which makes it necessary to improve its chemical composition and mechanical properties, reduce its thickness and weight, expand the range of new types of steel strip and increase its production. This entails a large number of technological operations dependent on precise temperature measurement and control. In some industrial plants, the steel strip is in continuous motion, which makes the use of contact measuring devices impossible. When using non-contact measuring devices such as pyrometers or thermal imaging cameras, the emissivity of the materials being measured is a problematic parameter, as setting an incorrect emissivity value to the measuring device results in inaccurate temperature readings. The essence of this research was to establish a measurement method and to perform experimental measurements of the emissivity of a polished steel strip used in a continuous annealing line, the subsequent processing of the data from these measurements and their evaluation. The emissivity measurements were carried out for 5 types of steel strip of different parameters, while the measurement itself was carried out in the long wavelength range of 7.5–14 μm and at strip temperatures of 100–300 °C. Depending on the type of steel strip, the mean emissivity values ranged from 0.0835– to 0.1143. The emissivity of the steel strip increased with increasing strip temperature, and it was not a linear dependence. The emissivity values determined in this research could be applied to measuring equipment in actual production, which could improve the accuracy of temperature measurement in the heat treatment of polished steel strip. Thermal camera measurements in the long wavelength range, taking thermal images and their processing and determining the emissivity value of polished steel strips are the parts of this research that make it different from other already published research.

Keywords: emissivity; temperature; steel strip; thermal image; experimental measurement; thermal imaging camera

1. Introduction

Steel strips or sheets have long been an important input semi-finished product for the production of products in various industries. Depending on the industry, these strips have different chemical compositions. Every year, there is a demand to increase the production of these strips, to extend the range with new types of strips, to improve their properties and to reduce their thickness and weight [1]. In order to achieve these requirements, it is necessary to improve and innovate areas of technological and production processes. A large part of these processes takes place in industrial furnaces, which reach high outputs; thus, their energy consumption is significant [2].

Continuous annealing lines are among the important industrial furnaces designed for the heat treatment of cold-rolled steel strips. These annealing lines consist of several chambers in which the steel strip undergoes a significant temperature change in a short time. The first step is the heating of the strip in the heating chamber, in which this heating takes place mainly through heat transfer by radiation and partly by convection. In the next step, the steel strip remains at the annealing temperature in the annealing chamber, followed by its cooling in the cooling chambers [3,4]. In each chamber of the continuous annealing line, a gaseous protective atmosphere flows to protect the steel strip from oxidation [5,6]. The velocity of the steel strip movement in all the furnace chambers is synchronized with the heating and cooling rates of the strip. To ensure continuous operation, 2 unwinders are used in the inlet section and 2 winders are used in the outlet section. Once the desired length of strip has been wound, the strip is separated using the splitting shears. After annealing in this line, the deformed grains of the steel strip are replaced with new ones and thus its properties including ductility are improved [7].

To achieve the desired mechanical properties of the steel strip and to achieve higher energy and economic efficiency in the annealing process, precise control and the measurement of the temperature in the furnace and on the steel strip throughout the annealing process is necessary [8,9]. In most industries, contact measuring devices such as thermocouples are used for accurate temperature measurement. Since in a continuous annealing line, the steel strip is in continuous motion, it is problematic to measure its temperature by using contact measuring devices. In this case, it is practical to use pyrometers or thermal imaging cameras [10].

Infrared thermography is a branch of science dealing with the acquisition of thermal data from non-contact measuring devices and their subsequent processing. The basis of this thermography is infrared radiation, also referred to as thermal radiation, which is a part of electromagnetic radiation in the wavelength range from 0.76 μm to 1000 μm. The wavelengths of infrared radiation are longer than the wavelengths of visible light. The field of infrared radiation also includes industrial heat sources such as iron or steel production [11–13].

Infrared radiation is divided into several ranges according to the size of the wavelengths, and their exact values vary depending on the reference. Table 1 gives an overview of the different wavelength ranges according to multiple references [11,13].

Table 1. Wavelength ranges by multiple references.

Wavelength Range	**Abbrevation**	**Value of Wavelengths by Reference (μm)**			
		Usamentiaga et al., 2014 [11]	**InfraTec [14]**	**RP Photonics Encyclopedia [15]**	**ISO 20473 [16]**
Near-infrared	NIR	0.8–1.7	0.78–1.4	0.75–1,4	0.78–3
Short-wavelength infrared	SWIR	1–2.5	1.4–3	1,4–3	-
Mid-wavelength infrared	MWIR	2–5	3–8	3–8	3–50
Long-wavelength infrared	LWIR	8–14	8–15	8–15	-
Far-infrared	FIR	-	15–1000	15–1000	50–1000

There are other divisions; for example, according to the International Commission on Illumination (CIE) and the German Institute for Standardisation (DIN), infrared radiation is divided into 3 ranges [13,14]:

- IR-A—range from 0.78 μm to 1.4 μm
- IR-B—range from 1.4 μm to 3 μm
- IR-C—range from 3 μm to 1000 μm

There is a dependence between the radiated power and the temperature of the object, with a higher temperature resulting in a higher radiated power. Pyrometers and thermal imaging cameras are used for the non-destructive and non-contact measurement of the temperature or temperature field on the surface of an object emitting infrared radiation [11]. However, in the case of these non-contact measuring devices, for proper temperature measurement, it is necessary to know several parameters affecting the measurement, among which is the exact value of emissivity, which is an important parameter of heat transfer via the radiation of real objects [17].

Emissivity is a dimensionless parameter determining the ability of an object to emit thermal radiation. It is defined as the ratio between the radiated power of a real object and the radiated power of an absolute blackbody at the same temperature. The emissivity value ranges from 0 to 1 or rather 0 to 100% [18,19]. Based on Kirchhoff's law, the emissivity of a material is equal to the absorptivity of the material, which is shown by Equation (1) [20]:

$$\varepsilon = \alpha \tag{1}$$

where ε represents emissivity and α represents absorptivity, which represents the ratio of radiation absorbed by the surface of the object.

For the absorptivity of non-transparent materials, Equation (2) is as follows:

$$\alpha = 1 - \rho \tag{2}$$

where ρ represents reflectivity, which represents the ratio of radiation reflected from the surface of the object [20].

The blackbody is an ideal emitter and represents 100% of the radiated power or intensity of thermal radiation [21]. The emissivity of the blackbody is equal to 1, which represents the maximum achievable value. In the case of a blackbody, it is assumed that at a steady temperature, all the absorbed radiation is emitted [11].

Emissivity has a significant impact on non-contact measurements by pyrometers or thermal imaging cameras. If the emissivity value set into the pyrometer or thermal imaging camera is different from that of the real measured object, then such a measurement would be accompanied by significant errors, which would distort, for example, the temperature measurement [22,23]. The correct emissivity value is also important for thermal engineering calculations and for creating mathematical models and process simulations. The current literature, scientific publications and other available resources provide emissivity values for various materials, including steel, with different types of steel having different emissivity values. In some cases, the emissivity value is different even for the same type of steel, or the emissivity value varies from steel strip to steel strip. As a result, there is a problem in selecting a suitable emissivity value and consequently, deviations arise in thermographic measurements and also in thermal engineering calculations and simulations. The emissivity of steel is affected by temperature, measurement wavelength range, chemical composition, surface roughness, oxidation, steel surface contamination, measurement angle and surface shape [24,25]. Another problem is that a large number of sources reporting the emissivity value for steel do not indicate under which conditions this value is valid, or the wavelength range and even the temperature are not specified. Steels with high chromium content mostly have a lower emissivity value than other types of steel; this is due to the protective chromium oxide layer [26].

The determination of emissivity using numerical simulations is very problematic and experimental measurement appears to be the most appropriate method for its determination [27]. The emissivity value of steel increases with increasing temperature, while the dependence in this case is not linear; this is evident from the results of experimental measurements reported in the scientific publications of the authors Roger et al., 1979 [28]; Wen, 2010 [26]; Wen, 2011 [10]; Woods et al., 2014 [29]; Zhang et al., 2015 [23]; Zareba et al., 2016 [30]; Deus et al., 2020 [31]; and Zhao et al., 2023 [32]. The emissivity of steel decreases with increasing wavelengths, and even in this case, it is not a linear dependence, as shown by the data in the scientific publications of the authors Wen, 2010 [26]; Wen, 2011 [10]; Xing et al., 2015 [33]; Fukuyama et al., 2019 [34]; Suleiman et al., 2021 [8]; Li et al., 2022 [35]; and Li et al., 2023 [36].

Table 2 shows the emissivity values of stainless steel at different wavelengths and temperatures.

Table 2. Emissivity of stainless steel.

Wavelength (µm)	Temperature (°C)	Emissivity	Reference
1	-	0.350	[37]
1	500	0.436	[23]
1	700	0.651	[23]
1.6	-	0.250	[37]
2	500	0.424	[23]
2	700	0.605	[23]
8	−60	0.131	[38]
10	−60	0.149	[38]
12	−60	0.125	[38]
14	−60	0.112	[38]
8–14	-	0.100	[37]

The aim of this research is as follows:

- Determination of the method of measuring the emissivity of steel strips;
- Experimental measurement of the emissivity of steel strips, subsequent processing and evaluation of the measurement data.

2. Materials and Methods

The material for this experimental measurement was a polished steel strip used in continuous annealing. The surface of the polished steel strip has greater reflectivity than ordinary steel strips. Five types of this steel strip were selected for the experimental measurement, and the differences in the parameters of each type of strip were in thickness and quality. The parameters of the selected steel strip types are written in Table 3.

Table 3. Parameters of polished steel strip.

Variable	Steel Strip 1	Steel Strip 2	Steel Strip 3	Steel Strip 4	Steel Strip 5
Steel strip grade	581	589	582	549	549
Steel strip thickness (mm)	0.48	0.47	1.38	0.52	0.80

Equipment and software used during the experimental measurements:

- Electric laboratory muffle furnace VEB Elektro Bad Frankenhausen LM 212.11: AREKO s.r.o., Bratislava, Slovakia, temperature range of operation 0 to 1200 °C, dimensions 170 × 270 × 90 mm;
- Thermal imaging camera Testo 868: K-Test s.r.o., Košice, Slovakia, infrared resolution 160 × 120 pixels, spectral range 7.5 to 14 µm, measuring range 0 to +650 °C, accuracy ±2 °C, ±2% of measured value, emissivity 0.01 to 1;
- Thermocouple type K: Meratex s.r.o., Košice, Slovakia, 3 thermocouples for measuring the surface temperature of the polished steel strip, 1 thermocouple for measuring the temperature inside the furnace;
- Data logger Ahlborn ALMEMO 5690-1: AREKO s.r.o., Bratislava, Slovakia, for measured data acquisition;
- Thermo Hygrometer Beurer HM 16: NAY a.s., Bratislava, Slovakia, for measuring the ambient temperature and relative humidity around the laboratory furnace;
- Software ALMEMO control: version 5.20, for the read-out of the data logger;
- Software testo IRSoft: version 5.0, for processing thermal images.

The experimental measurement method was based on heating a polished steel strip to the required temperature of 100–300 °C and measuring the temperature of this strip using thermocouples and a thermal imaging camera. On both measuring devices, it was necessary to reach the same temperature value of the steel strip. The thermocouples were attached to the steel strip or touched it during the whole measurement and were therefore considered as an objective indication of the temperature of the steel strip. On the other hand, when the temperature was measured using a thermal imaging camera, it was possible to measure different temperature values of the steel strip. In order for the thermal imaging camera to be able to measure the same temperature of the steel strip as the thermocouples measured, it was necessary to set the correct emissivity value on the thermal imaging camera. The procedure of this measurement methodology was described in scientific publications by Rakrueangdet et al. 2016 [39] and Zhu et al. 2017 [40]. Since the adjustment of the emissivity to the thermal imaging camera in the real-time measurement was time-consuming, the last part of the measurement methodology was modified. During the experimental measurements, thermal images were taken of the steel strip and the emissivity determination was performed when these thermal images were processed using the software. In this way, the emissivity value for each type of steel strip was determined. This was followed by data processing and evaluation.

A steel strip temperature of 300 °C was chosen because the emissivity of steel increases with oxidation and the experimental measurement conditions did not have the protective atmosphere normally found in a continuous annealing line. At higher temperatures, the strip measurement would have been affected just by the oxidation of the steel strip surface, which was undesirable for this experimental measurement.

2.1. Procedure of Experimental Measurement

In the first step of each of the experimental measurements of this research, a data logger with a computer was run and the temperatures in the furnace and the temperatures on the surface of the steel strip were measured before the actual heating. Since no other measurements were made on this furnace outside of the experimental measurements, this first temperature measurement took approximately the same temperature value on all thermocouples, which represented the ambient temperature. The next step was to close the furnace door and start the electric laboratory furnace and set the heating to 130 °C. Thermocouples measured and recorded the change in temperature inside the furnace and on the surface of the strip the entire time, and these temperature changes showed that shortly after the heating was started, the temperature inside the furnace began to rise, as did the temperature on the steel strip. After the time required for heating had elapsed, the furnace automatically stopped heating the spirals, and these spirals, heated to a high temperature, continued to transfer heat to the furnace space, the steel strip, and the furnace walls. The temperature of the furnace space reached its maximum temperature of 130 °C and thereafter thermal exchange took place between the furnace space, the steel strip and the furnace walls until the temperatures were equalized.

This first part of the measurement procedure is illustrated in Figure 1.

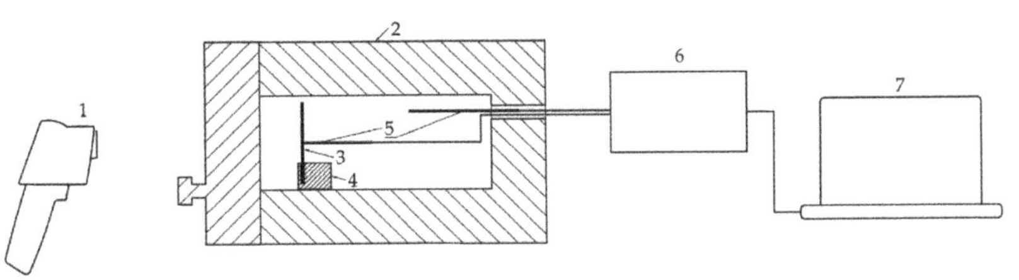

Figure 1. Schematic of the first part of the experimental measurement. (1) thermal imaging camera; (2) closed laboratory furnace; (3) polished steel strip; (4) bracket; (5) thermocouples; (6) data logger; (7) computer with temperature measurement software.

After the temperatures were equalized, or after the heat exchange between the furnace space, the steel strip and the furnace walls was completed, their temperature began to slowly decrease. When the temperature of the steel strip reached a value close to 100 °C, the furnace door was opened, and thermal images of the steel strip were taken with a thermal imaging camera; at the same time, the humidity and ambient temperature were recorded. The thermal imaging camera was placed on a tripod at a distance of 32 cm from the polished steel strip.

After the 100 °C thermal images were taken, two more heating periods followed, and thermal images of the steel strip were also taken at 200 °C and 300 °C. For each of these three temperatures, 3–5 thermal images were taken, which took 15–60 s to capture. At the end of the whole measurement, all the thermal images were transferred to the computer for processing. The measured temperatures from the data logger and the ambient temperatures were also stored.

Figure 2 shows a schematic of the second part of the measurement procedure.

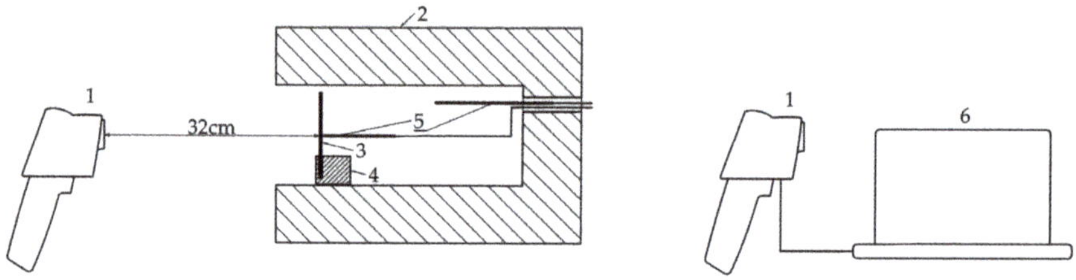

Figure 2. Schematic of the second part of the experimental measurement. (1) thermal imaging camera; (2) open laboratory furnace; (3) polished steel strip; (4) bracket; (5) thermocouples; (6) computer with software for processing of thermal images.

Figure 3 shows the time dependence of the temperature measured by the thermocouples inside the furnace and on the surface of the steel strip throughout the measurement. This figure is also divided into sections according to the points indicated in the procedure of experimental measurement.

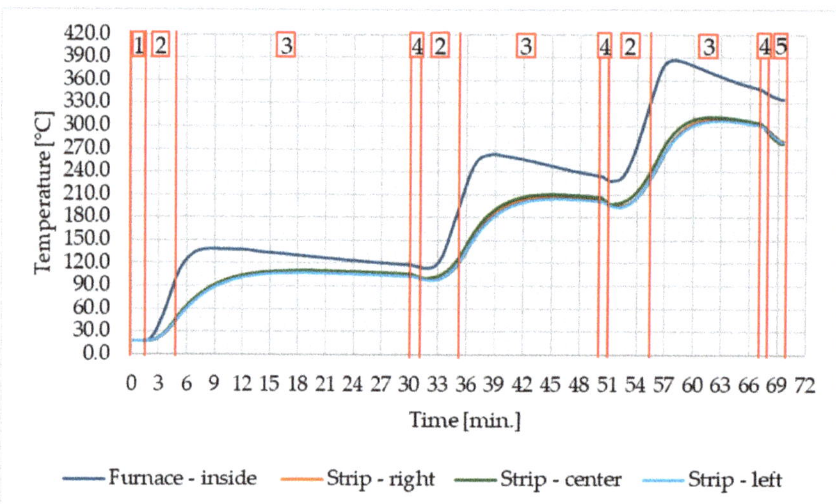

Figure 3. Temperatures in the furnace and on the steel strip throughout the measurement. 1. Switching on the data logger—recording the temperatures before heating; 2. closing the furnace door—starting the heating in the furnace; 3. switching off the furnace—transferring the heat to the temperature equilibration; 4. opening the furnace—taking thermal images; 5. leaving the door open and switching off the data logger.

From Figure 3, it can be seen that the entire run of one of the measurements took 69 min, with the most time-consuming of the individual sections being the sections marked with the number 3, which shows the heat transfer inside the furnace. At the beginning of these sections, there is a significant temperature difference between the thermocouple measuring the temperature inside the furnace and the thermocouples measuring the temperature on the surface of the steel strip. The difference in these temperatures gradually decreases, but by the end of Section 3, these temperatures have ceased to equalize and all begin to decrease. On this basis, it was considered that there was a steady state in the furnace and the thermal images were taken.

Figure 4 shows the dimensions of the steel strip, which were determined based on the dimensions of the laboratory furnace.

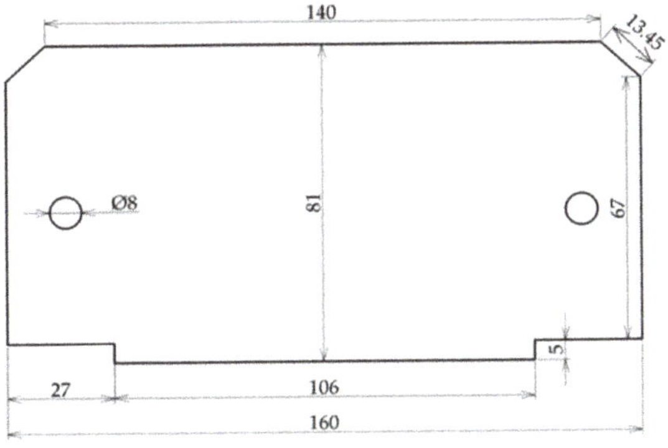

Figure 4. Dimensions of polished steel strip.

Figure 5 shows the actual shape of the polished steel strip and the location of the thermocouples measuring the temperature of this strip. The dimensions of the actual steel strip correspond to those shown in Figure 4 with an error of 1–2 mm due to the machining of the strip. The thermocouples were attached to the strip using screws placed in holes in the edges of the steel strip, and the thermocouple in the center of the strip was attached using a bracket.

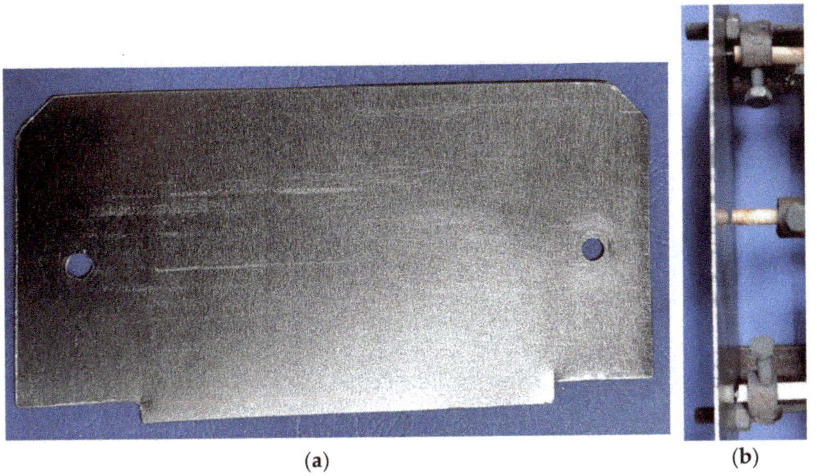

(a) (b)

Figure 5. Polished steel strip. (**a**) Actual shape; (**b**) thermocouple location.

Thermocouples attached to the steel strip had no protection and thus could be affected by the temperature of the hot air in the furnace. However, since a steady state condition was considered in the furnace just before the thermal images were taken, these thermocouple inaccuracies were neglected.

Figure 6 shows the placement of a polished steel strip in a laboratory furnace and the process of taking thermal images of this strip using a thermal imaging camera.

(**a**) (**b**)

Figure 6. Taking thermal images of the steel strip. (**a**) Placement of the strip in the furnace area; (**b**) thermal imaging camera.

2.2. Data Processing

One part of the data processing was to calculate the mean temperature of the steel strip as measured by the three thermocouples. The IRSoft software version 5.0 was used to process the thermal images, a screenshot of which is shown in Figure 7.

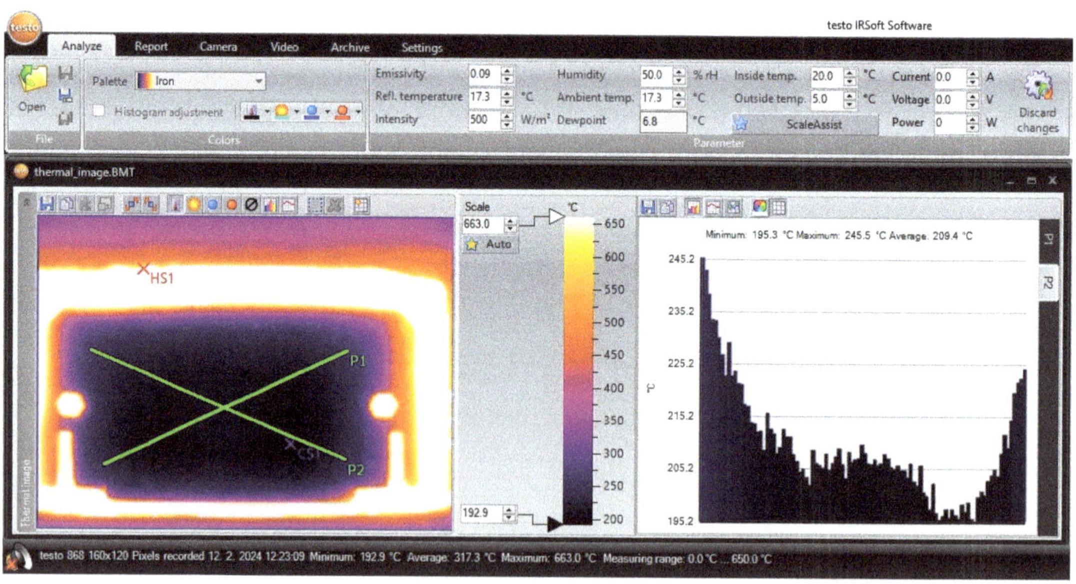

Figure 7. IRSoft software for processing thermal images from a thermal imaging camera.

In this software, two diagonals were drawn across the steel strip surface showing the temperature progression of the steel strip along the length of the diagonals, and the temperature values from these diagonals were displayed on a graph in a side window of the IRSoft software. This graph showed the minimum, maximum and average temperature values of the steel strip. The average temperature of the steel strip in the IRSoft software was compared to the mean temperature measured by the thermocouples at the corresponding date and time when the thermal image was taken. The emissivity value was then determined via the emissivity window so that the average temperatures from the software and the thermocouples were the same or the difference was as small as possible. In most cases, it happened to be that two emissivity values were assigned to the average temperature of the steel strip. When adjusting the emissivity value in the software, the average temperature displayed in the software graph was in one case lower and in another case higher than the mean temperature measured by the thermocouples, and in both cases, the deviation was of a similar value.

During the imaging of a polished steel strip with a thermal imaging camera, it was possible to observe traces of infrared radiation on the surface of this strip from surrounding objects in the room with the door of the electric laboratory furnace open. These traces of infrared radiation were therefore also transferred to the thermal images, thus creating a temperature distortion on the steel strip. Corrections of such distorted temperatures were made just with the IRSoft software through the determination of the temperature value in the window reflected temperature. At the time of the experimental measurement, no activities were performed on the surrounding objects in the room and thus these objects had a temperature close to the ambient temperature value, so that in the processing of the thermal images, the ambient temperature was also considered as the reflected temperature at the same time.

In Figure 7, a weak temperature homogeneity of the steel strip can be observed. In the middle of the strip, the temperatures are lower and, on the other hand, at the edges of the steel strip, the temperatures reach the maximum values. It is assumed that this is the reflected temperature, while in the middle of the steel strip, the temperature from the room is reflected, and at the edges of the strip, the temperature from the front of the furnace walls is reflected. Since the exact value of the temperature reflected from the front of the furnace was not known, the diagonals were drawn mainly in the middle part of the strip where the reflected ambient temperature was considered in the processing of the thermal images.

During the continuous processing of the thermal images using IRSoft software, it was found that the measured humidity had no effect on the temperature change of the steel strip and therefore the humidity change was neglected when further thermal images were processed.

3. Results

The emissivity values from IRSoft software, the average temperatures of the steel strip measured by thermocouples and the ambient temperature values were processed by descriptive statistics. Specifically, the number of measurements, mean, standard deviation, minimum value, maximum value, and median were determined. These results from descriptive statistics are written in Tables 4–8.

The following figures show the time dependence of the temperature or the temperature decrease of the steel strip during the thermal image capture with an open furnace door. For each type of strip, the temperature dependences of four measurements are shown, which deal with the temperature decrease from the opening of the furnace door or from the taking of the first thermal image to the taking of the last thermal image of that measurement. The duration of these dependencies is in the range of 15–60 s, with the resulting time value depending on the specific measurement, as different numbers of thermal images were taken for each measurement. The first of these figures is Figure 8.

The following figures marked with "a" show the dependence between the emissivity values from IRSoft software and the mean temperatures of the steel strip as measured by the thermocouples. The following figures marked with "b" show the dependence of the mean emissivity value and the grand average, or rather the mean of several means, of the steel strip temperature measured by the thermocouples. The first of these is Figure 9.

Figure 8 shows the temperature drop of steel strip 1 during thermal imaging.

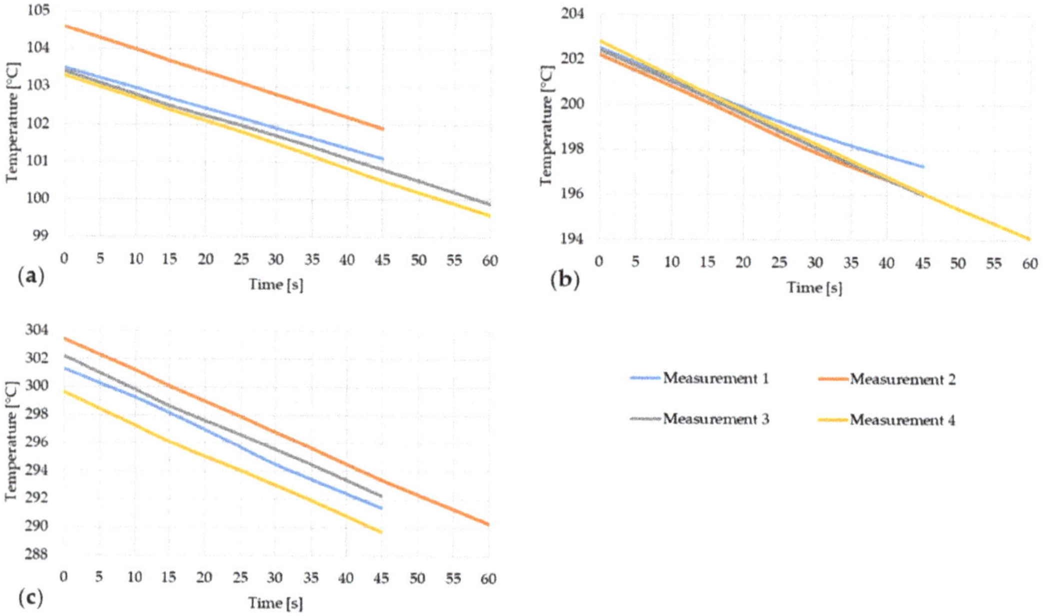

Figure 8. Temperature decrease of the steel strip during thermal image capture with the furnace door open for four measurements of the steel strip 1. (**a**) t ~ 100 °C; (**b**) t ~ 200 °C; (**c**) t ~ 300 °C.

Table 4. Descriptive statistics of measured data of steel strip 1. t1 ~ 100 °C; t2 ~ 200 °C; t3 ~ 300 °C.

Variable	*n*	Mean	Standard Deviation	Min	Max	Median
Steel strip emissivity t1	85	0.094	0.015	0.07	0.13	0.09
Steel strip temperature t1 (°C)	85	102.900	2.560	99.30	109.10	102.50
Ambient/reflected temperature t1 (°C)	85	17.480	1.589	13.80	18.80	18.00
Steel strip emissivity t2	85	0.098	0.009	0.08	0.12	0.10
Steel strip temperature t2 (°C)	85	195.297	6.372	178.70	203.95	196.90
Ambient/reflected temperature t2 (°C)	85	17.032	1.804	13.80	18.80	18.00
Steel strip emissivity t3	85	0.102	0.007	0.09	0.12	0.10
Steel strip temperature t3 (°C)	85	296.677	3.998	285.20	303.60	296.95
Ambient/reflected temperature t3 (°C)	85	17.103	1.756	13.80	18.80	18.00

Figure 9 shows the dependence of emissivity and temperature of steel strip 1.

In the graph, or rather in Figure 9a, a large range of measured emissivity values of 0.07–0.13 can be observed at 100 °C. This range decreases as the steel strip temperature increases, and at a strip temperature of 300 °C, the measured emissivity value is in the range of 0.09–0.12. The lower limit of the measured emissivity value increased with increasing temperature, and conversely, the higher limit of the measured emissivity decreased. Based on the literature review, it is assumed that this measurement instability at lower temperatures is due to the lower radiated power at which the thermal imaging camera was unable to measure with the necessary accuracy. As the temperature increased, the radiated power increased and thus the ability of the thermal imaging camera to measure a more accurate emissivity value increased. A similar trend in the measured emissivity values can be observed in the other graphs or figures marked with "a".

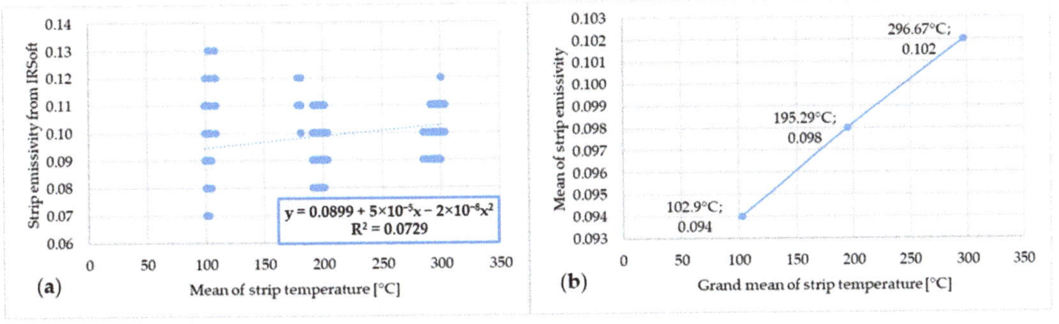

Figure 9. (**a**) Dependence of emissivity values from IRSoft software and mean temperature of steel strip 1; (**b**) dependence of mean emissivity value and grand mean temperature of steel strip 1.

Figure 10 shows the temperature drop of steel strip 2 during thermal imaging.

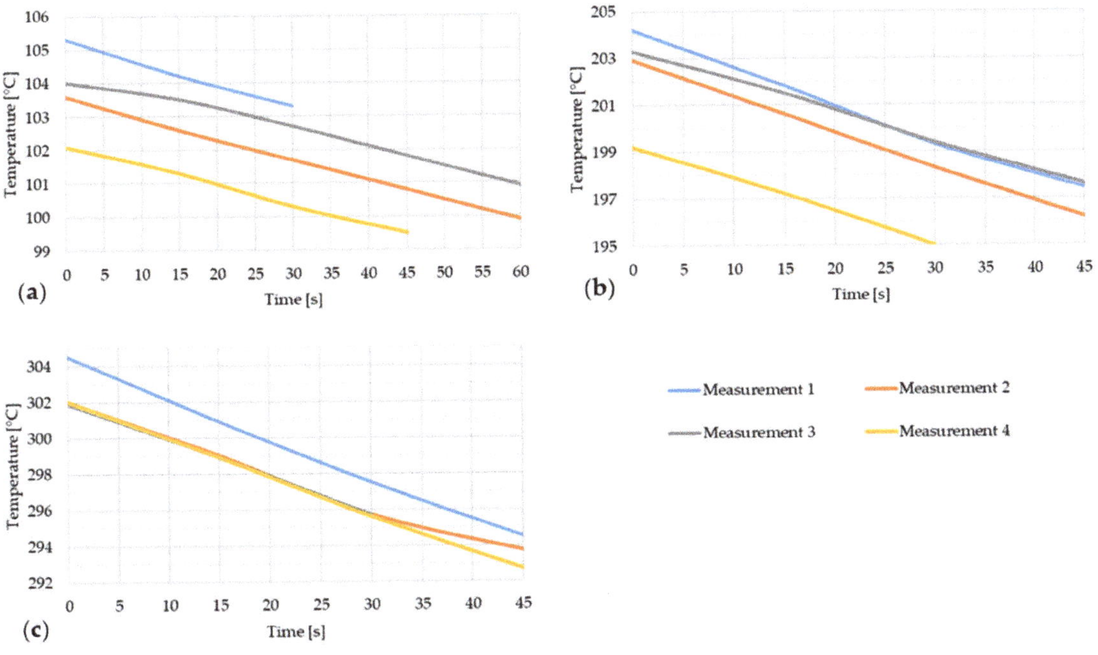

Figure 10. Temperature decrease of the steel strip during thermal image capture with the furnace door open for four measurements of the steel strip 2. (a) t ~ 100 °C; (b) t ~ 200 °C; (c) t ~ 300 °C.

Table 5. Descriptive statistics of measured data of steel strip 2. t1 ~ 100 °C; t2 ~ 200 °C; t3 ~ 300 °C.

Variable	n	Mean	Standard Deviation	Min	Max	Median
Steel strip emissivity t1	75	0.086	0.013	0.06	0.11	0.09
Steel strip temperature t1 (°C)	75	102.230	2.679	97.15	107.00	102.10
Ambient/reflected temperature t1 (°C)	75	17.966	1.195	14.30	18.90	17.90
Steel strip emissivity t2	75	0.088	0.008	0.07	0.11	0.09
Steel strip temperature t2 (°C)	75	197.614	3.567	189.55	202.75	198.15
Ambient/reflected temperature t2 (°C)	75	18.009	1.172	14.40	18.90	18.60
Steel strip emissivity t3	75	0.092	0.007	0.08	0.11	0.09
Steel strip temperature t3 (°C)	75	298.262	2.244	294.05	302.65	297.95
Ambient/reflected temperature t3 (°C)	75	17.860	1.323	14.30	18.90	17.90

Figure 11 shows the dependence of emissivity and temperature of steel strip 2.

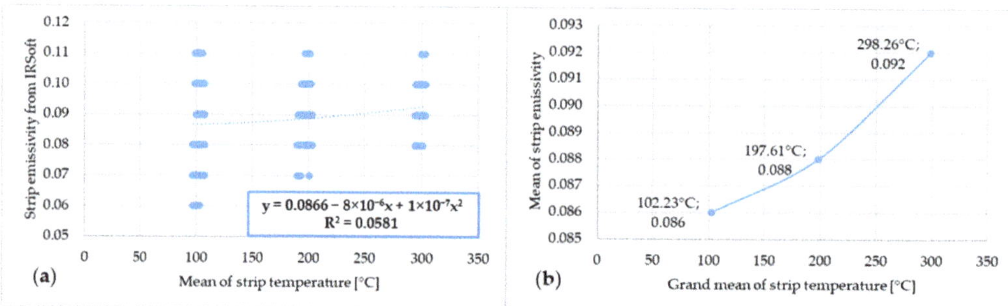

Figure 11. (a) Dependence of emissivity values from IRSoft software and mean temperature of steel strip 2; (b) dependence of mean emissivity value and grand mean temperature of steel strip 2.

Figure 12 shows the temperature drop of steel strip 3 during thermal imaging.

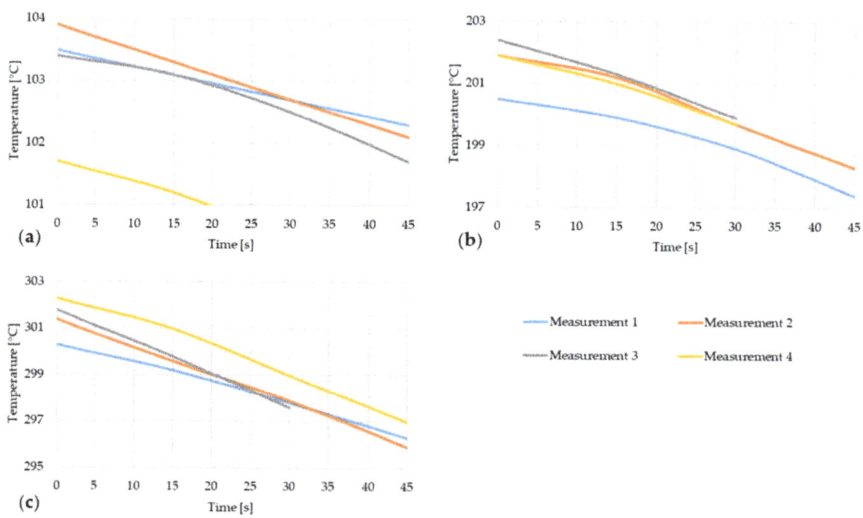

Figure 12. Temperature decrease of the steel strip during thermal image capture with the furnace door open for four measurements of the steel strip 3. (a) t ~ 100 °C; (b) t ~ 200 °C; (c) t ~ 300 °C.

Table 6. Descriptive statistics of measured data of steel strip 3. t1 ~ 100 °C; t2 ~ 200 °C; t3 ~ 300 °C.

Variable	n	Mean	Standard Deviation	Min	Max	Median
Steel strip emissivity t1	80	0.100	0.013	0.08	0.13	0.10
Steel strip temperature t1 (°C)	80	103.353	2.536	98.10	106.85	103.30
Ambient/reflected temperature t1 (°C)	80	19.711	0.356	18.90	20.00	19.90
Steel strip emissivity t2	80	0.108	0.010	0.09	0.13	0.11
Steel strip temperature t2 (°C)	80	201.400	2.411	197.30	206.45	200.57
Ambient/reflected temperature t2 (°C)	80	19.742	0.331	18.90	20.00	19.90
Steel strip emissivity t3	80	0.114	0.008	0.09	0.13	0.11
Steel strip temperature t3 (°C)	80	300.365	5.194	287.95	306.15	301.50
Ambient/reflected temperature t3 (°C)	80	19.707	0.355	18.90	20.00	19.90

Figure 13 shows the dependence of emissivity and temperature of steel strip 3.

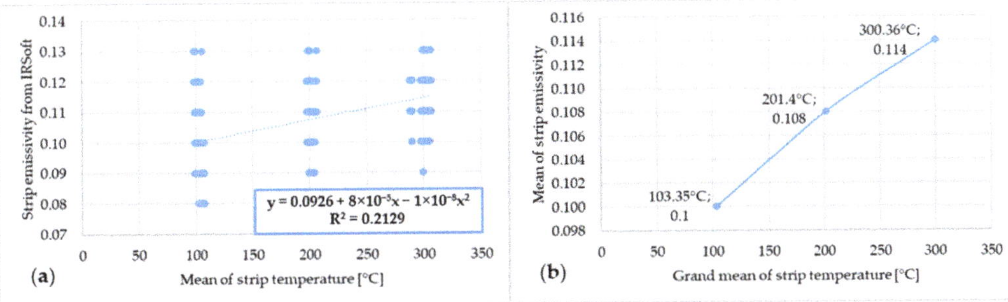

Figure 13. (**a**) Dependence of emissivity values from IRSoft software and mean temperature of steel strip 3; (**b**) dependence of mean emissivity value and grand mean temperature of steel strip 3.

Figure 14 shows the temperature drop of steel strip 4 during thermal imaging.

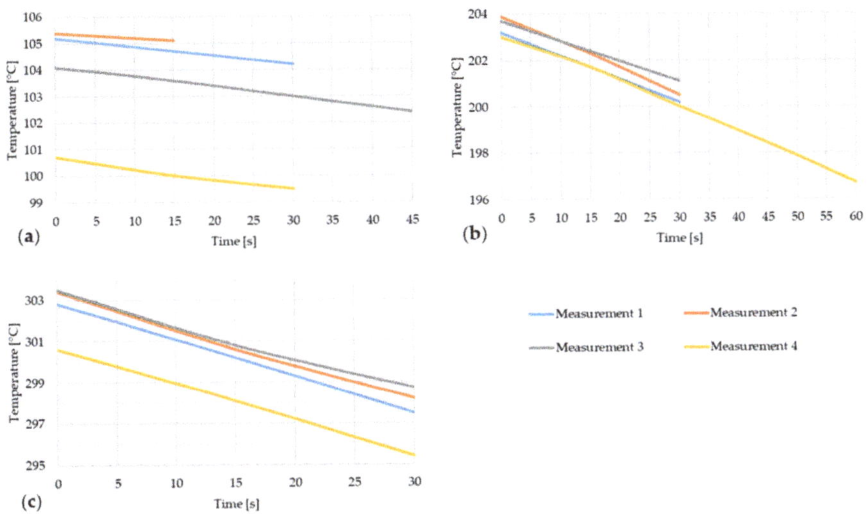

Figure 14. Temperature decrease of the steel strip during thermal image capture with the furnace door open for four measurements of the steel strip 4. (a) t ~ 100 °C; (b) t ~ 200 °C; (c) t ~ 300 °C.

Table 7. Descriptive statistics of measured data of steel strip 4. t1 ~ 100 °C; t2 ~ 200 °C; t3 ~ 300 °C.

Variable	*n*	Mean	Standard Deviation	Min	Max	Median
Steel strip emissivity t1	65	0.083	0.011	0.06	0.11	0.08
Steel strip temperature t1 (°C)	65	103.934	2.122	99.10	107.65	104.20
Ambient/reflected temperature t1 (°C)	65	16.661	1.053	14.90	18.70	17.30
Steel strip emissivity t2	65	0.087	0.008	0.07	0.10	0.09
Steel strip temperature t2 (°C)	65	202.150	2.201	197.00	206.45	201.95
Ambient/reflected temperature t2 (°C)	65	16.967	1.037	14.90	18.70	17.30
Steel strip emissivity t3	65	0.090	0.007	0.08	0.11	0.09
Steel strip temperature t3 (°C)	65	299.816	2.484	294.90	303.40	299.70
Ambient/reflected temperature t3 (°C)	65	16.967	1.070	14.90	18.70	17.30

Figure 15 shows the dependence of emissivity and temperature of steel strip 4.

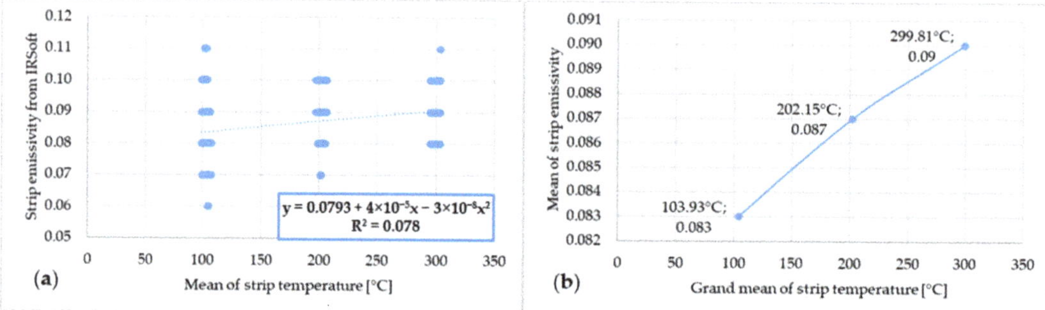

Figure 15. (**a**) Dependence of emissivity values from IRSoft software and mean temperature of steel strip 4; (**b**) dependence of mean emissivity value and grand mean temperature of steel strip 4.

Figure 16 shows the temperature drop of steel strip 5 during thermal imaging.

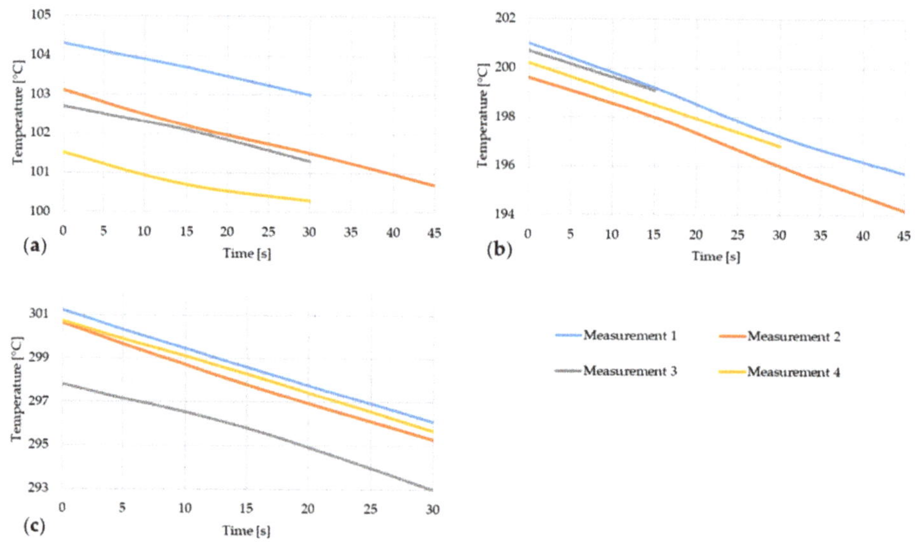

Figure 16. Temperature decrease of the steel strip during thermal image capture with the furnace door open for four measurements of the steel strip 5.

Table 8. Descriptive statistics of measured data of steel strip 5. t1 ~ 100 °C; t2 ~ 200 °C; t3 ~ 300 °C.

Variable	*n*	Mean	Standard Deviation	Min	Max	Median
Steel strip emissivity t1	60	0.086	0.012	0.06	0.11	0.09
Steel strip temperature t1 (°C)	60	103.315	2.309	98.10	107.60	103.42
Ambient/reflected temperature t1 (°C)	60	18.850	0.508	18.20	19.50	19.10
Steel strip emissivity t2	60	0.089	0.009	0.07	0.11	0.09
Steel strip temperature t2 (°C)	60	199.380	2.557	194.45	204.00	199.70
Ambient/reflected temperature t2 (°C)	60	18.820	0.505	18.20	19.50	19.10
Steel strip emissivity t3	60	0.094	0.005	0.08	0.11	0.09
Steel strip temperature t3 (°C)	60	296.976	2.092	292.85	300.15	297.05
Ambient/reflected temperature t3 (°C)	60	18.826	0.487	18.20	19.50	19.10

Figure 17 shows the dependence of emissivity and temperature of steel strip 5.

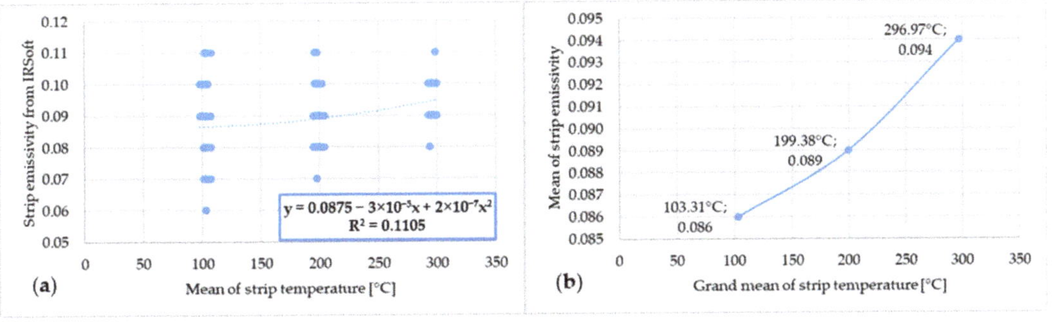

Figure 17. (**a**) Dependence of emissivity values from IRSoft software and mean temperature of steel strip 5; (**b**) dependence of mean emissivity value and grand mean temperature of steel strip 5.

4. Discussion

The measured emissivity values from this experimental measurement are constrained to thermal imaging camera wavelengths in the range 7.5–14 μm, which corresponds to the long wavelength range. The processing of the measured data using descriptive statistics yielded the following results:

- The mean temperature of the steel strip measured by the thermocouples is in the range 102.23–300.36 °C.
- The range of the mean ambient temperature is 16.66–19.74 °C.
- The mean emissivity value of the five steel strip types is in the range 0.083–0.114, with the lowest mean emissivity value of 0.083 with a standard deviation of 0.011 belonging to steel strip 4 and the highest mean emissivity value of 0.114 with a standard deviation of 0.008 from the measurements of steel strip 3. In terms of the general emissivity range of 0–1, or the range 0–100%, the emissivity results of the steel strip in this experimental measurement are near its lower limit, which represents a significantly low value.

In the figures showing the dependence of emissivity and temperature it is possible to see the increase in emissivity value with increasing temperature of the steel strip. This dependence between emissivity and steel strip temperature is not linear but polynomial. These results correspond to the results of the scientific publications presented in the introduction chapter of this article.

4.1. Emissivity Difference of the Same Grade of Steel

In Tables 4–8, different emissivity values can be observed between all types of steel strip. This is mainly due to the different steel grades of these strips, and the thickness of the strips also has some influence.

Steel strip 4 and steel strip 5 have the same grade value, so that in their case, the difference in emissivity is only due to the thickness of the strip. The difference in their emissivity at 100 °C is 0.003, at 200 °C it is 0.0016 and at 300 °C the difference in their emissivity is 0.0043. In all three cases, the higher emissivity value belongs to steel strip 5, which is 0.028 mm thicker than steel strip 4. From this comparison, it is clear that the strip with the greater thickness achieves a higher emissivity value than the thinner strip at the same grade and temperature. It is assumed that for the thicker strip to reach the same temperature as the thinner strip, it must accept a greater amount of heat. The same must also be true in reverse, so that when the furnace door was opened, the thicker strip cooled slower, and at the time when the thermal images were taken, it may have been at a slightly higher temperature than the thinner strip. However, in this case, there is no significant difference in emissivity.

4.2. Comparison of References and This Research

A common feature of the references and this research is the same basis of measurement methodology. Specifically, this research conducted the search for the same temperature of the object under investigation by using thermocouples and at the same time using a thermal imaging camera, whereby the same temperature can be achieved just by setting the appropriate emissivity value in the thermal imaging camera.

The difference is, for example, that in the case of the sources investigated, the emissivity measurements were in most cases made in the short wavelength range, whereas under the conditions of this research, the emissivity was measured in the long wavelength range. Another difference is that some sources measured the emissivity during the oxidation of the steel, but in this research, the oxidation of the steel strip was unacceptable, and the measurement was performed without oxidation. Under the conditions of this research, thermal images were taken during the measurement and processed as follows; this method was not presented in the sources reviewed. At the same time, during the search for sources, it was found that not many scientific sources deal with the issue of polished steel strips, and this is where the contribution of this research lies, as the results of the emissivity values of these strips are presented in this article.

5. Conclusions

Steel strips have long been an important input semi-finished product for several industrial areas; therefore, it is necessary to improve their chemical and mechanical properties, reduce their thickness and weight, expand the range of new types of steel strips and increase their production. In order to meet these requirements, the steel strip has to go through a large number of technological operations in which the accuracy of temperature measurement and control is crucial. In industrial plants where the steel strip is in constant motion, it is practically impossible to measure temperature using contact measuring devices. When using non-contact measuring devices such as pyrometers or thermal imaging cameras, the unknown emissivity value of the materials being measured is a problem, as setting the wrong emissivity value to the measuring device results in distorted temperature measurement data. Under the conditions of this research, the method of measurement was established and subsequently experimental measurements of the emissivity of a polished steel strip used in a continuous annealing line were carried out. This was followed by the processing of the data from these measurements and their evaluation. The experimental measurement of emissivity was carried out for five types of steel strips of different grades and thicknesses, the measurement itself being carried out in the temperature range 100–300 °C. Thermal images for emissivity determination were captured by a thermal imaging camera operating in the spectral range 7.5–14 μm, which represents the long wavelength range. After processing the measured data using descriptive statistics, the mean emissivity values ranged from 0.0835 to 0.1143 depending on the type of steel strip. The experimental results showed that the emissivity value of the steel strip increased as the temperature measured on the surface of this strip increased. The dependence between emissivity and strip temperature was not linear. In the case of actual production, the resulting emissivity values from this research could be applied to measuring equipment of the appropriate wavelength, which could improve the accuracy of temperature measurement of polished steel strips.

Author Contributions: Conceptualization, G.J., A.V. and J.K.; supervision, G.J., A.V. and J.K.; resources, G.J., R.D. and J.K.; methodology, Š.S. and G.J.; investigation, Š.S. and R.D.; formal analysis, Š.S., A.V. and R.D.; validation, Š.S., A.V. and R.D.; writing—original draft preparation, Š.S.; writing—review and editing, G.J. and J.K. All authors have read and agreed to the published version of the manuscript.

Funding: This research received no external funding.

Institutional Review Board Statement: Not applicable.

Informed Consent Statement: Not applicable.

Data Availability Statement: The original contributions presented in the study are included in the article, further inquiries can be directed to the corresponding author.

Conflicts of Interest: The authors declare no conflicts of interest.

References

1. Tang, B.; Kong, J.Y.; Wang, X.D.; Chen, L. Surface inspection system of steel strip based on machine vision. In Proceedings of the 2009 First International Workshop on Database Technology and Applications, Wuhan, China, 25–26 April 2009; pp. 359–362. [CrossRef]
2. Hao, Z.; Ren, F.; Kang, X.; Ni, H.; Lv, S.; Wang, H. Classification of steel strip surface defects based on optimized ResNet18. In Proceedings of the 2021 IEEE International Conference on Agents (ICA), Kyoto, Japan, 13–15 December 2021; pp. 61–62. [CrossRef]
3. Kang, Z.W.; Chen, T.C. Longitudinal residual warpage of strip in continuous annealing line. *Appl. Therm. Eng.* **2014**, *62*, 1–12. [CrossRef]
4. Chen, K.; Dai, M.; Xu, L.; Xu, S.; Xie, X.; Hu, X.; Huang, F.; Zhang, H. Inverse parameter identifications and forward strip temperature simulations of the continuous annealing line with physics-informed neural network and operation big data. *Eng. Appl. Artif. Intell.* **2024**, *127*, 107307. [CrossRef]
5. Kim, Y.; Lee, J.; Shin, K.S.; Jeon, S.H.; Chin, K.G. Effect of dew point on the formation of surface oxides of twinning-induced plasticity steel. *Mater. Charact.* **2014**, *89*, 138–145. [CrossRef]
6. Strommer, S.; Niederer, M.; Steinboeck, A.; Jadachowski, L.; Kugi, A. Nonlinear observer for temperatures and emissivities in a strip annealing furnace. *IEEE Trans. Ind. Appl.* **2017**, *53*, 2578–2586. [CrossRef]
7. Kang, Z.W.; Chen, T.C. Three-dimensional temperature distributions of strip in continuous annealing line. *Appl. Therm. Eng.* **2013**, *58*, 241–251. [CrossRef]
8. Suleiman, F.K.; Lin, K.; Daun, K.J. Development of a multivariate spectral emissivity model for an advanced high strength steel alloy through factorial design-of-experiments. *J. Quant. Spectrosc. Radiat. Transf.* **2021**, *271*, 107693. [CrossRef]
9. Lin, K.; Pourmajidian, M.; Suleiman, F.K.; McDermid, J.R.; Daun, K.J. Interpreting the radiative properties of advanced high strength steel strip using a hybrid thin film/geometric optics model. *J. Quant. Spectrosc. Radiat. Transf.* **2022**, *277*, 107963. [CrossRef]
10. Wen, C.D. Study of steel emissivity characteristics and application of multispectral radiation thermometry (MRT). *J. Mater. Eng. Perform.* **2011**, *20*, 289–297. [CrossRef]
11. Usamentiaga, R.; Venegas, P.; Guerediaga, J.; Vega, L.; Molleda, J.; Bulnes, F.G. Infrared thermography for temperature measurement and non-destructive testing. *Sensors* **2014**, *14*, 12305–12348. [CrossRef]
12. Schieke, S.M.; Schroeder, P.; Krutmann, J. Cutaneous effects of infrared radiation: From clinical observations to molecular response mechanisms. *Photodermatol. Photoimmunol. Photomed.* **2003**, *19*, 228–234. [CrossRef]
13. ICNIRP. Infrared Radiation. Available online: https://www.icnirp.org/en/frequencies/infrared/index.html (accessed on 13 June 2024).
14. InfraTec. Infrared Radiation. Available online: https://www.infratec.eu/sensor-division/service-support/glossary/infrared-radiation/ (accessed on 13 June 2024).
15. Paschotta, R. Infrared Light. *RP Photonics Encyclopedia*. Available online: https://www.rp-photonics.com/infrared_light.html (accessed on 13 June 2024). [CrossRef]

16. *ISO 20473*; Optics and photonics—Spectral bands. International Organization for Standardization: Vernier, Switzerland, 2007.
17. Shoji, E.; Takahashi, R.; Isogai, S.; Ito, N.; Kubo, M.; Tsukada, T.; Fukuyama, H. Compositional Dependence of Normal Spectral Emissivity of Molten Cu-Fe Alloy. *Metall. Mater. Trans. B* **2019**, *50*, 2454–2458. [CrossRef]
18. Ravindra, N.M.; Sopori, B.; Gokce, O.H.; Cheng, S.X.; Shenoy, A.; Jin, L.; Abedrabbo, S.; Chen, W.; Zhang, Y. Emissivity measurements and modeling of silicon-related materials: An overview. *Int. J. Thermophys.* **2001**, *22*, 1593–1611. [CrossRef]
19. Shi, D.; Liu, Y.; Sun, J.; Zhu, Z.; Liu, H. A new experimental apparatus for measurement of spectral emissivity of opaque materials using a reflector as the dummy light source. *Int. J. Heat Mass Transf.* **2012**, *55*, 3344–3348. [CrossRef]
20. Epsilon Engineer. Radiation, Emissivity, Absorbance and Reflectance. Available online: https://www.epsilonengineer.com/radiation.html (accessed on 14 May 2024).
21. Gao, Y.; Tian, G.Y. Emissivity correction using spectrum correlation of infrared and visible images. *Sens. Actuators A Phys.* **2018**, *270*, 8–17. [CrossRef]
22. Barreira, E.; Bauer, E.; Mustelier, N.; Freitas, V. Measurement of materials emissivity—Influence of the procedure. In Proceedings of the AITA—13th International Workshop on Advanced Infrared Technology and Applications, Pisa, Italy, 29 September–2 October 2015; pp. 242–245.
23. Zhang, F.; Yu, K.; Zhang, K.; Liu, Y.; Xu, K.; Liu, Y. An emissivity measurement apparatus for near infrared spectrum. *Infrared Phys. Technol.* **2015**, *73*, 275–280. [CrossRef]
24. Avdelidis, N.P.; Moropoulou, A. Emissivity considerations in building thermography. *Energy Build.* **2003**, *35*, 663–667. [CrossRef]
25. Xing, W.; Shi, D.; Sun, J.; Zhu, Z. Emissivity model of steel 430 during the growth of oxide layer at 800–1100 K and 1.5 μm. *Infrared Phys. Technol.* **2018**, *88*, 23–31. [CrossRef]
26. Wen, C.D. Investigation of steel emissivity behaviors: Examination of Multispectral Radiation Thermometry (MRT) emissivity models. *Int. J. Heat Mass Transf.* **2010**, *53*, 2035–2043. [CrossRef]
27. Švantner, M.; Honnerová, P.; Veselý, Z. The influence of furnace wall emissivity on steel charge heating. *Infrared Phys. Technol.* **2016**, *74*, 63–71. [CrossRef]
28. Roger, C.R.; Yen, S.H.; Ramanathan, K.G. Temperature variation of total hemispherical emissivity of stainless steel AISI 304. *JOSA* **1979**, *69*, 1384–1390. [CrossRef]
29. Woods, S.I.; Jung, T.M.; Sears, D.R.; Yu, J. Emissivity of silver and stainless steel from 80 K to 300 K: Application to ITER thermal shields. *Cryogenics* **2014**, *60*, 44–48. [CrossRef]
30. Zareba, S.; Wolff, A.; Jelali, M. Mathematical modelling and parameter identification of a stainless steel annealing furnace. *Simul. Model. Pract. Theory* **2016**, *60*, 15–39. [CrossRef]
31. Deus, V.S.D.; Castro, J.A.; Correa, S.R. Correlation among the input thermal parameters and thermography measurements data of the resistance seam welding. *Mater. Res.* **2020**, *23*, e20200029. [CrossRef]
32. Zhao, Y.; Li, X.; Zhang, H.; Shen, F.; Huang, C.; Liu, H.; Qi, H.; Huang, Z.; Geng, Z.; Xin, J.; et al. Preliminary development of emissivity measurement system at low temperature based on radiometric method. *Cryogenics* **2023**, *132*, 103671. [CrossRef]
33. Xing, J.; Cui, S.; Qi, W.; Zhang, F.; Sun, X.; Sun, W. A data processing algorithm for multi-wavelength pyrometry-which does not need to assume the emissivity model in advance. *Measurement* **2015**, *67*, 92–98. [CrossRef]
34. Fukuyama, H.; Higashi, H.; Yamano, H. Thermophysical properties of molten stainless steel containing 5 mass% B4C. *Nucl. Technol.* **2019**, *205*, 1154–1163. [CrossRef]
35. Li, J.; Huang, Y.; Fu, L.; Fu, T.; Zhang, J. Measurements of spectral emissivity, reflectance and transmittance at high temperatures using laser heating and auxiliary light source. *Int. J. Heat Mass Transf.* **2022**, *183*, 122092. [CrossRef]
36. Li, L.; Ruan, W.; Wang, W.; Yu, K.; Zhang, K.; Liu, Y.; Liu, Y. Predicting the directional spectral emissivity for rough surfaces polished by sandpaper. *Tribol. Int.* **2023**, *185*, 108557. [CrossRef]
37. Fluke. Infrared Technology, Emissivity–Metals, Emissivity Values for Metals. Available online: https://www.flukeprocessinstruments.com/en-us/service-and-support/knowledge-center/infrared-technology/emissivity-metals (accessed on 14 May 2024).
38. Ma, J.; Zhang, Y.; Wu, L.; Li, H.; Song, L. An Apparatus for Spectral Emissivity Measurements of Thermal Control Materials at Low Temperatures. *Materials* **2019**, *12*, 1141. [CrossRef]
39. Rakrueangdet, K.; Nunak, N.; Suesut, T.; Sritham, E. Emissivity measurements of reflective materials using infrared thermography. In Proceedings of the International MultiConference of Engineers and Computer Scientists (IMECS), Hong Kong, China, 16–18 March 2016.
40. Zhu, W.; Shi, D.; Zhu, Z.; Sun, J. Spectral emissivity model of steel 309S during the growth of oxide layer at 800–1100 K. *Int. J. Heat Mass Transf.* **2017**, *109*, 853–861. [CrossRef]

materials MDPI

Article

Applying Binder Jetting Technology to 316L Stainless Steel Materials and Testing Its Mechanical and Dimensional Properties Depending on the Printing Method

Ján Kráľ [1], Tibor Dzuro [1,*] and Hubert Debski [2]

[1] Department of Production Technology and Robotics, Faculty of Mechanical Engineering, Technical University of Kosice, 042 00 Kosice, Slovakia; jan.kral.2@tuke.sk

[2] Department of Fundamentals of Machine Design and Mechatronic Systems, Faculty of Mechanical Engineering, Lublin University of Technology, 20-618 Lublin, Poland; h.debski@pollub.pl

* Correspondence: tibor.dzuro@tuke.sk; Tel.: +421-55-602-2673

Abstract: This article discusses special additive technologies, with a particular focus on the innovative binder jetting technology used to create three-dimensional objects. The theoretical part of this article defines the production process–its shortcomings and benefits. Also, the article describes process parameters and individual steps that must be optimally set for the desired result. Further, the article characterizes the most influential factors that are indispensable in the printing process-metallic powder, binder, printing parameters, and finishing operations after the printing itself. The conclusion of the theoretical part deals with various material possibilities when using binder jetting technology. In the practical part of the article, the properties of the material, the chemical composition, and the resulting accuracy of the printed samples will be verified experimentally. The information obtained will subsequently be used to identify an economically advantageous application of binder jetting technology.

Keywords: 3D metal printing; additive manufacturing; additive technology; binder jetting; metal parts; metallic powder; non-fused printing technology; powder metallurgy

1. Introduction

Additive technologies make it possible to produce technologically unmanufacturable complex parts from various materials while reducing steps in the production process. Some additive technologies allow both design and production with excellent dimensional accuracy from materials such as polymers, ceramics, and metal [1].

Additive technologies represent production technologies whose principle is material layering in various ways, such as extruding, welding, curing, and others. As a rule, the process begins with the design of a three-dimensional (3D) object in a digital environment using CAD (Computer-aided design) systems. Using the machine and modeling program, individual layers and paths of the tool are then defined in order to achieve the desired geometry. In general, additive technologies are divided into two groups [2]:

1. Fusion-based methods—fuse-based principle (FBP):
   - DED—Direct energy deposition;
   - LPBF—Laser powder bed fusion;
   - EBM—Electron beam melting.
2. Non-fusion-based methods—no curing:
   - Extruding;
   - Binder jetting;
   - Lamination.

Depending on the material and the desired properties, individual processes differ slightly. Postprocessing of additive technologies can include finishing operations that are necessary to achieve the required tolerances.

The advantages of layer-by-layer production of components include a high level of freedom at the design stage, uniform production times, reduction of material waste through recycling, and the ability to create geometrically complex objects in a single operation. No rebuilding of the equipment is required for the production of another part; it is simply possible to manufacture object A alongside object B—if the capacity of the machine allows it. The nature of the technology predetermines its use also for the production of unique and complex parts, or for the production of prototypes, etc. By reusing unused materials, additive technologies are also suitable in sectors where scarce materials are used, such as aerospace.

Another advantage of additive technologies is that no specific tools are required for manufacturing, whereas conventional technologies, such as machining or forging, require a certain range of tools. Studies comparing the life cycle of additive technologies (especially binder jetting) and conventional technologies have shown that in addition to shortening production times, reducing steps in the production process, and minimizing tool needs, this technology also dominates in a more favorable environmental impact [3]. The mass adoption of additive technologies is currently hampered by several factors. The first is a smaller number of materials–especially metals and ceramics suitable for additive technologies. With regard to material properties, the relationships between process parameters and resulting properties have only recently begun to be characterized, resulting in lower confidence for structural applications. In addition, due to the novelty of additive technologies, machines and processes often differ, which also does not contribute to trust in the technology. The cost of additive technologies at this time is highly volatile due to the volatile prices of energy and input materials, which predetermines the use of additive technologies only in some high-value-added industrial sectors, such as aerospace and biomedical industries [4].

One of the methods of manufacturing components by additive technologies is the use of a powder bed, in which metallic powder is dispersed into a thin layer. This applied layer is subsequently glued together or joined together using a heat source. The joined layer is lowered, and another layer of powder is applied to it. The process runs until the printing of the entire part is complete. Just as additive technologies are divided, this method of manufacturing can also be divided into fusion technologies—those that use a heat source (laser, electron beam) and technologies that use a binder to bind layers—such as binder jetting [5]. It is this method of production that uses a binder to selectively bond the material in a powder bed. Ziaee and Crane [6] presented a broad review of technologies and approaches that have been applied in binder jet printing and pointed toward opportunities for future advancement. Li et al. [7] prepared a comprehensive review of the currently available literature sources on the jetting of metal binders. They summarized critical factors and their effects in the jetting of metal binders and divided them into two categories, namely material-related factors and process-related parameters. At the same time, they presented data on density, dimensional and geometric accuracy, and mechanical properties achieved by jetting a metal binder. By optimizing the parameters and choosing the sintering process appropriately, they showed that these materials achieve a relative density of more than 90%, with several types of stainless steel obtaining equivalent or better mechanical properties compared to cold forming. The above aspects require further scientific research, which includes understanding the powder coating process, binder-powder interactions, and shrinkage of parts. Shahed and Manogharan [8] investigated the powder-binder interaction for bimodal powders with actual binder jetting conditions over a wide range of packing densities and binder jetting conditions. Using Discrete Element Modeling (DEM—powder deposition)–VOF (Volume of fluid–binder interaction), they analyzed the interaction of powder and binder in terms of penetration depth, spreading time, and area per droplet on the powder bed. Their results provide a new understanding of the spatio-temporal characteristics of the binder-powder interaction, which helps in identifying the optimal printing parameters for the bimodal powder feedstock. Roberts et al. [9] investigated the maximum size of binder nanoparticles to fill the gaps between bed particles. Their results provide information on the selection of nanoparticle sizes required for binder composition, density optimization, and shrinkage reduction in jet binder printed components. Yang et al. [10] created a highly accurate computational model of liquid-solid interaction to simultaneously reproduce both binder flow and powder movement. The model accurately reproduces binder droplet impact, binder spreading and penetration, and powder particle spattering and agglomeration and provides in-depth insight into the fundamental mechanisms of binder-powder interactions as well as a basis for process and materials optimization. Mao et al. [11] developed an original binder and investigated the effect of layer thickness and binder saturation on the strength, dimensional accuracy, and surface roughness of raw 316L stainless steel samples produced by binder jetting additive manufacturing. They observed a decrease in strength and an increase in roughness with increasing layer thickness for raw printed samples. Increasing the binder saturation improved the strength and surface roughness of the raw samples. At the same time, they discussed the relationships between the binder and the mechanical performance of the raw material samples in terms of uneven distribution of the binder in depth. Yang et al. [12] simulated binder jetting and analyzed the joining mechanism of magnesium alloy. They used COMSOL Multiphysics simulation software (COMSOL 6.2) to create a simulation model of the binder movement and deposition process. They found that the higher the flow velocity, the larger the dispersion width of the binder drop after impacting the powder bed, which seriously affects the dimensional accuracy of the blank. Bae et al. [13] investigated the influence of binder and powder properties used in Binder Jet 3D Printing on Build-Up. They used an inorganic binder based on silicate (SiO_2–Na_2O) and powder in the form of sand. The results confirmed the increased print quality because the more hydrophilic the surface of the powder used for 3D printing, the higher the affinity between the surface and the inorganic binder, which increases the wettability. However, at the same time, the final strength of the product decreased. Lai et al. [14] investigated the effect of surface roughness on the fatigue behavior of 316L stainless steel produced by the binder jetting process. They performed Tension-compression fatigue tests on 316L stainless steel produced by binder jetting to assess the effect of surface roughness. Porosity was analyzed using an X-ray CT scan to study the distribution and the correlation to fatigue crack initiation. They found that the surface roughness lowered the fatigue strength by approximately 21%. The 316L material produced by binder jetting had lower yield strength than its fatigue strength.

This article deals with binder jetting technology for setting up mass production processes for 316L steel components. The properties of the material, the chemical composition, and the resulting accuracy of the printed samples will be verified experimentally. The information obtained will subsequently be used to identify an economically advantageous application of binder jetting technology.

2. Additive Binder Jetting Technology

Binder jetting (Figure 1) is an additive manufacturing technology that uses a binder to selectively (Figure 2) bind powder particles together, creating three-dimensional objects layer by layer. This technology is highly versatile and can produce objects from a variety of materials, including metals, ceramics, sand, and polymers. As the process progresses, layers of material are glued together to the desired geometry. After layering is completed, if necessary, the entire container is placed in the furnace, where hardening and settling of the binder takes place. The temperature and time in the oven depend on the binder used. After this step, the excess unbonded powder is sucked out and can be reused. Depending on the material, some components are finished at this stage and do not need any finishing operations. Furthermore, metallic and ceramic materials need to be subjected to heat treatment, such as sintering or infiltration, using an additional material to achieve the required mechanical properties [5].

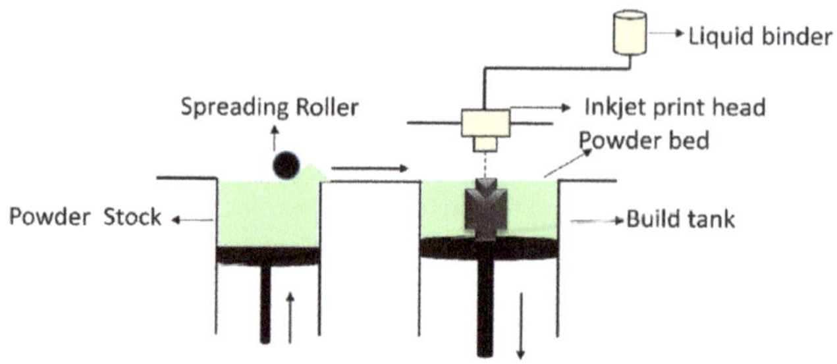

Figure 1. Schematic representation of the binder jetting [15].

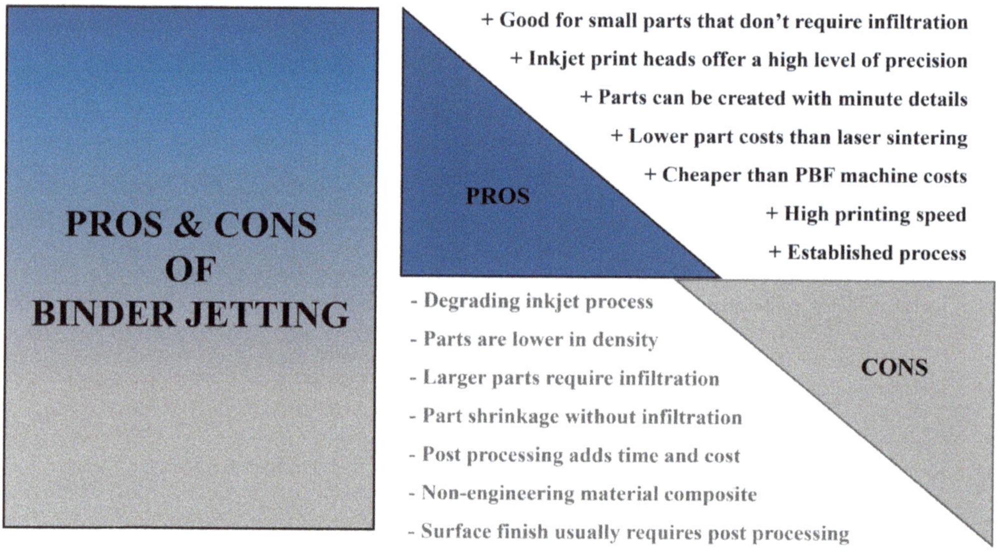

Figure 2. Advantages and disadvantages of binder jetting technology.

The most significant factor in binder jetting technology, currently limiting its potential, is the inability to estimate the level of deformations that occur during the sintering process. Recently, therefore, research has been focused on computational tools that minimize this problem [16]. Due to the similarity of technology with powder metallurgy and sintering, it is possible to use a wealth of knowledge directly from these sources. First of all, research is needed in the field of powder morphology, particle sizes, chemical composition, and their effects on the printing process itself. Other important parameters that affect density, strength, and other properties are the phase of the interaction of the binder and the powder, the thickness of the powder layer, the level of soaking of the binder, the drying time of the binder, and the printing speed. Post-processing operations, thanks to which we can achieve the required levels of density and surface quality, play an important role in the final properties and overall usability of products. In short, in addition to the knowledge drawn from powder metallurgy, casting, and other additive technologies, a number of studies are needed to optimize the process and further specify its parameters [5].

The following Figure 3. shows and describes examples of components made by binder jetting technology.

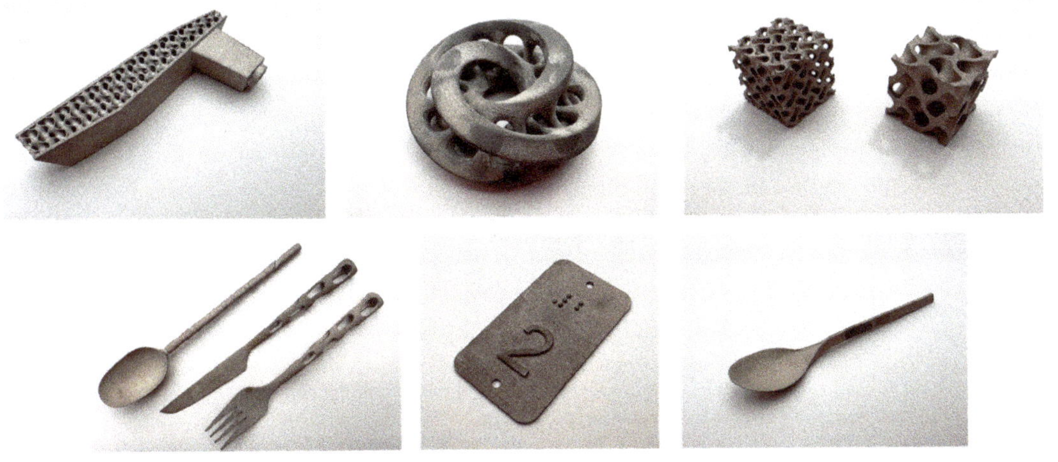

Figure 3. Examples of components made by binder jetting technology.

2.1. Processes in Binder Jetting Technologies

Processes in binder jetting technology include the following steps:

- Press

The current process of binder jetting is basically the same as it was at the birth of the technology. The fundamental change is mainly represented by new materials. ISO (International Organization for Standardization)/ASTM (American Society for Testing and Materials) 52900:2015 defines binder jetting as an additive manufacturing process in which a liquid binder is selectively applied to bind the powder into the desired shape [17].

To complete the geometric shape of a component, it is necessary to pay attention to the following steps [18]:

- A 3D model of a scanned, real, or digital object is required to create a CAD model. The model is digitally divided into layers and saved as an STL (Standard Triangle Language) file that is compatible with the machine;
- A thin layer of powder is spread in the container with a cylinder, which fills a thin gap and, with repeated opposite movements, pushes the excess powder away;
- A liquid binder (most often of polymeric origin in solvent or aqueous solution) is applied by the print head to a layer of powder at the points forming the geometry of the object [18]. Important is its ability to saturate into powder;
- After applying the binder, an electric heater passes through the powder container, which partially hardens and dries the binder. The heater also helps to maintain an even process temperature. The curing time after the application of individual layers is an important factor, as if the binder is insufficiently dried and bonded with the powder, cracking or sticking to the spreading roller could occur;
- After applying and drying the binder, the piston under the powder bed drops by a height of one layer (usually 50 to 200 μm). A further batch of powder is then prepared from the powder reservoir, which is spread, gently compacted, and smoothed to the correct height [5,19].

- Powder curing and suction

After printing is completed, some of the binder jetting technologies require additional operations to completely dry the binder, due to which the part acquires the required strength. Usually, the entire container is manually moved to the oven, where it is heated to a temperature between 180–200 °C. In the future, there will be an effort to automate this step. The curing time and conditions are derived from the binder properties and container volume. After passing this step, the parts have sufficient strength to move to the furnace. Before starting sintering, loose powder is removed with a vacuum cleaner. Depending on the complexity of the parts, manual cleaning with a brush or compressed air may also be required. Caution should be exercised when handling geometrically complex parts containing elements such as fine or overhanging parts. The cleaned part is then sintered or thickened with another material that enters the structure and provides the required mechanical properties [5].

- Sintering or infiltration with other material

The part has a relative density of 50–60% after the curing and powder removal phase. When observing the structure with a microscope, one can see individual grains of powder. In order to achieve the required density and mechanical properties, some of the methods already mentioned can be used. Regardless of the method of compaction of the structure, complete evaporation of the binder must first occur [5].

When choosing the right post-processing operation, it is important to start from factors such as material composition, powder particle size, sintering atmosphere, process time, and temperature [20]. As the characteristics of individual materials vary during sintering, it is sometimes advisable to adjust the process by using additives that help improve the compaction process. This can be achieved, for example, by mixing powders of different sizes or by using particles coated with different films [21]. Strategies differ from material types. Since ceramic materials have considerably higher sintering temperatures and lower densification potential compared to metals, metal infiltration of the ceramic matrix is a frequent combination to achieve the desired structure. On the other hand, when printing polymers that have considerably lower melting points, the polymerization of the structure occurs already when the individual layers are applied [22,23].

Depending on the material and the technology chosen, e.g., infiltration, a high degree of tolerances and accuracy can be achieved. Conversely, when sintering the entire volume of a part of some alloys, significant deformations occur. Therefore, it is important to consider the size and location of critical elements at the design stage. A factor that must not be overlooked is gravity, which can be the source of sinkholes formed during sintering. The correct orientation of the component is, therefore, also important [5].

- Finishing operations

A factor that directly affects the quality of the surface is the softness of the powder and the thickness of the layer. The thicker the layers and the larger the powder particles, the rougher the surface. However, very fine powders may become more difficult to apply due to their adherence to the dispenser. Uneven deposition of the layer can lead to an inhomogeneous structure in terms of the resulting density. From an economic point of view, it is more advantageous to print thicker layers, where the printing process is shortened [24].

2.2. Factors Affecting the Binder Jetting Process

As with other 3D printing technologies, there are several parameters that have a major impact on the properties of parts. They can be divided into the following groups [25]:

- Characteristics of powder

The characteristics of the powder have a major impact on the final product, process conditions, and economic aspects. A lack of knowledge and understanding of powder process and mechanics can greatly affect the resulting quality and cost of production. The most important features that are monitored in binder jetting technology include [26]:

- Geometry and morphology of the powder–most commonly defined by circularity;
- Powder spreadability and powder flow (defined by the Hall flow test);
- Mean particle size and PSD (Particle Size Distribution);
- Powder layering density;
- Surface and internal chemical composition.

Since binder jetting is a relatively new technology, it is not possible to directly apply all knowledge from other powder technologies (e.g., PM (Powder metallurgy) or MIM (Metal Injection Molding)) [27]. The methods of powder production also differ. Alignment and understanding of process requirements and materials engineering can open up further possibilities for product performance improvements and production improvements [28]. The nature of the binder jetting process makes it possible to maintain certain phases in particles that would be lost in fusion technologies. However, further research is needed into recycling and particle reliability, possibly developing new types of powders that would allow special properties to be achieved [29]. Table 1 describes the metallic powder materials currently in use.

Table 1. Currently used metallic powder materials [29].

Base Material	Colloquial/Commercial Name	Alloy Marking (According to American Standard ASME *)
Iron	Stainless steels	316, 316L, 304, 410, 420
	Maraging steel	18 MAR 300
	Hardening tool steel	15-5
	Hardening stainless steel	17-4PH, 13-8
	Tool steels	H13, S7
Nickel	Superalloys	IN-718, IN-625, Waspalloy, René 41
Cobalt	Superalloys	Co-Cr-Mo, Satellite class 6, and class 21
Titanium	Ti-rich	Ti-6Al-4V, Ti-6Al-7Nb, Ti-6Al-2Sn-4Zr-2Mo
	Commercially used pure Ti	CP
Copper	Bronze	Cu-Sn
Aluminum	Alloys for casting	Al-Si-10Mg, Al4047, Al-Si-7Mg-Cu, Al-12Si

Explanation: * American Society of Mechanical Engineers.

An important parameter of the process is viscosity. It depends not only on the reliable dosing of the binder but also on the resulting accuracy and quality of surfaces. The suitability of the binder for the process can be derived using the Weber (We) and Reynolds numbers (Re). The Formula (1) after modification looks like this [5]:

$$Oh = \frac{\sqrt{We}}{Re} = \frac{\eta}{\sqrt{\gamma \cdot \varsigma \cdot d}};. \quad (1)$$

where

Oh—Ohnesorg number;
γ—surface tension;
ς—density of the binder in the liquid state;
d—the diameter of the drop coming out of the nozzle;
η—dynamic viscosity of the binder in the liquid stat

- Print parameters

  Print parameters can be categorized as follows:

- Print thickness;
- The print height is a parameter defined as the thickness of the binder-wetted powder layer in the direction of the Z-axis [30];
- Print and powder application speed;
- A lower speed leads to higher accuracy and homogeneity of the structure but negatively affects the length of the process. On the other hand, at higher speeds, the risk of cavities and irregular structures increases [31];
- The ability of the binder to saturate powder particles—Saturation;
- The ability to correctly dose the binder is directly dependent on the capacity of the print head and the arrangement of the individual nozzles located in it. Based on this, the level of saturation of the powder with the binder and the overlap of the

binder between the rows are defined. Incorrect deposit of binder negatively affects the homogeneity of the structure as well as the dimensional inaccuracy of prints [32,33]. Similar to the speed of application of powder layers, the optimal setting of binder dosing is the result of trial and error by the user [5]. The wettability level can be approximated using Equation (2):

$$Uz = \frac{1000 \cdot \mathrm{V}}{\left(1 - \left(\frac{PR}{100}\right)\right) \cdot \mathrm{X} \cdot \mathrm{Y} \cdot \mathrm{Z}}; \tag{2}$$

where

V—volume of one drop of binder;
PR—powder density (powder bed density) [%];
X and Y—distances between nozzles [μm];
Z—coating thickness [μm].

- Drying time and heating element performance;
- The curing time depends on the selected saturation level, the type of binder and its chemical composition, the thickness of the layer, the wettability of the powder, and the properties of the powder bed (thermal conductivity, density, area, etc.);
- With no or short drying time, there is an increased risk of blockage of some parts of the print head, which is ultimately reflected in the properties of the component [33,34].

- Print orientation

For additive technologies, there are generally two terms related to print orientation that need to be defined, namely:

1. Layering orientation

It affects the porosity and mechanical properties of the component. In a study focused on print orientation, it was shown that samples printed at an angle of 45° had a compressive strength of 13.4 ± 4.6 MPa and a porosity of 37 ± 2%, while samples printed in a perpendicular direction showed a tensile strength of up to 45.1 ± 6.8 MPa and a porosity of 30 ± 2%. In another study on the influence of orientation on properties, it was reported that layering in the direction parallel to the pressure is approximately 48% stronger than in layers oriented in the direction perpendicular to the load [16];

2. Component orientation

The orientation of the component during printing also has a significant influence. It has been proven that the highest quality and accuracy is achieved when pushed to the primary plane *XY*. When printing primarily *Z*-oriented, components achieve lower surface quality and lower density in the raw state. This is due to the need for more layers compared to parts whose printing is oriented in the direction of the *X* or *Y* axis. Thus, it is clear that defects associated with orientation form primarily in weaker bound layers [35].

- Post-processing operations

After completing the printing stage, it is necessary to perform additional operations that will allow to achieve the desired properties. This includes [4,5,18,22,29,36,37]:

1. Desiccation/initial curing

The purpose of this step is to remove excess binder after the printing process. It usually takes place in a whole container with a printout and excess powder. In some methods, only the raw component is cured. The curing time and temperature depend on the binder used, geometry, wall thickness, component volume, and overall container size. The curing process ensures that the bonds between powder and binder are strengthened;

2. Removing unused powder

Removing unused powder from the printed part is necessary to achieve the required accuracy and purity of the resulting object.

3. Check the strength of the raw print

We consider a raw component to be a part that has undergone a curing process but has not yet undergone a compaction process by infiltration or sintering. Strength at this stage is very important, especially due to the manipulations when removing the remaining powder. If the strength of a raw component is too low, it will usually be damaged. Conversely, if a component is too durable, this may indicate an excessive amount of binder. Based on this, when setting process conditions, for example, in the case of a new type of powder or binder, the raw strength of the component is taken as a decisive parameter for proper process optimization. Since there is no prescribed standard for the strength control of a raw part for binder jetting technology yet, it is usually based on ASTM and MPIF standards;

4. Pyrolysis or firing of binder

Following the curing of the raw components, the binder firing step follows. This action takes place before the process of sintering or infiltration, usually in the same furnace. In order to achieve the right level of sintering or infiltration, it is necessary to remove the polymer binder that remains in the interparticle space. Most often, pyrolysis is used for this, in which thermal decomposition of organic materials occurs without the presence of oxygen-containing media. Firing occurs when heated above the decomposition temperature of the polymer binder or by reaction with gaseous components present;

5. Control of the content and impact of residues

The decomposition of the binder and the associated processes can result in the formation of residual carbon and oxygen. Residual substances most often form oxides in the case of oxygen and carbides in the case of carbon, which can negatively affect the properties of materials. The most important thing is to check for materials that are sensitive to even small changes in the content of certain substances, such as steel and carbon. Knowledge of the process and its optimal setting can have a significant impact on the final characteristics of products. If the binder firing process is properly managed and thus a minimum of impurities remains in the structure, a negligible negative effect on shrinkage can be expected;

6. Infiltration

This process involves the densification of the porous structure with a material with a lower melting point than the base material–such as steel and bronze. Thanks to this process, it is possible to achieve complete structure density with minimally affected dimensional accuracy. The driving force of the infiltration process is capillary phenomena. Infiltration in binder jetting technology favorably affects the mechanical properties of the final product, including hardness, tensile elasticity, yield strength, and others. Process optimization is essential to achieve the desired properties. Too high a level of infiltration can negatively affect dimensional accuracy. An insufficiently saturated base material has an inhomogeneous structure and worse mechanical properties. One of the innovations for achieving even better mechanical properties and structure is the process of infiltration in a vacuum;

7. Sintering

Sintering is another method of thickening a porous structure after printing. The sintering kinetics and structure thickening process depend on the chemical composition and surface characteristics of the powder, powder morphology, PSD, and sintering atmosphere. The most important aspects in binder jetting metal powder printing that affect shrinkage linearity, microstructure, structure phase formation, and porosity are PSD, morphology and chemical composition of the powder, heating parameters, cooling parameters, temperature, and sintering time.

2.3. Suitable and Researched Materials for Binder Jetting Technology

Binder jetting technology has a wide range of possible materials that can be used. Some of the most commonly used materials are, for example:

- Metals

Metals such as aluminum, stainless steel, and titanium, as well as various metal alloys, are popular for their high strength and durability [30,31]. Binder jetting makes it possible to create complex metal parts with relatively low production costs [38];

- Ceramics

Ceramic materials such as aluminum oxides, silicon oxides, and carbides are used in applications that require high temperature resistance and chemical stability [6]. Binder jetting allows you to create detailed and precise ceramic parts with complex geometries [38];

- Polymers

Various types of plastics and polymers, such as nylon, polycarbonate, and elastomers, are suitable for the production of prototypes, models, and functional parts. These materials are often affordable and have a wide range of properties [5,6];

- Composites

Composite materials that combine different types of fibers and matrixes are used to produce high-strength and lightweight parts [39]. Binder jetting allows you to create composite parts with precise fiber orientation and optimized properties [6,40].

3. Used Material and Equipment in the Practical Experiment

3.1. Used Material

In the experiment, 316L stainless steel powder was used, which is a low-carbon alloy known for high corrosion resistance and relatively good strength. In addition, it also has good formability and thermal properties within the 3D printing process. The material specifications given in Table 2. are directly provided by the ExOne printer manufacturer (Huntingdon, PA, USA). In Section 4, some properties of the material mentioned are verified through tests.

Table 2. Material properties declared by the manufacturer ExOne [41].

Material Property	Test Method	Values
Ultimate strength Rm Yield strength (0.2%) Extension to infringement Modulus of elasticity	ASTM E8	X, Y: 450–580 MPa Z: 450–520 MPa X, Y: 140–220 MPa Z: 140–220 MPa X, Y: 40–55% from 40–50% X, Y: 190–220 GPa W: 180–190 GPa
Hardness	ASTM E18	67–71 HRB *
Impact toughness	ASTM E23	55–75 J
Poisson number Relative density Density Surface roughness		0.28–0.30 96–99% 7600–7900 $kg{\cdot}m^{-2}$ Ra 3–12 μm

Explanation: * Rockwell Hardness Test on the B Scale.

The chemical composition of the 316L steel material is described in the following Table 3.

The influence of parameters such as temperature, thickness of powder, glue and others have a major impact on the resulting shrinkage rate. These enumerated effects were assessed on the optical device KEYENCE IM-8000 (Osaka, Japan) (see Section 4.2).

Table 3. Chemical composition of steel material 316L [41].

Chemical Composition	
Element	**[%]**
Nickel	10–14
Chromium	16–18
Carbon	max. 0.03
Molybdenum	2–3
Manganese	max. 2.0
Silicon	max. 1.0
Iron	remainder

3.2. Equipment Used

For experimental purposes, a device from ExOne was selected and subsequently used, namely the model DesktopMetal X160Pro®, on which all samples were printed.

The X160Pro® (Gersthofen, Germany) provides a proven 3D printing method using binder jetting technology, capable of printing ultrafine metallic, sand, ceramic and specialty material powders. This is the largest device of this type, which has a capacity of up to 160 L. This device can be used for a wide range of production, whether it is prototyping or mass production. The greatest benefit of the device is the free hand at the design stage and the ability to easily produce even shape-complex parts for a single operation [42].

In addition to the machine itself, other parts are also necessary for the operation of the technology, such as the powder hopper, the furnace in which the binder solidifies, the powder removal device, and the sintering furnace.

After the binder has hardened and sufficient rigidity of the prints has been achieved for their handling, the platform is moved from the furnace to the equipment where the primary dedusting process takes place. In this step, the unsaturated powder is sucked out.

The working area of the machine is closed during the dedusting process–this is a dust-free process. The inside of the machine is equipped with an extraction head, the pressure of which is adjusted so as not to damage the integrity of the green parts. However, damage may occur with complex parts.

The stay of the parts in the sintering furnace ensures complete bonding of the material. The time during which the parts are in the furnace depends entirely on the material, powder, and binder. For 316L steel, the so-called 12-h recipe was used, which consists of heating to a sintering temperature of $\approx$1380 °C, holding at this temperature for 130 min, and then cooling to 22 °C. Due to the laminar heat flow in the furnace, the manufacturer states that the difference between the different zones is $\pm$1 °C. The parts are inserted into the furnace chamber on special inserts.

4. Practical Experiment—Optical Measurement of Samples and Material Tests

4.1. Print Parameters

Even before the experiment itself, first of all, it is necessary to set the parameters of the press as well as the powder used. The following Table 4 describes the parameters of printing and powder used.

PSD refers to the particle size distribution in a given material. PSD is important because it affects material properties such as flow, density, surface area, and reactivity.

D10, D50, D90

These parameters are quantile values that describe the particle size distribution within the PSD.

Table 4. Parameters of printing and powder used.

Device	ExOne DesktopMetal X160Pro®
Powder layer height	50 μm
The speed of movement of the recoater	100 mm/s
Roller speed (opposite)	800 rpm
Material	Steel 316L
PSD	
D10	3–4 μm
D50	11–12 μm
D90	21–22 μm
Particle shape	Spherical powder (balls) Produced by gas atomization

- D10: Provides information about the smallest particles in the mixture. It is important for applications where fine particles can affect the reactivity or flow resistance of the powder. This value means that 10% of the particles in the powder have a diameter smaller than or equal to this value;
- D50: Also known as the median particle size, it indicates that 50% of the particles have a diameter smaller than or equal to this value. It is the median value of the particle size distribution. D50 is often used to assess the overall average particle size in the powder. In the case of 3D printing, it can affect the quality of the printed product and the homogeneity of the material;
- D90: Provides information about the largest particles in the mixture, which can be critical for mechanical properties or for ensuring a uniform surface appearance. This value means that 90% of the particles have a diameter smaller than or equal to this value.

The device allows you to adjust the ratio between new powder and recycled powder obtained in dedusting processes.

In Figure 4, one can see the arrangement of the cubes during the printing process. In addition to scrolling, printed parts can also be rotated freely. The correct orientation of the parts has a significant influence on the resulting deformation.

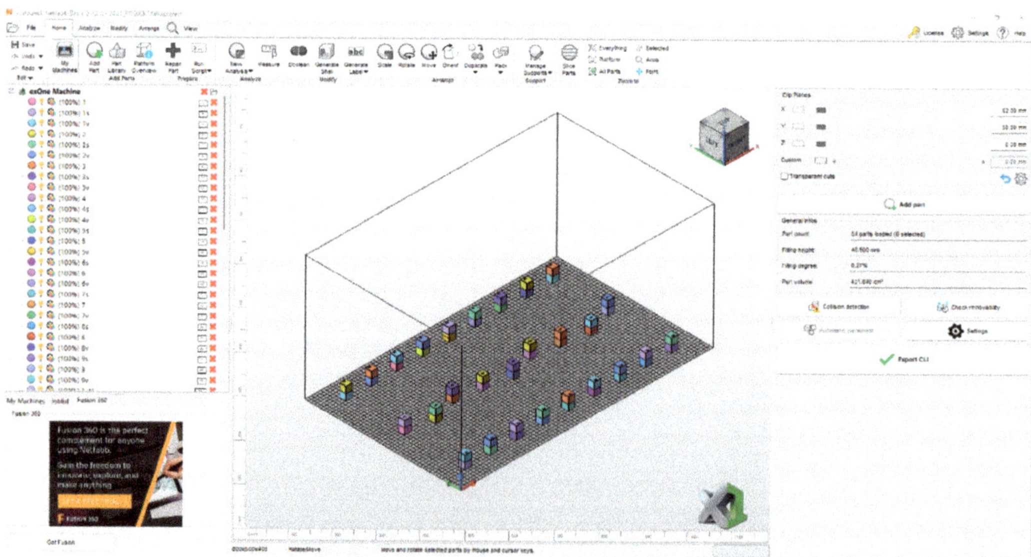

Figure 4. The work area of Autodesk Netfabb.

4.2. Estimation of Shrinkage in Individual Directions

Measurements were performed on a KEYENCE IM-8000 device (Figure 5). The test samples for measuring dimensions and shrinkage after sintering have the form of a cube with an initial side dimension of 20 mm (Figure 6a–c). Test specimens of 316L material are shown in Figure 6d.

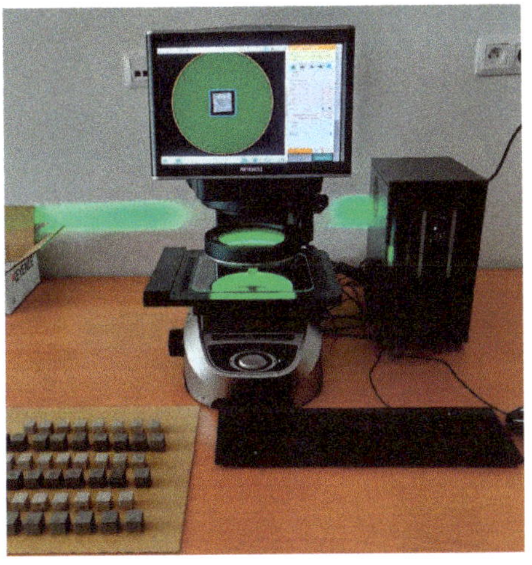

Figure 5. Measuring device KEYENCE IM-8000.

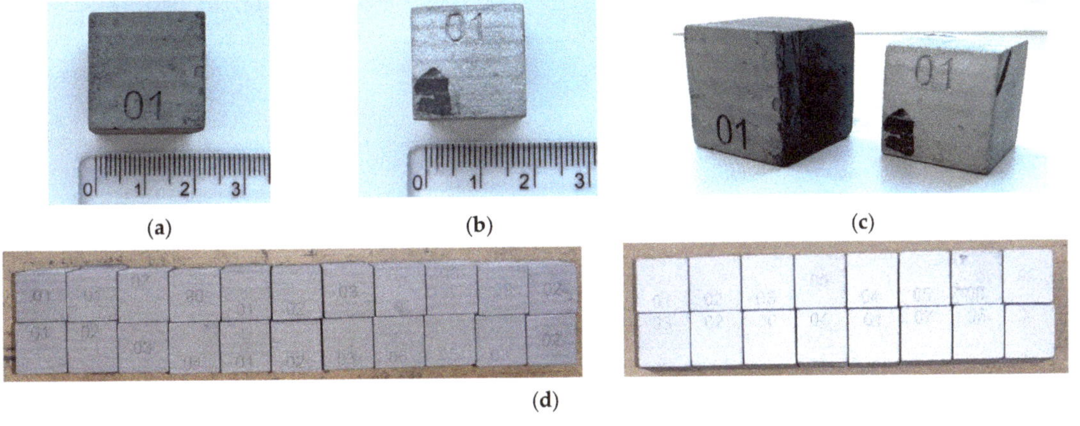

Figure 6. Printouts of test samples. (**a**) cube before sintering, (**b**) cube after sintering, (**c**) isometric view, (**d**) test samples of 316L.

The following Tables 5 and 6. present the results of measurements of the first samples in different directions. These are cubes that have been pushed to the starting size of a 20 mm raw component, but this can vary from one direction to another due to factors such as gravity or printing inaccuracy. The measurements were carried out in three views (1, 2, 3), according to which we find and compare the relative shrinkage in the direction of the *X* and *Y* axes (Figure 7) in the upper and lower bases of the cube and in the direction of the *Z* axis in the side wall.

Table 5. Comparison of dimensional changes before and after sintering in the direction of the *X* and *Y* axes on the upper base.

Measured Wall					Upper Base			
Measurement Orientation								
	X				Y			
	Dimension before Sintering [mm]	Dimension after Sintering [mm]	Shrinkage Rate [mm]	Shrinkage Rate [%]	Dimension before Sintering [mm]	Dimension after Sintering [mm]	Shrinkage Rate [mm]	Shrinkage Rate [%]
	20.007	17.591	2.416	12.08	20.334	17.763	2.571	12.64
	19.923	17.585	2.338	11.74	20.381	17.828	2.553	12.53
	20.071	17.547	2.524	12.58	20.331	17.877	2.454	12.07
	20.001	17.544	2.457	12.28	20.347	17.804	2.543	12.50
	19.983	17.579	2.404	12.03	20.238	17.856	2.382	11.77
	20.076	17.608	2.468	12.29	20.254	17.806	2.448	12.09
	20.010	17.518	2.492	12.45	20.067	17.788	2.279	11.36
	19.914	17.636	2.278	11.44	19.902	17.685	2.217	11.14
	20.066	17.662	2.404	11.98	20.113	17.838	2.275	11.31
	20.030	17.489	2.541	12.69	20.154	17.797	2.357	11.69
	20.033	17.613	2.420	12.08	20.268	17.785	2.483	12.25
	20.006	17.617	2.389	11.94	20.304	17.882	2.422	11.93
	19.998	17.654	2.344	11.72	19.993	17.868	2.125	10.63
	20.038	17.577	2.461	12.28	20.269	17.777	2.492	12.29
	20.052	17.552	2.500	12.47	20.331	17.728	2.603	12.80
	19.987	17.667	2.320	11.61	20.105	17.682	2.423	12.05
	20.039	17.598	2.441	12.18	20.289	17.791	2.498	12.31
	20.002	17.611	2.391	11.95	20.381	17.833	2.548	12.50
	20.051	17.651	2.400	11.97	20.305	17.814	2.491	12.27
	20.062	17.597	2.465	12.29	20.274	17.782	2.492	12.29
Min.	19.914	17.489	2.278	11.44	19.902	17.682	2.125	10.63
Max.	20.076	17.667	2.541	12.69	20.381	17.882	2.603	12.80
AP	20.017	17.595	2.423	12.10	20.232	17.799	2.433	12.02
SD	0.04311	0.04681			0.12961	0.0543		

Table 6. Comparison of dimensional changes before and after sintering in the direction of the *X* and *Y* axes on the lower base.

Measured Wall					Lower Base			
Measurement Orientation								
	X				Y			
	Dimension before Sintering [mm]	Dimension after Sintering [mm]	Shrinkage Rate [mm]	Shrinkage Rate [%]	Dimension before Sintering [mm]	Dimension after Sintering [mm]	Shrinkage Rate [mm]	Shrinkage Rate [%]
	19.976	17.571	2.405	12.04	20.238	17.809	2.429	12.00
	19.916	17.491	2.425	12.18	20.278	17.826	2.452	12.09
	19.989	17.473	2.516	12.59	20.268	17.814	2.454	12.11
	19.937	17.497	2.440	12.24	20.232	17.833	2.399	11.86
	19.961	17.585	2.376	11.90	20.810	17.884	2.926	14.06
	20.204	17.553	2.651	13.12	20.008	17.77	2.238	11.19
	19.942	17.425	2.517	12.62	20.006	17.755	2.251	11.25
	19.879	17.587	2.292	11.53	20.053	17.621	2.432	12.13
	19.983	17.593	2.390	11.96	20.166	17.812	2.354	11.67
	19.891	17.480	2.411	12.12	20.088	17.807	2.281	11.36
	19.945	17.558	2.387	11.97	20.265	17.791	2.474	12.21
	19.933	17.634	2.299	11.53	20.135	17.781	2.354	11.69
	19.912	17.545	2.367	11.89	20.009	17.772	2.237	11.18
	19.978	17.574	2.404	12.03	20.091	17.766	2.325	11.57
	19.926	17.512	2.414	12.11	20.112	17.694	2.418	12.02
	20.012	17.650	2.362	11.80	20.233	17.676	2.557	12.64
	19.921	17.411	2.510	12.60	20.198	17.813	2.385	11.81
	19.993	17.513	2.480	12.40	20.214	17.762	2.452	12.13
	19.917	17.539	2.378	11.94	20.017	17.669	2.348	11.73
	19.946	17.501	2.445	12.26	20.195	17.715	2.48	12.28
Min.	19.879	17.411	2.292	11.53	20.006	17.621	2.237	11.18
Max.	20.204	17.65	2.651	13.12	20.810	17.833	2.557	12.64
AP	19.958	17.535	2.423	12.14	20.148	17.762	2.385	11.84
SD	0.06607	0.06135			0.17133	0.06301		

From Table 5, it can be read that the change in dimensions after sintering in the directions *X* and *Y* (Figure 8) is similar at the upper base. In the direction of the *X*-axis, the average shrinkage was 2.423 mm. The percentage is 12.1%. In the direction of the *Y*-axis, the average shrinkage is similar—2.433 mm. As a percentage, it is 12.02%. However, the difference can be seen in the values of the dimensions before sintering between the two directions. In the direction of the *X*-axis, the average value measured on the device is 20.017 mm, while in the direction of the *Y*-axis, we measured an average of 20.232 mm. The shrinkage ratio after sintering is almost identical.

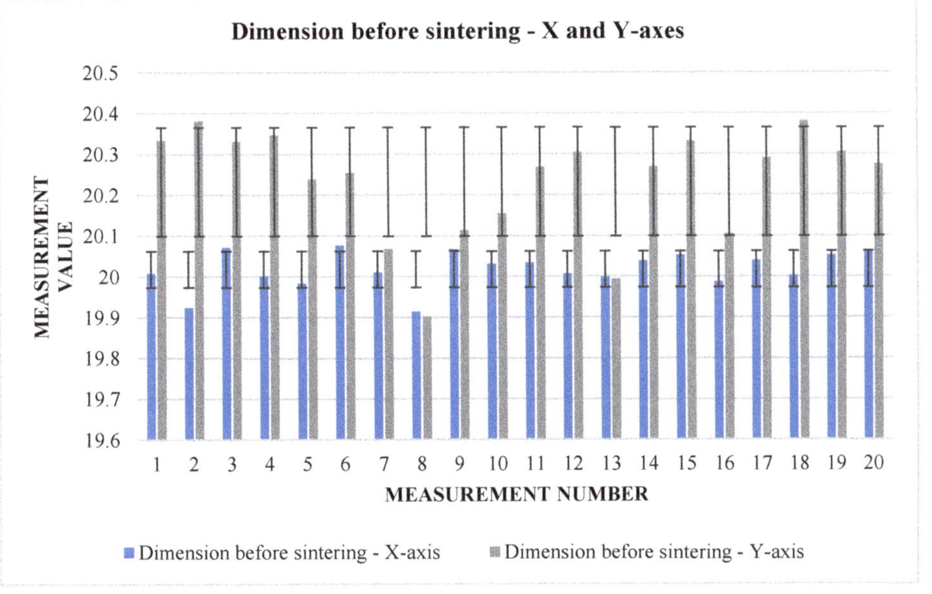

Figure 7. Graphical representation of Table 5—dimension before sintering—*X* and *Y*-axes (upper base).

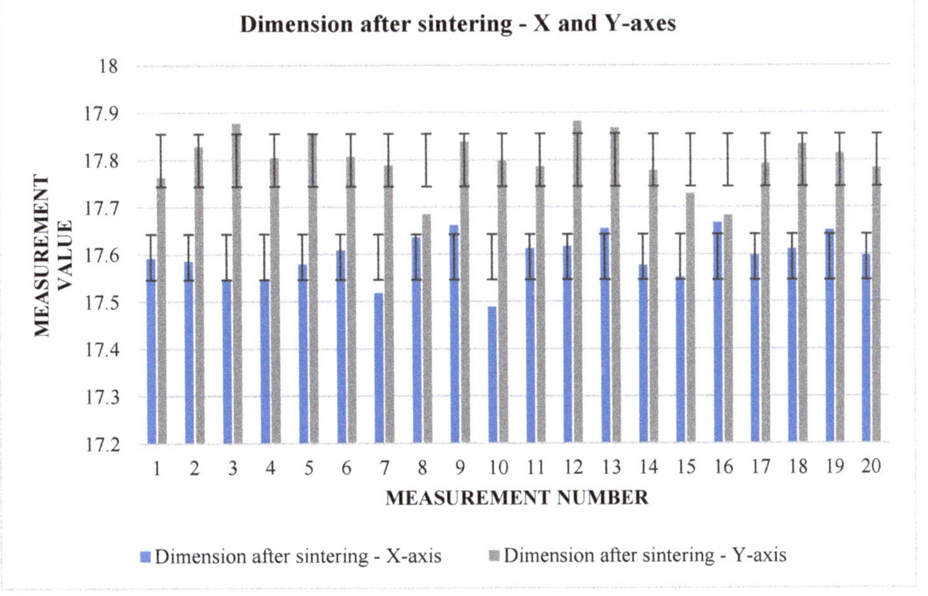

Figure 8. Graphical representation of Table 5—dimension after sintering—*X* and *Y*-axes (upper base).

When measuring the lower base of the printed cubes (Table 6), we can observe a change in dimensions from the opposite base already before sintering. This can be caused by the combination of gravity factors and layering of the binder between the individual layers of the powder, where the binder can "overflow" along the edges of the cube. As for shrinkage, in the direction of the X-axis it is almost identical to the upper base. In the Y direction, it differs by two-tenths of a millimeter.

The graphs (Figures 9 and 10) indicate some variation between individual cubes in both X and Y directions. There are also considerable rebounds between dimensions on both axes (Figures 11 and 12), the origin of which is not yet known.

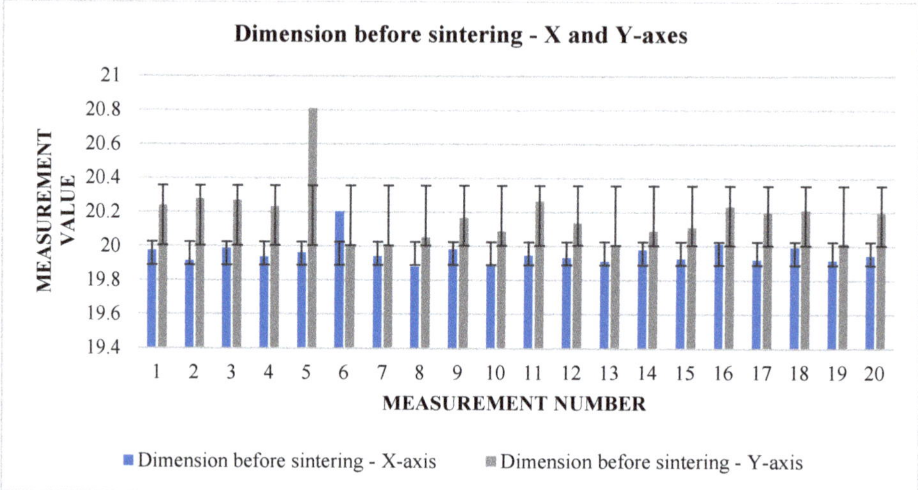

Figure 9. Graphical representation of Table 6—dimension before sintering—X and Y-axes (lower base).

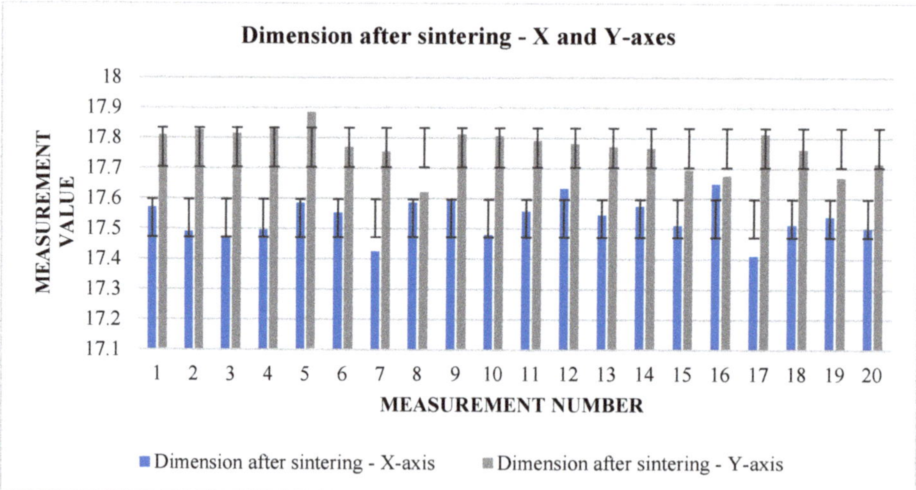

Figure 10. Graphical representation of Table 6—dimension after sintering—X and Y-axes (lower base).

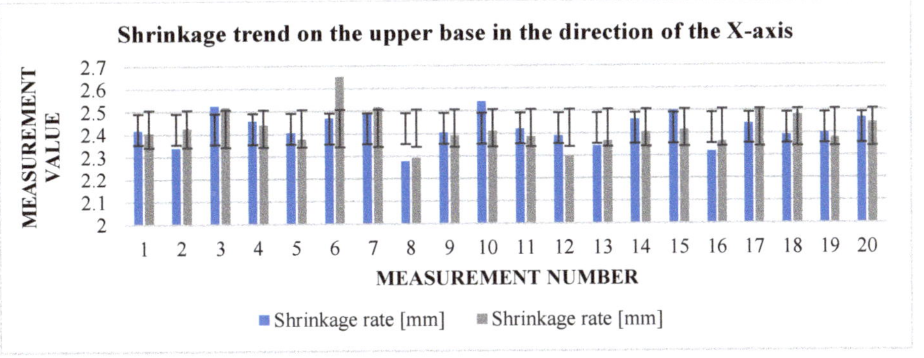

Figure 11. Graphical representation of Tables 5 and 6—shrinkage in the direction of the *X*-axis.

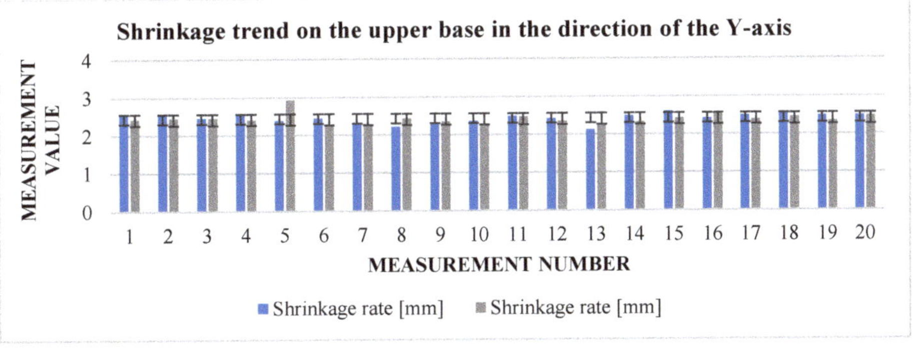

Figure 12. Graphical representation of Tables 5 and 6—shrinkage in the direction of the *Y*-axis.

From the measured values obtained in the vertical direction, it is possible to immediately see the influence of gravity and the weight of the powder on the values of the resulting shrinkage. Compared to area shrinkage (*X*, *Y*), vertical shrinkage (*Z*) (Figure 13) is more than 3% higher (Table 7).

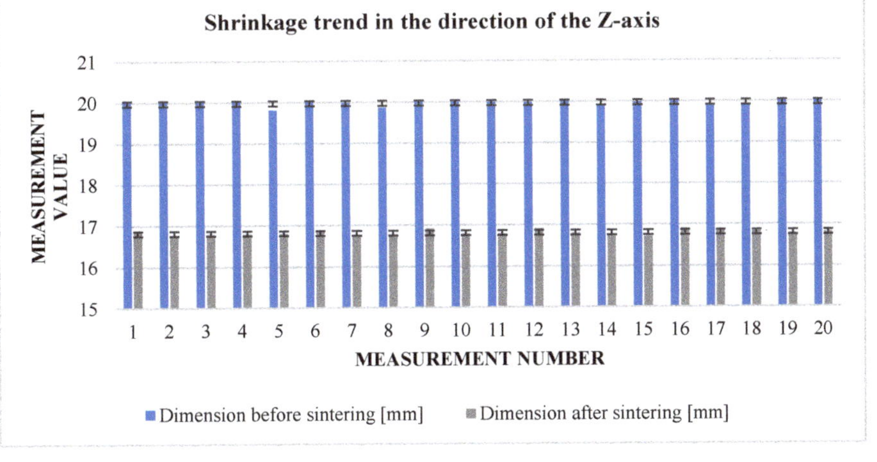

Figure 13. Graphical representation of Table 7—*Z*-axis shrinkage.

Table 7. Comparison of dimension changes before and after sintering in the direction of the *Z*-axis.

Measured Wall			Side Wall	
Measurement Orientation				
From				
Dimension before Sintering [mm]		Dimension after Sintering [mm]	Shrinkage Rate	
			[mm]	[%]
	20.009	16.865	3.144	15.71
	20.006	16.767	3.239	16.19
	20.037	16.788	3.249	16.22
	20.029	16.795	3.234	16.15
	19.806	16.801	3.005	15.17
	20.04	16.849	3.191	15.92
	19.985	16.73	3.255	16.29
	19.859	16.784	3.075	15.48
	20.012	16.920	3.092	15.45
	20.003	16.789	3.214	16.07
	19.997	16.793	3.204	16.02
	19.985	16.912	3.073	15.38
	20.032	16.796	3.236	16.15
	19.895	16.771	3.124	15.70
	19.982	16.714	3.268	16.35
	20.011	16.906	3.105	15.52
	19.881	16.871	3.010	15.14
	19.933	16.836	3.097	15.54
	20.007	16.793	3.214	16.06
	20.012	16.811	3.201	16.00
Min.	19.806	16.714	3.005	15.14
Max.	20.040	16.92	3.268	16.35
AP	19.976	16.815	3.162	15.83
SD	0.064	0.05564		

Based on the results obtained, we can estimate the rate of shrinkage and, thus, the need for additions to achieve the required dimensions in individual directions at approximately *X*: 12%, *Y*: 12%, and *Z*: 15.5%.

Thanks to this measurement, we have approached the shrinkage of cubes in different directions and the influence of gravity on printing accuracy, which will benefit the dimensional accuracy of future prints.

4.3. Optical Measurement of Samples with Set Shrinkage Rate

From previous experiments and our experiment, shrinkage rates of *X*: 13%, *Y*: 13%, and *Z*: 16% were determined for further printing. Again, cubes were made to the size of the green part a = 20 mm. Unlike in the previous experiment, the placement of these samples in the powder bed will also be compared in this chapter, which may also affect the final deformation and accuracy. We will distinguish three hints according to the following Figure 14. From an economic point of view, there are also other parts in the powder bed to make printing as productive as possible. Arrows 1, 2, and 3 indicate the rows in which the dice are located. Optical measurement was performed on the same device—KEYENCE IM 8000.

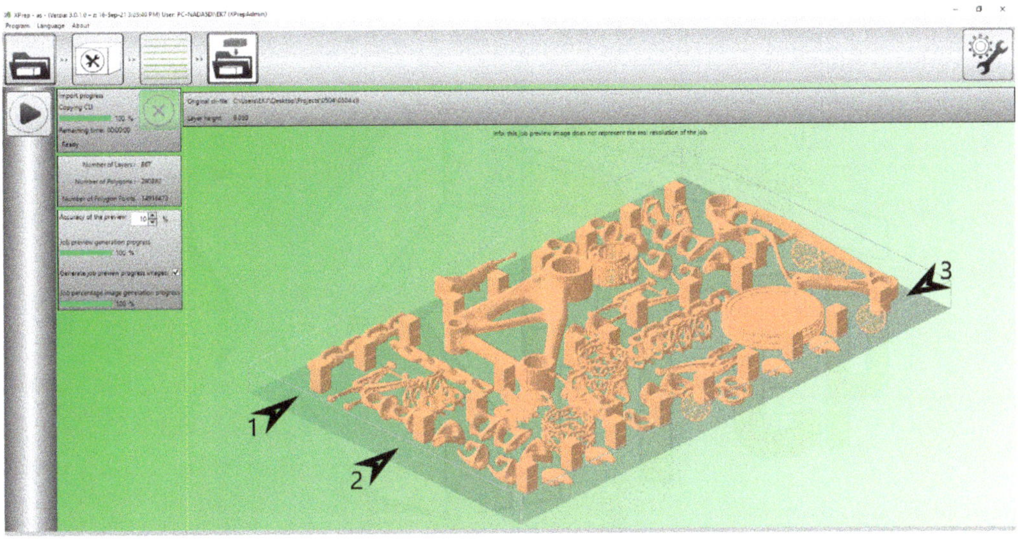

Figure 14. Distribution of cubes in the powder bed at the preprocessing stage.

From Table 8, we can read that the average shrinkage in the direction of the *X* and *Y* axes is slightly higher than the specified 13% in both. Therefore, a slight correction can be considered for further experiments so that the resulting dimensions are as close as possible to the requirements. Deviations between individual dimensions in the pre-sintering phase, but also after sintering, can be associated with printing parameters but also with the dedusting process, where fine layers can be removed when handling cubes. Fluctuations between dimensions can also be seen in the graphs (Figures 15 and 16), showing shrinkage in both axes.

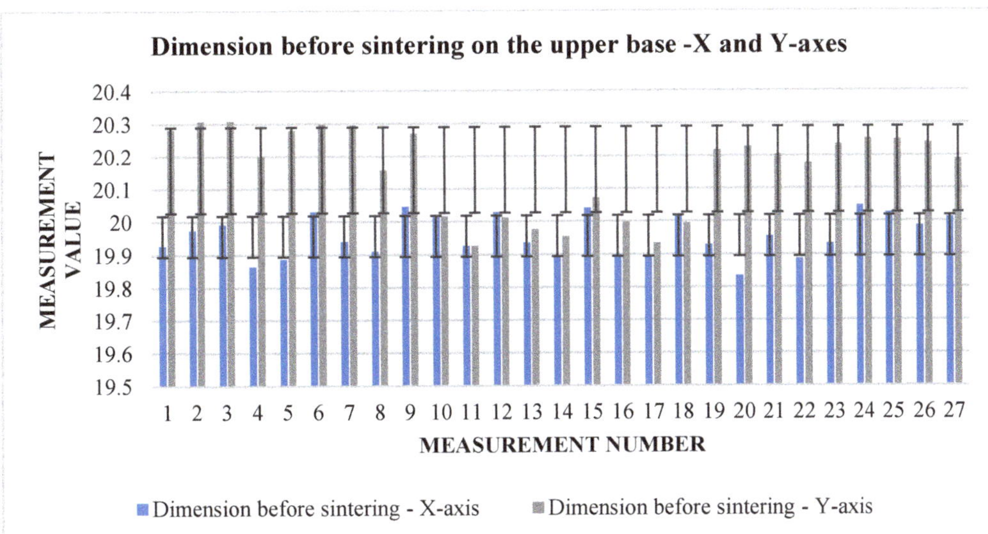

Figure 15. Graphical representation of Table 8—dimension before sintering—*X* and *Y*-axes (upper base).

Table 8. Comparison of dimensional changes before and after sintering in the direction of the *X* and *Y* axes on the upper base.

Measured Wall					Upper Base					
Measurement Orientation										
X					*Y*					
	Dimension before Sintering [mm]	Dimension after Sintering [mm]	Shrinkage Rate [mm]	Shrinkage Rate [%]	Dimension before Sintering [mm]	Dimension after Sintering [mm]	Shrinkage Rate [mm]	Shrinkage Rate [%]	
	19.927	17.222	2.705	13.57	20.281	17.473	2.808	13.85	
	19.974	17.217	2.757	13.80	20.307	17.441	2.866	14.11	
	19.992	17.35	2.642	13.22	20.308	17.554	2.754	13.56	
	19.863	17.36	2.503	12.60	20.201	17.668	2.533	12.54	
	19.886	17.308	2.578	12.96	20.279	17.754	2.525	12.45	1st series
	20.030	17.38	2.650	13.23	20.298	17.596	2.702	13.31	
	19.940	17.268	2.672	13.40	20.295	17.446	2.849	14.04	
	19.910	17.349	2.561	12.86	20.157	17.494	2.663	13.21	
	20.045	17.133	2.912	14.53	20.269	17.368	2.901	14.31	
	20.013	17.157	2.856	14.27	20.016	17.425	2.591	12.94	
	19.926	17.246	2.680	13.45	19.926	17.272	2.654	13.32	
	20.028	17.275	2.753	13.75	20.012	17.315	2.697	13.48	
	19.936	17.382	2.554	12.81	19.976	17.018	2.958	14.81	
	19.897	17.370	2.527	12.70	19.954	17.335	2.619	13.13	2nd series
	20.040	17.296	2.744	13.69	20.072	17.167	2.905	14.47	
	19.899	17.411	2.488	12.50	19.998	17.025	2.973	14.87	
	19.896	17.268	2.628	13.21	19.934	17.226	2.708	13.58	
	20.014	17.091	2.923	14.60	19.994	16.982	3.012	15.06	
	19.929	17.367	2.562	12.86	20.217	17.577	2.64	13.06	
	19.834	17.407	2.427	12.24	20.228	17.627	2.601	12.86	
	19.954	17.439	2.515	12.60	20.204	17.697	2.507	12.41	
	19.885	17.406	2.479	12.47	20.177	17.874	2.303	11.41	
	19.932	17.413	2.519	12.64	20.235	17.648	2.587	12.78	3rd series
	20.047	17.424	2.623	13.08	20.253	17.649	2.604	12.86	
	20.026	17.407	2.619	13.08	20.249	17.629	2.62	12.94	
	19.986	17.373	2.613	13.07	20.24	17.533	2.707	13.37	
	20.011	17.376	2.635	13.17	20.19	17.501	2.689	13.32	
Min.	19.834	17.091	2.427	12.24	19.926	16.982	2.525	12.45	
Max.	20.045	17.411	2.923	14.60	20.308	17.754	3.012	15.06	
AP	19.949	17.293	2.656	13.31	20.136	17.388	2.748	13.65	
SD	0.06104	0.09301			0.12895	0.22521			

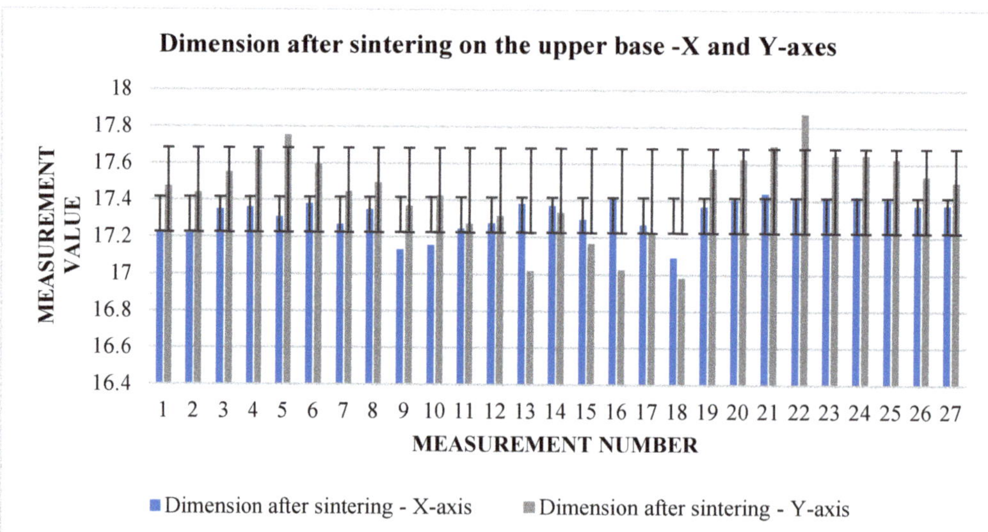

Figure 16. Graphical representation of Table 8—dimension after sintering—*X* and *Y*-axes (upper base).

When comparing the values of shrinkage rates between cubes (Table 9), which were printed in different rows, it can be seen that in both directions on the upper base, the change in dimensions in the third row after sintering is always below 13%, although the average value of the dimensions after printing is not the lowest. The question is whether this difference originates from the printer or the layout of the cubes in the sintering furnace.

Table 9. The difference between shrinkage in different rows.

Shrinkage X [%]		Shrinkage Y [%]	
1st series	13.35	1st series	13.49
2nd series	13.44	2nd series	13.96
3rd series	12.80	3rd series	12.78

From Table 10, we can read that in the direction of the *X*-axis we are approaching the values we expected. The deviation is only 0.11%. In the *Y*-axis direction, this is almost 0.77%–which should be considered in further experiments.

Table 10. Comparison of dimensional changes before and after sintering in the direction of the *X* and *Y* axes on the lower base.

Measured Wall					Lower Base				
Measurement Orientation									
	X				*Y*				
	Dimension before Sintering [mm]	Dimension after Sintering [mm]	Shrinkage Rate [mm]	Shrinkage Rate [%]	Dimension before Sintering [mm]	Dimension after Sintering [mm]	Shrinkage Rate [mm]	Shrinkage Rate [%]	
	19.991	17.219	2.772	13.87	20.321	17.498	2.823	13.89	1st series
	19.944	17.345	2.599	13.03	20.267	17.589	2.678	13.21	
	19.985	17.437	2.548	12.75	20.266	17.644	2.622	12.94	
	19.972	17.301	2.671	13.37	20.287	17.632	2.655	13.09	
	19.994	17.364	2.630	13.15	20.318	17.682	2.636	12.97	
	19.948	17.417	2.531	12.69	20.258	17.62	2.638	13.02	
	19.955	17.407	2.548	12.77	20.281	17.637	2.644	13.04	
	20.005	17.344	2.661	13.30	20.24	17.526	2.714	13.41	
	20.075	17.235	2.840	14.15	20.289	17.459	2.83	13.95	
	19.967	17.288	2.679	13.42	19.972	16.855	3.117	15.61	2nd series
	19.964	17.173	2.791	13.98	20.111	17.211	2.9	14.42	
	20.018	17.397	2.621	13.09	20.007	17.352	2.655	13.27	
	19.954	17.295	2.659	13.33	20.046	16.79	3.256	16.24	
	19.95	17.327	2.623	13.15	20.074	17.331	2.7425	13.66	
	20.003	17.394	2.609	13.04	19.974	17.376	2.598	13.01	
	19.574	17.392	2.182	11.15	20.097	16.896	3.201	15.93	
	19.94	17.247	2.693	13.51	20.168	17.244	2.924	14.50	
	19.983	17.294	2.689	13.46	19.968	17.143	2.825	14.50	
	19.867	17.459	2.408	12.12	20.08	17.61	2.47	12.30	3rd series
	19.978	17.402	2.576	12.89	20.239	17.647	2.592	12.81	
	20.019	17.47	2.549	12.73	20.222	17.676	2.546	12.59	
	19.979	17.404	2.575	12.89	20.216	17.737	2.479	12.26	
	19.956	17.458	2.498	12.52	20.224	17.692	2.532	12.52	
	20.121	17.479	2.642	13.13	20.018	17.668	2.35	11.74	
	19.986	17.484	2.502	12.52	20.252	17.687	2.565	12.67	
	20.145	17.399	2.746	13.63	19.959	17.661	2.298	11.51	
	19.933	17.31	2.623	13.16	20.133	17.459	2.674	13.28	
Min.	19.574	17.173	2.182	11.15	19.968	16.790	2.470	12.30	
Max.	20.075	17.459	2.84	14.15	20.321	17.682	3.256	16.24	
AP	19.953	17.337	2.617	13.11	20.163	17.387	2.776	13.77	
SD	0.09531	0.08295			0.11979	0.26799			

It can be seen from the graphs (Figures 17 and 18) that the differences between dimensions are slightly stable in series 1 and 3 compared to the upper base. There are significant differences in the second row. When comparing lower base shrinkage in individual rows Table 11, an improvement can be seen in the third row compared to the upper base. However, the total shrinkage in the Y direction is constant (Figures 19 and 20). This is due to rather high deviations in the second row.

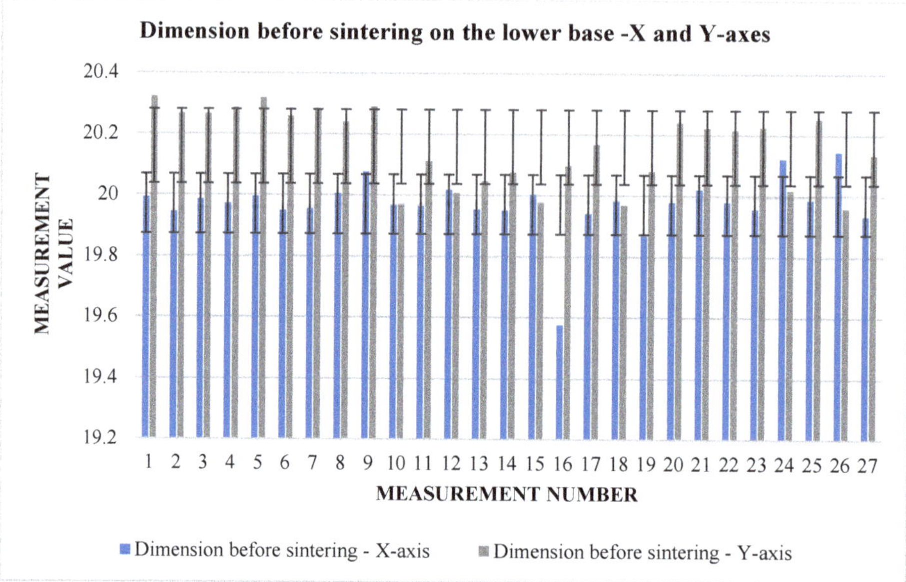

Figure 17. Graphical representation of Table 10—dimension before sintering—X and Y-axes (lower base).

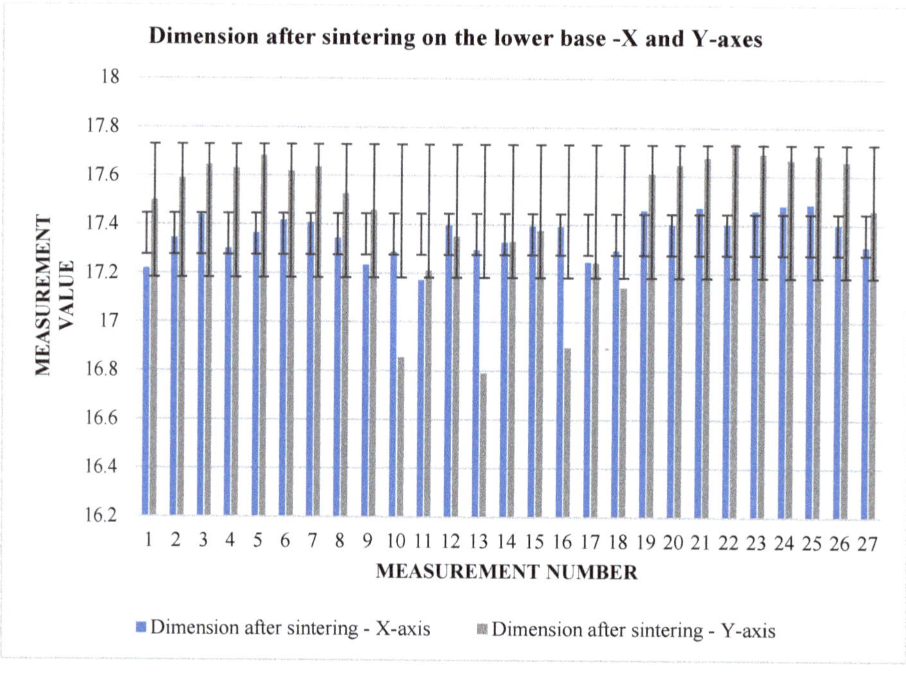

Figure 18. Graphical representation of Table 10—dimension after sintering—X and Y-axes (lower base).

Table 11. The difference between shrinkage in different rows.

Shrinkage X [%]		Shrinkage Y [%]	
1st series	13.23	1st series	13.28
2nd series	13.12	2nd series	14.53
3rd series	12.84	3rd series	12.41

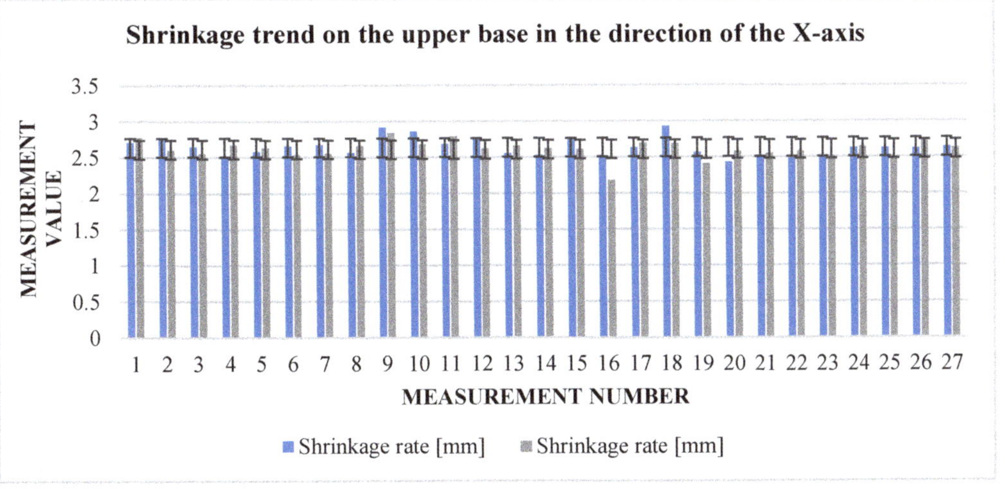

Figure 19. Graphical representation of Tables 8 and 10—shrinkage in the direction of the *X*-axis.

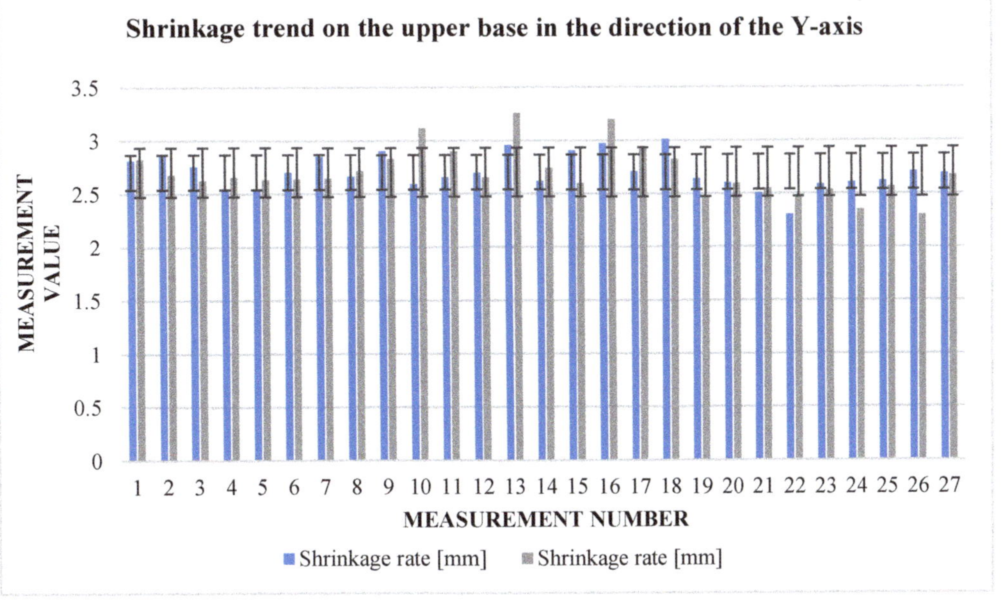

Figure 20. Graphical representation of Tables 8 and 10—shrinkage in the direction of the *Y*-axis.

In the direction of the Z-axis, the addition was set at 16%. The actual situation is 18.9%. Thus, the cubes were much more deformed than expected under the influence of gravity and other parameters, and the loss of dimensions could already be seen at a glance. It is, therefore, important to pay special attention to this direction in further attempts. The following graph shows the relative shrinkage values from Table 12.

Table 12. Comparison of dimension changes before and after sintering in the direction of the Z-axis.

	Measuring Wall	**Side Wall**			
		Measurement Orientation			
		From			
	Dimension before Sintering [mm]	**Dimension after Sintering [mm]**	**Shrinkage Rate [mm]**	**Shrinkage Rate [%]**	
	20.046	16.186	3.860	19.26	
	19.975	16.258	3.717	18.61	
	19.971	16.409	3.562	17.84	
	20.012	16.338	3.674	18.36	
	20.019	16.44	3.579	17.88	1st series
	19.97	16.421	3.549	17.77	
	19.948	16.282	3.666	18.38	
	20.03	16.179	3.851	19.23	
	20.05	16.069	3.981	19.86	
	19.953	15.914	4.039	20.24	
	19.991	16.046	3.945	19.73	
	19.944	16.260	3.684	18.47	
	19.961	16.183	3.778	18.93	
	20.002	16.223	3.779	18.89	2nd series
	19.957	16.147	3.810	19.09	
	20.02	16.230	3.790	18.93	
	19.946	16.000	3.946	19.78	
	19.927	15.868	4.059	20.37	
	19.837	16.246	3.591	18.10	
	19.9	16.266	3.634	18.26	
	20.007	16.274	3.733	18.66	
	19.886	16.307	3.579	18.00	
	19.964	16.339	3.625	18.16	3rd series
	19.984	16.335	3.649	18.26	
	19.991	16.279	3.712	18.57	
	19.911	16.256	3.655	18.36	
	20.005	16.002	4.003	20.01	
Min.	19.837	15.868	3.549	17.77	
Max.	20.050	16.44	4.059	20.37	
AP	19.982	16.198	3.775	18.90	
SD	0.04885	0.14475			

From Table 13, it can be seen that the greatest rate of shrinkage in the direction of the Z-axis (Figure 21) was in the second row. Overall, however, the figures in this regard are well above the planned 16%. This is due to the stacking printing method, where cubes printed at a lower level are affected by the weight of the cubes above them.

Table 13. The difference between shrinkage in different rows.

Shrinkage from [%]	
1st series	18.57
2nd series	19.38
3rd series	18.49

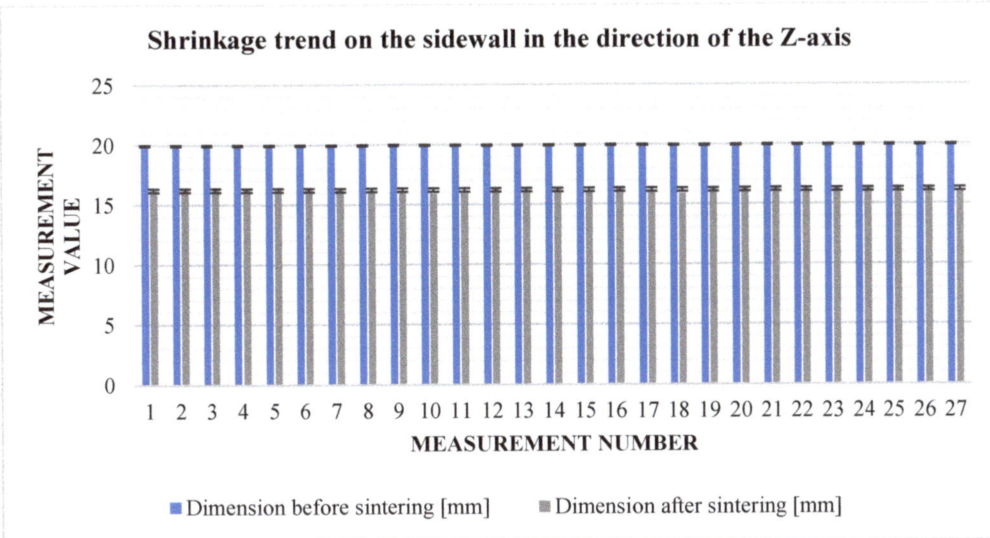

Figure 21. Graphical representation of Table 12—Z-axis shrinkage.

4.4. Hardness Tests

The hardness test was carried out according to Vickers (Table 14), carried out according to STN EN ISO (Slovak Technical Standard International European Standard Organization for Standardization) 6507-1 [43]. The punctures were made on a ground surface in the middle of the individual samples. The measured puncture diagonals are recalculated according to the device manufacturer's tables. Hardness was measured on WPM HPO-250 (Leipzig, Germany).

Table 14. Vickers hardness test values.

	u1 [mm]	u2 [mm]	Diameter [mm]	HV10	HRB
1. Measurement	0.350	0.360	0.3550	147	-
2. Measurement	0.360	0.349	0.3545	148	-
3. Measurement	0.351	0.350	0.3505	151	-
4. Measurement	0.350	0.350	0.3500	151	-
5. Measurement	0.358	0.352	0.3550	147	-
Diameter				148.8	77.9

In the material sheet from the manufacturer ExOne, the hardness according to Rockwell 67–71 HRB is indicated. After converting the units from our test, we obtain a value of 77.9 HRB. This means that we measured a hardness higher than the value stated by the manufacturer.

4.5. Static Tensile Test

Tensile tests were carried out according to STN EN (Slovak Technical Standard International European Standard) 10002-1. The proposed shape of the samples was circular in shape. The following Figure 22 shows deformed tensile samples.

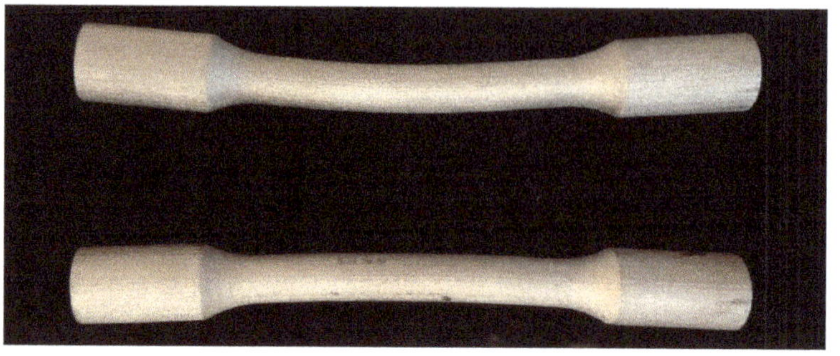

Figure 22. Deformed tensile samples.

The first printed samples were highly deformed not only in the axis of the samples but also in cross-section—it was not a circle but an ellipse. This deformation occurred during the sintering process. To print correction samples, a program was applied that predicts the rate of shrinkage and the direction of deformation during sintering. The result of the program is the design of such a geometry of the product, which is deformed into the desired shape during sintering. In addition to adjusting the geometry, the program (Figure 23) can also design supporting elements.

Figure 23. Samples printed using the Live Sinter simulation program.

The following Figure 24 shows samples with supporting elements suggested by Live Sinter.

Figure 24. Samples with support elements suggested by Live Sinter.

The first attempts in the tensile test (Figure 25) resulted repeatedly in the bars slipping out of the fixture. The material deformed above yield strength, but at values ≈ 455–460 MPa, it repeatedly slipped.

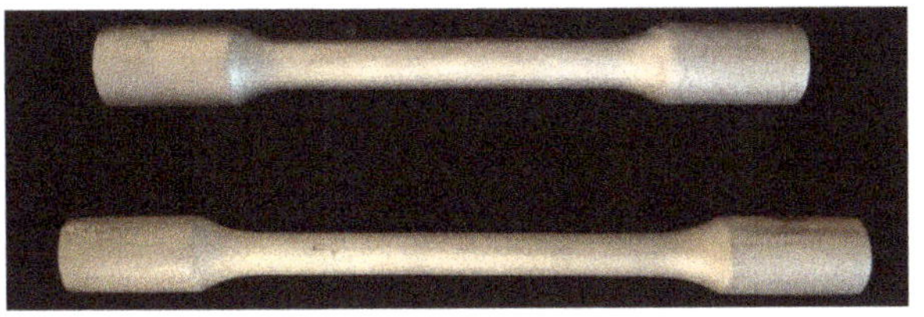

Figure 25. Samples after tensile test.

The bottom sample in Figure 25. with considerable elongation and change in section thickness. However, slipping (Figure 26) from the fixture prevented it from breaking, and therefore, a different geometry will be chosen for further tests.

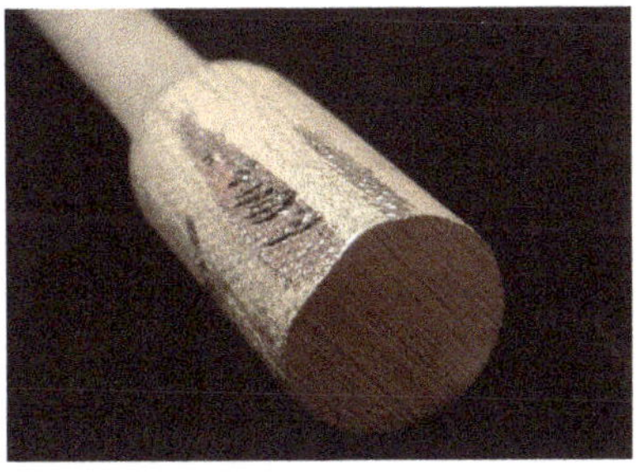

Figure 26. Traces of slipping out of the preparation.

To eliminate the extraction of samples from the preparation, new rods were designed which were terminated with a thread (Figure 27). The samples were made by turning printed rods. These rods were then screwed into the special fixture in which the test was conducted.

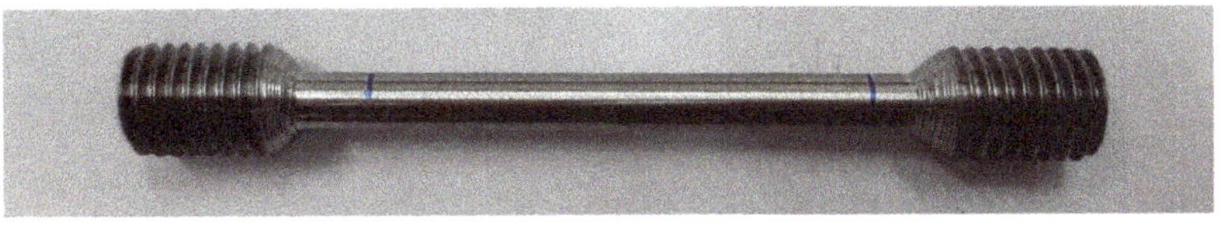

Figure 27. M12 threaded samples used in tensile test corrective experiments.

Turned samples were successfully tested. The tensile test results are shown in the diagram (Figure 28).

The testing used 10 samples and averaged the results. The following Table 15 compares the values measured by tests and the material sheet supplied with ExOne's 316L steel.

Table 15. Comparison of values obtained by experiment and material sheet.

	Manufacturer's Declared Value	The Value Obtained by the Experiment
Rm [MPa]	450–580	530–545
Rp 0.2 [MPa]	140–220	245–259
A * [%]	40–50	39.5–41

Explanation: * Elongation at break.

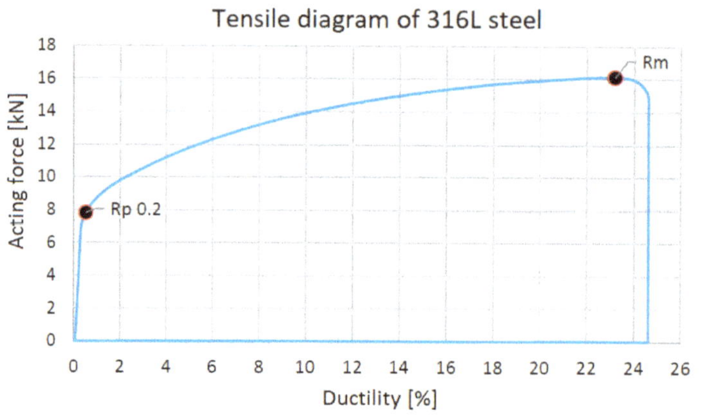

Figure 28. Tensile test conduct.

When comparing the results, we can say that the characteristics declared by the manufacturer agree with the values obtained by the experiment (Figure 29). The test samples were pushed to plane *XY*. The current state of optimization of the printer does not yet allow printing a similar sample in the vertical direction.

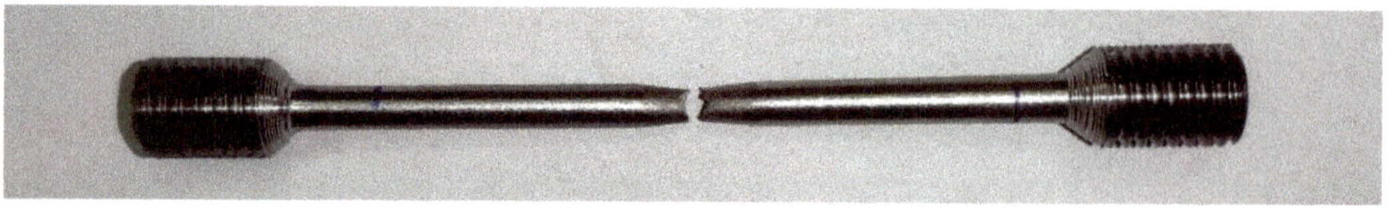

Figure 29. Sample after rupture.

4.6. Spectrometry-Verification of Chemical Composition

Verification of the chemical composition of the samples was performed on the BELEC COMPACT PORT spectrometer. The principle of chemical analysis consists of discriminating the surface of the sample with an argon gun. The following Figure 30 shows the sample after the test.

Figure 30. The surface of the sample after discussing the surface.

Spectrometry determined the content not only of the elements declared by the manufacturer but also of others, such as P, S, Cu, and Nb (Table 16). The result of spectrometry showed that the powdered material has values that fall into the material sheet from the manufacturer ExOne. The only element that is outside the interval is carbon (Table 17). Based on the chemical composition, the spectrometer estimated steel grade 1.4401, marked X5CrNiMo17-12-2, in US (United States) standard (AISI (American Iron and Steel Institute)) designated 316. Stainless austenitic chromo-nickel-molybdenum steel has the same properties as 316L steel but is not resistant to intergranular corrosion.

Table 16. Spectrometry measurement results.

Measurement	Fe [%]	C [%]	Si [%]	Mn [%]	P [%]	S [%]	Cu [%]	
1	67.760	0.052	0.870	1.310	0.016	<0.002	0.231	
2	67.670	0.044	0.924	1.352	0.017	<0.002	0.225	
3	67.580	0.040	0.924	1.357	0.017	<0.002	0.224	
SD	0.07348	0.00498	0.02254	0.02107	0.00047	0	0.00309	
Measurement	**Al [%]**	**Cr [%]**	**Mo [%]**	**Ni [%]**	**V [%]**	**Ti [%]**	**Nb [%]**	**Co [%]**
1	0.024	17.030	2.232	10.280	0.059	0.005	0.017	0.108
2	0.022	16.750	2.173	10.580	0.076	0.006	0.032	0.122
3	0.022	16.700	2.142	10.720	0.082	0.006	0.037	0.125
SD	0.00094	0.14522	0.03733	0.18354	0.00974	0.00047	0.00849	0.00740

Table 17. Comparison of spectrometry results and material sheet.

	Fe [%]	Ni [%]	Cr [%]	C [%]	Mo [%]	Mn [%]	Si [%]
Diameter	67.670	10.527	16.827	0.045	2.182	1.340	0.906
ExOne	remainder	10–14	16–18	max. 0.03	2–3	max. 2.0	max. 1.0

4.7. Density

An important parameter that has a significant impact on the quality and mechanical properties of the printed components is density. The manufacturer ExOne indicates a density in the range of 7.6–7.9 g/cm^3 for 316L. In our experiment, we will measure the relative density of cubes before and after sintering, with an indicator of 7.97 g/cm^3. The measured samples were randomly selected from the first, second, and third series of the desktop of the X160Pro printer.

From Table 18. we can see that the cubes entered the process with an input relative density ≈ 61.21%. The highest values were achieved in the third row. Further research could include a deeper comparison of the influence of sample placement on the resulting density.

Table 18. Relative density of samples before sintering.

		X [mm]	Y [mm]	Z [mm]	V [mm^3]	Weight [g]	Density [g/cm^3]	Relative Density [%]
	Sample 1	19.927	20.281	20.046	8101.380	38.254	4.72	59.77
Series 1	Sample 2	19.974	20.307	19.975	8102.100	38.994	4.81	60.92
	Sample 3	19.992	20.308	19.971	8108.177	39.761	4.90	62.07
	Sample 4	20.013	20.016	19.953	7992.777	37.454	4.69	59.32
Series 2	Sample 5	19.926	19.926	19.991	7937.336	37.781	4.76	60.25
	Sample 6	20.028	20.012	19.944	7993.562	38.533	4.82	61.02
	Sample 7	19.929	20.217	19.837	7992.418	39.070	4.89	61.88
Series 3	Sample 8	19.834	20.228	19.9	7983.923	39.746	4.98	63.02
	Sample 9	19.954	20.204	20.007	8065.834	39.927	4.95	62.66
SD							3490.93	
			Average					61.21

Table 19 shows the effect of shrinkage on the finite relative density. The resulting relative density is 98.7% (Figure 31). Again, we can see that in the three series, the values are the highest. The value achieved by us is in accordance with the value declared by the manufacturer ExOne in the material sheet for steel 316L.

Table 19. Relative density of samples after sintering.

		X [mm]	Y [mm]	Z [mm]	V [mm3]	Weight [g]	Density [g/cm^3]	Relative Density [%]
	Sample 1	17.222	17.473	16.186	4870.691	37.901	7.78	98.50
Series 1	Sample 2	17.217	17.441	16.258	4881.980	38.416	7.87	99.61
	Sample 3	17.35	17.554	16.409	4997.556	39.082	7.82	98.99
	Sample 4	17.157	17.425	15.914	4757.661	35.560	7.47	94.61
Series 2	Sample 5	17.246	17.272	16.046	4779.669	37.186	7.78	98.48
	Sample 6	17.275	17.315	16.26	4863.636	38.207	7.86	99.44
	Sample 7	17.367	17.577	16.246	4959.250	39.106	7.89	99.81
Series 3	Sample 8	17.407	17.627	16.266	4990.949	39.227	7.86	99.49
	Sample 9	17.439	17.697	16.274	5022.449	39.417	7.85	99.34
SD							4436.13	
			Average					98.70

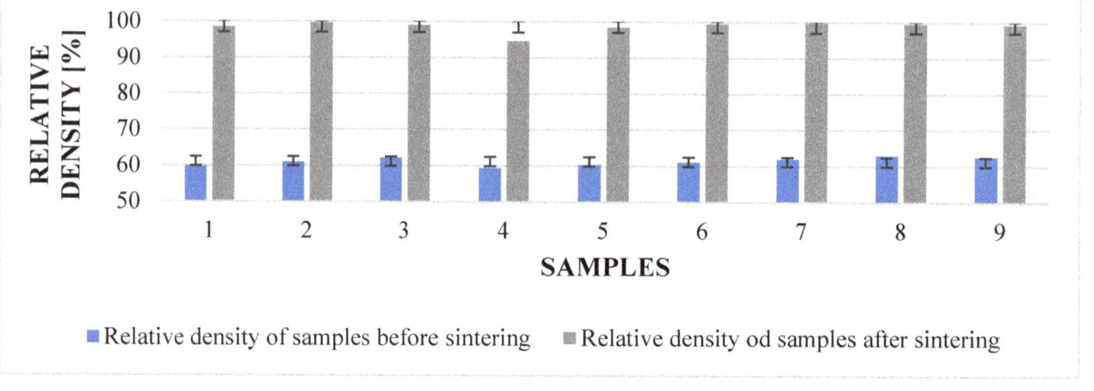

Figure 31. Comparative graph between the densities obtained before and after sintering.

4.8. *Macrostructure of Steel 316L*

The microstructure of the sample made of 316L steel was assessed on the KEYENCE VH-Z100R digital microscope.

Figure 32 shows a comparison between the microstructure given in the material specification and the microstructure captured on a digital microscope.

We can say that with more thorough polishing of the sample, the resulting microstructure would be closer to the one stated by the manufacturer ExOne. Our sample shows a slightly higher degree of porosity. However, it was not significantly reflected in the resulting density and mechanical properties.

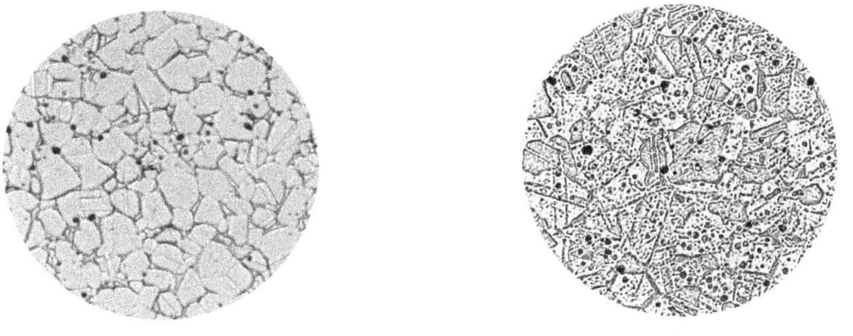

(a) (b)

Figure 32. Samples: (**a**) comparison of the microstructure given in the material specification [6], (**b**) microstructure captured on a digital microscope. Scale bar: 1:500.

4.9. Cost-Effective Binder Jetting Technology

For an approximate estimate of the economic advantage of binder jetting technology, we compared the production of a six-edge screw M12 x 100 made of stainless steel, similar in properties to 316L steel. The calculation was carried out using an online calculator offered by the manufacturer ExOne [42]. In the application, it is necessary to enter the type of machine on which the printing will take place, the material of the powder, the dimensions of the final component, its approximate volume in mm^3, and the method of distribution or percentage of filling the powder bed. The output values include approximate prices for powder, binder, and detergents.

The resulting values calculated by the calculator are shown in Table 20.

Table 20. Estimated cost of production.

CapEx	0.05 €
Material	2.20 €
Other	1.95 €
Total cost of 1 part	4.20 €

The average price of a finished M12 x 100 screw, made using conventional technology, is approximately 1.48 € [44]. It is necessary to take into account the fact that, at the moment, we cannot produce a screw of the required precision and quality using binder jetting technology without finishing modifications of turning, grinding, or polishing. These would significantly increase the resulting cost of one component. It is also important to mention current electricity prices. The most energy-intensive process is the sintering phase in a sintering furnace, consuming approximately 1.5 MWh per sintering cycle.

Based on these facts, we can say that finding a product suitable for series production for 316L steel is difficult. A more suitable application of binder jetting technology is piece and prototype production. It is especially suitable for components where we can take advantage of the fact that the component is printed as a whole and thus eliminate the need to use a combination of conventional technologies (e.g., bending and welding), which would shorten production time and reduce production costs. One of the proposed parts for the use of binder jetting technology is a pressurized hydrogen tank (Figure 33) with special cooling channels.

Figure 33. Hydrogen tank made by binder jetting technology.

By using binder jetting technology, several technological operations are eliminated. The inside of the tank contains special channels that were made during one operation together with the tank itself. This eliminates the technology needed to make channels and connect them to the tank. The tank itself is also printed as a single unit without the need for welding. By producing the tank and its interior as a whole, the unsaturated powder is enclosed in the cavity of the tank. Therefore, it is necessary to remove it with compressed air or other technology that will not damage the geometry of the component before sintering.

In addition to minimizing the technologies required to produce certain components, the great benefit of additive binder jetting is the reduction in the weight of parts. The fact that the technology theoretically does not know an obstacle in terms of the manufacturability of geometrically complex parts opens up space for the use of so-called topological optimization and generative design. The principle of topological optimization (Figure 34) is based on the design of a structurally and geometrically ideal shape of a component by gradual material removal (Figure 35). Generative design, on the other hand, designs a geometric shape by gradually adding material. Suitable software generates a part based on boundary conditions such as material, load, and degrees of freedom that would often not be possible to produce with conventional technologies or would be economically disadvantageous. The final geometry of the product is as lightweight as possible while meeting the strength and mechanical requirements. The minimum weight of the currently ever-increasing fuel prices is particularly suitable in the automotive industry. In addition to the automotive industry, it is also used in the construction, medical, and aerospace industries.

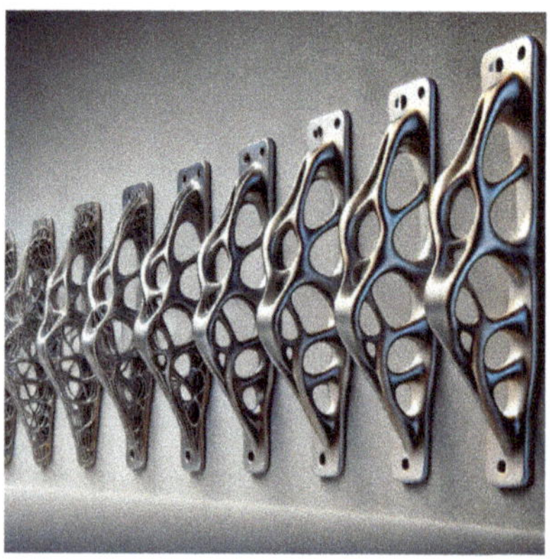

Figure 34. Gradual removal of material from the structural element by topological optimization.

Figure 35. Piston made by 3D printing technology.

The piston (Figure 35) was designed by Porsche using topological optimization. Additive technology was used to produce it, reducing weight by up to 10% on such a small component [45].

5. Discussion

Binder jetting technology has enormous potential, and current trends indicate its growing importance in industrial applications. It now enables the printing of the largest metal parts, which puts it at the forefront among other additive technologies. It is extremely suitable for the automotive, marine, aerospace, and even aerospace industries, requiring minimal finishing technological modifications, known as post-processing.

The advantages of this technology lie in its ability to create complex geometries that would be unfeasible or costly with conventional manufacturing methods. In addition, binder jetting allows the use of a wide range of materials, which opens the door to innovative applications and new business opportunities.

However, there is much room for improvement in this technology, not only in terms of hardware but also in terms of software. Improving data pre-preparation for devices without design interventions is a key aspect that can significantly impact print quality and efficiency. The authors of the article have two devices with binder jetting technology at their disposal and plan to optimize them in the future based on data preparation. This step is extremely important, as optimizing the data to eliminate material density depending on the location of the print area can lead to higher material homogeneity. Homogeneity is very important for the mechanical properties and reliability of the final products.

The expansion of binder jetting technology and the acceleration of its deployment can lead to a reduction in production costs, which will increase its competitiveness against other additive technologies as well as traditional production methods. In addition, advances in software tools for data preparation can bring better control over the printing process, allowing for even higher product quality and accuracy.

Another important aspect to consider is the environmental impact of binder jetting technology. This technology has the potential to reduce waste and material consumption compared to traditional production methods, which is in line with the global pressure for more sustainable production processes. In addition, reducing costs and increasing efficiency can lead to wider adoption of this technology, which could have a positive impact on the entire industrial ecosystem.

As experiments with the 316L material have shown, achieving the specified product properties implies the perfect alignment of several stages of the production process. The most important influence on the final accuracy of the product is the sintering process in the sintering furnace. In this step, the inaccuracies achieved at the powder and binder layering stage are multiplied as a result of shrinkage. The resulting geometry is also significantly influenced by the powder's own weight. This was also confirmed by optical measurement of samples, where the vertical direction appeared to be the most problematic direction of printing. In this axis, shrinkage reached its highest values. In addition to dimensional accuracy, compliance with geometric tolerances is also a challenge. A simulation program predicting deformation offers a potential solution to deviations that occur during sintering. A deeper investigation of the parameters and their effects in the printing, dedusting, and sintering phases is a must.

Although binder jetting technology already offers significant advantages today, its future lies in continuous development and optimization. Continued investment in research and development, collaboration between academia and industry, and a focus on improving software and hardware solutions are key to unlocking the full potential of this technology. This not only improves the quality and efficiency of production but also makes the industrial process more sustainable and economically advantageous.

The economic advantage of technology depends on several factors. The most important factor is the manufacturability of another technology. Another important factor is the material, which predetermines the field of application. The 316L steel we examine is the most widely used material in PM. Its practical use in terms of economic return is inefficient when compared to the mass production of conventionally produced components. A negative factor affecting economic profitability is the production time and the sintering itself. The sintering process is both an energy-consuming and time-consuming process that negatively affects economic returns. An unmistakable factor is that binder jetting is currently unable to produce parts without the need for finishing operations. The production flexibility of binder jetting is particularly suitable for piece or prototype production. To achieve a precisely defined geometry, it is necessary to use the addition of material for finishing operations, which increases production costs. The production of similarly complex parts requires considerable research into the effects of process parameters that directly affect the final density, accuracy, and properties of the final product.

6. Conclusions

The additive approach of binder jetting technology, coupled with the wide range of materials available, is a potential solution even for conventionally non-manufacturing components.

The optimal production setup has a significant impact on the overall economy of the process binder jetting technology. The influence of parameters such as powder layering rate, binder deposition, powder and binder thickness, sintering time and sintering temperatures, automation of dedusting processes, parts handling, and others are other areas of research whose benefits for binder jetting will open up new possibilities of application.

The measurement found that the shape in the vertical direction is significantly influenced by gravity, which deforms the printed parts differently depending on the vertical location in the printing area. The difference between the printed samples in the upper and lower base in the vertical direction was 3%. During the data design, it must be calculated according to the location of the printed part.

The properties of the material declared by the manufacturer have been verified by static tensile test and hardness test. With a relative density of almost 99%, the resulting sample properties were comparable to conventionally treated 316L stainless steel.

The results of the tensile test proved that the material is able to withstand a stress of more than 530 MPa.

The hardness of the structure was above the manufacturer's declared value—77.9 HRB. At a relative density above 99% or using infiltration, these values would be even higher. The chemical composition of metallic powder has been verified by spectrometry. When observing the microstructure with a digital microscope, porosity could be seen, which, however, did not affect the resulting material properties too negatively.

Sintering in a sintering furnace is one of the most expensive items from the cost of making binder jetting. One cycle of sintering in a sintering furnace consumes 1.6 MWh of electricity, which, at current astronomical prices in Europe, makes it impossible to use this technology for commercial use. At a price of €0.50 per 1kWh of electricity without VAT, only the cost of sintering is €800; with a price of powder of €12 per kilogram, the cost of electricity is higher than 65 kg of material. It is, therefore, necessary to use the maximum capacity of the sintering furnace for efficiency because the price of the consumed electricity is distributed over a larger amount of sintered material. As it has emerged from measurements, when the powder itself is heavily loaded on the lower part of the printing base, there is a more significant deformation and different deformation compared to the parts printed in the upper part.

The authors are currently working on optimizing the data for printing-better determining deformations in the vertical direction, which were also reflected in the measurements already presented. Another research is the mechanical improvement of the application parts of the device, which is already the subject of a patent procedure.

Author Contributions: Methodology, T.D.; Software, H.D.; Validation, T.D.; Formal analysis, J.K.; Writing—original draft, J.K.; Writing—review & editing, T.D.; Supervision, H.D. All authors have read and agreed to the published version of the manuscript.

Funding: This research received no external funding.

Institutional Review Board Statement: Not applicable.

Informed Consent Statement: Not applicable.

Data Availability Statement: The original contributions presented in the study are included in the article, further inquiries can be directed to the corresponding author.

Acknowledgments: The article was written in the framework of Grant Projects: KEGA 054TUKE-4/2022 "Innovation of production technology study programs focused on the internationalization of education using virtual or augmented reality methods". Also, with the support of the Operational Programme Integrated Infrastructure for the project Center for the Development of Textile Intelligence and Antimicrobial Technologies co-financed by European Regional Development Fund. ITMS2014+ 313011AVF5 Center for the Development of Textile Intelligence and Antimicrobial Technologies.

Conflicts of Interest: The authors declare no conflict of interest.

References

1. Murr, L.E.; Martinez, E.; Amato, K.N.; Gaytan, S.M.; Hernandez, J.; Ramirez, D.A.; Shindo, P.W.; Medina, F.; Wicker, R.B. Fabrication of Metal and Alloy Components by Additive Manufacturing: Examples of 3D Materials Science. *J. Mater. Res. Technol.* **2012**, *1*, 42–54. [CrossRef]
2. Sames, W.J.; List, F.A.; Pannala, S.; Dehoff, R.R.; Babu, S.S. The Metallurgy and Processing Science of Metal Additive Manufacturing. *Int. Mater. Rev.* **2016**, *61*, 315–360. [CrossRef]
3. Cerdas, F.; Juraschek, M.; Thiede, S.; Herrmann, C. Life Cycle Assessment of 3D Printed Products in a Distributed Manufacturing System. *J. Ind. Ecol.* **2017**, *21*, S80–S93. [CrossRef]
4. Ford, S.; Despeisse, M. Additive Manufacturing and Sustainability: An Exploratory Study of the Advantages and Challenges. *J. Clean. Prod.* **2016**, *137*, 1573–1587. [CrossRef]
5. Mostafaei, A.; Elliott, A.M.; Barnes, J.E.; Li, F.; Tan, W.; Cramer, C.L.; Nandwana, P.; Chmielus, M. Binder Jet 3D Printing—Process Parameters, Materials, Properties, Modeling, and Challenges. *Prog. Mater. Sci.* **2021**, *119*, 100707. [CrossRef]
6. Ziaee, M.; Crane, N.B. Binder Jetting: A Review of Process, Materials, and Methods. *Addit. Manuf.* **2019**, *28*, 781–801. [CrossRef]
7. Li, M.; Du, W.; Elwany, A.; Pei, Z.; Ma, C. Metal Binder Jetting Additive Manufacturing: A Literature Review. *J. Manuf. Sci. Eng. Trans. ASME* **2020**, *142*, 090801. [CrossRef]
8. Shahed, K.S.; Manogharan, G. Powder-Binder Interaction in Binder Jetting Process: A Simulation Study on Bimodal Powders. In Proceedings of the ASME 2023 18th International Manufacturing Science and Engineering Conference, New Brunswick, NJ, USA, 12–16 June 2023; ASME: New Brunswick, NJ, USA, 2023; Volume 1, pp. 1–10.
9. Roberts, J.W.; Sutcliffe, C.J.; Green, P.L.; Black, K. Modelling of Metallic Particle Binders for Increased Part Density in Binder Jet Printed Components. *Addit. Manuf.* **2020**, *34*, 101244. [CrossRef]
10. Yang, Z.; Zhang, Y.; Yan, W. High-Fidelity Modeling of Binder–Powder Interactions in Binder Jetting: Binder Flow and Powder Dynamics. *Acta Mater.* **2023**, *260*, 119298. [CrossRef]

11. Mao, Y.; Li, J.; Li, W.; Cai, D.; Wei, Q. Binder Jetting Additive Manufacturing of 316L Stainless-Steel Green Parts with High Strength and Low Binder Content: Binder Preparation and Process Optimization. *J. Mater. Process. Technol.* **2021**, *291*, 117020. [CrossRef]
12. Yang, Q.; Li, M.; Zhao, Z.; Liao, X.; Li, J. Simulation of Binder Jetting and Analysis of Magnesium Alloy Bonding Mechanism. *3D Print. Addit. Manuf.* **2024**, *11*, 751–763. [CrossRef]
13. Bae, M.A.; Kim, K.H.; Baek, J.H. Effect of the Properties of Binder and Powder Used in Binder Jet 3D Printing on Build-Up. *Int. J. Met.* **2023**, *17*, 2500–2507. [CrossRef]
14. Lai, W.-J.; Ojha, A.; Luo, Z. Effect of Surface Roughness on Fatigue Behavior of 316L Stainless Steel Produced by Binder Jetting Process. In Proceedings of the TMS 2022 151st Annual Meeting & Exhibition Supplemental Proceedings, Anaheim, CA, USA, 27 February–3 March 2022; Springer International Publishing: Cham, Switzerland, 2022; pp. 164–176.
15. Navaf, M.; Sunooj, K.V.; Aaliya, B.; Akhila, P.P.; Sudheesh, C.; Mir, S.A.; George, J. 4D Printing: A New Approach for Food Printing; Effect of Various Stimuli on 4D Printed Food Properties. A Comprehensive Review. *Appl. Food Res.* **2022**, *2*, 12. [CrossRef]
16. Gibson, I.; Rosen, D.; Stucker, B.; Khorasani, M. *Additive Manufacturing Technologies*, 3rd ed.; Springer Nature: Cham, Switzerland, 2021; ISBN 978-3-030-56126-0.
17. Laa, D. ASTM and ISO Publish New Standard Terminology for Additive Manufacturing Technologies—ISO/ASTM 52900:2015. Available online: https://www.3printr.com/astm-standard-terminology-for-additive-manufacturing-technologies-withdrawn-5133486/ (accessed on 1 January 2020).
18. Allen, S.M.; Sachs, E.M. Three-Dimensional Printing of Metal Parts for Tooling and Other Applications. *Met. Mater. Int.* **2000**, *6*, 589–594. [CrossRef]
19. Mostafaei, A.; Behnamian, Y.; Krimer, Y.L.; Stevens, E.L.; Luo, J.L.; Chmielus, M. Effect of Solutionizing and Aging on the Microstructure and Mechanical Properties of Powder Bed Binder Jet Printed Nickel-Based Superalloy 625. *Mater. Des.* **2016**, *111*, 482–491. [CrossRef]
20. Fang, Z.Z. *Sintering of Advanced Materials*, 1st ed.; Woodhead Publishing: Sawston, UK, 2010; ISBN 9780081014776.
21. German, R.M.; Suri, P.; Park, S.J. Review: Liquid Phase Sintering. *J. Mater. Sci.* **2009**, *44*, 1–39. [CrossRef]
22. de Oro Calderon, R.; Gierl-Mayer, C.; Danninger, H. Fundamentals of Sintering. In *Encyclopedia of Materials: Metals and Alloys*; Elsevier Inc.: Wien, Austria, 2021; pp. 481–492, ISBN 978-0-12819-733-2.
23. Yap, Y.L.; Wang, C.; Sing, S.L.; Dikshit, V.; Yeong, W.Y.; Wei, J. Material Jetting Additive Manufacturing: An Experimental Study Using Designed Metrological Benchmarks. *Precis. Eng.* **2017**, *50*, 275–285. [CrossRef]
24. Kumbhar, N.N.; Mulay, A.V. Post Processing Methods Used to Improve Surface Finish of Products Which Are Manufactured by Additive Manufacturing Technologies: A Review. *J. Inst. Eng. Ser. C* **2018**, *99*, 481–487. [CrossRef]
25. Guo, N.; Leu, M.C. Additive Manufacturing: Technology, Applications and Research Needs. *Front. Mech. Eng.* **2013**, *8*, 215–243. [CrossRef]
26. Astm, B. Automatic Hall Flowmeter Calibrated Funnel Hall Flowmeter for Metal Powder. 2020. Available online: https://www.aaamachine.com/partner/tsutsui/pdf/Brochure_Automatic_Hall_Flowmeter_Tsutsui_AAAmachine.pdf (accessed on 20 January 2024).
27. Utela, B.; Storti, D.; Anderson, R.; Ganter, M. A Review of Process Development Steps for New Material Systems in Three Dimensional Printing (3DP). *J. Manuf. Process.* **2008**, *10*, 96–104. [CrossRef]
28. McCarthy, D.L.; Williams, C.B. Creating Complex Hollow Metal Geometries Using Additive Manufacturing and Electroforming. In Proceedings of the 23rd Annual International Solid Freeform Fabrication Symposium—An Additive Manufacturing Conference, SFF 2012, Austin, TX, USA, 6–8 August 2012; Seepersad, C.C., Bourell, D.L., Marcus, H.L., Beaman, J.J., Crawford, R.H., Eds.; University of Texas at Austin: Austin, TX, USA, 2012; pp. 108–120.
29. Anderson, I.E.; White, E.M.H.; Dehoff, R. Feedstock Powder Processing Research Needs for Additive Manufacturing Development. *Curr. Opin. Solid State Mater. Sci.* **2018**, *22*, 8–15. [CrossRef]
30. Sutton, A.T.; Kriewall, C.S.; Leu, M.C.; Newkirk, J.W. Powders for Additive Manufacturing Processes: Characterization Techniques and Effects on Part Properties. In Proceedings of the Solid Freeform Fabrication 2016: Proceedings of the 27th Annual International Solid Freeform Fabrication Symposium—An Additive Manufacturing Conference, SFF 2016, Austin, TX, USA, 8–10 August 2016; Bourell, D.L., Ed.; University of Texas at Autin: Austin, TX, USA, 2016; pp. 1004–1030.
31. Mostafaei, A.; Stevens, E.L.; Hughes, E.T.; Biery, S.D.; Hilla, C.; Chmielus, M. Powder Bed Binder Jet Printed Alloy 625: Densification, Microstructure and Mechanical Properties. *Mater. Des.* **2016**, *108*, 126–135. [CrossRef]
32. Schmutzler, C.; Stiehl, T.H.; Zaeh, M.F. Empirical Process Model for Shrinkage-Induced Warpage in 3D Printing. *Rapid Prototyp. J.* **2019**, *25*, 721–727. [CrossRef]
33. Chen, H.; Zhao, Y.F. Process Parameters Optimization for Improving Surface Quality and Manufacturing Accuracy of Binder Jetting Additive Manufacturing Process. *Rapid Prototyp. J.* **2016**, *22*, 527–538. [CrossRef]
34. Meteyer, S.; Xu, X.; Perry, N.; Zhao, Y.F. Energy and Material Flow Analysis of Binder-Jetting Additive Manufacturing Processes. *Procedia CIRP* **2014**, *15*, 19–25. [CrossRef]
35. Salehi, M.; Gupta, M.; Maleksaeedi, S.; Sharon, N.M.L. *Inkjet Based 3D Additive Manufacturing of Metals*; Materials Research Forum LLC: Millervile, PA, USA, 2018; ISBN 978-1-945291-44-9.

36. Do, T.; Shin, C.S.; Stetsko, D.; Vanconant, G.; Vartanian, A.; Pei, S.; Kwon, P. Improving Structural Integrity with Boron-Based Additives for 3D Printed 420 Stainless Steel. *Procedia Manuf.* **2015**, *1*, 263–272. [CrossRef]
37. Movrin, D.; Luzanin, O.; Guduric, V. Using Statistically Designed Experiment to Optimize Vacuum-Assisted Post-Processing of Binder Jetted Specimens. *Rapid Prototyp. J.* **2019**, *25*, 653–663. [CrossRef]
38. Kernan, B.D.; Sachs, E.M.; Oliveira, M.A.; Cima, M.J. Three Dimensional Printing of Tungsten Carbide-Cobalt Using a Cobalt Oxide Precursor. In Proceedings of the 14th International Solid Freeform Fabrication Symposium, Austin, TX, USA, 4–6 August 2003; University of Texas at Austin: Austin, TX, USA, 2003; pp. 616–631.
39. Wang, H.R.; Cima, M.J.; Kernan, B.D.; Sachs, E.M. Alumina-Doped Silica Gradient-Index (GRIN) Lenses by Slurry-Based Three-Dimensional Printing (S-3DP^{TM}). *J. Non. Cryst. Solids* **2004**, *349*, 360–367. [CrossRef]
40. Cho, W.; Sachs, E.M.; Patrikalakis, N.M.; Troxel, D.E. A Dithering Algorithm for Local Composition Control with Three-Dimensional Printing. *CAD Comput. Aided Des.* **2003**, *35*, 851–867. [CrossRef]
41. ExOne: Metal 3D Printers—Materials & Binders. Available online: https://exone.com/Admin/getmedia/92ff696e-1b4e-4c7e-a45c-ae45b04340da/PSC_X1_MaterialData_316L_10192020_V6.pdf (accessed on 20 January 2024).
42. Desktopmetal: X160Pro® Specifications. Available online: https://www.desktopmetal.com/uploads/20231015_Spec-Sheet_X160Pro_2024-04-19-185222_zcan.pdf (accessed on 25 August 2024).
43. *STN EN ISO 6507-1*; Metallic Materials: Vickers Hardness Test. Part 1, Test Method. International Organization for Standardization: Geneva, Switzerland, 2005.
44. Metric Screw DIN 933 M12X100 6HR Full Thread Stainless Steel. Available online: https://techmat-sk.eu/obchod/skrutka-metricka-din-933-m12x100-6hrh-celozavit-nerez/p-4380398.xhtml (accessed on 26 January 2024).
45. Porsche 3D-Printing Innovative Pistons for Increased Power and Efficiency. Available online: https://www.greencarcongress.com/2020/07/20200714-porsche.html (accessed on 7 February 2024).

materials

MDPI

Article

Electron Beam Welding of Dissimilar Stainless Steel and Maraging Steel Joints

Matúš Geľatko *, Radoslav Vandžura, František Botko and Michal Hatala

Faculty of Manufacturing Technologies, Technical University of Košice with a Seat in Prešov, 080 01 Prešov, Slovakia
* Correspondence: matus.gelatko@tuke.sk

Abstract: The incorporation of additive manufactured (AM) metal parts to real assemblies is a crucial issue for the increasing of their industrial utilization. The presented research is devoted to the electron beam welding (EBW) of dissimilar steel joints. Dissimilarity is defined by the various types of steel and manufacturing processes used for the creation of specimens. Conventional AISI 316 stainless steel, selective laser melted (SLM) SS 316L stainless steel, and SLM M300 maraging steel were welded at variable parameters in the form of a welding current and a welding velocity. EBW joints were evaluated considering the macroscopic and microscopic characteristics, as well as a reached microhardness. The obtained preliminary results represent important input data for the follow-up experiments focused on the setting of optimal EBW parameters of welding the dissimilar joints including SLM products, with the consideration of their basic macroscopical and microscopical characteristics, mechanical properties, and residual stresses.

Keywords: fusion welding; additive manufacturing; selective laser melting; AISI 316; SS 316L; M300

1. Introduction

Additive manufacturing (AM) is sophisticated, highly innovative and currently well-known technology that brings many positives into the industrial sector, mainly in the form of the freedom in designing the shape of final components [1]. The dimensions of AM parts are still restrictive, as they are defined by the possibilities of 3D printing machine parameters, concretely by their chamber volume. This phenomenon consequently leads to the issue of their incorporation into the real functional assemblies within the connection of AM products with conventional parts (dissimilar joints) and equally their mutual connection [2]. It also includes components created using laser powder bed fusion (LPB-F) technologies working with the input material in the form of a metal powder [3], whereas some practical applications include design of lattice structures that were experimentally verified in the case of numerically modeled L-PBF AlSi10Mg aluminum alloy with promising results [4]. Another similar experiment was conducted on the LPB-F AISI 316L stainless steel, where roughness was evaluated on different configurations of structures [5]. Considering the requirements on the creation of inseparable joints, metal-based materials demand the necessity of applying the welding technologies for their connection [6].

Within recently conducted experiments, more sophisticated welding technologies have been used for the creation of joints composed of AM materials. Tungsten inert gas (TIG) welding was used for the joining of tubes made of SLM (selective laser melting) AISI 316L stainless steel [7]. The experiment showed that higher temperatures have positive influence on the microstructure, due to its increasing homogeneity in the direction of the weld axis, in which the microhardness decreases. Also, porosity was shown to be significantly influential on the mechanical properties of the weld joint. The same technology and similar material were used in the study by Huysmans, S. et al. [8], but within dissimilar joints (conventional 316L and AM 316L), where the corrosion testing is worth mentioning, as it showed that pitting corrosion is situated mainly in the AM material and the heat-affected zone (HAZ). AM AISI 316L was the material of interest in the study [9], where its microstructure and tensile mechanical properties were compared with the same cold-rolled plates after the laser welding process. Defect-free welds were reached with the ferritic–austenitic mode solidification, and a fracture occurred in the weld metal (WM). For dissimilar joints (AISI 316L), input energy of a laser beam is a critical parameter for the weldability of used materials, as was stated in a study by Matilainen, V.P. et al. [10]. During the comparison of TIG and laser welding of AM and conventional aluminum parts (AlSi10Mg) [11], the most significant factor was shown to be the susceptibility of AM components to the creation of porosity, mainly in the case of a TIG welding. The lowest microhardness was reached on the fusion line of TIG weld composed of two AM parts and in the weld metal center of laser weld. The fracture occurred in the mentioned places during tensile tests. The positive influence of electron beam welding (EBW) technology on the final mechanical properties of specimens made of the same material (AlSi10Mg) was observed [12]. Electron beam welding was also used for the joining of LPB-F Inconel 718 specimens, where sufficient penetration depth was reached (proportional to the heat input), with nail-shaped welds of lower hardness in comparison with the base metal (BM) [13]. EBW was one of the technologies applied on the specific SLM CLF-1 material with good joint formation without observed defects [14].

Considering the AISI 316/316L stainless steel, various EBW currents were tested with the objective of evaluating the microstructure and mechanical properties of welds, where fine austenite grains in base metal and fine equiaxed grain in combination with elongated columnar austenite grains in weld metal were observed [15]. Technology showed high potential during the welding of the same material of 160 mm thickness at one-time penetration using the appropriate technological parameters [16]. Residual stresses in EBW 316L joints were analyzed due to the verification of finite element models [17] and also their sensitivity to the beam oscillation, where influenced corrosion properties were taken into account [18]. EBW is also highly effective for the creation of dissimilar joints of various materials, such as titanium alloys (TC4/TA7), where various microstructures were observed in various areas of a weld, with dislocations in a weld metal [19]. The same technology was effective for the welding of dissimilar nickel-based superalloys (GH3039/IC10) in reaching the sufficient joint without any defects, and the microstructure was composed of cellular dendrites and columnar crystals in the weld metal, with the average value of microhardness being 300 HV in this area [20]. 316L stainless steel was part of the dissimilar joint with the RAFM (reduced-activation ferritic/martensitic) steel [21]. The reached microstructure of a weld metal was martensitic with a small amount of a ferrite, whereas the weld metal of two 316L steels (similar joint) was composed of austenite. The heat-affected zone was not observed on the side of the 316L material. Another study evaluated the EBW joint of AISI 316L steel with the Inconel 718 alloy using X-ray tomography [22]. Welding parameters were tested based on the occurrence of porosity and undercuts.

The maraging steel was experimentally tested for welding using the TIG/GTAW (gas tungsten arc welding) method. Concretely, maraging steel grade 250 was welded, and the microstructure was found to change in both the fusion and heat-affected zone due to the high input of heat. The microhardness decreased in the direction of the weld axis, as the consequence of coarsening the grain [23]. Similar hardness development was described in the study [24], where the same technology was used for the maraging steel grade 300 and the embrittlement fracture behaviors with other mechanical properties within HAZ were tested. Laser welding was used for the joints composed of maraging steel 250, in order to create the optimization of welding parameters and numerical simulation, considering tensile strength, hardness, and penetration depth. The observed microstructure was martensitic in the WM and HAZ, with the highest microhardness in the HAZ [25]. Subashini, L. et al. [26] compared the application of the laser-MIG (metal inert gas) hybrid welding and autogenous laser welding on the same material and considered the influence of wire filler metal on microstructure and mechanical properties. A maraging steel was also the main subject of the critical review study focused on the susceptibility of its weldments to the stress corrosion cracking [27]. A dissimilar joint composed of C300 maraging steel and SS 316L stainless steel was reached using a DED (direct energy deposition) method, whereas the microstructure in the fusion zone was without intermetallic phases, and the evaluated mechanical properties of joints were similar to the stainless steel [28]. During the EBW application on maraging steels, WMs are represented by a dendritic microstructure [29], with the presence of a certain HAZ with microstructure similar to BM and the observation of a dark band region [30].

The experiment described is devoted to the application of EBW technology for the creation of dissimilar steel joints, whereas a dissimilarity is represented by both different manufacturing technology of base metals (SLM and conventional) and different steel kinds (stainless and maraging). The motivation of the research comes out directly from the unusual combination of the mentioned base metals, their production technologies, and obviously the welding technology, within which it is necessary to find the appropriate technological parameters for the creation of sufficient joints. Within this first stage of research, welds are evaluated from the macroscopic and microscopic view, and microhardness is measured, all based on various parameters in the form of a welding current and a welding velocity. The results are a necessary steppingstone for the comprehensive research, focused on the evaluation of created EBW welds of dissimilar SLM materials, with the objective of sufficient incorporation of metal AM components to greater assemblies, such as demanding engineering parts or specific cutting tools.

2. Materials and Methods

Three materials were selected as the experimental specimens: conventional stainless steel AISI 316, AM stainless steel SS 316L, and AM maraging steel M300. The following Tables 1–3 include the chemical compositions of the mentioned base metals. The main physical, mechanical, and thermal properties of the experimental materials are included in Table 4.

Experimental specimens of SS 316L and M300 materials were prepared using selective laser melting (SLM) technology. Specimens were prepared in the Technical University in Ostrava (Ostrava, Czech Republic)–Center of 3D printing Protolab. For SS 316L stainless steel and for M300 maraging steel, RenAM500S Flex and AM400 systems by Renishaw company (Wotton-under-Edge, UK) were used, respectively. Both machines provide production speed at 5–20 $cm^3 \cdot h^{-1}$, scanning speed up to 2000 $mm \cdot s^{-1}$, positioning speed up to 7000 $mm \cdot s^{-1}$, and layer thickness between 20 and 100 μm. The first can create parts with laser power up to 500 W and the second up to 400 W in chambers of 250 × 250 × 350 mm and 250 × 250 × 300 mm, respectively. Particle distribution for both powders is in the range of 15–45 μm. Technological parameters for SLM preparation of experimental materials are included in Table 5.

$$\varepsilon = P/(v \times d \times t_L) \quad (1)$$

Table 1. Chemical composition of AISI 316 stainless steel [31].

Element	Fe	Cr	Ni	Mo	Mn	Si	N	P	C	S
Mass (%)	Balance	16–18	10–14	2–3	≤2	≤1	≤0.1	≤0.045	≤0.03	≤0.03

Table 2. Chemical composition of SS 316L stainless steel [32].

Element	Fe	Cr	Ni	Mo	Mn	Si	N	O	P	C	S
Mass (%)	Balance	16–18	10–14	2–3	≤2	≤1	≤0.1	≤0.1	≤0.045	≤0.03	≤0.03

Table 3. Chemical composition of M300 maraging steel [33].

Element	Fe	Ni	Co	Mo	Ti	Si	Mn	C	P	S
Mass (%)	Balance	17–19	7–10	4.5–5.2	0.3–1.2	≤0.1	≤0.15	≤0.03	≤0.01	≤0.01

Table 4. Mechanical and physical properties of the experimental materials [31–33].

Parameter	Direction	AISI 316	SS 316L (As Built)	M300 (As Built)
Density [g·cm^{-3}]	-	8.03	7.99	8.10
Hardness [HV]	XY Z	187	198 ± 8 * 208 ± 6 *	350 ± 15 * 357 ± 12 *
Yield strength [MPa]	XY Z	205	547 ± 3 494 ± 14	976 ± 17 794 ± 22
Elongation at break [%]	XY Z	40	43 ± 2 35 ± 8	15 ± 1 10 ± 2
Modulus of elasticity [GPa]	XY Z	193	197 ± 4 190 ± 10	185 ± 9 189 ± 6
Thermal conductivity [W·mK^{-1}]	-	16.3	16.2	14.2
Melting point [°C]	-	1375–1400	1371–1399	1413

* Vickers hardness HV0.5.

Table 5. SLM technological parameters.

Parameter	SS 316L	M300
Laser power P [W]	200	400
Scanning velocity v [mm·s^{-1}]	650	820
Hatch distance d [µm]	110	95
Layer thickness t_L [µm]	50	40
Energy input [J·mm^{-3}] (1) [34]	55.94	128.37
Scanning strategy (Figure 1)	Chessboard	Meander
Gas protection	Argon (Ar)	

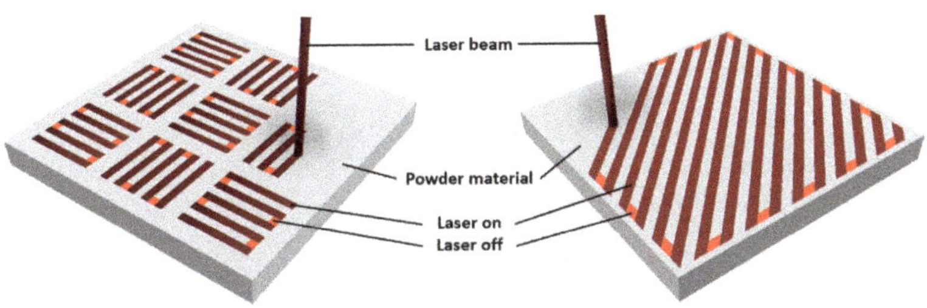

Figure 1. Stripe (**left**) and meander (**right**) scanning strategies [35].

Experimental materials (50 mm long, 20 mm wide, and 10 mm thick) were joined by the fusion welding using the EBW technology that uses a high concentration of electrons arising in the focal point of the electron beam, which causes local heating and the melting of influenced materials. Welds were created in First Welding Company, Inc., in Bratislava (Slovakia) using the automated EBW device PZ ELZA UNI 2G, with a vacuum chamber (10^{-2} Pa) in dimensions of 1300 × 1500 × 1550 mm and motion range up to 451 mm on the X and Y axes. Accelerating voltage was between 25 and 60 kV, electron beam power was in the range of 30 W to 30 kW, and the range of the welding current was 1–500 mA. The characteristic EBW weld bead profile of the experimental specimens was reached by the influence of the electron beam and its motion in a longitudinal direction of welded specimens, when the keyhole was created during the melting, which defines the weld bead profile after solidification. Its shape was also influenced by the EBW technological parameters. The following Table 6 summarizes the parameters of welding the experimental specimens, which were divided into three groups. The first group includes mutual joints of conventional AISI 316 steels for adjusting the main welding parameters. The second group includes joints of AISI 316 and SLM SS 316L steels for the evaluation of welding the stainless steels with different manufacturing processes. The third group includes joints of AISI 316 and SLM M300 steels for the evaluation of possibilities in welding of dissimilar joints composed of various steels, considering different manufacturing technologies and compositions. All specimens were welded using accelerating voltage (U_a) at 55 kV, focusing current (I_f) at 833 mA, and the oscillation diameter of 1 mm at 555 Hz frequency (f_o).

Table 6. EBW technological parameters.

Group	Specimen (Weld)	Materials	Welding Current (I_w) [mA]	Welding Velocity (v_w) [mm.s^{-1}]
1	W1	AISI 316–AISI 316	135	35
	W2	AISI 316–AISI 316	135	30
	W3	AISI 316–AISI 316	135	25
	W4	AISI 316–AISI 316	150	20
2	W5	AISI 316–SS 316L	135	25
	W6	AISI 316–SS 316L	150	30
3	W7	M300–AISI 316	135	25
	W8	M300–AISI 316	150	20
	W9	M300–AISI 316	150	20
	W10	M300–AISI 316	150	25

Welded specimens were subjected to the observation of macrostructure and microstructure in the cross-section due to the evaluation of morphologic geometry of the weld joint; rate of metallographic fusion of base metals; and characteristics of WM, HAZ, and BM. The prepared metallographic specimens were macroscopically analyzed using the Stemi 508 stereo microscope with an AxioCam ERc 5s camera (Carl Zeiss AG, Oberkochen, Germany). Microscopic analysis was conducted using the Axioskop 2 MAT optical microscope (Carl Zeiss AG, Oberkochen, Germany) with an AxioCam ERc 5s camera and a Keyence VHX-7000 digital microscope (Keyence, Osaka, Japan). The measurement of microhardness was conducted using the Micro-Vickers Hardness Tester CV-400DAT (CV Instruments, Borgharenweg, The Netherlands) at 2.942 N force and dwell time for 15 s.

3. Results

3.1. Macroscopic Evaluation

In order to adjust EBW parameters for experimental specimens, four welds were created of conventional AISI 316 stainless steels (Figure 2). All these joints included typical characteristics of the EBW process with certain discrepancies. The first weld (W1) was made using the 135 mA current at 35 mm·s^{-1} velocity. The whole thickness of base metals was sufficiently melted and fused in the form of a nail-shaped joint. However, the width (1.33 mm) of the weld joint was not quite auspicious. The steep narrowing right below the reinforcement was observed, as the consequence of the combination of the lower welding current with higher velocity. This phenomenon was suppressed by lowering the velocity to 30 mm·s^{-1} (W2). The nail joint was thicker with stable dimensions throughout the whole depth. Also, the excess of reinforcement was suppressed. In this case, a steep transition into the root area was observed. A wedge-shaped weld (W3) was reached at lowering the velocity to 25 mm·s^{-1}, with the thickness of 2.69 mm below the reinforcement and 1.38 mm above the root, whereas its shape was smooth through the whole depth without any steep transitions. A significant discontinuity was observed in the form of an undercut on the left side of the face. The last weld (W4) of the first group was created using the higher current (150 mA) and a velocity of 20 mm·s^{-1}. In comparison with the W3 weld, the reached width was higher, mainly in the area above the root (1.71). The reached width of the face was 3.78 mm, and the root was 1.73 mm wide; hence, the typical texture of a weld in a welding direction was visible. It is worth mentioning the significant excess of reinforcement, which was probably caused by the combination of a higher welding current and lower welding velocity.

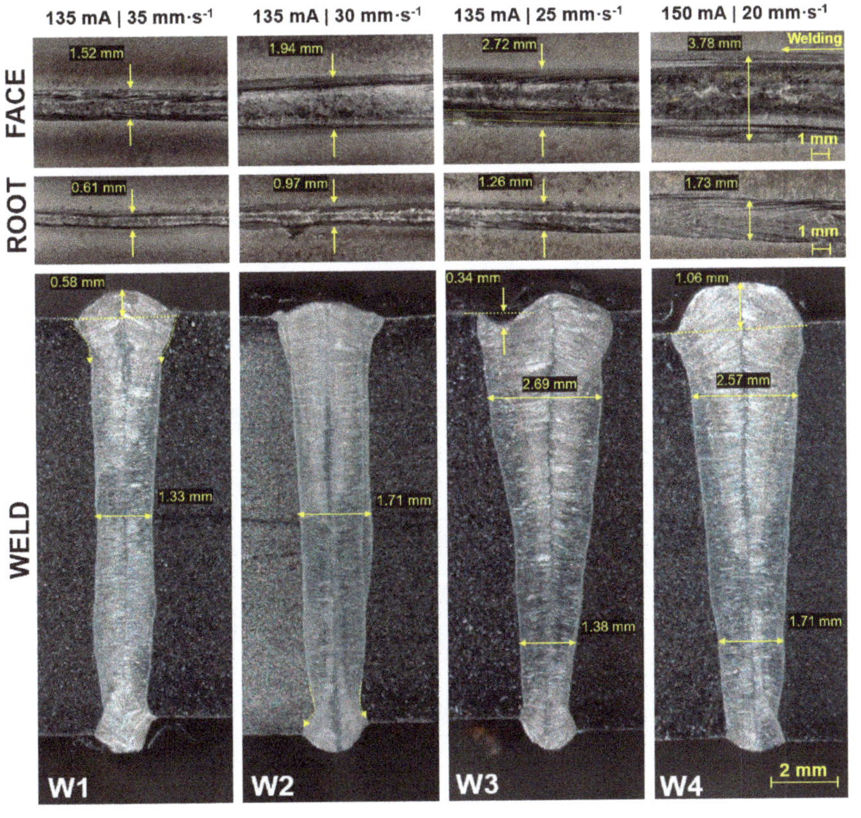

Figure 2. Macroscopic view of AISI 316–AISI 316 welds.

Similar welding parameters were used for the creation of dissimilar joints within the second group (Figure 3). For the first weld (W5) of this group, the same parameters of W3 weld were used (135 mA, 25 mm·s^{-1}). The reached basic shape characteristics of the wedge joint were also similar to the W3 weld, with the sufficient thickness of reinforcement and sequential narrowing in a direction to the root. A dissimilarity of base metals (different manufacturing processes) caused various dimensions of the weld to interface to its center axis, which was longer on the side of the SLM material in this case (1.33 mm). Also, the undercut was observed on this side, as well as the sudden variation of weld metal texture in the area of the highlighted curve, which could have been the consequence of the mixing of liquid metals in the fusion zone. On the other side of the weld (conventional base metal), a steep transition was observed in both below the reinforcement and above the root. For the W6 weld, the same current (150 mA) was used as in the case of the W4 weld, but the velocity was set at 30 mm·s^{-1}, due to the excess of reinforcement on the W4 weld. The weld metal was fused to a sufficient extent. The phenomenon of various weld dimensions to its center axis was not present through the whole depth of the wedge joint but only at the area below the reinforcement. In comparison with the W5 weld, a higher input energy caused a more significant impact of the electron beam, which resulted in the non-negligible root dropout in the evaluated cut of a weld.

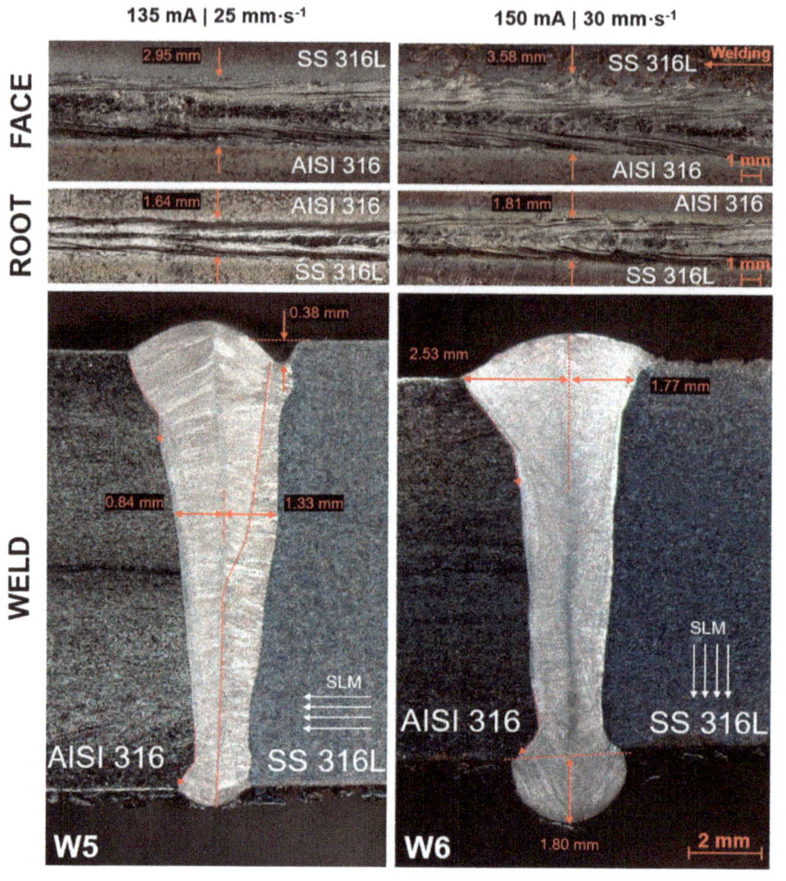

Figure 3. Macroscopic view of AISI 316–SS 316L welds.

Using the welding current at 135 mA and welding velocity at 25 mm·s^{-1} was also verified for the creation of the W7 weld (Figure 4). It is evident from the shape of the joint that the impact of the electron beam at these parameters was insufficient. The upper part of this wedge weld fulfils basic requirements. However, low welding parameters caused steep transition of its width below the reinforcement on both sides and a significant narrowing of joint, with the insufficient fusion of base metals on the root surface. Increasing the current to 150 mA and decreasing the velocity to 20 mm·s^{-1} caused the improvement in the weld thickness uniformity in a direction of depth (2.09 mm) within the straight W8 weld with the sufficient creation of weld root (1.14 mm wide). Similarly, steep transitions below the reinforcement were present, with a considerable underfill of the weld face. Vortex textures, caused by the formation of keyhole due to the sufficient energy density, were observed on both sides of a depth direction. The upper vortex is the consequence of the molten metal flow under the influence of surface tension, and the lower vortex is the consequence of the influence of the electron beam and the bottom surface tension. Within previously described welds (W1–W7), molten metal flows upward around the weld axis, caused by the recoil pressure arising during the metal vaporization and the flow on the surface of the weld driven by the hydrostatic pressure arising during the motion of the weld beam [36]. The energy input was lower, and hence the flows of molten metal below the upper surface and above the bottom surface were not present. The application of the selected EBW parameters on the W9 weld caused the creation of a straight weld with similar thickness (2.40 mm).

However, the reinforcement highness is satisfactory, but the root dropout occurred in this case. The root width was stated at 0.97 mm, but the view was distorted due to the excess of SLM support structures. Increasing the velocity (W10 weld) caused the creation of a joint with similar characteristics. Differently, the weld thickness was smaller in the place of the cut, and the epitaxial growth of the substructure composed of columnar crystals was observed above the root area; thus, the metal flow above the root surface was not present due to the insufficient impact of the electron beam in a deeper volume of the weld. However, face and root width were greater (2.83 mm and 1.54 mm, respectively). A serious defect was observed in the lower part of the weld through its whole thickness. This cavity seemed to be the consequence of higher speed and consequently lower energy density that led to insufficient time for solidification that could not stand the impact of cutting during the preparation of specimens for the microstructure analysis. Another source can be the porosity cluster, which caused a division of larger amount of material after its yield to cutting during the preparation of specimens. A residual pit after the pore presence is highlighted in the left upper corner of the cavity. Such porosity could be transferred from base metals, mainly M300, which could be porous after the SLM process.

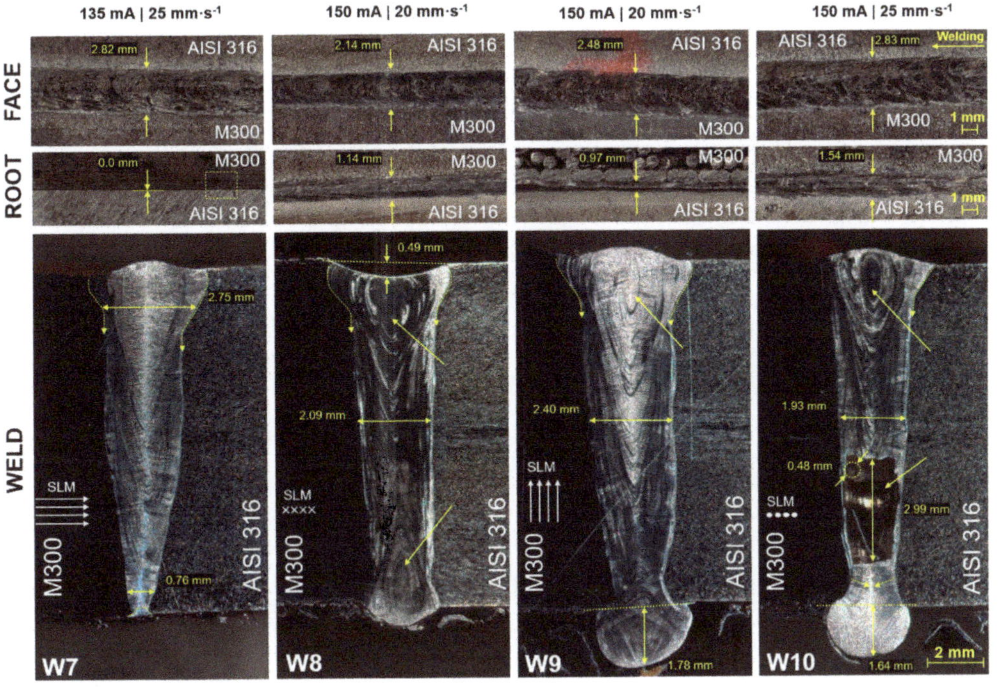

Figure 4. Macroscopic view of M300–AISI 316 welds.

The parameters used for the welding of the W7 specimen caused insufficient root penetration. Detailed analysis on the bottom surface of a specimen showed that the root penetration was different along the length of the weld (highlighted area of the root in Figure 4). The following figures depict various profiles within this area. Figure 5 includes the profile of the weld joint, where base metals were not fused on the bottom surface. The created crater was in the recorded depth around the level of 0.040 mm, which was approximately 0.035 mm below the analyzed surface. It is worth mentioning that surface profiles of base metals are different, mainly due to the various manufacturing processes. The profile of AISI 316 stainless steel was smoother within the evaluated area. A texture is defined by the traces in a diagonal direction as the consequence of the cutting process.

The base metal of the M300 maraging steel was more articulated without a defined pattern, caused by the higher roughness reached after the SLM process. The profile in the evaluated area included more deviations.

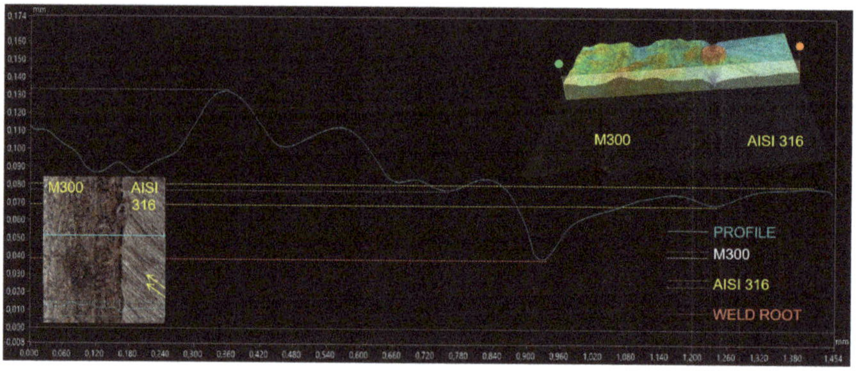

Figure 5. Profile of the root area on the W7 weld—insufficient fusion.

The following Figure 6 includes the profile on the line where root dropout occurred. Similarly, the profile on the side of the AISI 316 base metal was smoother with negligible deviations. On the side of the M300 base metal, variations of profile were more significant, but in comparison with the area in the previous Figure 5, the range was smaller. In this case, the peak was reached on the weld center axis, where the mentioned root dropout was present. Besides the fact that the used welding parameters were not able to fuse base metals on the bottom surface, the molten metal reached the root in some points. The absence of solidified material around this point on the weld axis caused an overflow of the mentioned molten metal, which caused the root dropout after the solidification.

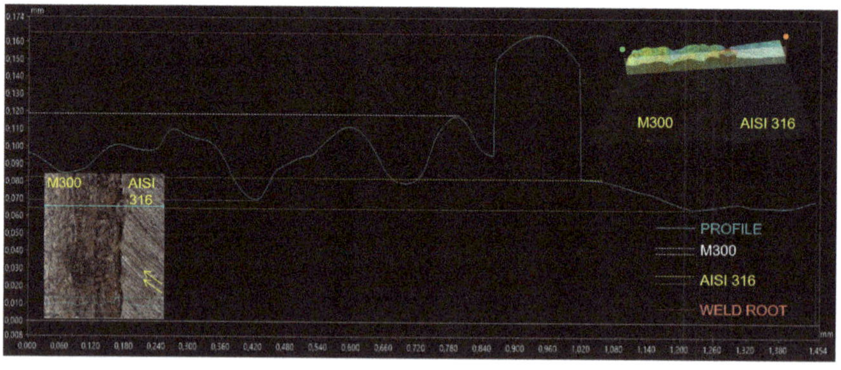

Figure 6. Profile of the root area on the W7 weld—root dropout.

The profile depicted in Figure 7 includes the previously mentioned characteristics of conventional stainless steel (AISI 316) and SLM maraging steel (M300). However, the variance of deviations was the lowest in comparison with the other evaluated lines (0.08–1.120 mm). Furthermore, base metals seemed to be fused from the view on the bottom of the weld, but the profile curve depicts a certain valley on the center axis of the weld, and hence it can be stated that the used welding parameters were highly insufficient for these experimental specimens.

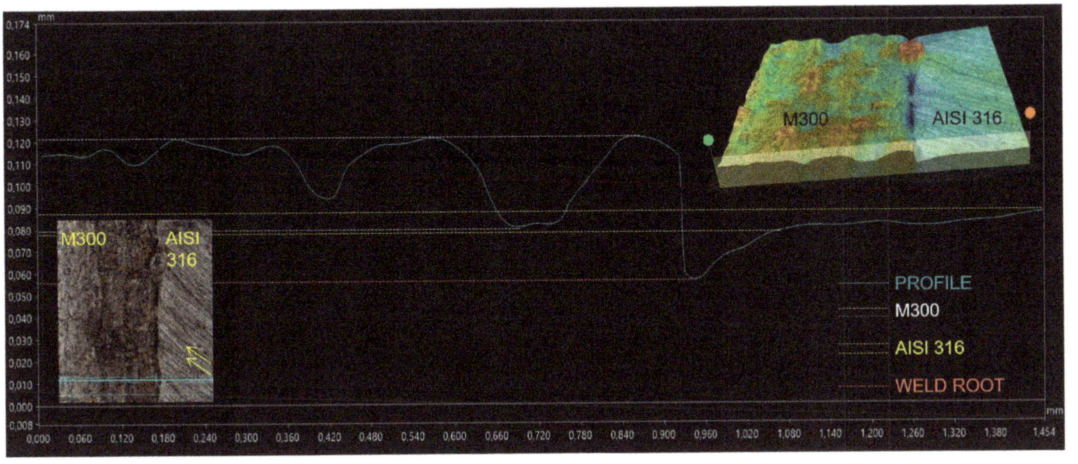

Figure 7. Profile of the root area on the W7 weld—insufficient fusion.

3.2. Microscopic Evaluation

The microstructure of the conventionally made base metal (AISI 316) within welds was polyhedral–characteristic for stainless steels, including austenitic grains with equiaxed, variously oriented shapes sized between 10 and 100 μm (Figure 8). The shape morphology of grains stayed unchanged after the welding with the occurrence of annealing twins. The matrix also included unorganized dispersing of morphologically heterogeneous precipitates of alloying elements' (Cr, Ni, etc.) carbide phases. Any discontinuities or defects, such as porosity or metallurgical inclusions, were not identified during the observation of AISI 316 base metal microstructure.

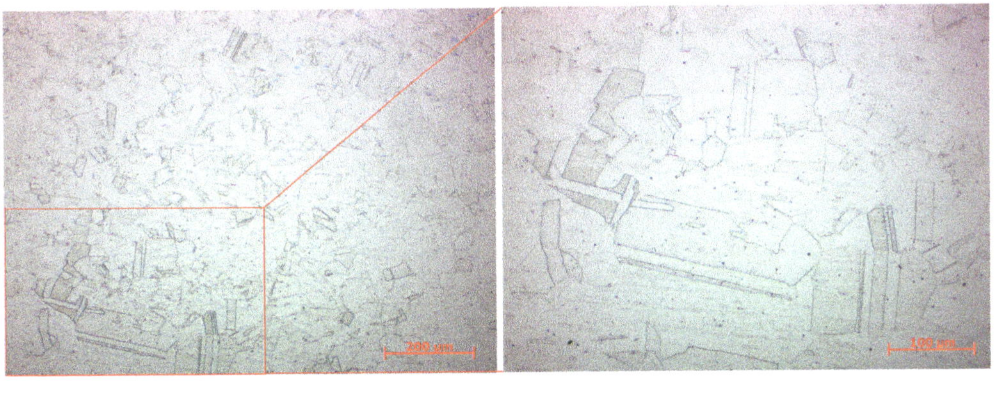

Figure 8. Microstructure of the conventional AISI 316 base metal (W3 specimen).

The microstructure of SLM SS 316L base metal is characterized by the spherical substructure of grains representing the solidified melt pools, defined by a technological process, during which the laser beam impulses melt a material layer by layer (Figure 9). The size of grains is approximately 100 μm in a perpendicular (laser motion) direction and 20–100 μm in a printing (layering) direction, whereas this range is defined by the penetration of individual melt pools and their mutual interaction and bonding during the layering process. The structure of grains itself, growing in the direction of the layering process, is dendritic and identical with a direction of heat flow during the solidification. A discrete and heterogeneous segregation was observed on the grain boundaries, copying the shape of the substructure, due to the faster solidification, and hence more rapid recrystallization. In comparison with conventional material, precipitates of carbide phases of steel were not monitored in the matrix, which can be ascribed to the causal impact of SLM technology.

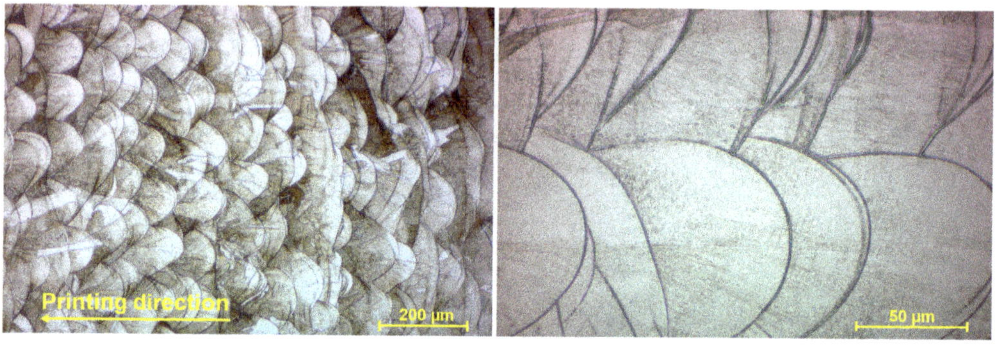

Figure 9. Microstructure of the SLM SS 316L base metal (W5 specimen).

The base metal represented by the SLM M300 maraging steel exhibited a typical lath martensite single-phase microstructure (Figure 10a). The melt pools created during the SLM process were not so expressive, being adjusted using a different etchant (Figure 10b). The spherical substructure of grains was more evident, and individual printing layers could be distinguished. In the perpendicular direction, the size of grains was approximately 100–150 µm, and the size in the printing direction was mostly approximately 10–50 µm. Similarly to the SLM SS 316L material, the range was defined by the overlapping of layers, and hence a smaller range in the printing direction was the consequence of smaller layer thickness. The grains were composed of a dendritic structure, with the growing trend and solidification in the direction of the heat flow. A segregation was present on melt pool boundaries by the same previously mentioned reason. Certain discrepancies were observed in the SLM M300 base metal, in the form of black dots. It can be the individual and randomly occurred porosity, which arises under the influence of gas entrapped in the molten material and stays as a gap in the solidified material.

A weld joint of two AISI 316 stainless steels (Figure 11a) was fully fused with base metals throughout the whole thickness of a single created welding layer. Its microstructure was mainly dendritic as a consequence of rapid heat removal during the solidification of liquid weld metal after the EBW process, with the epitaxial growth of substructure in the form of columnar crystals mutually oriented at a dihedral angle (Figure 11b). In addition was the influence of the previously described molten metal flow up in the direction of the weld axis in combination with the decrease in solidification speed in a direction from the BM to the WM. Furthermore, a microstructure is composed of cellular crystals and equiaxed grains present in the center of the weld metal without the significant segregation or drains. A finer microstructure is the consequence of the EBW characteristics, where a thin fusion zone and subsequent more rapid solidification does not cause the creation of such coarsened grains. The same description of WM microstructure can be applied for a dissimilar weld composed of AISI 316 and SLM SS 316L (Figure 12), reporting similar characteristics. However, the angle of columnar crystals was the opposite due to the analyzed area closer to the lower part of the weld joint, where they tended to direct towards the root surface. Considering the frequently occurring defects, no cracks, cavities, pores, inclusions, or incomplete fusion with base metals were observed in these two analyzed joints.

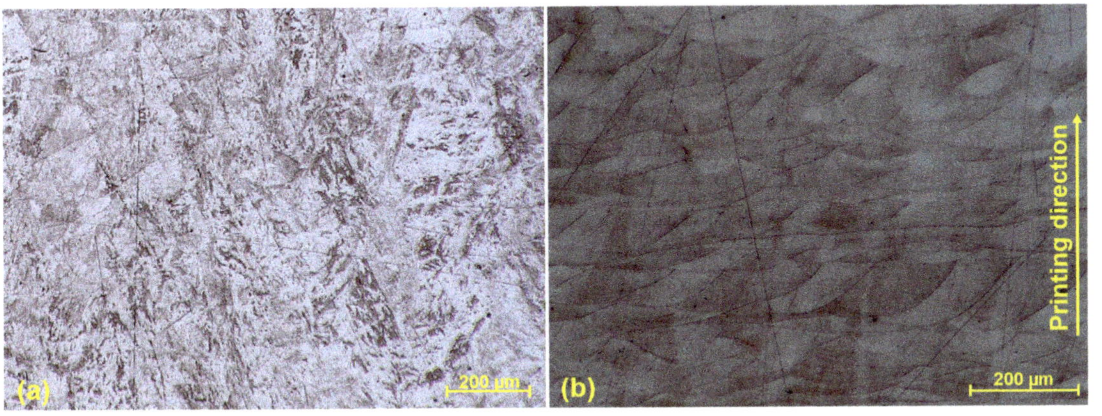

Figure 10. Microstructure of the SLM M300 base metal (W9 specimen); (**a**) lath martensite single-phase microstructure; (**b**) dendritic structure (SLM melt pools).

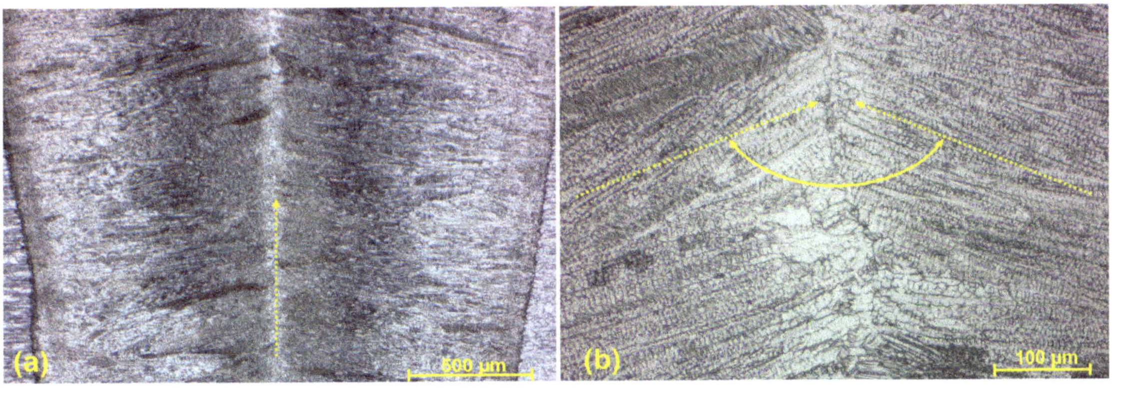

Figure 11. Microstructure of the AISI 316–AISI 316 weld metal (W3 specimen); (**a**) WM with epitaxial growth of substructure;(**b**) columnar crystals oriented at dihedral angle.

The weld composed of AISI 316 stainless steel and M300 maraging steel was created in a sufficient fusion of both metals without the observation of a significant lack of material mixture and related weld defects. A fine dendritic microstructure was similarly created. Columnar crystals were mostly oriented perpendicularly to the weld axis. However, the texture of WM in the upper part of the weld was notably influenced by the molten metal flow in combination with the presence of surface tension, resulting in the vortex shape (Figure 13a). A typical epitaxial growth of a substructure (columnar crystals) was observed right below the tip of the vortex (Figure 13b). Characteristically for the EBW, the fine microstructure reached was caused by a thin fusion zone and fast solidification, no matter what kind of steels were welded. The fusion zone exhibited darker zones on the left side of the WM, being influenced by the higher presence of martensite on the side of the M300 material. The etching adapted to the maraging steel also influenced this phenomenon and caused the blackening of AISI 316 BM. The observed discontinuities in the form of porosity were related to those described in M300 BM, which could be transferred to the weld metal. Any cracks or incomplete fusion with BM were not observed in this joint.

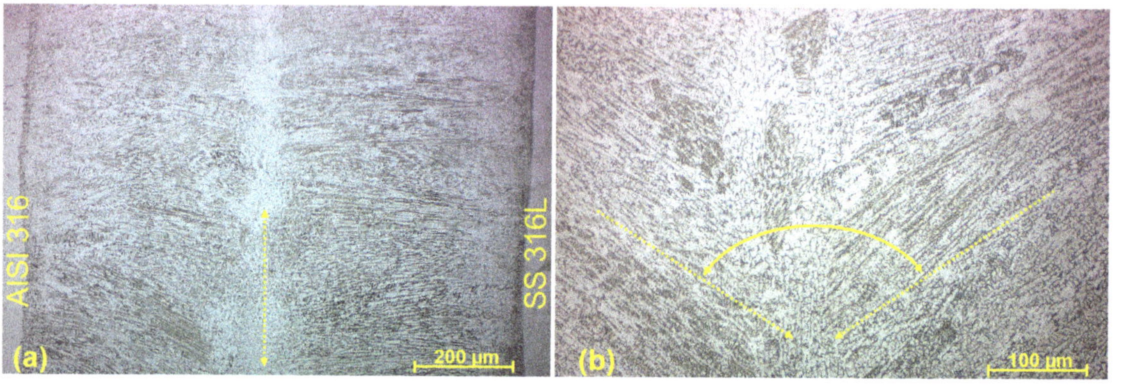

Figure 12. Microstructure of the AISI 316–SS 316L weld metal (W5 specimen); (a) WM with epitaxial growth of substructure; (b) columnar crystals oriented at dihedral angle.

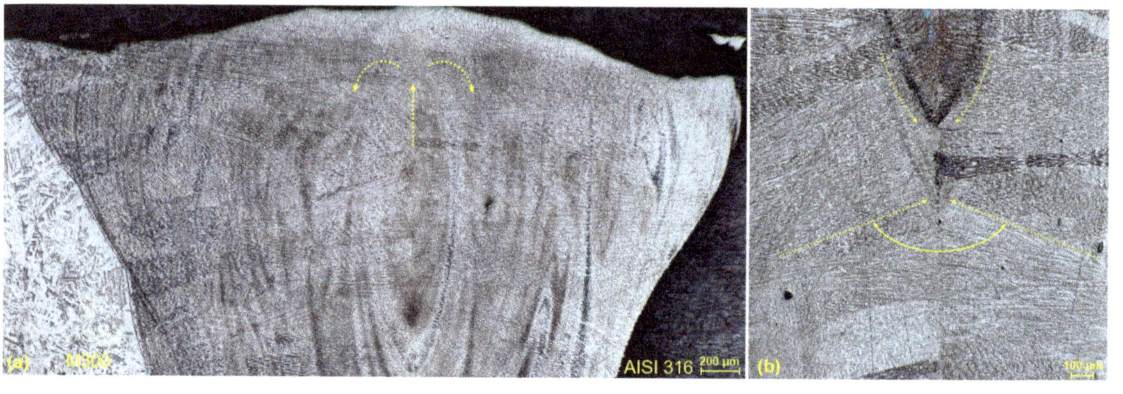

Figure 13. Microstructure of the M300–AISI 316 weld metal (W9 specimen); (a) vortex shape in the upper part of WM; (b) tip of the vortex with columnar crystals oriented at dihedral angle.

The heat-affected zone of conventional stainless steel (AISI 316) was composed of a microstructure identical to the base metal, with the rapid transformation to the weld microstructure on their boundary (Figure 14a). Any recrystallization of the austenitic matrix was not observed, wherein carbide phases precipitates were structurally and morphologically retained. Similarly, the heat-affected zone of SLM stainless steel (SS 316L) resembled typical spherical-shaped grains of a base metal up to the weld boundary, with the sufficient fusion of the weld metal and a fluent transition (Figure 14b). Hence, any morphological transformation did not occur after the solidification process within HAZ. From the side of the weld metal, WM microstructure, represented by the columnar crystals, was present right after the boundary in both cases. Such heat-affected zones are characteristic for the EBW process, where the energy of beam penetrates more significantly into the depth of a material than into the weld width direction. Also, the quite rapid solidification plays its role when there is not enough time for the gradual material recrystallization between WM and BM.

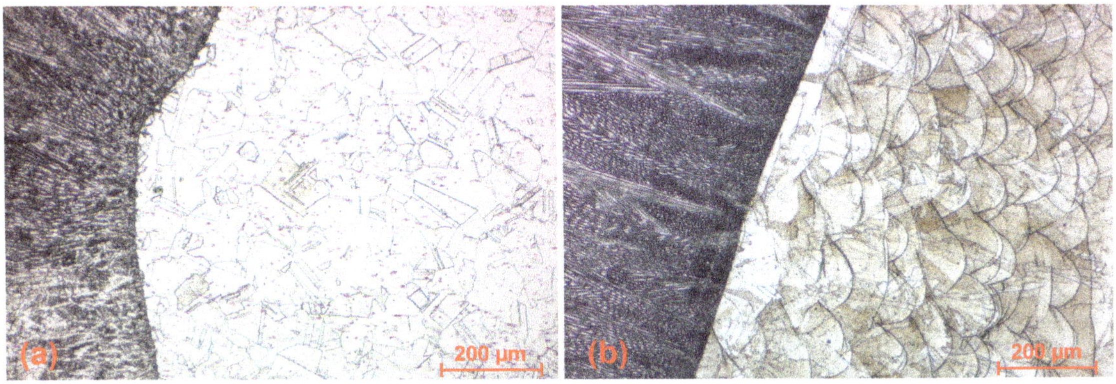

Figure 14. Heat-affected zone: (**a**) AISI 316 (W3 specimen); (**b**) SS 316L (W5 specimen).

In comparison with previous negligible HAZs, the one of the M300–AISI 316 weld was noticeable throughout the whole thickness of the joint (approximately 1.5 mm). The etching was adapted to the depiction of all three microstructures (WM, HAZ, and BM) at a sufficient level (Figure 15a). The finer lath martensite single-phase microstructure of BM was unchanged during the transition of material through the HAZ, up to the WM, where a dendritic microstructure was present right after the fusion line. From the side of the BM, the HAZ was bounded with a dark band zone, being closer to the line character (approximately 150 µm wide). A dark band zone is represented by the two-phase microstructure, composed of martensite and austenite. The reverted austenite is created in a martensite matrix due to the influence of high temperatures on martensite stage [25]. After the etching for a depiction of the characteristic SLM melt pools (Figure 15b), a dark band zone is a more significant, and an unchanged spherical microstructure of a BM in a HAZ is more evident. The previously described porosity observed in M300 BM and WM were also present in the HAZ.

Figure 15. Heat-affected zone of the M300–AISI 316 weld (W9 specimen); (**a**) lath martensite single-phase microstructure in HAZ; (**b**) dark band zone with SLM melt pools.

3.3. Microhardness Measurement

Three specimens (one of each group) were selected for the microhardness measurement, and each specimen was measured in three lines composed of 21 impressions, by which the reaching of sufficient distance in base metal from the weld was secured. The zero point within each line was in the center of the weld metal, and 10 impressions were made in both directions of base metals, with their mutual distance of 0.25 mm. The mutual distance between lines was 3 mm and was 2 mm from the upper and lower surfaces. The graphs do not include HAZ regions due to their absence (minimal size) at the EBW welds. The dividing lines of weld regions are highlighted separately for all three measurement lines due to the irregular width and shape of the weld in a depth direction. The following Figure 16 includes the microhardness of the W4 specimen. The microhardness of the weld can be considered as being consistent in the evaluated area of measurement lines, with the values around a BM value. Overall, the measured microhardness values were defined by the austenitic microstructure of the BM and a fine microstructure of the WM. The upper part of the weld (Line 1) reached 189 HV in the center with its increasing in both directions and a stabilization in both base metals. Line 2 from the middle part of the weld showed similar microhardness in the center of the weld (186.1 HV), with a similar increasing and stabilization trend. Microhardness within the lower part of the weld (Line 3) reached a higher value in the center (205.3 HV), with a decrease in both directions before stabilization. On the right side, stabilization occurred already in the weld metal.

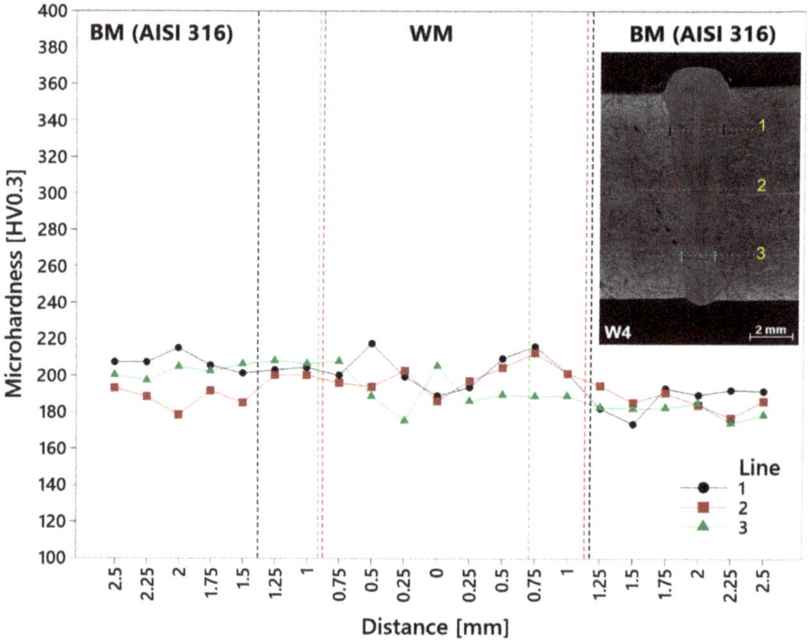

Figure 16. Microhardness of the AISI 316–AISI 316 weld (W4 specimen). Fusion lines (dashed): black—Line 1; red—Line 2; green—Line 3.

The lines of a microhardness measurement on the W6 specimen were situated lower from the upper surface, due to the not negligible width of a reinforcement (Figure 17). The overall microhardness of a weld is on the level of conventionally made AISI 316 stainless steel with the increase on the side of the SLM SS 316L stainless steel, a little higher than a data sheet value. Nevertheless, the AISI 316 stainless steel included higher carbon content, which caused the increasing of microhardness, and the SS 316L material reached higher values. It is the consequence of the SLM process, which creates finer austenitic microstructure with dislocations of higher density situated on grain boundaries on the substructure level, which lead to the higher resistance of material to the indentation [37,38]. The upper part (Line 1—below the reinforcement) reached 202.3 HV in the center of the weld with the following decrease in both directions. In the direction of the AISI 316 material, values were stable. In the opposite direction, the gradual increase to the maximum occurred in the BM (229.3 HV). The microhardness of the middle part (Line 2) progressed with a similar trend, where the 210.5 HV in the center of the weld decreased to lower values with a stabilization in the direction of conventional base metal. In the direction of the SLM material, the increasing of stable values started right after the weld boundary. The overall maximum of the weld was reached in the SLM material on Line 3 (253.9 HV). The microhardness in a weld metal is the most stable in this area.

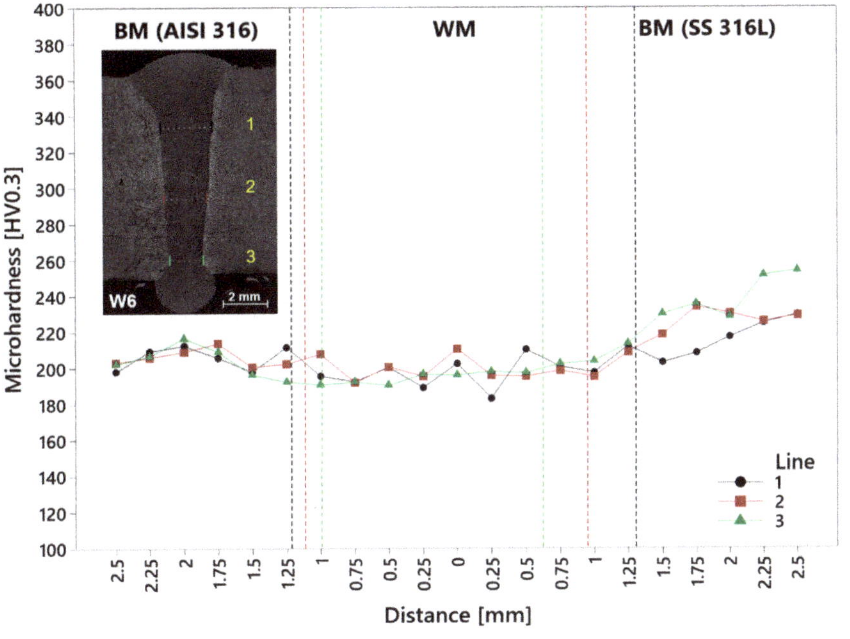

Figure 17. Microhardness of the AISI 316–SS 316L weld (W6 specimen). Fusion lines (dashed): black—Line 1; red—Line 2; green—Line 3.

The microhardness of the specimen W9 was not so consistently distributed in the evaluated area of the weld (Figure 18). The curves represent minimal values in the area of a weld metal, caused by the mixing of different base metals of various compositions in a fusion zone. The slight increase was recorded in the direction of AISI 316 stainless steel and significantly in the direction of SLM M300 maraging steel, approximately to the level of data sheet values. The high microhardness of a maraging steel is provided by its specific chemical composition, even though the natural hardness-increasing carbon (C) content is low ($\leq$0.03). The final martensite microstructure is generally characterized by higher hardness, whereas the important factor is higher nickel (Ni) content, which suppresses the creation of ferrite and pearlite, as well as cobalt (Co), increasing the hardness to high values through the acceleration of precipitation and subsequently reducing the duration of ageing [39]. Furthermore, higher titanium (Ti) content also has the ability to increase the hardness of maraging steel [40], which can be ascribed to the influence of higher dislocation density and intergranular zones including Ti and precipitates of CoNi [41]. Worth mentioning are the non-negligible deviations, mainly in the areas of weld metal and M300 base metal. The upper part of the weld (Line 1) reached 185 HV in its center, which can be considered stable in the whole weld metal. Right beyond the border, microhardness rapidly increased. On the other side of the weld, a slight increase in microhardness occurred in the AISI 316 base metal, which stayed stable in points situated further from the weld. The overall minimum of the whole weld was reached in the zero point of the middle part (Line 2) of the weld (141.8 HV). Microhardness increased in both directions, slightly towards the AISI 316 material with its stabilization in a base metal and significantly on the boundary of the weld with the M300 base metal. A rapid increase was observed in the lower area of the weld (Line 3), where the second (0.5 mm) point was impressed in the M300 HAZ due to the shape of the weld joint. The same as in the case of points within other lines, the slight increase occurred in the direction of AISI 316 stainless steel with a stable state in a base metal. A difference in microhardness between a HAZ and a M300 BM was not observed due to the unchanged microstructure. The boundary line represented by a dark band zone is highlighted with blue lines in Figure 18.

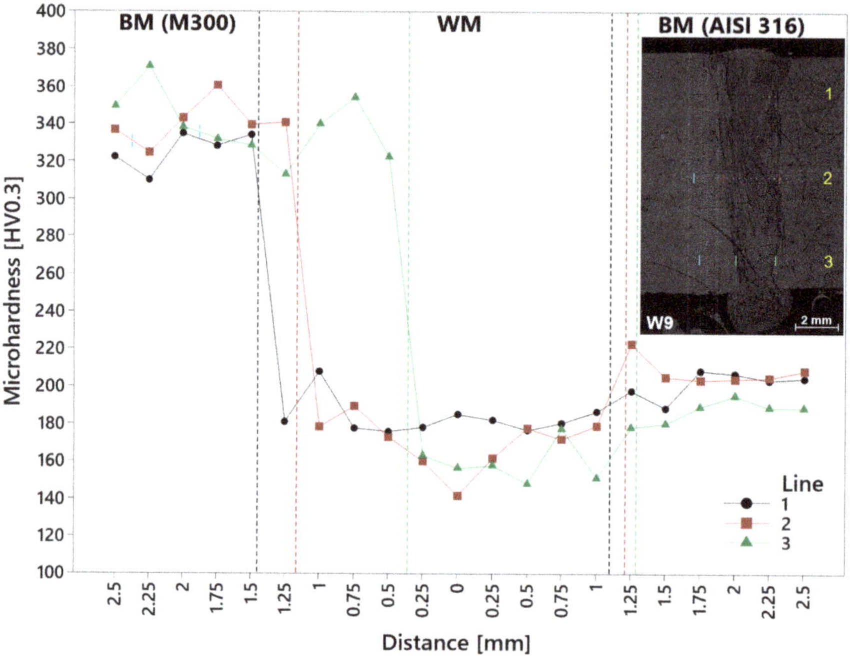

Figure 18. Microhardness of the M300–AISI 316 weld (W9 specimen). Fusion lines (dashed): black—Line 1; red—Line 2; green—Line 3; blue—HAZ (BM M300).

4. Discussion

The described characteristics of welds within this study are in a certain correlation with other studies from the field. The used welding parameters (adjusted mainly according to similar AISI 316 welds) were different from other studies [15,16,18] at first sight. However, considering the various thicknesses of experimental specimens within the mentioned experiments, the used parameters can be regarded as similar. Considering the macroscopical parameters of the reached joints, the fusion zones of welds in all three groups were significantly larger in a depth direction than in the width direction, which is characteristic for EBW, no matter what the thickness of the weld is, having been proven in the case of 160 mm welds with the large weld depth–width ratio [16]. The overall shapes of welds were nail- and wedge-shaped, similarly stated in the study [17] (wine-glass shape) and the study [21], where also no significant weld defects were observed and HAZ was negligible in the case of AISI 316L stainless steel joints. Additionally, the same shape characteristics were observed within dissimilar AISI 316–SS 316L and M300–AISI 316 welds in this experiment. The microstructure of stainless steel welds (similar and dissimilar) was transformed from the austenitic equiaxed grains in the BM to the columnar and partly equiaxed grains in the WM right after the fusion line without the significant HAZ, which corresponds to results of related studies [15,16,21]. In the case of the SS 316L material, the same transition of the microstructure in the direction of the weld axis was observed, despite the fact that the BM austenitic microstructure is influenced by the SLM technology. The martensitic microstructure of M300 BM was retained within the observed HAZ, and a dendritic microstructure was present in the WM, which was also described in the case of other studies focused on the EBW welding of maraging steels [29,30]. HAZ is divided with a BM by a dark band zone, being thinner in comparison with other welding methods [30], whereas it was shown to also be related to the etching of the analyzed specimens. The HAZ of the AISI 316 material was also not present within a dissimilar weld with the M300 material. A microhardness in all welds was on the level of the ASI 316 base metal, which was also described in other studies [15,21], with the exception of [16], where a little increase was recorded in the weld center (15–20 HV higher than BM). Within dissimilar joints, the microhardness increased in the direction of the SS 316L base metal, which is the consequence of a different manufacturing technology (SLM), and a significant increase was recorded in the direction of the M300 base metal, whereas the presence of M300 HAZ did not have an influence on the microhardness of a joint. A similar distribution of microhardness in similar materials was also described by using the other welding technologies, such as TIG welding of SLM stainless steel [7] or GTAW/TIG welding [23] and laser welding of maraging steels [25]. Additionally is a depiction of microhardness distribution within dissimilar EBW joints, including SLM SS 316L stainless steel and SLM M300 maraging steel, within the presented study. A description of characteristics of dissimilar joints including unusual used materials representing stainless steels (AISI 316 and SLM SS 316L) and mainly SLM M300 maraging steel expresses the overall added value of a described experiment.

5. Conclusions

The presented experiment focuses on the evaluation of possibilities in the electron beam welding of dissimilar welds composed of different types of steels that were produced using different manufacturing processes. Three groups of specimens were divided according to welded base metals, wherein the first included two conventional AISI 316 stainless steels, the second included conventional AISI 316 with SLM SS 316L stainless steels, and the third included M300 maraging steel with conventional AISI 316. The created welds, reaching variable parameters (welding current and welding velocity), were evaluated regarding three main aspects (macrostructure, microstructure, and microhardness), based on which the following statements can be concluded:

- The selected EBW parameters were able to reach sufficient weld joints in all three groups of materials without the significant lack of fusion. The combination of the 135 mA current and the velocity in the range of 25–35 mm·s^{-1} was appropriate for the creation of sufficient joints within the first and second group of specimens, and hence it can be stated that materials on the similar base (316 stainless steel) can be welded with the same parameters. The current of 150 mA was suitable for both groups and more appropriate for the third group of specimens using a certain velocity. The influence of SLM orientation on the creation of welds was not recorded.
- Within the first group, the application of lower current and higher velocity created nail-shaped joints, whereas a decrease in velocity caused the reaching of wedge-shaped joints, which also applies for the second group. The same effect on the shape of a joint was reached using the higher welding current. Appropriate lower-current parameters (135 mA and 25 mm·s^{-1}) for stainless steels were not sufficient for the third group joints, where the lack of root penetration occurred. A 150 mA current with the appropriate velocity (20 and 30 mm·s^{-1}) created straight welds of a sufficient width and the presence of a vortex, acting as an indicator of good molten metal flow. From macroscopical defects, excesses of reinforcement, undercuts, root dropouts, underfill of the face, and a cavity were observed.
- The microstructure of AISI 316 BM was austenitic with equiaxed grains and annealing twins. In the case of the SS 316L BM, a microstructure was defined by the spherical grains with dendritic substructure, as the consequence of the SLM process. Within the

M300 BM, lath martensite single-phase microstructure and a spherical microstructure with a dendritic substructure were observed, according to applied etchant. The microstructure of WM in all three groups was mainly dendritic with the epitaxial growth of substructure in the form of columnar crystals. In the case of the third group, the texture was influenced by the molten metal flow and the creation of a vortex. The excessive HAZ was only present in the third group with a thin dark band zone (M300), without significant recrystallization. Discontinuities were observed in M300 BM and their welds.

- The microhardness of WM was on the level of AISI 316 BM in all three analyzed welds, whereas it increased in the direction of SS 316L BM due to the finer microstructure and more significantly in the direction of M300 BM for the same reason, whereas the microhardness in the HAZ was unchanged due its similar microstructure with the BM. Furthermore, it must be mentioned that it is appropriate to adapt the number of measuring points to the width of the present HAZ.

According to the examined aspects, it can be stated that AISI 316 stainless steel has the potential to be a satisfactory base metal for the connection of SLM SS 316L and mainly SLM M300 maraging steel with subassemblies. However, it is necessary to conduct subsequent experiments focused on the evaluation of other mechanical properties. The application of scanning electron microscopy can be helpful for better microstructural analysis, or X-ray diffraction for the evaluation of residual stresses, which has an important influence on weld. Considering the selection of EBW parameters, the design of experiment (DoE) method with follow-up analysis could bring a better understanding of their influence on the evaluated characteristics of weld joints.

Author Contributions: Conceptualization, M.G. and R.V.; methodology, F.B. and M.H.; software, R.V. and M.G.; validation, F.B.; formal analysis, R.V.; investigation, M.G.; resources, R.V. and M.G.; data curation, F.B. and M.G.; writing—original draft preparation, M.G. and F.B.; writing—review and editing, M.H. and R.V.; visualization, M.H.; supervision, F.B.; project administration, R.V. and M.H.; funding acquisition, M.H. All authors have read and agreed to the published version of the manuscript.

Funding: This work was funded by the Slovak Research and Development Agency under contract No. APVV-21-0228 and the projects VEGA 1/0391/22 and KEGA 017TUKE-4/2023 were granted by the Ministry of Education, Science, Research and Sport of the Slovak Republic. This article is the result of the following project implementation: Development of excellent research capacities in the field of additive technologies for the Industry of the 21st century, ITMS: 313011BWN5, supported by the Operational Program Integrated Infrastructure funded by the ERDF.

Institutional Review Board Statement: Not applicable.

Informed Consent Statement: Not applicable.

Data Availability Statement: Data are contained within the article.

Acknowledgments: The authors would like to thank the Center of 3D Printing Protolab (VSB-TU Ostrava) for the preparation of experimental specimens and the First Welding Company, Inc. (Bratislava), for the welding of experimental specimens.

Conflicts of Interest: The authors declare no conflicts of interest. The funders had no role in the design of the study; in the collection, analyses, or interpretation of the data; in the writing of the manuscript; or in the decision to publish the results.

References

1. Gibson, I.; Rosen, D.; Stucker, B.; Khorasani, M. Design for additive manufacturing. In *Additive Manufacturing Technologies*; Springer: Cham, Switzerland, 2021; pp. 555–607. [CrossRef]
2. Gill, M.; Terry, E.; Abdi, Y.; Hawkes, S.; Rindler, J.; Schick, D.; Ramirez, A.; Herderick, E.D. Joining technologies for metal additive manufacturing in the energy industry. *Jom* **2020**, *72*, 4214–4220. [CrossRef]

3. Chowdhury, S.; Yadaiah, N.; Prakash, C.; Ramakrishna, S.; Dixit, S.; Gupta, L.R.; Buddhi, D. Laser powder bed fusion: A state-of-the-art review of the technology, materials, properties & defects, and numerical modelling. *J. Mater. Res. Tech.* **2022**, *20*, 2109–2172. [CrossRef]
4. Alaimo, A.; Del Prete, A.; Mantegna, G.; Orlando, C.; Panella, F.W.; Primo, T.; Tumino, D.; Vindigni, C.R. Modified beam modeling of powder bed fusion manufactured lattice structures. *Int. J. Mech. Sci.* **2023**, *259*, 108599. [CrossRef]
5. Nogueira, P.; Silva, M.B.; de Deus, A.M.; Cláudio, R.; Carmezim, M.; Santos, C.; Marginho, J.; Reis, L.; Alves, J.L.; Oliveira, L.; et al. Evaluation of the roughness of lattice structures of AISI 316 stainless steel produced by laser powder bed fusion. *Eng. Manuf. Let.* **2024**, *2*, 39–44. [CrossRef]
6. An, F.; Liu, X.; Zhang, H.; Zhang, L.; Ning, J.; Na, S.J. Comparison of performance of laser powder bed fusion thin-walled TC11 alloy samples welded via laser welding and electron beam welding. *Opt. Laser Technol.* **2024**, *170*, 110256. [CrossRef]
7. Mohyla, P.; Hajnys, J.; Sternadelová, K.; Krejčí, L.; Pagáč, M.; Konečná, K.; Krpec, P. Analysis of welded joint properties on an AISI316L stainless steel tube manufactured by SLM technology. *Materials* **2020**, *13*, 4362. [CrossRef]
8. Huysmans, S.; Peeters, E.; De Bruycker, E.; De Prins, K. Weldability study of additive manufactured 316L austenitic stainless steel components—Welding of AM with conventional 316 L components. *Weld. World.* **2021**, *65*, 1415–1427. [CrossRef]
9. Mokhtari, M.; Pommier, P.; Balcaen, Y.; Alexis, J. Laser welding of AISI 316L stainless steel produced by additive manufacturing or by conventional processes. *J. Manuf. Mat. Process.* **2021**, *5*, 136. [CrossRef]
10. Matilainen, V.P.; Pekkarinen, J.; Salminen, A. Weldability of additive manufactured stainless steel. *Physcs. Proc.* **2016**, *83*, 808–817. [CrossRef]
11. Zhang, C.; Bao, Y.; Zhu, H.; Nie, X.; Zhang, W.; Zhang, S.; Zeng, X. A comparison between laser and TIG welding of selective laser melted AlSi10Mg. *Opt. Laser Technol.* **2019**, *120*, 105696. [CrossRef]
12. Nahmany, M.; Hadad, Y.; Aghion, E.; Stern, A.; Frage, N. Microstructural assessment and mechanical properties of electron beam welding of AlSi10Mg specimens fabricated by selective laser melting. *J. Mater. Process. Tech.* **2019**, *270*, 228–240. [CrossRef]
13. Sali, A.R.; Patel, V.; Hyder, J.; Hyder, D.; Corliss, M.; Hung, W. Electron-beam welding of laser powder-bed-fused Inconel 718. *Int. J. Eng. Mater. Manuf.* **2021**, *6*, 209–224. [CrossRef]
14. Wang, Z.; Liu, Z.; Ma, J.; Su, Y.; Deng, H.; Zhou, N.; Zhai, H.; Wu, J. Investigation on microstructure and mechanical properties of electron-beam-welded joint of reduced activation ferritic/martensitic steel fabricated by selective laser melting. *Mat. Sci. Eng. A-Struct.* **2023**, *881*, 145333. [CrossRef]
15. Li, H.; Huang, F.; Li, J.; Li, J.; Zhao, G.; Chang, Y. Electron beam welding beam current on microstructure and mechanical properties of AISI 316 austenitic stainless steel. *Arch. Civ. Mech. Eng.* **2023**, *23*, 259. [CrossRef]
16. Ma, J.; Wang, Z.; Liu, Z.; Ji, H.; Xia, X.; Li, C.; Liu, Z.; Zhang, Z.; Wu, J. Investigation on microstructure and mechanical properties of 160mm thick 316L stainless steel electron-beam-welded joint. *Fusion Eng. Des.* **2023**, *195*, 113973. [CrossRef]
17. Mokhtarishirazabad, M.; McMillan, M.; Vijayanand, V.D.; Simpson, C.; Agius, D.; Truman, C.; Knowles, D.; Mostafavi, M. Predicting residual stress in a 316L electron beam weld joint incorporating plastic properties derived from a crystal plasticity finite element model. *Int. J. Pres. Ves. Pip.* **2023**, *201*, 104868. [CrossRef]
18. Kar, J. Effect of beam oscillation on residual stress and corrosion properties of AISI 316L electron beam welds. *Metal. Micro. Anal.* **2021**, *10*, 652–660. [CrossRef]
19. Li, L.; Wang, S.; Huang, W.; Jin, Y. Microstructure and mechanical properties of electron beam welded TC4/TA7 dissimilar titanium alloy joint. *J. Manuf. Process.* **2020**, *50*, 295–304. [CrossRef]
20. Sun, W.; Wang, S.; Xin, J.; Tan, G.; Hong, M.; Wu, M.; Ke, L. Microstructure and mechanical properties of the IC10/GH3039 dissimilar electron beam welded joint. *Vacuum* **2020**, *181*, 109592. [CrossRef]
21. Fan, X.; Shen, X.; Zhang, Y.; Liu, Z.; Wu, J.; Liu, Z.; Ma, J. Microstructure and mechanical properties of similar and dissimilar joints of RAFM and 316L by electron beam welding. *Fusion Eng. Des.* **2021**, *162*, 112130. [CrossRef]
22. Kar, J. X-Ray Tomography Studies of Electron Beam Welded Inconel-718 TO AISI-316L Dissimilar Joints. *Trans. Indian Inst. Met.* **2024**, *77*, 229–236. [CrossRef]
23. Gope, D.K.; Kumar, P.; Chattopadhyaya, S.; Wuriti, G.; Thomas, T. An investigation into microstructure and mechanical properties of maraging steel weldment. *IOP Conf. Ser. Mater. Sci. Eng.* **2021**, *1104*, 012014. [CrossRef]
24. Takahashi, A.; Toyohiro, T.; Segawa, Y.; Kobayashi, M.; Miura, H. Embrittlement fracture behavior and mechanical properties in heat-affected zone of welded maraging steel. *Materials* **2024**, *17*, 440. [CrossRef]
25. Ch, S.R.; Rajesh, S. Process parameters optimisation and numerical simulation of laser beam welded butt joints of maraging steel. *Int. J. Automot. Mech. Eng.* **2022**, *19*, 9709–9720. [CrossRef]
26. Subashini, L.; Prabhakar, K.P.; Ghosh, S.; Padmanabham, G. Comparison of laser-MIG hybrid and autogenous laser welding of M250 maraging steel thick sections—Understanding the role of filler wire addition. *Int. J. Adv. Manuf. Tech.* **2020**, *107*, 1581–1594. [CrossRef]
27. Bagchi, A.; Gope, D.K.; Chattopadhyaya, S.; Wuriti, G. A critical review on susceptibility of stress corrosion cracking in maraging steel weldments. *Mater. Today Proc.* **2020**, *27*, 2303–2307. [CrossRef]
28. Ben-Artzy, A.; Reichardt, A.; Borgonia, J.P.; Dillon, R.P.; McEnerney, B.; Shapiro, A.A.; Hosemann, P. Compositionally graded SS316 to C300 Maraging steel using additive manufacturing. *Mater. Design.* **2021**, *201*, 109500. [CrossRef]
29. Hu, Z.F.; Mo, D.F.; Wang, C.X.; He, G.Q.; Chen, C.S. Different behavior in electron beam welding of 18 Ni co-free maraging steels. *J. Mater. Eng. Perform.* **2008**, *17*, 767–771. [CrossRef]

30. Murthy, C.V.S.; Krishna, A.G.; Reddy, G.M. Dissimilar welding of maraging steel (250) and 13-8 Mo stainless steel by GTCAW, LBW and EBW processes. *T. Indian I. Metals.* **2019**, *72*, 2433–2441. [CrossRef]
31. AISI 316 Stainless Steel, Austenitic. Available online: https://www.matweb.com/search/datasheet.aspx?matguid=172060d502d846dd8cc18785ec397bd0 (accessed on 14 September 2024).
32. Data Sheet: SS 316L-0407 Powder for Additive Manufacturing. Available online: https://www.renishaw.com/resourcecentre/en/details/data-sheet-ss-316l-0407-powder-for-additive-manufacturing--90802 (accessed on 14 September 2024).
33. Data Sheet: Maraging Steel M300 for 400 W Powder for Additive Manufacturing. Available online: https://www.renishaw.com/resourcecentre/en/details/Data-sheet-Maraging-steel-M300-for-400-W-powder-for-additive-manufacturing--96326?lang=en&srsltid=AfmBOoqLIFedyFO-AKFdX97rrxD7ezHFaPPSUfYtomjslvsNQ2jK3aJo (accessed on 14 September 2024).
34. Gu, D.; Shen, Y. Balling phenomena in direct laser sintering of stainless steel powder: Metallurgical mechanisms and control methods. *Mater. Design.* **2009**, *30*, 2903–2910. [CrossRef]
35. Meander Stripe and Chessboard Hatching Pattern. Available online: https://www.renishaw.com/resourcecentre/en/details/Meander-stripe-and-chessboard-hatchings-pattern--95492?lang=en (accessed on 14 September 2024).
36. Wu, H.Y.; Jiang, F.; Wang, S.L.; Chen, Y.H. Flowing model of weld pool during vacuum electron beam welding. *J. Phys. Conf. Ser.* **2021**, *1748*, 052026. [CrossRef]
37. Mohd Yusuf, S.; Chen, Y.; Boardman, R.; Yang, S.; Gao, N. Investigation on porosity and microhardness of 316L stainless steel fabricated by selective laser melting. *Metals* **2017**, *7*, 64. [CrossRef]
38. Saeidi, K.; Gao, X.; Zhong, Y.; Shen, Z.J. Hardened austenite steel with columnar sub-grain structure formed by laser melting. *Mater. Sci. Eng. A-Struct.* **2015**, *625*, 221–229. [CrossRef]
39. Mouritz, A.P. *Introduction to Aerospace Materials*; Woodhead Publishing Limited: Sawston, UK, 2012.
40. Strakosova, A.; Roudnická, M.; Šafka, J.; Ackermann, M.; Dvorský, D.; Školáková, A.; Vronka, M.; Svora, P.; Drahokoupil, J.; Pinc, J.; et al. Effect of titanium on microstructure and mechanical behaviour of additively manufactured 1.2709 maraging steel. *Addit. Manuf.* **2024**, *88*, 104264. [CrossRef]
41. Dehghahi, S.; Ghoncheh, M.H.; Hadadzadeh, A.; Sanjari, M.; Amirkhiz, B.S.; Mohammadi, M. The role of titanium on the microstructure and mechanical properties of additively manufactured C300 maraging steels. *Mater. Des.* **2020**, *194*, 108965. [CrossRef]

materials

MDPI

Article

Influence of Microstructure on Music Properties of SWP-B Music Steel Wire Under Different Annealing Treatments

Xinru Jia [1,2], Qinghua Li [1,2,*], Fuguo Li [1,2,3,*], Xiaohui Fang [1,2], Junda You [1,2], Qian Zhao [1,2], Xia Wang [4] and Jinhua Lu [3]

1 State Key Laboratory of Solidification Processing, School of Materials Science and Engineering, Northwestern Polytechnical University, Xi'an 710072, China; jiaxinru@mail.nwpu.edu.cn (X.J.); fxh@mail.nwpu.edu.cn (X.F.); yjd@mail.nwpu.edu.cn (J.Y.); zq2019@mail.nwpu.edu.cn (Q.Z.)

2 Shaanxi Key Laboratory of High-Performance Precision Forming Technology and Equipment, Northwestern Polytechnical University, Xi'an 710072, China

3 Joint R&D Center for Metallic Materials, Metallic Wire and Metallic Card Clothing, Xi'an 710021, China; 13773865632@163.com

4 Department of Engineering Mechanics, College of Civil & Architecture Engineering, Xi'an Technological University, Xi'an 710021, China; wangxiaxit@126.com

* Correspondence: qinghual@nwpu.edu.cn (Q.L.); fuguolx@nwpu.edu.cn (F.L.); Tel.: +86-180-9252-1022 (F.L.)

Abstract: The mechanical properties of music wire are contingent upon its microstructure, which in turn influences its applications in music. Chinese stringed instruments necessitate exacting standards for comprehensive performance indexes, particularly with regard to the strength, resilience, and rigidity of the musical steel wires, which differ from the Western approach to musical wire. In this study, SWP-B music wire was selected for investigation through metal heat treatment, which was employed to regulate its microstructure characteristics. Furthermore, a spectral analysis was conducted to evaluate the musical expression, encompassing attributes such as pitch and timbre. In conclusion, the governing law of the impact of the microstructure of music wire on its musical expression was established. The results demonstrate that steel wire subjected to a 200 °C annealing treatment for cementite spheroidization can effectively reduce stress concentration, thereby reducing the probability of fracture and consequently improving tonal uniformity and richness while increasing tensile strength from 2578 MPa to 2702 MPa. Conversely, the high-temperature annealing treatment alters the crystalline structure of the material and refines the grain structure, thereby improving the material's performance and sound quality. The fine microstructure of the music steel wire displays enhanced uniformity. As the annealing temperature increases, the strength of the ferrite phase <110>//ND (<010>//ND, indicating that the <010> direction of the crystal is parallel to the normal direction of the material) and the cementite phase <010>//ND demonstrates a gradual decline. However, this also results in a more pronounced harmonic performance, which, in turn, affects the overall music expression.

Keywords: music steel wire; microstructure; cementite spheroidization; spectral properties; musical expression

1. Introduction

As an important component of stringed instruments, the evolution of music steel wire reflects the close integration of music and materials engineering. From the initial utilization of natural materials, such as sheep gut strings, to the advent of modern high-performance steel wires, the evolution of stringed instruments not only aligns with the trajectory of technological advancement but also exemplifies the uninterrupted enhancement of musical expression. In classical Chinese music, each instrument can express a unique set of emotions. The guqin, a seven-stringed plucked instrument in some ways similar to the zither, is an instrument that can convey profound emotions in a reserved and distant manner. The erhu (two-string fiddle) is particularly adept at expressing sadness and tenderness, rendering it an appropriate instrument for conveying delicate emotions. The pipa (pear-shaped lute/Chinese lute), with its dexterous fingering and variable speed, is a highly expressive musical instrument. As music steel wire has become more widely used in musical instruments, the tone of these instruments has become brighter, allowing them to meet the diverse demands of modern musical expressions. The advancement of classical Chinese musical instruments and the evolution of string materials reflect the ongoing transformation of musical culture, while simultaneously illustrating the multifaceted nature of musical expression. The advent of music steel wire represents a pivotal moment in this evolution, enabling traditional instruments to flourish once more in the context of contemporary music.

The selection of material for music steel wire is primarily based on its acoustic characteristics, including pitch accuracy, low internal damping, high elasticity, and toughness. In addition to these acoustic properties, cost and durability are also important considerations. The forming process exerts a significant influence on the properties of music steel wire. A variety of drawing and processing techniques can be employed to produce strings with varying diameters and properties. In recent years, research on forming processes has focused on the optimization of drawing speed, annealing temperature, and cooling method. For example, the implementation of continuous drawing and multistage annealing processes has been demonstrated to enhance the uniformity and elasticity of strings. From the perspective of practical engineering, drawn steel wires are subjected to additional heat treatment during the manufacturing process. Such processes may include bluing at 300 °C to 400 °C or hot-dip galvanizing at 450 °C [1–3]. These temperatures are high enough to modify the mobility of dislocations and alloying elements within the deformation structure, thereby influencing the mechanical properties of the material. In light of these considerations, significant effort has been dedicated to elucidating the effects of annealing on the deformation microstructure [4]. Li et al. observed that, during high-temperature annealing, spheroidized cementite particles were situated at triple junctions between coarse hexagonal ferrite substructure grains [2]. The findings revealed that the mechanical properties of strings are markedly influenced by the specific heat treatment parameters. In this field, a number of heat treatment protocols have been proposed for different materials. For example, a variety of heat treatment techniques have been proposed for high-carbon and alloy steels with the objective of attaining the most favorable balance between strength and modulus of elasticity. The advent of high-quality stainless steel and alloy materials has resulted in a notable enhancement in the toughness, stability, and acoustic quality of the musical steel wire [5].

Modifications to the microstructure of music wire have a direct impact on its modulus of elasticity, yield strength, fatigue life, and acoustic properties, which subsequently influence the range of frequencies and timbre it is capable of producing [6]. The presence of finer grains typically results in enhanced material strength and resilience. Furthermore, the anisotropy of the metal is closely associated with its crystal orientation. An optimal crystal orientation facilitates enhanced sound conduction efficiency and sound quality. Recent studies have demonstrated that the phase composition and interactions within the microstructure of music steel wire exert a significant influence on determining the material's acoustic properties of the material. This is illustrated by the observation that grain refinement can enhance the strength and elasticity of the steel wire [7]. In the event of high strain, cementite is observed to undergo chemical decomposition and act as a carbon source through mechanical alloying. This process results in the segregation of carbon to ferrite boundaries, thereby enhancing the tensile strength of the material [8,9]. The optimization of the microstructure through heat treatment and forming processes has been demonstrated to enhance the fatigue resistance and the service life of the strings. Zhou et al. observed that annealing at temperatures ranging from 210 °C to 350 °C did not alter the morphology of cementite in the wire, but did result in a structural change in the cementite from an almost amorphous state to a nanocrystalline form.

The steel wire utilized in musical instruments is subjected to meticulous regulation with regard to a multitude of factors, including surface finish, tensile strength, stiffness, and other mechanical and fatigue-related properties [10]. The mechanical properties of music steel wire strings serve as the fundamental indicators of their suitability for performance. The evaluation of string performance necessitates the consideration of a number of pivotal parameters, including tensile strength, fatigue resistance, and Young's modulus. The mechanical properties of the material are largely contingent upon the dimensions and configuration of the pearlitic lamellae. In general, a reduction in the spacing of the pearlitic lamellae results in an increase in the yield strength, tensile strength, hardness, and elongation of the material [11]. The optimal layer spacing is conducive to optimizing the wire's vibration characteristics, thereby enhancing the stability of pitch and the clarity of sound. An excessive layer spacing can result in an uneven tonal response of the wire during vibration, whereas a smaller layer spacing helps to produce a more uniform sound.

The expressiveness of music steel wire is contingent upon three primary factors: timbre, tone intensity, and range. An acoustic analysis revealed that the timbre and tone intensity of a given instrument can be significantly influenced by the material from which it is constructed and the characteristics of its microstructure. These factors can consequently affect both the performance of the player and the experience of the listener. Hidekazu Kodama conducted a comprehensive analysis of the impact of diverse string materials on timbre and performance characteristics, uncovering notable discrepancies in resonance frequency, sound quality, and aftertaste across different string materials. In particular, the elasticity and density of the strings exert a direct influence on the characterization of the sound propagation [12]. Jonathan Paul Crosmer conducted a comprehensive analysis of the factors influencing the quality of string tone, integrating performance tests, recordings, and harmonic frequency analysis. His findings revealed that the primary determinants of string tone quality are the composition and tension of the strings, the type of instrument being used, vibrations transmitted from the strings, bow speed, bow pressure, and the contact point between the bow and the strings [13].

The tension in strings is typically increased in order to achieve a more robust sound, but this is limited by the tensile strength of the music steel wire. There are three kinds of common steel wires used for strings: the first category, designated as SWP-A, exhibits a diameter range of 0.08 mm to 10.0 mm. These wires find primary application in general dynamic load springs, encompassing various types of pressure, tension, and torsion springs. The second category, designated as SWP-B, features a diameter range of 0.08 mm to 8.00 mm, suitable for dynamic load springs, with a relatively higher working stress, commonly used in automobile engine valve springs, precision instruments, electrical equipment, production machinery, brake springs, clutch springs, refrigerator compressor springs, etc. SWP-V, with a diameter range of 1.00~6.00 mm, is used specifically for valve springs or similar applications with higher requirements for springs. For a given material, the only means of increasing the available tension is to increase the diameter of the wire [14]. To augment the available tension for a given string material, it is essential to augment the diameter of the music steel wire. Nevertheless, an increase in wire diameter precipitates an enhancement of the bending stiffness of the wire. Excessive bending stiffness of the strings can have a deleterious impact on the sound quality [15,16]. The impact of string stiffness on the timbre and tuning methodology of the instrument is well documented [17]. All genuine strings exhibit a specific degree of stiffness, which is contingent upon the diameter and other intrinsic material properties of the string. It has been demonstrated that the discordance of the vibration of the music steel wire can be reduced when the tension is increased to the ultimate tensile strength (T_S), as shown in the following equations:

$$f_n = n \cdot f_1 (1 + \alpha \cdot n^2) \tag{1}$$

$$\alpha = \frac{\pi^3 \cdot Er^4}{8L^2T} \tag{2}$$

In Equation (1), $f = \sqrt{c/L}$ is the fundamental frequency of the string (here c is the wave velocity, $c = \sqrt{T/\rho}$, L is the effective vibration length of the string, and ρ is the linear density of the string). α is the dissonance factor (the deviation from the harmonic is measured by the value of the dissonance factor α), T is the tensile force, and Er is the Young's modulus. An increase in the ultimate tensile strength of the music steel wire allows for the creation of smaller dissonances [18]. The enhanced strength and ductility of the wire enable it to withstand higher tensile forces without rupturing, thereby increasing its lifespan while maintaining optimal elasticity. This quality contributes to pitch stability and consistent aural quality throughout the playing process. Additionally, the use of this wire in industrial production settings can result in cost savings, making it a cost-effective solution for industrial applications.

A substantial body of research has been conducted on the microstructure of steel wires subjected to heat treatment. Nevertheless, the microstructural evolution of steel wires during the annealing process, as well as the correlation between the changes in tensile strength and their relationship to music steel wires, remain unclear. In this study, the spectral information of Fourier-transformed sound waves generated by the vibration of string wires subjected to different annealing treatments is employed to investigate the musical expression. Scanning electron microscopy (SEM), X-ray diffraction (XRD), and electron backscatter diffraction (EBSD) were employed to investigate the microstructure of music steel wire in diverse annealing conditions. The transformation in mechanical properties of music steel wire at varying annealing temperatures was delineated through the utilization of nanoindentation hardness and tensile properties. Furthermore, the Fourier transform of sound waves generated by the vibration of music steel wire was analyzed using MATLAB R2023b to examine the vibration characteristics and spectral response of the wire. Additionally, the parameters of Young's modulus, tensile strength, nanoindentation hardness, and decay time of the music steel wire were measured and calculated, respectively, to analyze the musical music expression parametrically from three perspectives: tone, amplitude, and loudness. Moreover, the impact of varying microstructures, mechanical properties, and string wrapping on the musical music expression of the music steel wire was evaluated. A comprehensive analysis of the large-strain SWP-B music wire was conducted through additional annealing of cementite spheroidization, ferrite deformation, and other phenomena, with the aim of systematically exploring the relationship between mechanical properties and string composition, manufacturing process, and mechanical properties of music steel wire. This was accomplished through an examination of the music steel wire following the annealing of disparate internal cementite spheroidization processes. The findings indicate that the internal cementite spheroidization of music steel wire has a considerable impact on the musical expression, thereby advancing the existing conclusions.

2. Experimental Materials and Methods

The materials employed in this experiment were guzheng strings procured from the Xi'an Conservatory of Music Musical Instrument Factory. The guzheng is one of the oldest folk musical instruments in China. It is a 21- or 25-stringed plucked instrument. The strings were stripped of the outer coil wire, nylon wire, and copper wire. The resulting decomposition diagram is illustrated in Figure 1a, while the sectional diagram is shown in Figure 1b. The steel wire located within the core was retained for the purpose of undergoing experimental testing, and its designated model is SWP-B. The initial microstructure of the music steel wire is illustrated in Figure 2. The microstructure comprises layered pearlitic structures, comprising ferrite and cementite, which are markedly elongated along the axes of the strings and exhibit a fine fiber texture. The chemical composition of the music steel wire was determined in accordance with the Chinese Industrial Standard YB/T 5218-2017, as detailed in Table 1. A differential scanning calorimetry (DSC) test was conducted on the music steel wires using a simultaneous thermal analyzer (STA 449C) from NETZSCH (Selb, Germany), as shown in Figure 3, and the peak of the heat flow was identified within the range of approximately 700–800 °C, indicating that the phase transition point of the music steel wires is 741 °C.

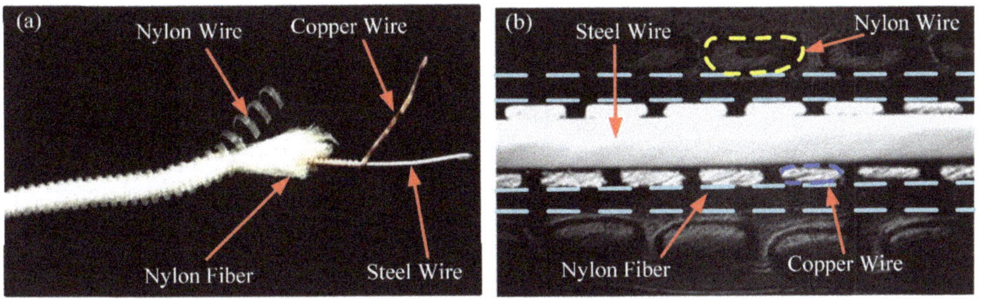

Figure 1. (**a**) String breakdown diagram; (**b**) String longitudinal section.

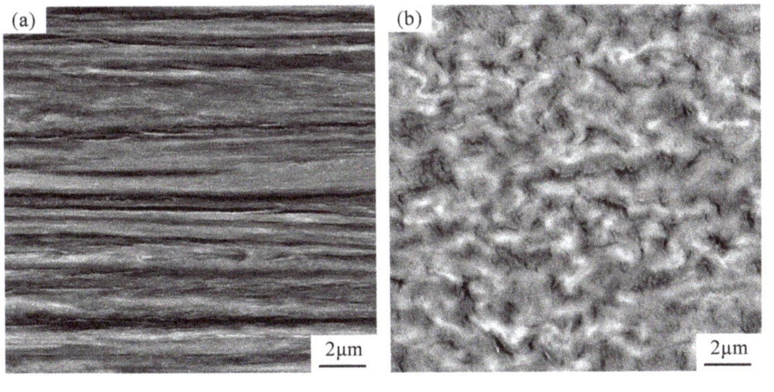

Figure 2. (**a**) Longitudinal section SEM graph of the music wire; (**b**) Cross-section SEM graph of the music wire.

Table 1. Chemical Composition of High Carbon Music Wire of SWP-B (wt.%).

Element	Fe	C	Si	Cr	Mn	Cu	P	S	Al
Music wire	Bal	0.781	0.21	0.06	0.67	0.04	0.001	0.004	0.005

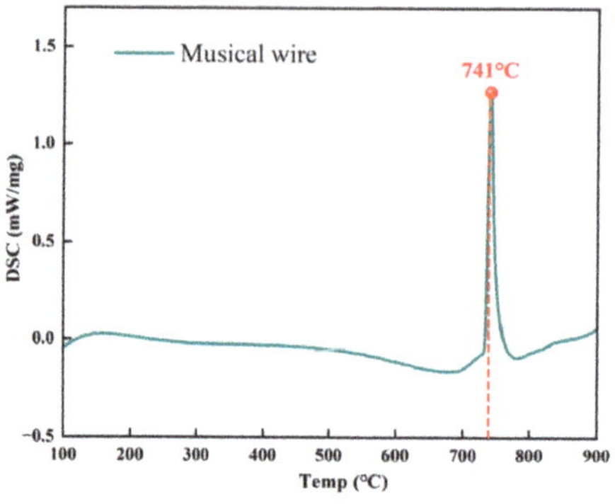

Figure 3. The phase transition curve for DSC of SWP-B music wire.

The elemental composition of high-carbon steel wire is of the utmost importance, particularly the carbon content. Steel wire exhibits a strong magnetic property and, due to the low atomic number of carbon and sulfur, cannot be scanned accurately using an energy dispersive spectrometer (EDS) electronic probe. This paper utilizes the CS844 Carbon and Sulfur Analyzer (LECO, St. Joseph, MO, USA) to measure the carbon and sulfur content. The method is based on the standard GB/T 20123-2006, which details the determination of the total carbon and sulfur content of iron and steel. The combustion of the sample after being poured into the infrared detection cell is performed using a high-frequency induction furnace, and the infrared absorption method is used to calculate the content. This method ensures high accuracy. A UV–visible near-infrared spectrophotometer is employed to measure the silicon, manganese, and phosphorus content. The sample is dissolved in acid and subsequently reacted with reagents to create color. The measurement of the sample's absorption at distinct wavelengths enables the calculation of each element's content.

2.1. Microstructure and Mechanical Property Test

A differential scanning calorimetry (DSC) test was conducted on the music steel wires using a simultaneous thermal analyzer (STA449C) from NETZSCH (Selb, Germany) to perform the DSC thermal analysis of SWP-B wires. The tests were conducted using an aluminum crucible with a diameter of 4 mm and a crimp seal. Prior to the commencement of the DSC analysis, the ultra-fine steel wire was meticulously divided into segments of 3 mm in length. This procedure ensured that the aluminum crucible and the wire's test surface were thoroughly devoid of any extraneous contaminants. The weight of the steel wire sample was then meticulously regulated within the range of 30 to 50 mg using an analytical balance. Subsequent to this, the aluminum crucible underwent a process of compaction through the utilization of a press. During the test, the heating rate was set to 20 K/min, the empty aluminum pan was kept at 40 °C for 1 min, and then the SWP-B steel wire was placed into the pan. The temperature interval was set from 40 °C to 900 °C. The utilization of flowing nitrogen as a protective gas during the course of the tests is also a notable aspect of the experimental design. The collected DSC test data were then subjected to thorough analysis using NETZSCH Proteus software 4.2.1. Heat flow was measured in relation to time and can be converted linearly to temperature. Heat flow in the positive lead was obtained by subtracting the heat flow in the pan from the heat flow in the pan and the lead.

The string steel wire has been subjected to a heat treatment process. This process effectively enhances the wire's toughness and elasticity, thereby ensuring its stability under tension. It is distinguished by high tensile strength and high hardness. In this study, the music steel wire was subjected to a low-temperature annealing treatment at 100 °C, 150 °C,

200 °C, 250 °C, 300 °C, 350 °C, 400 °C, 450 °C, and 500 °C at temperatures below the phase transition point, with a holding time of 30 min.

Given the high carbon content of music steel wire, it can be classified as hypereutectoid steel wire. Consequently, spheroidization annealing, isothermal annealing, and incomplete annealing were employed for music steel wires above the phase transition point. The resulting heat treatment curve is illustrated in Figure 4. Spheroidal annealing, as illustrated in Figure 4a, involves heating the steel wire to 750 °C, maintaining this temperature for one hour, and then the wire is cooled in the furnace to 710 °C (FC). This temperature is maintained for one hour, after which the furnace is used to cool the wire to 500 °C for one hour. Finally, it is removed from the furnace and cooled in air to room temperature (AC). To ensure uniformity in the spheroidization microstructure and precise control over the hardness of the annealed material, the objective is to achieve a uniform distribution of spherical or granular carbides within the ferrite matrix. Isothermal annealing is illustrated in Figure 4b: the music steel wire is heated to 780 °C for one hour to achieve incomplete austenitizing. It is then cooled more rapidly to 600 °C (the pearlitic temperature range) for one hour in an isothermal holding process. This allows the austenite to be completely transformed into pearlite. The material is then air-cooled to room temperature and isothermally annealed. The resulting microstructure and hardness are more uniform than those achieved through complete annealing. Incomplete annealing (Figure 4c) involves heating the music steel wire to 780 °C, maintaining the temperature for one hour to achieve incomplete austenitization, and then slowly cooling the wire to room temperature. This process can refine the microstructure and reduce the hardness of the material. The evolution of the microstructure and mechanical properties of the annealed music steel wires and the respective differences in the corresponding musical expression were observed and analyzed.

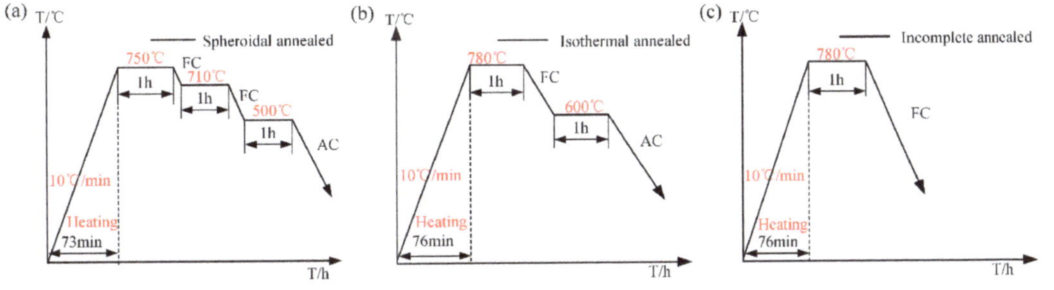

Figure 4. Annealing heat treatment curves of music wire: (**a**) Spheroidal annealing curve; (**b**) Isothermal annealing curve; (**c**) Incomplete annealing curve (FC: furnace cooling; AC: air cooling).

To examine the microscopic morphology of music steel wire following annealing treatment at varying temperatures, the heat-treated samples were inlaid. The music steel wire with a diameter of 0.5 mm and a length of 15 mm was affixed to the XQ-2B metallographic specimen inlay machine with double-sided adhesive tape from Shanghai Optical Machinery Factory. Thereafter, the wire was embedded with conductive resin and powder, resulting in a cylindrical sample with a diameter of 20 mm and a length of 10 mm. Once the resin had cured, polishing was conducted along the longitudinal axis of the wire, with the objective of achieving a polished surface that was as close as possible to the central axis and a width of the polished sample surface greater than 96% of the wire diameter. Subsequently, the samples were ground with sandpaper in a coarse-to-fine sequence, beginning with 1000# and progressing to 3000#. Following grinding, the samples were subjected to fine polishing with W2.5 diamond polishing paste on a tweed polishing cloth until all the scratches on the surface of the samples were removed, resulting in a bright surface. Subsequently, the samples were etched with a 3~4% nitric acid solution for 30s to observe the microstructure.

In this experiment, a metallographic optical microscope (Axiovert 40 MAT, Carl Zeiss, Oberkochen, Germany) was used to observe the metallographic microstructure of the music steel wire after corrosion at a magnification of 1000. Additionally, an ultra-high-resolution field emission scanning electron microscope (Clara GMH, Tescan, Czech Republic) with an electron-beam acceleration voltage of 15–20 kV was employed to observe the microstructure of the heat-treated music steel wire for SEM analysis. Finally, ImageJ2, an open Java image processing and analysis program, was used to observe the microstructure of the heat-treated music steel wire. The volume percentage of ferrite/pearlitic lamella spacing was determined from multiple optical micrographs using ImageJ2. Additionally, various crucial parameters, such as lamella spacing, were obtained from several high-resolution FE-SEM images through the linear intercept method. A field emission scanning electron microscope (ZEISS Sigma 300, Carl Zeiss, Oberkochen, Germany) equipped with a Nordlys Nano Detector was employed to obtain the Kikuchi pattern (TKD) of the music steel wire. Subsequently, the polar plots were generated using Aztec Crystal 2.1.2, EBSD analysis software, to analyze the textile structure orientation and grain size changes of the music steel wire after heat treatment.

In this experiment, X-ray diffractometer (D8 DISCOVER, Bruker, Billerica, MA, USA) was utilized to obtain the diffraction peak data of SWP-B steel wire for XRD analysis. The incoming ray method generator was the most common Cu target (λ = 0.1542 nm) with a wavelength of λ = 1.5406 Å. Prior to instrument use, calibration was conducted using a Si standard. It is imperative to note that since XRD is performed by scanning an area to obtain the diffraction peaks, it is essential to ensure that the area of the specimen is larger than 10 mm × 10 mm. The experiment was conducted with a tube voltage U = 40 kV, a tube current I = 200 mA, and a diffraction angle of 2θ in the scanning range of 30° to 130°.

In this experiment, the objective was to ascertain the impact of distinct annealing treatments on the hardness of music steel wire. Nanoindentation hardness testing (HIT) was conducted using a nanoindentation instrument (Hysitron TI 980, Bruker, Billerica, MA, USA) equipped with a Berkovich indenter, crafted from diamond with a radius of 100 nm. An A3 × 3 matrix of nanoindentation points was collected from the center to the edges of the longitudinal section of the mirror-polished music steel wire samples with a uniform spacing of 30 µm, forming a grid-like pattern. A load of 10 mN was applied, and the loading, holding, and unloading times were set to 5 s, 2 s, and 5 s, respectively. The final hardness values were averaged. Nanoindentation hardness was converted to Vickers hardness in accordance with the standards set forth in ISO 14577-1:2015(E), where 1 GPa is equivalent to 102.04 HV.

The objective of this study is to examine the relationship between alterations in microstructure and mechanical properties, with a particular focus on the tensile strength of music steel wire, which is a critical factor in determining its service life. As the diameter of music steel wire is insufficient for easy clamping, a tensile testing machine (INSTRON 3382, Boston, MA, USA) was employed to ascertain the tensile strength and Young's modulus of music steel wire following various annealing treatments. The wire was twisted and affixed to fixtures at both ends, with node readings recorded at a step size of 10 N. The data were collected from multiple measurements, and the final fitting of the force and displacement change curves was converted into stress–strain curves. To ensure the reliability of the results, the data from the five tensile tests conducted on the music steel wires after each annealing condition were averaged.

2.2. Music Expression Test

In order to conduct a visual analysis of the musical expressiveness of the guzheng's music steel wires, excitation was applied to the four steel wires with different heat treatments with the objective of inducing vibration. To record the vibration frequencies of the music steel wires in their playing state, an audio recording device was positioned near the tensioned music steel wires, with the objective of capturing the emitted sound frequencies. The audio recording device employed was a Multi-Track Linear PCM Recorder (LS-100, Olympus, Tokyo, Japan), which is capable of high-fidelity PCM recording at a maximum frequency of 96 kHz and a maximum of 24 bits. Following the excitation of the strings, the audio signals were recorded and the requisite program code was written in MATLAB software R2023b for the calculation of the Fourier transform (FFT) spectra of the acoustic signals. This was performed in order to analyze the frequency domain and time domain characteristics of the music steel wires after different treatments, as described in references [19,20]. The vibration of the string can be classified into four main categories: transverse vibration, longitudinal vibration, torsional vibration, and octave vibration. As the transverse vibration represents the primary form of vibration, this paper will concentrate on it. The fundamental vibration frequency formula proposed by Mason can be used to calculate the transverse vibration [21]:

$$f = \frac{1}{2L}\sqrt{\frac{T}{\rho_l}} \tag{3}$$

In this equation, T represents tension (N), L denotes the effective string length (m), and the string's line density is represented by ρ_l (kg/m), $\rho_l = \rho_v$ A = ρ_v (πd^2)/4. The density of the steel wire string (density of high-carbon steel wire) is 7850 kg/m^3. It can be observed that the frequency of the transverse vibration is inversely proportional to the length of the string, directly proportional to the square root of the tension, and inversely proportional to the square root of the wire density.

The duration of an instrument's sound signal articulation is significantly influenced by the decay time of its fundamental frequency. The term "decay time" is defined as the time required for the maximum amplitude of the fundamental frequency harmonic to decay to 10% of the maximum amplitude. This value is employed in the calculation of the duration of the interval of the fundamental frequency component of the sound signal generated by the vibration of the strings under disparate plucking strengths. The formula for this calculation is as follows:

$$T = |T_{LPmax} - T_{LPmax\cdot 0.1}| \tag{4}$$

where T_{LPmax} is the time from the onset of oscillation of the fundamental frequency signal to its maximum amplitude, $T_{LPmax\cdot 0.1}$ is the time for the fundamental frequency signal to decay from peak to 10% maximum amplitude.

The tension of each string of the guzheng (simplified model in Figure 5a) was measured using a combination of the model DYZL-107 string tension tester and the three-roller tension sensor DY220-K1T2, manufactured by Anhui Bengbu Dayang Sensing Co., Ltd. (Bengbu, China) (see Figure 5b). The tension of the various heat-treated music steel wires was adjusted on a manual tensile tester (K-100HB-H, Fuzhou Aipu Instrument Co., Fuzhou, China), and the vibration frequency was subsequently measured after the application of excitation at varying tensions (T) and different vibration lengths (L). Subsequently, the audio data underwent Fourier transform analysis using MATLAB software R2023b, resulting in the generation of frequency domain and time domain plots of the music steel wires.

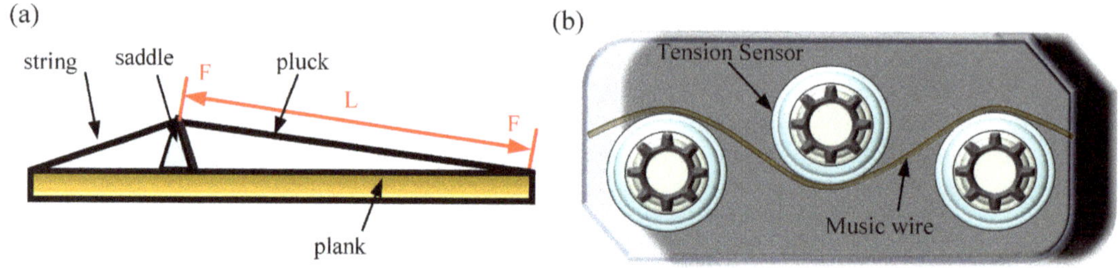

Figure 5. (**a**) Simple model of music wire oscillation; (**b**) Three-roller tension sensor measuring scheme. (Note: F: tension; L: length).

3. Experimental Results and Discussion

3.1. Microstructure Analysis of Music Wire After Low-Temperature Annealing

Figure 6 shows the XRD patterns of music steel wires and various post-annealing treatments. The calibration of the indices for ferrite is based on card number PDF#06-0696, while that for cementite is based on card number PDF#34-0001. Figure 6a depicts the XRD patterns of music steel wire in various annealed states, whereas Figure 6b illustrates the ferrite (110) crystal diffraction peaks of the wire in various annealed states.

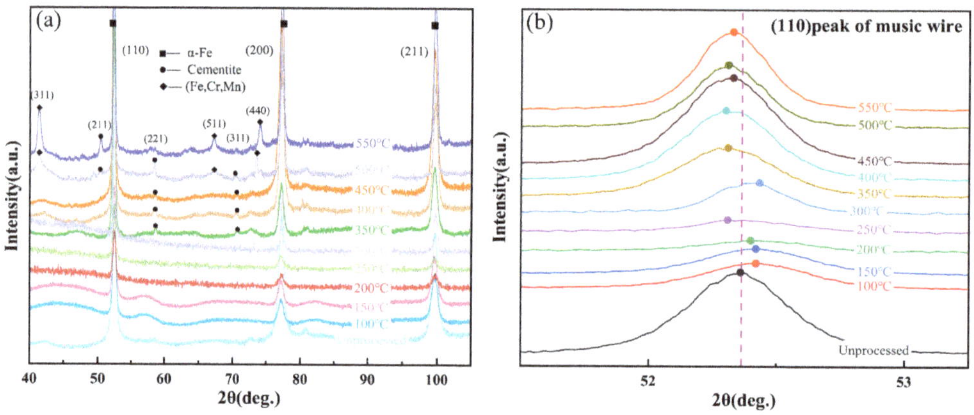

Figure 6. XRD patterns of music wires after different annealing treatments: (**a**) X-ray diffraction patterns; (**b**) Ferrite (110) crystal plane diffraction peaks.

As shown in Figure 6a, at temperatures below 300 °C, the annealed music steel wire does not exhibit the diffuse cementite phase. Instead, C atoms are dissolved in the ferrite. As the annealing temperature increases, cementite is gradually precipitated. At 400 °C, precipitation of the Cr and Mn phases occurs. The diffraction peaks of the ferrite (110) appeared shifted, broadened, and significantly weakened, which is due to the nanostructures and residual internal stresses caused by the large deformation of the music steel wire after cold drawing, so that the internal ferrite lamella refinement and dislocation density of the music steel wire increased, and the occurrence of cementite lamella deformation and decomposition and the formation of carbide nanocrystalline particles and part of the amorphous cementite, which is reflected in the music steel wire having a higher tensile strength [22].

Figure 7 presents the SEM images of music steel wire magnified at 20,000× and localized SEM images at 50,000× magnification under varying annealing conditions. The longitudinal microstructure morphology of music steel wire was observed under low-temperature annealing at a conventional temperature for 30 min. It can be observed that, with the increase in annealing temperature, the internal microstructure of music steel wire is gradually spheroidized after annealing at 350 °C, and the internal microstructure of the steel wire is significantly spheroidized when it is annealed at 550 °C.

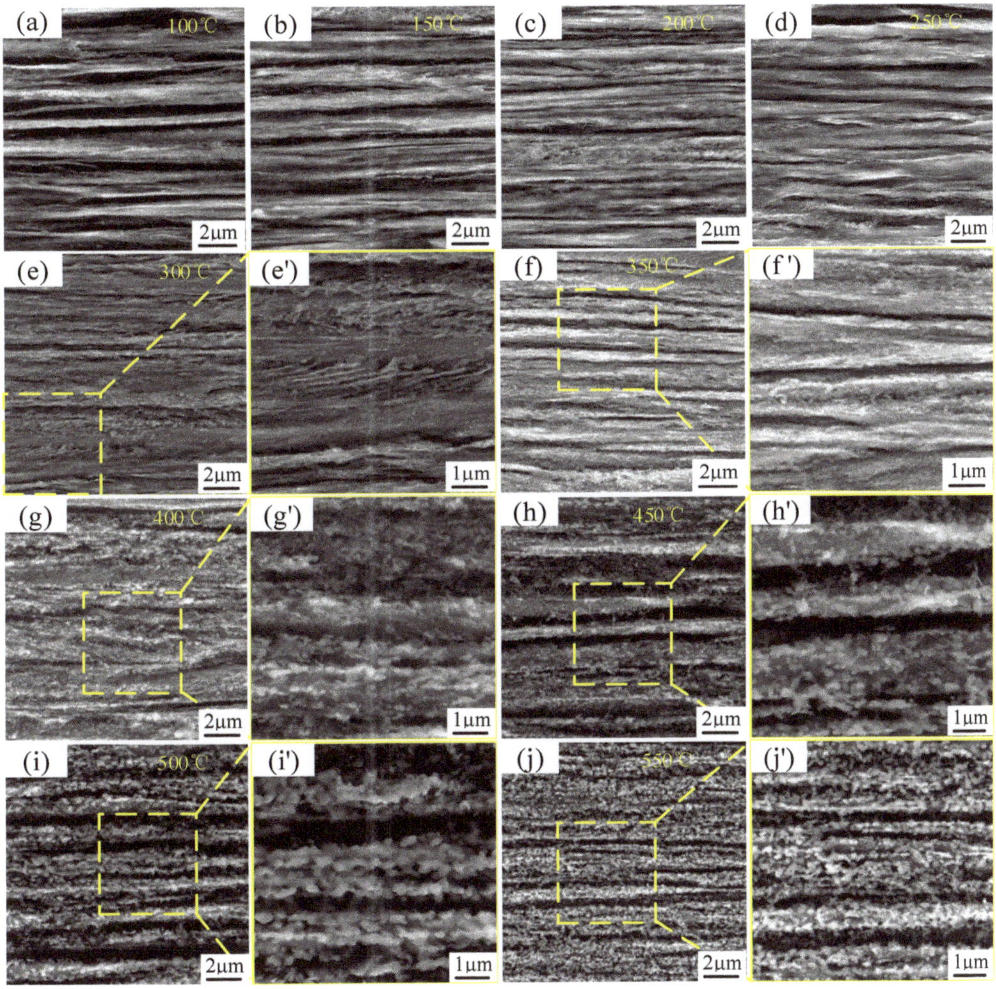

Figure 7. SEM micrographs of the music wires with different annealing treatments: (**a**–**j**) Annealed at 50-degree intervals from 100 °C to 550 °C for 30 min each at a magnification of 20,000 times; (**e′**–**j′**) Localized areas corresponding to 50,000× magnification of each.

As shown in Figure 7, the microstructure does not differ much from the original cold-drawn condition after annealing at 150 °C for holding 30 min. Although fragmentation of the cementite flakes can be observed, the lamellar structure is still dominant in the large-strain music steel wire after annealing at 250 °C for holding 30 min (Figure 7d), and the lamellar structure remains stable, there is no spheroidization of the cementite and no recrystallization. A small amount of detached cementite was observed in the annealed wire at 350 °C (Figure 7f). The internal microstructure of the music steel wire is still clearly lamellar-oriented before annealing below 400 °C (Figure 7g) and, after heat treatment at 400 °C for 30 min, substantial changes in the microstructure are observed, accompanied by a notable reduction in lamellar spacing structure of the original lamellar pearlitic phase. In addition, the cementite flakes undergo spheroidization, mainly located in the interlayer spacing junctions. The internal lamellar structure of the music steel wire after annealing from 450 °C (Figure 7h) is still visible in the axial direction, while cell/sub-boundaries were formed in the ferrite, and the cementite flakes started to spheroidize, with the appearance of small-diameter spheroidized carbides. Since the <110> fiber textile structure formed during wire drawing limits further sliding in each lamella to ordinary strain, the compatibility of adjacent grains can only be maintained by bending of the grains toward each other [23], so that the cementite lamellae curl and fragment into numerous short segments or even round grains. This also proves that cementite flakes are capable of plastic deformation and even strain-induced fragmentation [24]. Meanwhile, according to Y.J. Li [25], it was shown that atom redistribution was not observed after annealing at 250 °C for 30 min, and during annealing at 450 °C (Figure 7h), due to the uncertainty of the existence of layered cementite before annealing, there may be interactions between carbon atoms and dislocations, causing carbon atoms dissolved in ferrite to migrate to the pristine carbide. This migration is driven by a reduction in the ferrite/cementite phase boundary driven by a reduction in the ferrite/cementite phase boundary area, resulting in the formation of amorphous cementite crystallization generation and spheroidization of the cementite [2,26]. With increasing temperature, the spheroidization process becomes more pronounced and reaches its maximum at 550 °C (Figure 7j), where the internal microstructure is completely transformed into a distinct spheroidal cementite phase, coexisting with the formation of highly deformed ferrite flakes which recrystallize and agglomerate into larger ferrite grains, with a blurred and almost invisible/cementite phase boundary.

3.2. Microstructure Analysis of Music Wire After High-Temperature Annealing

Figure 8 depicts the SEM microstructure of SWP-B music steel wire at 20,000× magnification following a high-temperature annealing treatment. The treatment processes of spheroidization annealing (Figure 8b,c), isothermal annealing (Figure 8e,f), and incomplete annealing (Figure 8d) were employed for temperatures exceeding the phase transition point of music steel wire, respectively.

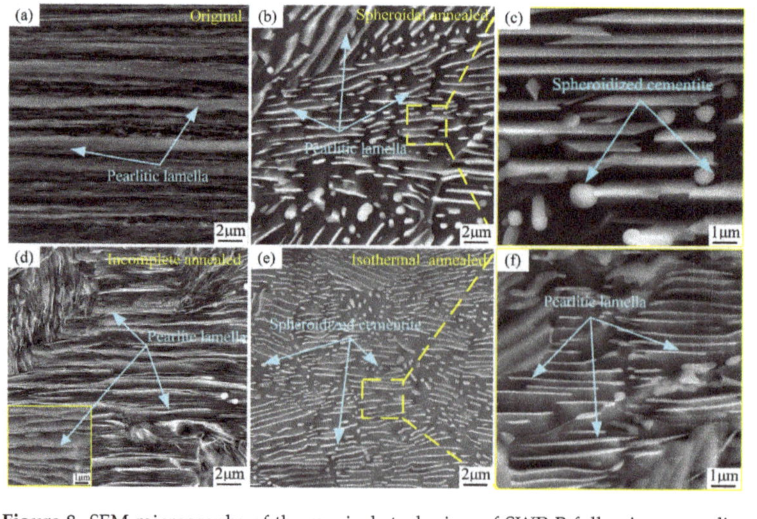

Figure 8. SEM micrographs of the musical steel wires of SWP-B following annealing treatments: (**a**) Original music wire; (**b**) Spheroidal annealed; (**c**) Spheroidal annealed partial enlargement; (**d**) Incomplete annealed; (**e**) Isothermal annealed; (**f**) Isothermal annealed partial enlargement.

As illustrated in Figure 8a, the microstructure of the music steel wire under 20 k magnification exhibits distinct axial lamellar pearlite spacing, as evidenced by the clear separation of the lamellae. Figure 8b,c depicts the formation of multiple pearlite clusters with varying orientations resulting from the decomposition of eutectic austenite after reaching the spheroidization and annealing temperatures. Figure 8c presents a magnified view of the interior of the clusters. It can be observed that the ferrite lamellae and cementite lamellae are randomly oriented, and the crystal orientations of the ferrite exhibit no strong tendency towards continuity or discontinuity. These are surrounded by minute particles of cementite [27]. Figure 8d illustrates the formation of multiple pearlite clusters with varying orientations, derived from eutectic austenite decomposition following the attainment of an incomplete annealing temperature in the music steel wire. The pearlite lamella spacing within these clusters is considerable, exhibiting no spheroidization. Figure 8e depicts the formation of multiple orientations of pearlite clusters in the music steel wire upon reaching the isothermal annealing temperature through eutectic austenite decomposition. The clusters of internal cementite plates have undergone a transformation, assuming the form of either short rods or spherical particles. The influence of cementite on the deformation of music steel wires is considered to be anisotropic due to the axially aligned laminar structure of the wire [28]. The cementite plate is regarded as a barrier that prevents dislocations from passing through the ferrite, resulting in the deformation direction along the carbide plate being designated as the "soft" orientation of the plastic deformation [29,30]. Conversely, when stress is applied at an angle to the cementite plate, the plate is prone to fracturing, which in turn gives rise to the formation of high-density dislocation entanglement in the ferrite region in close proximity to the fractured cementite. This phenomenon gives rise to the bending of high-density dislocations within the ferrite. As a result, the internal microstructure of the music steel wire, following spheroidization annealing and isothermal annealing of the steel wire, displays a distinctive cementite and ferrite plate-bending phenomenon.

Figures 9a–d and 10a–d depict the long-axis EBSD color plots and IPF maps and the phase distribution of SWP-B music steel wire subjected to four distinct treatments: original music wire (Figures 9a and 10a), spheroidal annealing (Figures 9b and 10b), isothermal annealing (Figures 9c and 10c), and incomplete annealing (Figures 9d and 10d).

As illustrated in Figure 9a, the original music steel wire exhibits a fiber texture that is nearly parallel to the axial direction, with fine grains displaying an anisotropic structure. Following three distinct high-temperature annealing treatments, the longitudinal cross-section IPF images of the music steel wire, as shown in Figure 9b–d, reveal that, upon annealing of the music steel wire above the phase transition point, there is a notable shift in the internal organization of the material, specifically in the tensile direction of the parallel textile structure segmentation, which generates large-angle grain boundaries and multicrystalline grains. These observations suggest a tendency towards isotropy in each of these structures. Moreover, the recrystallized grains, which diverge from the original organization, exhibit pearlitic lamellar spacing and a drawing direction parallel to the anisotropic state. The latter display a random distribution of pearlitic crystallographic orientation, as the grains continue to grow, resulting in the formation of a fine pearlitic lamellar spacing structure within the grains, as illustrated in Figure 8b–e. The grain size is more uniform following spheroidization annealing. Additionally, a considerable number of minute grain boundaries emerged within the metal wire following isothermal annealing. Furthermore, a combination of minute grain boundaries and larger grains was observed within the steel wire after incomplete annealing. As seen in Figure 10, the content of the cementite phase during spheroidization annealing is the highest under different annealing methods and then decreases continuously, followed by an increase in the content of other precipitated phases.

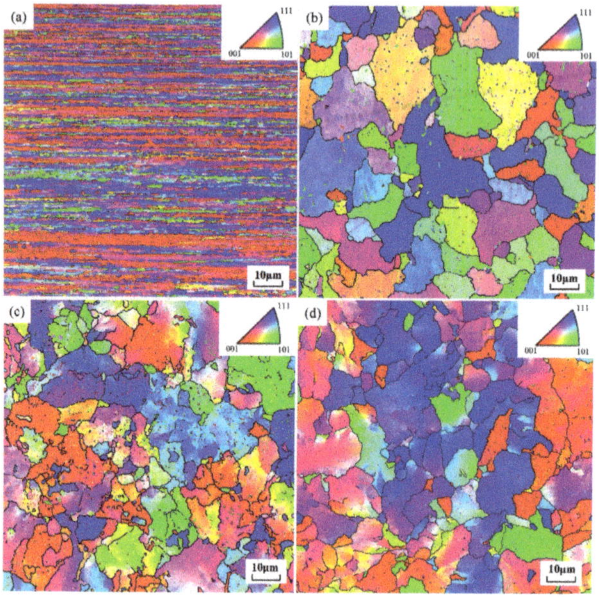

Figure 9. Inverse pole figure (IPF) of ferrite in the pearlite colonies: (**a**) Original music wire; (**b**) Spheroidal annealed; (**c**) Isothermal annealed; (**d**) Incomplete annealed.

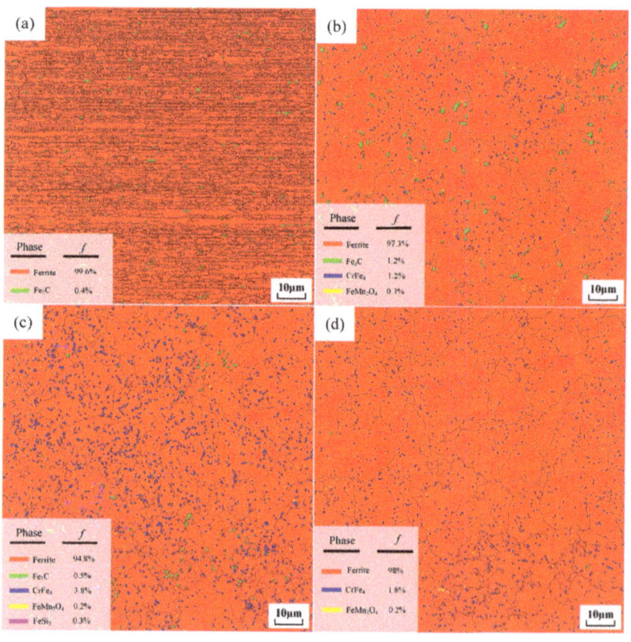

Figure 10. The phase distribution by EBSD: (**a**) Original music wire; (**b**) Spheroidal annealed; (**c**) Isothermal annealed; (**d**) Incomplete annealed.

Figure 11a–d,a′–d′ illustrate the PF plots of the ferrite phase and cementite phases of the original music steel wire, as well as the spheroidal annealing, isothermal annealing, and incomplete annealing treatments. The distribution of the phases can be clearly observed from the PF phase distribution diagrams. Additionally, the evolution of the internal microstructure of the music steel wire deformation texture under different annealing treatment levels is depicted by the polar diagrams of the ferrite and cementite phases.

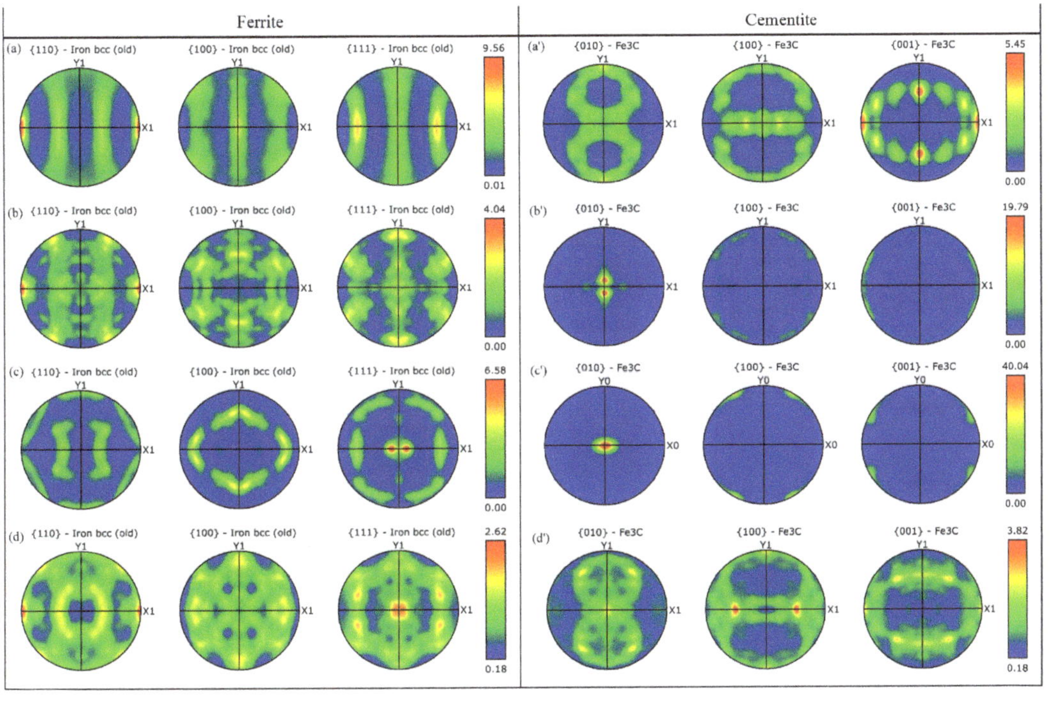

Figure 11. Pole figures of the ferrite and cementite phases of the musical wires with different annealing treatments: Ferrite phase, (**a**) Original music wire; (**b**) Spheroidal annealed; (**c**) Isothermal annealed; (**d**) Incomplete annealed. Cementite phase, (**a′**) Original music wire; (**b′**) Spheroidal annealed; (**c′**) Isothermal annealed; (**d′**) Incomplete annealed.

The strengths of the ferrite <110> fiber texture of the unannealed music steel wire and the spheroidally annealed, isothermally annealed, and incompletely annealed music steel wires are quantified and presented in Figure 11a–d. Figure 11a illustrates that the strength of the ferrite <110> fiber texture of the original music steel wire is 9.56, which corresponds to a markedly robust strength arising from the situation. As the annealing temperature increases, the axial pearlitic lamella spacing inside the steel wire is destroyed, the ferrite phase is decomposed, and the strength of the <110> fiber texture is significantly reduced. This reduction in strength is observed until the incomplete annealing treatment, at which point the strength of the ferrite phase <110> fiber texture decreases to 2.62. Figure 11a′–d′ demonstrate that the formation of spheroidized cementite results in an enhancement of the strength of the <110> phase of cementite. The isothermally annealed steel wire microstructure exhibits the highest strength of the cementite <010> phase, reaching 40.04, which can be attributed to its highest spheroidization rate after isothermal treatment.

Figure 12a–h illustrate the pearlite clusters of unannealed, spheroidally annealed, isothermally annealed, and incompletely annealed steel wires. Given that the music steel wire is a high-carbon steel wire, it follows that the mechanical properties of the high-carbon steel wire and the pearlitic lamellar spacing are inextricably linked. Pearlite is a mixture of ferrite and cementite, and the pearlite organization offers the benefits of high strength and good plasticity [31]. The pearlitic rate serves as a crucial metric for assessing the organization of the steel wire. To ensure accurate measurement of the pearlitic lamellar spacing of the steel wires following different annealing treatments, four groups of pearlitic clusters were selected from each of the four groups visible in the 20k times scanning electron microscope images. The intercept method was employed to ascertain the pearlitic lamella spacing (ILS) [32], with the red line representing an arbitrary intercept line, as illustrated in Figure 12a–h. The measured results are presented in Table 2.

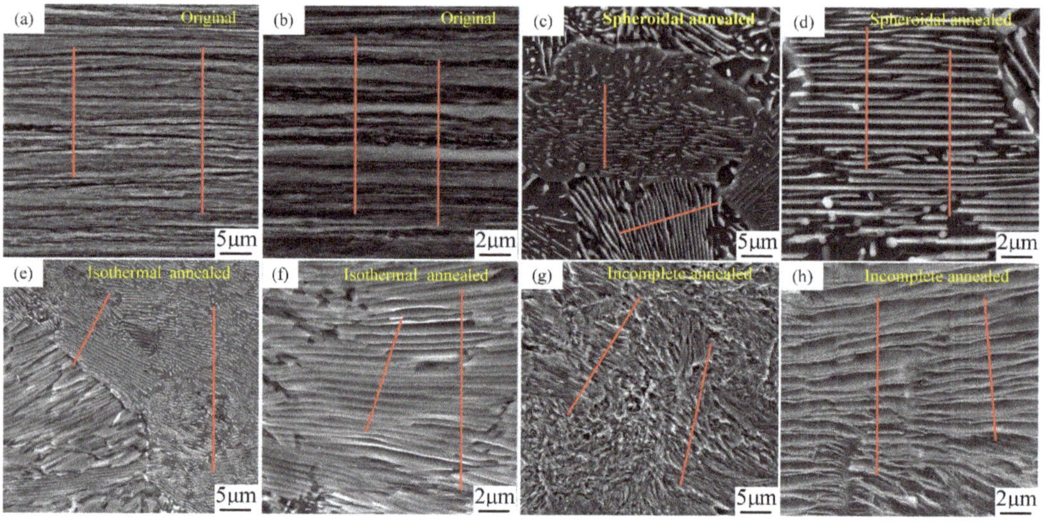

Figure 12. SEM images of pearlite colonies at 10 k times and 20 k times after various annealing treatments: (**a**,**b**) Original music wire; (**c**,**d**) Spheroidal annealed; (**e**,**f**) Isothermal annealed; (**g**,**h**) Incomplete annealed.

Table 2. The pearlitic lamella spacing measurements of music wire after various annealing treatments.

	Lamella Spacing (μm)				
	1	**2**	**3**	**4**	**Average Value**
Original music wire	0.75966	0.86182	0.85458	0.87536	0.83786
Spheroidal annealed	0.37564	0.38747	0.37470	0.36415	0.37549
Isothermal annealed	0.22160	0.26875	0.26760	0.26469	0.25566
Incomplete annealed	0.53573	0.53024	0.56706	0.57031	0.55084

The micrographs obtained via EBSD (Figure 9) demonstrate that the size of the pearlite clusters remains consistent following both spheroidization annealing and isothermal annealing treatments. This is attributed to the comparable heat treatments administered prior to isothermal transformation (Figure 4). Figure 9b–d illustrate that the cluster size diminishes with a reduction in isothermal transformation temperature. The refinement of the pearlite cluster size is typically accompanied by an increase in the number of nucleation sites in the cementite at higher supercooling [33]. The transformation of music steel wires at lower temperatures is distinguished by a reduction in interlayer spacing, which can be attributed to the increase in nucleation sites during supercooling and a concomitant decrease in the diffusion rate. The interlayer spacing of pearlite in SWP-B music steel wire exhibits the smallest value following isothermal annealing treatment. In comparison to the spheroidization annealing heat treatment process at an isothermal temperature of 710 °C, the isothermal annealing process at an isothermal temperature of 600 °C is relatively low (Figure 4b), which results in a reduction in the ability of grain boundaries to move, a slowing down of the growth rate of pearlite, and a refinement of the final pearlite lamella spacing in comparison to the spheroidal annealing [34]. The reduction of pearlitic lamellar spacing results in a slight increase in tensile strength for isothermally annealed music steel wire in comparison to spheroidally annealed wire. This phenomenon can be attributed to the fact when the wire is subjected to an external force, plastic deformation occurs within the ferrite. It is the cementite layer that plays a role in preventing this deformation, and its lamellar spacing represents the maximum slip distance. It can be reasonably deduced that a reduction in lamellar spacing will result in a reduction in slip distance, as well as in ferrite and cementite obstruction. Furthermore, the steel wire will demonstrate an increase in resistance to plastic deformation, which will consequently result in an increase in hardness.

3.3. Mechanical Properties Analysis

Figure 13 illustrates the variation in nanoindentation hardness and tensile strength of SWP-B musical steel wire following different heat treatments. The microstructure of the SWP-B musical steel wire is predominantly lamellar pearlite. The observed decline in hardness can be attributed to the alteration in the structural composition of pearlite and cementite as a consequence of elevated annealing temperatures.

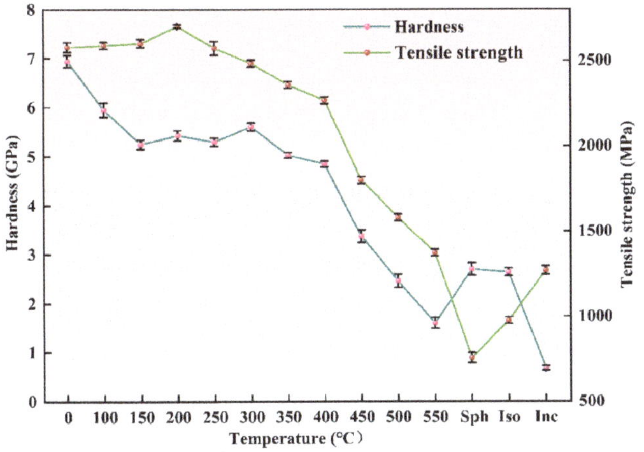

Figure 13. Hardness and tensile strength of music wire after different annealing treatments.

The cementite present in the music steel wire is capable of withstanding the same degree of plastic strain as the ferrite during cold drawing, despite the inherent brittleness of the cementite itself [27,35,36]. No notable reduction in strength was observed following annealing at 150 °C for 30 min (Figure 13). Instead, a slight increase in strength was observed, reaching a peak of 2.7 GPa after annealing at 200 °C for 30 min. Following annealing and holding at 250 °C for 30 min, a slight decline in strength was observed. However, the level remained at approximately 2.5 GPa, indicating that no significant alterations in the microstructure were apparent. Moreover, it has also been demonstrated that, during annealing at relatively low temperatures (150–300 °C), reprecipitation, nanocrystals, and roughening of severely strained cementite occur [2,4]. Accordingly, the prevailing view in the literature is that the formation of cementite nanocrystals after a period of low-temperature annealing at approximately 250 °C results in wire hardening and a catastrophic deterioration of the torsional toughness of the wire [37,38]. This is evidenced by the fluctuating indentation hardness of the music steel wire between 150 °C and 300 °C as the annealing temperature increases, followed by a continuous decrease, and significant decomposition of the cementite occurs, while the lamellar structure remains dominant [39]. At temperatures exceeding 400 °C, the original layered structure is replaced by a supersaturated ferrite substructure comprising grains [8]. Furthermore, the coarsening of the ferrite substructure grains gradually becomes the dominant structure, exerting control over the strength of the material during subsequent annealing. This is the primary factor responsible for the observed decrease in strength during annealing. While the hardness measurements are predominantly surface-based, the observed decrease in hardness can be attributed to the dissolution of the amorphous mixed cementite structure, which was induced by the large deformation of the drawn music steel wires after annealing at 400 °C for 30 min, resulting in a significant reduction in their surface hardness [40]. It is noteworthy that the spheroidization of cementite begins to predominate during annealing at 400 °C (Figure 7). Following annealing at 450 °C for 30 min, the spheroidization of cementite intensifies, and the hardness of the music steel wires appears to decline, reaching 3.37 GPa (approximately 343 HV), which is about half of the hardness of the unannealed music steel wires, which is 6.94 GPa (approximately 707 HV). The internal lamellar structure of the music steel wire is no longer present after annealing at temperatures up to 550 °C, and the hardness also decreases to 1.6 GPa (approximately 163 HV).

Above the phase transition point, the high percentage of spheroidization of the isothermally spheroidized and isothermally annealed cementite, as well as the presence of the pearlitic lamellar spacing structure (Figure 8b,c), resulted in a slightly higher hardness of the isothermally spheroidized and isothermally annealed microstructure than that of the incompletely annealed microstructure and that of the low-temperature annealed microstructure at 500 °C and 550 °C. In particular, the incompletely annealed microstructure of SWP-B music steel wire exhibits minimal spheroidization (Figure 8d) and, consequently, the material exhibits the lowest recorded hardness value of 0.68 GPa (69 HV).

Figure 13 also demonstrates the correlation between tensile strength and annealing temperature. The tensile strength of the music steel wire at room temperature is 2578 GPa, which indicates that the wire is of a robust and high-strength construction. Following annealing at 200 °C for 30 min, the tensile strength is maintained with a slight increase to 2702 MPa. Following annealing at 250 °C for 30 min, a slight reduction in tensile strength is observed, although it remains at a high level. For annealing temperatures above this threshold, the relationship between tensile strength and annealing temperature is approximately linear, exhibiting a decrease in tensile strength. In high-temperature annealing above the phase transition point, the isothermal annealing microstructure and spheroidal annealing microstructure form large-angle grain boundaries, and their internal cementite spheroidization rate accounts for a significant percentage (Figure 9b,c). Consequently, the tensile strength is significantly reduced. However, in the case of an incompletely annealed microstructure, the internal pearlitic lamellar spacing structure of the grain accounts for a considerable proportion of the grain, and the spheroidization rate is relatively low (Figure 9d), resulting in a tensile strength of 1309 MPa. This is higher than that observed in the microstructure after spheroidal annealing and isothermal annealing.

The observed increase in tensile strength of the steel wire following annealing at temperatures below 200 °C for 30 min can be attributed to either static strain aging during low-temperature annealing or even during the storage of cold-deformed material at room temperature. The increase in wire strain serves to mitigate the effects of low-temperature annealing and subsequent strain aging in large-strain music steel wires. This phenomenon can be attributed to the supersaturation of ferrite with carbon during the cold drawing of large-strain music steel wires and the slightly higher temperatures that occur during the drawing process. These conditions allow for the intense polarization of carbon into individual dislocations in the presence of intense ferrite strain during wire drawing [41]. This effect constrains the number of available sites on the carbon-modified dislocations during static storage at room temperature or annealing at low temperature. Furthermore, given that the formation of the ferrite dislocation cell/substructure is also decorated with carbon, the density of individual dislocations is observed to decrease with continued drawing [25].

Interactions between interstitial solute atoms and dislocations also serve to determine a multitude of mechanical properties inherent to steel [42]. As proposed by Y.J. Li, the formation of the cell/substructure grain in large-strain drawn steel wires after annealing at temperatures above 200 °C is not attributable to recrystallization. Rather, it can be attributed to pronounced recovery of the deformed microstructure. Upon annealing at temperatures above 250 °C, softening commences, accompanied by a pronounced decline in tensile strength. This phenomenon is primarily attributed to the static recovery of the deformed microstructure, manifested in the annihilation, rearrangement, or polygonization of ferrite dislocations [2,43]. Furthermore, additional microstructural modifications during annealing, including spheroidization of cementite, redistribution of alloying elements, recovery of point defects (vacancies and interstitials), and relatively minor coarsening of the cell/substructure grain, may also contribute to a reduction in tensile strength. The application of intense plastic deformation, such as cold drawing, results in the formation of a high dislocation density within the material. The majority of dislocations generated during cold drawing are stored at the interface between ferrite and cementite, where they form cell/substructure grains superimposed on the phase boundary [25]. Concurrently, as a consequence of cementite decomposition, dislocations from adjacent ferrites may converge at the site of the original carburization, thereby forming dipole configurations [44]. In conjunction with the accelerated diffusion process across the phase boundary resulting from severe plastic deformation, the rapid annihilation of dislocation dipoles during annealing becomes a viable phenomenon.

The inhibition of recrystallization in large-deformation SWP-B music wires can be attributed to their fiber texture [45], which reduces the drive for recrystallization and decreases the mobility of grain boundary boundaries, as well as the formation of small-angle grain boundaries. The presence of a pearlitic lamellar spacing microstructure in music steel wires contributes to an increase in strength and hardness compared to wires containing spheroidized cementite. This is attributed to the fact that the latter material exhibits a smaller heterogeneous interfacial area fraction. During annealing at temperatures above 200 °C, the phase boundaries diminished by spheroidized cementite are replenished by the cell/substructure grain, which inhibits dislocation motion. Furthermore, the ferrite cell/substructure grain undergoes continuous coarsening in conjunction with the spheroidization of cementite throughout the annealing process [2]. In the present case, it is challenging to quantify the reduction in tensile strength due to the loss of phase boundary density resulting from the spheroidization of the cementite. It is evident that the softening effect resulting from phase boundary loss is concurrent with the softening effect associated with cell/substructure grain coarsening.

3.4. *Analysis of Musical Expression of Annealed Musical Wire*

Without loss of generality, Figures 14–17 illustrate the selected nylon–steel composite wires of the guzheng, which are located on the 3rd string (pitched at 880 Hz) and 6th string (pitched at 493.8 Hz) in the treble region, the 9th string (pitched at 329 Hz) in the middle region, the 12th string (pitched at 220 Hz) and the 15th string (pitched at 146.8 Hz) in the bass region, and the 18th string (pitched at 110 Hz) in the double bass region. The strings are tuned in such a way that they are susceptible to excitation by the sound waves themselves, once they have been tuned. The frequency and time domain plots of the sound waves generated following a Fourier transform are presented herein.

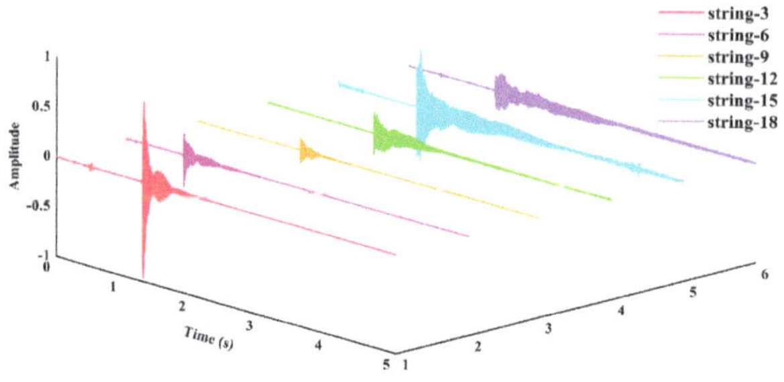

Figure 14. Time domain plot of treble strings 3 and 6, alto string 9, bass strings 12 and 15, and double bass string 18 at standard tones.

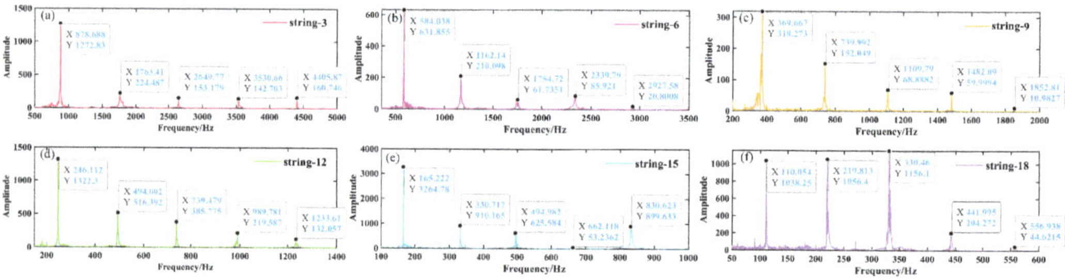

Figure 15. Frequency domain plot of treble strings 3 and 6, alto string 9, bass strings 12 and 15, and double bass string 18 at standard tones. (**a**) Frequency domain plot of string 3; (**b**) Frequency domain plot of string 6; (**c**) Frequency domain plot of string 9; (**d**) Frequency domain plot of string 12; (**e**) Frequency domain plot of string 15; (**f**) Frequency domain plot of string 18.

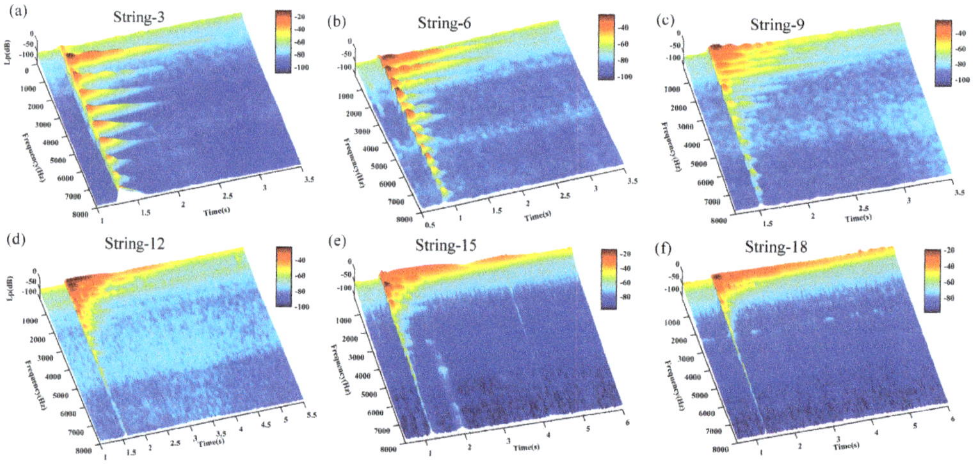

Figure 16. The 3D spectrograms of treble strings 3 and 6, alto string 9, bass strings 12 and 15, and double bass string 18 at standard tones: (**a**) Time and frequency domain plots of string 3; (**b**) Time and frequency domain plots of string 6; (**c**) Time and frequency domain plots of string 9; (**d**) Time and frequency domain plots of string 12; (**e**) Time and frequency domain plots of string 15; (**f**) Time and frequency domain plots of string 18.

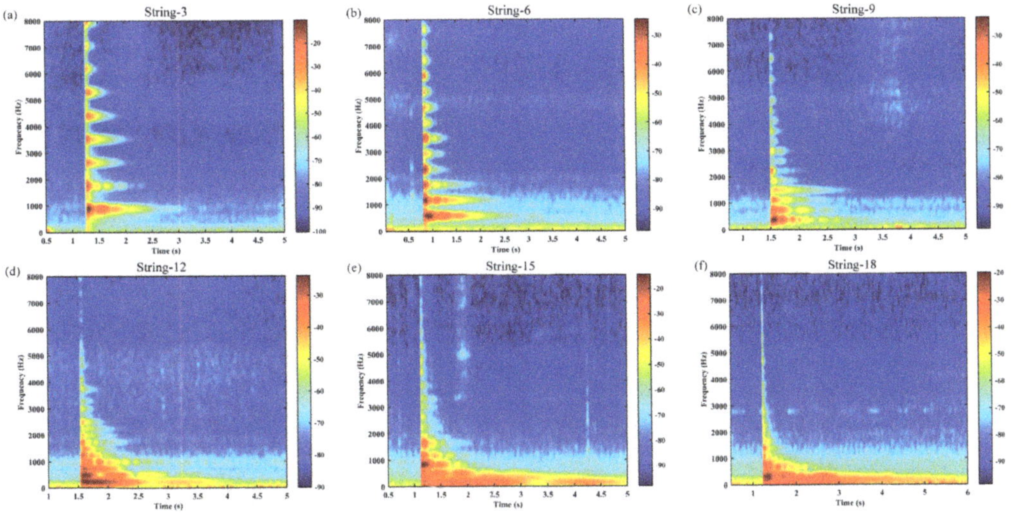

Figure 17. The 2D spectrograms of treble strings 3 and 6, alto string 9, bass strings 12 and 15, and double bass string 18 at standard tones. (**a**) Time and frequency domain plots of string 3; (**b**) Time and frequency domain plots of string 6; (**c**) Time and frequency domain plots of string 9; (**d**) Time and frequency domain plots of string 12; (**e**) Time and frequency domain plots of string 15; (**f**) Time and frequency domain plots of string 18.

As illustrated in Figures 14–17, the guzheng strings are composed of nylon–steel composite wire, which produces discernible vibration signals and a prolonged aftertone. These strings are distinguished by a full and rounded timbre, a purity and softness of tone, brightness, and clarity. As the vibration of the guzheng string is a compound vibration, it exists in two distinct regions, namely the center and one third of the string. This results in the production of a compound tone, which includes both the fundamental frequency and overtones. The low-frequency overtones are of greater amplitude, resulting in a mellower tone that is more suitable for individuals with more sensitive hearing. However, the high-frequency output is less pronounced, and the tone may be perceived as more monotonous. Figure 17 illustrates that the bass strings 12 and 15, in addition to the double bass string 18, exhibit a prolonged sustain and a gradual decay. In contrast, string 9 exhibits a more robust mid-frequency overtone, resulting in a tone that is more rounded and harmonious. In contrast, strings 3 and 6 exhibit augmented high-frequency overtones, which contribute to a brighter and more discernible tone. In the performance, the tone situated in the vicinity of the front yue-shan (a musical instrument part, specifically a horizontal wooden piece used to hold the strings on the forehead of the zither in ethnic musical instruments) is more robust, while the tone in the middle position of the zither code and the front yue-shan is comparatively softer. The entire vibration process of the guzheng's articulation and the quantity and strength of the overtones generated have become the essential elements that define the guzheng's distinctive tone. The tone is significantly influenced by the unique characteristics of the guzheng, including a clear and bright soprano, a round and full mid-range, a deep and dense bass, and a long aftertone.

Table 3 illustrates the mechanical property parameters that were evaluated for each of the selected nylon–steel composite wires, in addition to the bare steel wire, the 200 °C annealed steel wire, and the incomplete annealed steel wire.

Table 3. Mechanical performance parameters of music wires after varying annealing treatment conditions.

Music Wires	Diameter/mm	Modulus of Elasticity/(GPa)	Tensile Strength/(MPa)
Nylon–steel composite wire	1.75	6.4 ± 0.7	460 ± 24
Bare steel wire	0.6	24.1 ± 1.3	2578 ± 37
200 °C annealed bare steel wire	0.6	32.0 ± 1.8	2702 ± 53
Incompletely annealed bare steel wire	0.6	7.3 ± 1.4	1270 ± 48

As illustrated in Table 3, the tensile strength of nylon–steel composite wire with a core steel wire diameter of φ = 0.6 mm, bare steel wire with a diameter of φ = 0.6 mm, 200 °C annealed steel wire, and incompletely annealed steel wire is 460 MPa, 2578 MPa, 2702 MPa, and 1269 MPa, respectively. Furthermore, the Young's modulus is 6.4 GPa, 24.1 GPa, 32.0 GPa, and 7.3 GPa, respectively. Following annealing at 200 °C, the tensile strength of the steel wire reached a maximum of 2.7 GPa. Subsequent low-temperature annealing treatment resulted in an even higher tensile strength of 1.2 GPa in the case of the incompletely annealed steel wire. The enhanced tensile strengths permit a greater degree of inharmonicity in the utilization of the musical steel wire [18,46], thus the steel wire strings that have undergone annealing at 200 °C and the incomplete annealed steel wire strings that have been subjected to high temperatures were selected for the assessment of the musical expression of the strings. The most significant factor influencing the functionality of a string is its stiffness. This term does not refer to the hardness of the string but rather to the external force required to produce a displacement within the string. The material properties and geometry of the string contribute to the overall stiffness of the string [47]. In general, there is a proportional relationship between the stiffness of a material and its Young's modulus, E. Among the properties of a string, stiffness is of particular significance, given the strong correlation between the level of mechanical and physical properties and acoustic parameters. Moreover, the vibrational properties of the string are frequently employed to generate the sound of stringed instruments [7,47,48].

Materials with high stiffness facilitate the transmission of sound waves with greater efficiency, resulting in a more robust and clearer audio output. It is evident that the stiffness of a material may be subject to alteration as a consequence of tensile forces. This alteration is of particular consequence in audio apparatus, as it can influence the velocity of sound wave propagation and the frequency response of the audio [47,49]. To illustrate, 200 °C annealed steel wire (E = 24.1 GPa, TS = 2702 MPa) and bare steel wire (E = 32.0 GPa, TS = 2578 MPa) exhibit increased modulus and tensile strength, allowing them to withstand greater stresses under tension. This results in superior tone retention and a richer, more articulate tone due to their capacity to efficiently conduct high-frequency vibrations. The audio spectral clarity and richness of high-frequency harmonics directly influence the musical expression of the instrument [48]. In contrast, softer materials, such as nylon–steel composite wire (E = 6.4 GPa, TS = 460 MPa) and incomplete annealed steel wire (E = 7.3 GPa, TS = 1270 MPa), display greater susceptibility to vibration during stretching. This has an impact on the fullness of the sound quality and can result in a reduction in performance at higher pitches, which in turn affects the overall sound expression. Steel wire that has undergone annealing at 200 °C (with a Young's modulus of 32 GPa) exhibits the highest Young's modulus, which provides enhanced rigidity and stability. This makes it an optimal choice for generating higher frequencies and harmonics, which contribute to a more precise and clear sound quality. In comparison, incompletely annealed steel wire (with a Young's modulus of 7.3 GPa) displays diminished rigidity, which gives rise to instability in its high-frequency expression and a mellifluous, warm tone that may be more conducive to specific applications. However, this is frequently accompanied by a reduction in volume and loudness.

Figures 18–20 illustrate the time domain, frequency domain plots, as well as the 3D spectrograms of four different string materials played under excitation for the A6 tone (standard fundamental frequency of 1760 Hz) after Fourier transformations for comparison.

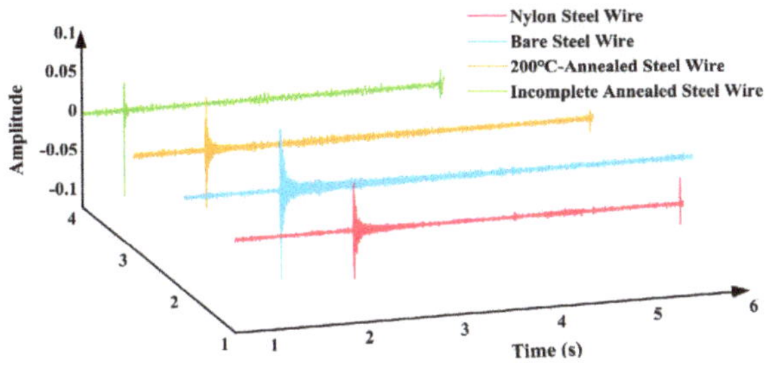

Figure 18. Time domain plot of various string materials after Fourier transform of the A6 tone (standard fundamental frequency of 1760 Hz) played under plucking.

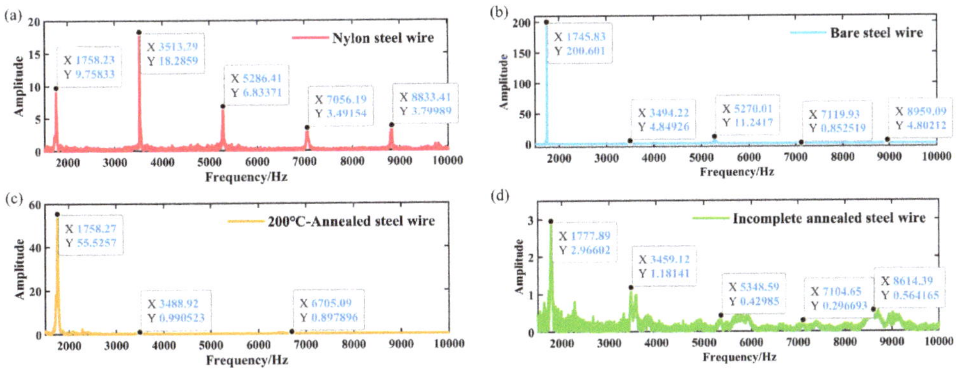

Figure 19. Frequency domain representation of various string materials after Fourier transform of the A6 tone (standard fundamental frequency of 1760 Hz) played under plucking. (**a**) Frequency domain plot of nylon–steel composite wire; (**b**) Frequency domain plot of bare steel wire; (**c**) Frequency domain plot of 200 °C annealed bare steel wire; (**d**) Frequency domain plot of incompletely annealed bare steel wire.

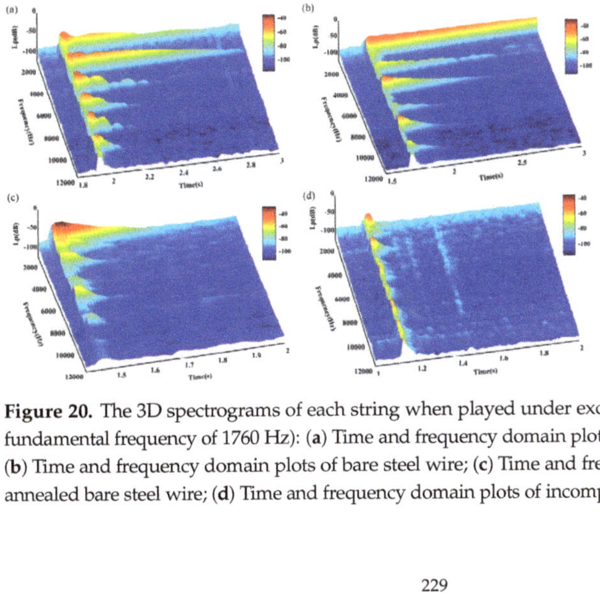

Figure 20. The 3D spectrograms of each string when played under excitation of an A6 tone (standard fundamental frequency of 1760 Hz): (**a**) Time and frequency domain plots of nylon–steel composite wire; (**b**) Time and frequency domain plots of bare steel wire; (**c**) Time and frequency domain plots of 200 °C annealed bare steel wire; (**d**) Time and frequency domain plots of incompletely annealed bare steel wire.

Table 4 presents the frequencies (amplitudes) of the first five harmonics of the A6 tone when played by the four wires that underwent distinct treatments.

Table 4. Amplitudes of the first five harmonics of the A6 tone played on four different strings.

	Nylon–Steel Composite Wire	Bare Music Wire	200 °C Annealed Bare Music Wire	Incompletely Annealed Bare Music Wire
1st harmonic	1758 Hz (9.75)	1745 Hz (200.6)	1758 Hz (55.52)	1777 Hz (2.96)
2nd harmonic	3513 Hz (18.28)	3494 Hz (4.85)	3488 Hz (0.99)	3459 Hz (1.18)
3rd harmonic	5286 Hz (6.83)	5270 Hz (11.24)	-	5348 Hz (0.43)
4th harmonic	7056 Hz (3.49)	7119 Hz (0.85)	6705 Hz (0.89)	7104 Hz (0.30)
5th harmonic	8833 Hz (3.80)	8961 Hz (4.93)	-	8614 Hz (0.56)

Note: The audio signal is processed by Fourier transform to ensure the amplitude value is relative and dimensionless.

As illustrated in Figures 19 and 20 and Table 4, musical wires composed of different materials demonstrate disparate frequency shifts between the various harmonic orders [50]. The first harmonic (fundamental frequency) exhibits a range of approximately 1745 Hz to 1777 Hz, with the highest frequency (1777 Hz) observed in the incompletely annealed steel wire, resulting in a brighter tone. The second harmonic exhibits a relatively narrow frequency range, with the nylon–steel composite wire (3513 Hz) demonstrating superior stability compared to the other materials, indicative of enhanced frequency retention. The spectra of the various wire material distributions demonstrate that the annealing treatment has a considerable impact on the harmonic frequencies [13]. In particular, the third harmonic lacks data for the bare steel wire annealed at 200 °C. However, the nylon–steel composite wire (5286 Hz) and incomplete annealed steel wire (5348 Hz) exhibit higher frequencies, indicating that both may be more effective in high-frequency timbres. Additionally, the fourth and fifth harmonics demonstrate a notable variation in frequency. The steel wire annealed at 200 °C (6705 Hz) exhibited relatively low frequencies at the fourth harmonic, which may contribute to a more profound aftertone. The amplitudes of the harmonics exhibited variability across the different materials. The amplitude characteristics of the bare steel wire and the nylon–steel composite wire are more pronounced in the higher harmonics, particularly the 5th harmonic. This may result in a richer tone and improved layering. The amplitude of the incomplete annealed wire is lower (8614 Hz), which affects the roundness and subtlety of the tone.

Figure 19 demonstrates that the steel wire that has undergone incomplete annealing displays a greater number of harmonics and a less discernible frequency amplitude. Upon the application of an excitation, the stabilization of the fundamental frequency of the nylon–steel composite wire occurs at a slower rate, while the stabilization of the fundamental frequency of the bare steel wire occurs at a faster rate. In conjunction with the Young's modulus test, the nylon–steel composite wire and the incompletely annealed steel wire exhibited the lowest Young's modulus value among all the strings, indicating that these materials possess relatively low stiffness and high deformability. This resulted in an instability and a less uniform vibration pattern when the strings were subjected to playing stimulation. Furthermore, the fundamental frequency demonstrated a prolonged stabilization period. The nylon–steel composite wire displays a greater richness of frequency components, a relatively weaker presence of high frequencies, and a notable prominence of spectral peaks in the low-frequency range. The fundamental frequency (1758 Hz) is robust and exhibits a gradual decay, while its harmonics (3513 Hz, 5286 Hz, 7056 Hz, 8833 Hz) are somewhat diminished and decay more rapidly, allowing for clear differentiation between scales. The timbre of nylon–steel composite wires is typically mellifluous with a warm quality. The phase response of these strings may result in relatively weak volume peaks at specific frequencies. These frequencies are distinguished by lower energy in the high frequencies and a more even distribution of harmonics, often with a greater proportion of low-frequency components. This can be attributed to the enhanced richness and layering of the string's timbre resulting from the wrapping of the nylon and copper wires, which also increases the durability and abrasion resistance of the strings while maintaining pitch stability and reducing intonation fluctuations [51].

In comparison to nylon–steel composite wire, bare steel wire displays enhanced stiffness, which results in a diminished amplitude and concentrated energy in the lower-frequency region. The fundamental frequency component in the spectrum is markedly amplified, accompanied by a more pronounced and powerful high-frequency component. Furthermore, the harmonics are more pronounced and robust, contributing to a brighter and clearer tone with a wider frequency distribution. The resonance effect and transient response of bare steel wire are typically more pronounced, with the potential for additional high-frequency peaks to appear within the frequency spectrum. Bare steel wires typically exhibit higher power and a more pronounced volume than nylon–steel composite wires and 200 °C annealed steel wires under the same playing force. This characteristic results in the generation of more penetrating acoustic effects. The spectral characteristics of 200 °C annealed steel wire are analogous to those of bare steel wire, albeit with a diminution in harmonics and a slight attenuation of the high-frequency range. The fundamental frequency and its harmonic energy are diminished, resulting in a less robust sound than that of ordinary steel wire strings. The initial stages of harmonic attenuation are accelerated, and the overall timbre is reminiscent of that nylon–steel composite wires. Nevertheless, the distinctive attributes of bare steel wire characteristics remain discernible. The spectrum of the incompletely annealed steel wire exhibits the most moderate characteristics. The application of high-temperature annealing can facilitate the release of internal stress, thereby enhancing the resonance and high-frequency gain. The spectrum displays a substantial high-frequency component and a complex harmonic structure. However, the energy is low and the sound quality is relatively poor, with a noticeable decrease in high-frequency power.

Figure 21 illustrates the time–frequency plots of the acoustic signals generated by each string of the aforementioned guzheng under plucking excitation following distinct heat treatments. These plots demonstrate the regularity of the changes in harmonic components over time.

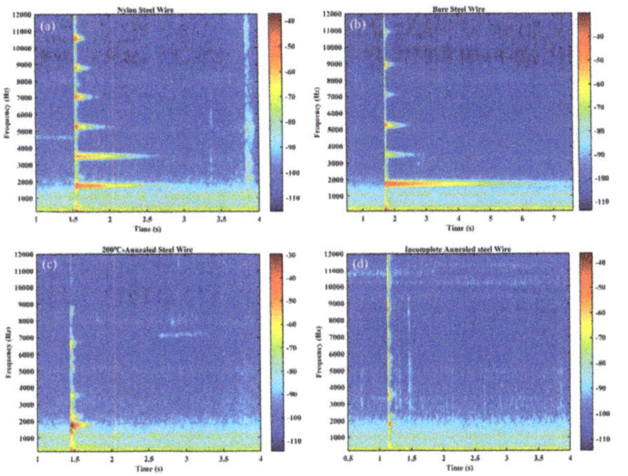

Figure 21. The 2D spectrograms of the acoustic signals produced by each string under plucking excitation: (**a**) Time–frequency plot of nylon wire composite wire; (**b**) Time–frequency plot of bare steel wire; (**c**) Time–frequency plot of 200 °C annealed steel wire; (**d**) Time–frequency plot of incompletely annealed steel wire.

Table 5 presents the calculated decay times of the fundamental frequency harmonic components of the acoustic signals generated during the vibration of four typical guzheng strings under excitation.

Table 5. Time required for the 1st harmonic of each string to decay to 10% of its peak value.

	Nylon Steel Wire	Bare Steel Wire	200 °C Annealed Steel Wire	Incompletely Annealed Steel Wire
Time	96.5 ms	319.7 ms	41.6 ms	2.31 ms

As illustrated in Figure 21 and Table 5, the duration of the intervals for all four strings of the guzheng exhibits variability to varying degrees. Consequently, the duration of the resulting sound signals is also subject to alteration. The fundamental wave components of all four strings of the guzheng reach their peak values with remarkable swiftness after the application of the plucking force and then undergo a rapid decay until they cease to exist [12].

The various categories of wires exhibited notable differences in their respective decay times. The bare steel wire exhibited the longest decay time (319.7 ms), while the incompletely annealed steel wire demonstrated a markedly shorter decay time of 2.3 ms, indicating a rapid decay of vibration energy. These findings suggest that there are notable differences in vibration retention among various materials. This phenomenon is associated with the impact of attenuation on sound vibration, which is contingent upon the specific microstructure of the material in question [52]. The decay time of the steel wire that has undergone a 200 °C annealing process is 41.6 ms, which indicates that it is better able to retain its acoustic properties. In contrast, the nylon–steel composite wire shows better sound retention in practical applications (96.5 ms), indicating that the material properties of the incompletely annealed steel wire are more favorable than those of the aforementioned wire, but not as optimal as those of the bare steel wire. The incorporation of a winding layer into the latter results in an increase in the weight of the string, the tension of the string, and the vibration characteristics of the steel wire core. This ensures a suitable amplitude and markedly improves the sound quality of the string.

The rest of the nylon–steel composite wire is shorter than that of bare steel wire. The material has a more readily dissipated sound, a relatively low sound beta value, a smaller volume, and is suitable for the expression of delicate, soft, and lyrical music. However, the material's softness constrains the emotional richness that it is capable of expressing. The fundamental frequency of the initial excitation exhibits enhanced stability, however, it is susceptible to fluctuations resulting from fingering. The fundamental frequency stabilization of bare steel wire occurs more rapidly, allowing it to quickly reach a stable state. The pitch accuracy is higher, with a decibel value that is greater, a volume that is more pronounced, rich harmonics, an aftertaste that is longer, a relatively long decay time, and a slower auditory disappearance. Bare steel wire is capable of expressing a wide range of emotions, rendering it an optimal selection for fast-paced, high-intensity musical compositions. The bright sound quality of this instrument allows for the clear conveyance of a multitude of musical emotions.

Table 5 lends further support to the hypothesis that the frequency of annealed steel wire at 200 °C is analogous to that of nylon–steel composite wire, with the latter displaying a more pronounced first-order frequency (1758 Hz). However, the first harmonic decays more rapidly, resulting in a notable decline in the amplitude of high-order frequencies. Although these changes may be discernible to the human ear, the resulting lack of rhythm and lack of tone, along with the loss of musical beauty, are less perceptible. The microstructure of the 200 °C annealed material is more uniform, which results in enhanced energy propagation characteristics, improved mechanical properties, and a fuller, relatively warm tone. This renders it more conducive to the expression of emotionally rich music. The annealing process has the effect of reducing internal stresses, increasing the toughness and elasticity of the string, and enhancing its responsiveness when played. The annealing process alters the crystalline structure of the material, resulting in a refinement of the grain structure. This improvement in the material's structure enhances the musical expression and sound quality of the string. A more uniform vibration results from a superior microstructure, which in turn gives rise to a richer harmonic characteristic and an enhanced overall acoustic effect. Consequently, the frequency of incompletely annealed steel wire is relatively high, and the amplitude is significantly lower in the low- and middle-frequency bands. This may result in a lack of warmth in the fundamental tone, rendering it suitable for the expression of certain high-frequency characteristics. Nevertheless, there are some limitations in regard to emotional expression.

4. Conclusions

In this study, SWP-B music wire was subjected to a series of heat treatments at varying temperatures, including low-temperature annealing, spheroidal annealing, isothermal annealing, and incomplete annealing. Subsequently, the evolution of the wire's microstructural properties was examined through a combination of microstructure analysis techniques. This study aimed to investigate the effect of microstructural evolution on the musical expression of SWP-B music wire. The following conclusions can be drawn:

(1) A low-temperature annealing process at 200 °C for 30 min can facilitate the formation of carbon atoms at the sub-boundaries of significant bias polymerization in music steel wire. This is accomplished by reducing the interfacial energy and stabilizing the sub-boundaries, which can then function as a barrier for dislocation movement. Moreover, the precipitation of fine interstitial carbides can impede dislocation movement, thereby conferring an additional strengthening effect on the wire. The tensile strength of SWP-B piano wire was increased from 2578 MPa to 2702 MPa.

(2) As the temperature of the low-temperature annealing process is increased, the precipitation of some intermediate carbides (ε-Fe_3C) occurs at the interfacial position of adjacent ferrite lamellae and in the ferrite matrix. Following low-temperature tempering at 450 °C, notable precipitation changes were observed for the elements Cr, Mn, and Si.

(3) Subsequent to spheroidizing, isothermal, and incomplete annealing treatments, the structure of SWP-B music wire undergoes a transformation from anisotropic to isotropic, exhibiting a near-random texture. This process results in a notable decline in both hardness and tensile strength. The strength of the ferrite phase <110>//ND and the cementite phase <010>//ND exhibited a gradual decline with the increase in annealing temperature.

(4) Nylon–steel composite wire strings are suitable for soft-toned musical instruments, although further improvement may be necessary in order to achieve optimal high-frequency characteristics. The annealed steel wires, subjected to a 30 min treatment at 200 °C, display superior frequency retention and sound expression characteristics in comparison to other materials. Consequently, they are well-suited for musical instruments that require high-quality tone and high-frequency characteristics. It has been demonstrated that incomplete annealing and other high-temperature annealing treatments result in a notable reduction in both tensile strength and Young's modulus of steel wire. Such alterations frequently result in unstable frequency changes, which have a detrimental impact on the tone. Furthermore, the resulting tone and performance are inadequate for the high standards required for musical instrument manufacturing.

Author Contributions: X.J.: Writing—review and editing, Data curation, Investigation, Methodology, Software. Q.L.: Project administration, Supervision, Visualization. F.L.: Conceptualization, Methodology, Data curation, Supervision. X.F.: Software, Visualization. J.Y.: Data curation. Q.Z.: Investigation. X.W.: Software, Validation. J.L.: Investigation, Data curation. All authors have read and agreed to the published version of the manuscript.

Funding: The authors are very grateful for the support received from the National Key Research and Development Program of China (Grant No. 2021YFB3400902), the National Natural Science Foundation of China (Grant No. 51275414, 52205418), the Fundamental Research Funds for the Central Universities with Grant No. 3102015BJ (II) ZS007, the Key Research and Development Program of Shaanxi Province (No. 2020ZDLGY12-07).

Data Availability Statement: The original contributions presented in this study are included in the article. Further inquiries can be directed to the corresponding authors.

Conflicts of Interest: The authors declare that they have no known competing financial interests or personal relationships that could appear to have influenced the work reported in this paper.

References

1. Maruyama, N.; Tarui, T.; Tashiro, H. Atom probe study on the ductility of drawn pearlitic steels. *Scripta Mater.* **2002**, *46*, 599–603. [CrossRef]
2. Li, Y.; Choi, P.; Goto, S.; Borchers, C.; Raabe, D.; Kirchheim, R. Evolution of strength and microstructure during annealing of heavily cold-drawn 6.3 GPa hypereutectoid pearlitic steel wire. *Acta Mater.* **2012**, *60*, 4005–4016. [CrossRef]
3. Lee, J.; Lee, J.; Lee, Y.; Park, K.; Nam, W. Effects of post-deformation annealing conditions on the behavior of lamellar cementite and the occurrence of delamination in cold drawn steel wires. *J. Mater. Process Tech.* **2009**, *209*, 5300–5304. [CrossRef]
4. Zhou, L.; Fang, F.; Wang, L.; Chen, H.; Xie, Z.; Jiang, J. Torsion delamination and recrystallized cementite of heavy drawing pearlitic wires after low temperature annealing. *Mat. Sci. Eng. A-Struct.* **2018**, *713*, 52–60. [CrossRef]
5. Goodway, M. Metals of music. *Mater. Charact.* **1992**, *29*, 177–184. [CrossRef]
6. Leitner, T.; Trummer, G.; Pippan, R.; Hohenwarter, A. Influence of severe plastic deformation and specimen orientation on the fatigue crack propagation behavior of a pearlitic steel. *Mat. Sci. Eng. A-Struct.* **2018**, *710*, 260–270. [CrossRef]
7. Baraz, V.R.; Rundkvist, N.A.; Belov, A.V. Piano wire and strings with high acoustic properties. *Met. Sci. Heat Treat+* **1992**, *34*, 283–287. [CrossRef]
8. Li, Y.; Raabe, D.; Herbig, M.; Choi, P.-P.; Goto, S.; Kostka, A.; Yarita, H.; Borchers, C.; Kirchheim, R. Segregation stabilizes nanocrystalline bulk steel with near theoretical strength. *Phys. Rev. Lett.* **2014**, *113*, 106104. [CrossRef] [PubMed]
9. Jaya, B.N.; Goto, S.; Richter, G.; Kirchlechner, C.; Dehm, G. Fracture behavior of nanostructured heavily cold drawn pearlitic steel wires before and after annealing. *Mat. Sci. Eng. A-Struct.* **2017**, *707*, 164–171. [CrossRef]
10. Chen, J.M.; Chew, C.H.; Seah, K.H.W. Mechanical characterization of cryogenically treated music wire. *J. ASTM Int.* **2006**, *3*, JAI13510. [CrossRef]
11. He, W.; Yinli, C.; Lan, S. Effect of cooling rate and isothermal temperature on the phase transformation and microstructure evolution in SWRH82B high-carbon steel. *Mater. Res. Express* **2018**, *5*, 086506. [CrossRef]
12. Kodama, H.; Sato, M.; Hoshika, S. Attenuation characteristics of tones and vibrations in guitars with nylon, fluorocarbon, and phosphor bronze strings pressed down against fret. *Acoust. Sci. Technol.* **2023**, *44*, 218–229. [CrossRef]
13. Crosmer, J.P. *A Comparison of Viola Strings with Harmonic Frequency Analysis*; University of Nebraska-Lincoln: St. Lincoln, NE, USA, 2011.
14. Giordano, N. Evolution of Music Wire and Its Impact on the Development of the Piano. In *Proceedings of Meetings on Acoustics*; AIP Publishing: Melville, NY, USA, 2011.
15. Fletcher, N.H.; Rossing, T.D. *The Physics of Musical Instruments*; Springer Science & Business Media: Berlin/Heidelberg, Germany, 2012; pp. 3–4.
16. Fletcher, H.; Blackham, E.D.; Stratton, R. Quality of piano tones. *J. Acoust.* **1962**, *34*, 749–761. [CrossRef]
17. Schuck, O.H.; Young, R.W. Observations on the vibrations of piano strings. *J. Acoust.* **1943**, *15*, 1–11. [CrossRef]
18. Martin, D.W.; Ward, W.D. Subjective evaluation of musical scale temperament in pianos. *J. Acoust.* **1961**, *33*, 582–585. [CrossRef]
19. Stoica, P.; Moses, R.L. *Spectral Analysis of Signals*; Pearson Prentice Hall: Upper Saddle River, NJ, USA, 2005; Volume 452.
20. Tarasiuk, T. Hybrid wavelet-Fourier spectrum analysis. *IEEE Trans. Power Deliv.* **2004**, *19*, 957–964. [CrossRef]
21. Handžić, I.; Reed, K.B. The musical kinetic shape: A variable tension string instrument. *Appl. Acoust.* **2014**, *85*, 143–149. [CrossRef]

22. Li, Y.; Kostka, A.; Choi, P.; Goto, S.; Ponge, D.; Kirchheim, R.; Raabe, D. Mechanisms of subgrain coarsening and its effect on the mechanical properties of carbon-supersaturated nanocrystalline hypereutectoid steel. *Acta Mater.* **2015**, *84*, 110–123. [CrossRef]
23. Raabe, D.; Heringhaus, F.; Hangen, U.; Gottstein, G. Investigation of a Cu-20 mass% Nb in situ composite, part I: Fabrication, microstructure and mechanical properties. *Z. Fur Met.* **1995**, *86*, 405–415. [CrossRef]
24. Embury, J.; Fisher, R. The structure and properties of drawn pearlite. *Acta Metall.* **1966**, *14*, 147–159. [CrossRef]
25. Li, Y.; Choi, P.; Borchers, C.; Westerkamp, S.; Goto, S.; Raabe, D.; Kirchheim, R. Atomic-scale mechanisms of deformation-induced cementite decomposition in pearlite. *Acta Mater.* **2011**, *59*, 3965–3977. [CrossRef]
26. Bi, Y.; Wang, Q.; He, T.; Liu, X.; Liu, Y.; Wang, Y.; Yao, Z. The significant impact of the crystallization of cementite in high carbon steel wire on its torsional performance. *Mater. Charact.* **2024**, *211*, 113885. [CrossRef]
27. Zhang, X.; Godfrey, A.; Hansen, N.; Huang, X. Hierarchical structures in cold-drawn pearlitic steel wire. *Acta Mater.* **2013**, *61*, 4898–4909. [CrossRef]
28. Zhou, L.; Fang, F.; Wang, L.; Hu, X.; Xie, Z.; Jiang, J. Torsion performance of pearlitic steel wires: Effects of morphology and crystallinity of cementite. *Mat. Sci. Eng. A-Struct.* **2019**, *743*, 425–435. [CrossRef]
29. Zhang, X.; Hansen, N.; Godfrey, A.; Huang, X. Dislocation-based plasticity and strengthening mechanisms in sub-20 nm lamellar structures in pearlitic steel wire. *Acta Mater.* **2016**, *114*, 176–183. [CrossRef]
30. Zhang, X.; Godfrey, A.; Liu, W.; Liu, Q. Study on dislocation slips in ferrite and deformation of cementite in cold drawn pearlitic steel wires from medium to high strain. *Mater. Sci. Technol.* **2011**, *27*, 562–567. [CrossRef]
31. Su, X.; Zhou, M.; Zhu, M.; Wang, H.; Zhang, Q.; Tian, J.; Xu, G. Microstructure and mechanical properties of Nb microalloyed high-carbon pearlitic steels subjected to isothermal transformation. *Mater. Charact.* **2023**, *202*, 113013. [CrossRef]
32. Bakshi, S.D.; Shipway, P.; Bhadeshia, H. Three-body abrasive wear of fine pearlite, nanostructured bainite and martensite. *Wear* **2013**, *308*, 46–53. [CrossRef]
33. Aita, C.R.; Weertman, J. Fatigue crack bifurcation in a Fe-0.64 C alloy. *Scr. Metall.* **1980**, *14*, 425–429. [CrossRef]
34. Babasafari, Z.; Pan, A.V.; Pahlevani, F.; Hossain, R.; Sahajwalla, V.; Du Toit, M.; Dippenaar, R. Effects of austenizing temperature, cooling rate and isothermal temperature on overall phase transformation characteristics in high carbon steel. *J. Mater. Res. Technol.* **2020**, *9*, 15286–15297. [CrossRef]
35. Zhou, L.; Fang, F.; Kumagai, M.; Pickering, E.; Yu, T.; Zhang, X. Structure restoration and coarsening of nanocrystalline cementite in cold drawn pearlitic wire induced by low temperature annealing. *Scripta Mater.* **2022**, *215*, 114696. [CrossRef]
36. Sevillano, J.G. Room temperature plastic deformation of pearlitic cementite. *Mater. Sci. Eng.* **1975**, *21*, 221–225. [CrossRef]
37. Zhou, L.; Fang, F.; Zhou, X.; Tu, Y.; Xie, Z.; Jiang, J. Cementite nano-crystallization in cold drawn pearlitic wires instigated by low temperature annealing. *Scripta Mater.* **2016**, *120*, 5–8. [CrossRef]
38. Xiang, L.; Liang, L.; Wang, Y.; Chen, Y.; Wang, H.; Dai, L. One-step annealing optimizes strength-ductility tradeoff in pearlitic steel wires. *Mat. Sci. Eng. A-Struct.* **2019**, *757*, 1–13. [CrossRef]
39. Languillaume, J.; Kapelski, G.; Baudelet, B. Cementite dissolution in heavily cold drawn pearlitic steel wires. *Acta Mater.* **1997**, *45*, 1201–1212. [CrossRef]
40. Guo, W.; Meng, Y.; Zhang, X.; Bedekar, V.; Bei, H.; Hyde, S.; Guo, Q.; Thompson, G.B.; Shivpuri, R.; Zuo, J.-m. Extremely hard amorphous-crystalline hybrid steel surface produced by deformation induced cementite amorphization. *Acta Mater.* **2018**, *152*, 107–118. [CrossRef]
41. Li, Y.; Choi, P.; Borchers, C.; Chen, Y.; Goto, S.; Raabe, D.; Kirchheim, R. Atom probe tomography characterization of heavily cold drawn pearlitic steel wire. *Ultramicroscopy* **2011**, *111*, 628–632. [CrossRef] [PubMed]
42. Clouet, E.; Garruchet, S.; Nguyen, H.; Perez, M.; Becquart, C.S. Dislocation interaction with C in α-Fe: A comparison between atomic simulations and elasticity theory. *Acta Mater.* **2008**, *56*, 3450–3460. [CrossRef]
43. Jafari, M.; Bang, C.W.; Han, J.C.; Kim, K.-M.; Na, S.-H.; Park, C.-G.; Lee, B.-J. Evolution of microstructure and tensile properties of cold-drawn hyper-eutectoid steel wires during post-deformation annealing. *J. Mater. Res. Technol.* **2020**, *41*, 1–11. [CrossRef]
44. Hausselt, J.; Blum, W. Dynamic recovery during and after steady state deformation of Al-11wt% Zn. *Acta Metall.* **1976**, *24*, 1027–1039. [CrossRef]
45. Zelin, M. Microstructure evolution in pearlitic steels during wire drawing. *Acta Mater.* **2002**, *50*, 4431–4447. [CrossRef]
46. Järveläinen, H.; Välimäki, V.; Karjalainen, M. Audibility of the timbral effects of inharmonicity in stringed instrument tones. *ARLO* **2001**, *2*, 79–84. [CrossRef]
47. Kakegawa, Y.; Asakura, T. Effect of Physical Characteristics of Strings on the Instrumental Sound of Violin. In Proceedings of the INTER-NOISE and NOISE-CON Congress and Conference Proceedings, Washington, DC, USA, 1–5 August 2023; Institute of Noise Control Engineering: Reston, VA, USA; pp. 4319–4324.
48. Railsback, O. Scale temperament as applied to piano tuning. *J. Acoust.* **1938**, *9* (Suppl. S3), 274. [CrossRef]
49. Kakegawa, Y.; Asakura, T. Effect of physical characteristics of strings on the subjective impression of violin sound. *Acoust. Sci. Technol.* **2023**, *44*, 396–398. [CrossRef]
50. Gunawan, D.; Sen, D. Spectral envelope sensitivity of musical instrument sounds. *J. Acoust.* **2008**, *123*, 500–506. [CrossRef]

51. Schoonderwaldt, E.; Guettler, K.; Askenfelt, A. Effect of the Width of the Bow Hair on the Violin String Spectrum. In Proceedings of the Stockholm Music Acoustics Conference, Stockholm, Sweden, 14–15 June 2023; SMAC: Carlsbad, CA, USA, 2003.
52. Alm, J.F.; Walker, J. Time-frequency analysis of musical instruments. *SIAM Rev.* **2002**, *44*, 457–476. [CrossRef]

materials

MDPI

Article

Comparison of the Performance of Nonlinear Time-Dependent Constitutive Models Calibrated with Minimal Test Data Applied to an Epoxy Resin

Rui Miranda Guedes [1,2,*] and José Lopes Morais [3]

1 LAETA—Associate Laboratory of Energy, Transports and Aerospace, Department of Mechanical Engineering, Faculty of Engineering, University of Porto, 4200-465 Porto, Portugal
2 LABIOMEP—Porto Biomechanics Laboratory, University of Porto, 4200-465 Porto, Portugal
3 CITAB—Centre for the Research and Technology of Agro-Environmental and Biological Sciences, School of Science and Technology, University of Trás-os-Montes and Alto Douro, 5000-801 Vila Real, Portugal; jmorais@utad.pt
* Correspondence: rmguedes@fe.up.pt; Tel.: +351-22-508-1400

Abstract: Epoxy resins are extensively employed as adhesives and matrices in fibre-reinforced composites. As polymers, they possess a viscoelastic nature and are prone to creep and stress relaxation even at room temperature. This phenomenon is also responsible for time-dependent failure or creep fracture due to cumulative strain. Several constitutive equations have been used to describe the mechanical time-dependent response of polymers. These models have been proposed over the past six decades, with minimal direct and practical confrontation. Each model is associated with a specific application or research group. This work assesses the predictive performance of four distinct time-dependent constitutive models based on experimental data. The models were deemed sufficiently straightforward to be readily integrated into practical engineering analyses. A range of loading cases, encompassing constant strain rate, creep, and relaxation tests, were conducted on a commercial epoxy resin. Model parameter calibration was conducted with a minimum data set. The extrapolative predictive capacity of the models was evaluated for creep loading by extending the tests to five decades. The selected rheological models comprise two viscoelastic models based on Volterra-type integrals, as originally proposed by Schapery and Rabotnov; one viscoplastic model, as originally proposed by Norton and Bailey; and the Burger model, in which two springs and two dashpots are combined in a serial and parallel configuration. The number of model parameters does not correlate positively to superior performance, even if it is high. Overall, the models exhibited satisfactory predictive performance, displaying similar outcomes with some relevant differences during the unloading phases.

Keywords: epoxy resin; rheology properties; viscoelasticity; viscoplasticity; stress relaxation; multi-step; creep; constant strain rate

1. Introduction

Epoxy resins have a wide range of applications, from the electronics industry, where they are used as encapsulants, to the building, automotive and aerospace industries, where they are used as structural adhesives and as matrices in fibre-reinforced composites. Its success is attributable to its excellent mechanical properties, high adhesion, heat resistance and electromagnetic insulation. The type of epoxy resin, hardener, and manufacturing process determine the resulting properties [1]. The long-chain molecular structure that is characteristic of polymers is the determining factor in their time-dependent response, a phenomenon referred to as "fading" memory [2]. Therefore, understanding and quantifying this phenomenon is relevant whenever dimensional stability is a concern or creep strain needs to be kept below critical levels under continuous load to avoid creep failure [3,4]. Time-dependent experimental characterisation is ordinarily an expensive and time-consuming process. A novel method using machine learning (ML) techniques has been developed to estimate the long-term creep modulus of thermoplastics using the open online database CAMPUS®. This method avoids the need for lengthy creep tests, thereby reducing the cost of obtaining the long-term creep modulus, an important design parameter for selecting polymers for long-term use [5]. However, ML techniques for predicting the long-term thermo-mechanical properties of materials are still in their infancy.

Due to the inherent properties of polymers, the use of viscoelastic or viscoplastic models is essential when analysing their time-dependent mechanical response. Therefore, it is important to have a comprehensive understanding of their constitutive behaviour under a range of loading conditions. Only then can an adequate prediction of their mechanical response under more general loading conditions be achieved.

In general, time-dependent materials are linear viscoelastic for sufficiently small strains, with the upper limit of linearity typically around 0.5% for many stiff polymers [6]. Above this threshold, the material starts displaying nonlinear behaviour, i.e., the response depends on the stress or strain state.

Different constitutive equations describe polymers' mechanical response's stress–strain time-dependency. These models have been put forth over the past six decades, with only a limited number of experimental and direct comparative assessments. Each model is associated with a specific application or research group. The constitutive models developed are generally expressed through integral or differential formulations. Classic examples of integral formulations for viscoelastic materials are the free-volume theory of Knauss and Emri [7] and the thermodynamic-based model of Schapery [8]. Differential formulation is the alternative type of constitutive equation easily adapted to finite element analysis [9,10]. Despite six decades of extensive research, there is no time-dependent constitutive model with global acceptance capable of predicting the mechanical response a polymer under complex loading histories [11–14]. Truly, the models denominated as unified viscoplastic theories were proposed for more complex loading conditions than creep or stress relaxation [15–17]. A double overstress (VBO) element configuration was developed for simulating strain rate sensitivity, creep, relaxation, and recovery behaviour, and addressing the non-monotonic changes in creep and relaxation that occur when a loading history involves some degree of unloading [14]. These advanced viscoplastic models involve large numbers of material parameters and naturally fit the experimental data better than the material models available in commercial finite element (FE) codes [16]. Since these approaches are not readily available in commercial FE code, it has inhibited their practical adoption [17].

Among several possibilities, four models were selected for this study: the Schapery single integral nonlinear constitutive model (SM) [8,11], the Burger's model (BM) [18], the Norton–Bailey viscoplastic model (NB-SH) [19,20], and the nonlinear approach proposed by Rabotnov (Rabotnov) [21,22].

The integral formulation developed by Schapery [8], based on the irreversible thermodynamic approach for nonlinear viscoelastic materials, is one of the most quoted integral constitutive formulations [13]. The classical mechanical elements of mass, spring, and linear viscous dashpot have been employed as fundamental components in searching for more comprehensive mechanical models. The governing differential equations for these mechanical models provide differential constitutive equations that offer a phenomenological description of the viscoelastic response of polymers. The classical Burger's model [18], i.e., the Maxwell and Kelvin elements connected in series, were used to describe the response of materials like asphalt [23] and soil beds [24]. The stress dependence of the model parameters is due to the nonlinear material response. Majda and Skrodzewicz [25] proposed Burger's model, with the coefficients dependent on stress level, to represent the creep behaviour of an epoxy adhesive. However, the value of the modulus of elasticity (the spring of the Maxwell element) was assumed to be constant.

Another study [26] proposed modelling bonded joints using standard models supported by commercial Finite Element Method software (ANSYS version 12). In that context, the Bailey–Norton law [19,20] was selected to represent the creep response of an epoxy-based adhesive. A modified version of the Norton–Bailey model was proposed by Majda and Skrodzewicz [25]. The model parameters were assumed stress-dependent, to increase the degree of curve fitting of the experimental curves.

This work compares different nonlinear time-dependent constitutive models when applied to an epoxy resin. The selection criteria comprised two conditions. These models should be simple enough for industrial applications and present a universal character to capture the main features of the nonlinear viscoelasticity response. This experimental investigation was limited to the uniaxial stress state at room temperature.

The performance of the selected models was evaluated through a series of distinct loading conditions, encompassing constant strain rate, creep, and relaxation tests, applied to a commercial epoxy resin. Furthermore, the model parameters' calibration was achieved with a minimum of experimental data. The number of material parameters differed between the models, with four for NB-SH and Rabotnov, seven for SM and eight for BM. Although the SM model achieved the best performance, the NB-SH model showed a good performance with few parameters.

2. Materials

The resin system used in this work was the SR 1500 epoxy, cured with SD 2503 hardener, and supplied by SICOMIN, Châteauneuf les Martigues, France. Accordingly, to the manufacturer, the formulation bases of SR 1500 epoxy are bisphenol A and F. This is an ambient temperature curing resin system, specifically developed for hand lay-up and vacuum bagging production. Plates with 200 $\times$ 200 $\times$ 2 mm were cast in a mould. The resin was cured by maintaining the polymer at 20 °C for 24 h, followed by 16 h at 60 °C. Later, the plates were subjected to a thermal post-cure, at 120 °C for 3 h, aiming at full curing. FTIR analysis confirmed no residual curing agent, or unreacted epoxy groups in the post-cured resin. The material was stored in a dry room for more than 12 months before being tested, long enough to reach the equilibrium at the glassy state. Briefly, cooling the epoxy resin from its rubbery state to a temperature lower than the glass transition temperature results in a non-equilibrium glassy state. This glassy state spontaneously evolves toward a temporally distant equilibrium, via slow molecular motions, under constant external conditions. Such a phenomenon is denominated as structural relaxation or physical ageing [27].

The test specimens with 80 $\times$ 6 $\times$ 2 mm^3 were cut out from plates and the gauge length was 50 mm. The average tensile strength previously measured was 74 MPa. Three repetitions of the mechanical responses were obtained for each loading case performed at room temperature.

3. Experimental Tests

The test machine was an Instron® ElectroPuls E1000 (High Wycombe, UK) with a 2 kN load cell. The data acquisition rate (displacement and load) was fixed at 2.5 Hz. Tests were performed at room temperature.

Eleven different types of tests were executed, as described in Table 1. These are grouped as, creep, constant strain rate (CSR) and stress relaxation. Three different multi-step loading cases were applied, as indicated in Table 2. The rise time varied from 0.4 to 1.0 s, imposed by a strain rate of 0.01/s. The last case was a multi-step relaxation with step strain evolution described in Figure 1.

Table 1. Index of all tests that have been carried out.

Test Type	Case Nr.	Test Description
Creep	1	Test at 6 MPa during 24 h
	2	Multi-step loading case A
	3	Multi-step loading case B
	4	Multi-step loading case C
CSR	5	Constant Strain Rate, 0.0001/s
	6	Constant Strain Rate, 0.001/s
	7	Constant Strain Rate, 0.01/s.
Stress Relax	8	Constant Strain, 0.0025 mm/mm
	9	Constant Strain, 0.005 mm/mm
	10	Constant Strain, 0.01 mm/mm
Multi-step Relaxation	11	Described in Figure 1

Table 2. Multi-step loading tests description.

Case A	Stress (MPa)	6	24	6	
	Dwell Time (s)	300	300	300	
Case B	Stress (MPa)	24	6	24	
	Dwell Time (s)	300	300	300	
Case C	Stress (MPa)	6	12	24	30
	Dwell Time (s)	300	300	300	300

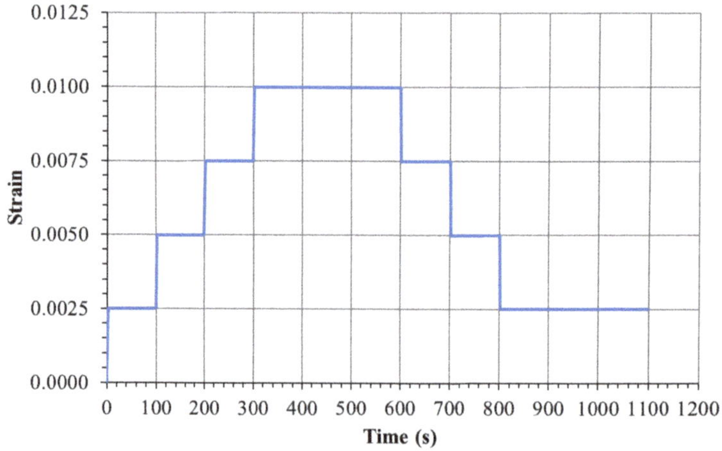

Figure 1. Step strain sequence (2 s ramp between each strain level) for the multi-step stress relaxation.

4. Constitutive Time-Dependent Models

A description of the constitutive models is provided. Since we restricted this work to uniaxial loading cases, all the following descriptions are made accordingly. Nevertheless, the unidirectional model constitutive equations can be readily extended to the isotropic 3-dimensional formulation.

An important point must be made about the present approach. All the relationships presented are based on creep compliance, which predicts the strain response to variable uniaxial input stress. The alternative would have been to base the relationships on the relaxation modulus. Creep compliance and relaxation modulus are related for linear viscoelastic materials by an analytical expression. Schapery [8] claimed that there is no fundamental reason for a material to conform to both representations unless it is approximately elastic or linearly viscoelastic. However, the observed behaviour for a variety of materials is consistent with one or the other. Nevertheless, determining nonlinear parameters from relaxation tests is challenging and complex [28]. In response, several research papers have proposed methods to analytically or numerically determine the stress relaxation response directly from creep tests [28–31]. As Pupure et al. point out, the situation becomes more complicated when the viscoplastic strain becomes relevant [32].

In our case, stress relaxation tests are predicted from a limited amount of creep data. The experimental data used to calibrate the models was limited to creep at 6 and 24 MPa measured over 300 s. The CSR test at the highest strain rate was used to calibrate the nonlinear elastic parameters, assuming that the time-dependent effects are negligible at this rate. This test was not used for (NB-SH) as it does not include the nonlinear elastic behaviour. All other experimental data were used to evaluate the predictions of the models.

4.1. Schapery Model (SM) and Burger's Model (BM)

These models have been described before but for completeness, a brief description of these two models is given in Appendix A. The SM [8] does not include damage and viscoplastic strain components. The latter are assumed to be non-recoverable, as discussed by Schapery in an updated analysis [33]. Nevertheless, the dashpot in the BM, which belongs to the Maxwell element, behaves as a viscoplastic component, as its strain is not recoverable unless an opposing stress is applied. To model the viscoelastic/viscoplastic behaviour of wood materials, the dashpot (of the Maxwell element in the Burger model) has been replaced by the Bingham element, which allows for a stress plasticity threshold [34]. The choice of components was made according to experimental observations. The aim was to reduce the number of parameters to a minimum.

4.2. Norton–Bailey Viscoplastic Model (NB-SH)

Bailey and Norton [19,20] developed a comprehensive model to describe the primary and secondary creep of metals at elevated temperatures. Singh and Mitchell proposed an enhanced expression for modelling the creep rate of soils, which was subsequently known as the Singh–Mitchell model [35]. Marry and Bray adopted this model to represent the time-dependent mechanical response of geomembranes made of high-density polyethylene, yielding favourable outcomes [36].

In this case, under a creep loading case, the total strain is given by

$$\varepsilon = \frac{\sigma_0}{E} + Ae^{m\frac{\sigma(\tau)}{s_0}} \frac{1}{n+1}\left(\frac{t}{\tau_0}\right)^{n+1}, \tag{1}$$

where t represents the time, E is the elastic modulus and A, m, n are the viscoplastic parameters. The parameters s_0 and t_0 are employed to assure a dimensionally coherent expression, in this case s_0 = 1 MPa and τ_0 = 1 h or τ_0 = 1 s, i.e., depending on the unit of time used. Despite its empirical foundation, power law has been successfully employed in the analysis of polymers [37–39]. Fractional models provide a deeper meaning of the power law [40,41]. Modification of the Maxwell model, comprising a spring and a dashpot in series, by replacing the dashpot with a spring-pot (fractional model) leads to the power law for the creep loading condition [42].

The constitutive equations for viscoplastic strains under a general loading condition, depending on the strain state of the material at the time of the stress change, becomes

$$\varepsilon_{vp}(t) = \frac{A}{n+1}\left\{\frac{1}{\tau_0}\int_0^t e^{\frac{m}{(n+1)}\frac{\sigma(\tau)}{s_0}}d\tau\right\}^{n+1}, \tag{2}$$

which is a strain-hardening formulation. Further details on these matters can be found in a previous work [43]. This model permits the incorporation of viscoplastic strain recovery during unloading stages. As observed by [44,45], the same viscoplastic model can represent the "positive" viscoplastic strains during loading and "negative" viscoplastic strains during unloading phases, noting that the direction of a viscous force is always opposite to the direction of displacement. While loading, a "positive" viscoplastic strain generates a viscous force acting in the opposite direction. Conversely, during unloading, the viscous force causes a "negative" viscoplastic strain. This effect explains the viscoplastic recovery, which may be partial.

4.3. Rabotnov Viscoelastic Model (Rabotnov)

Rabotnov [21] proposed an approach to construct a nonlinear equation as duly explained and tested using specimens made of polyoxymethylene [22]. Rabotnov assumed that all the nonlinearity may be gathered on the left-hand side of the equation

$$\varphi(\varepsilon) = \sigma(t) + \int_0^t \frac{K}{n}\left(\frac{t-\tau}{\tau_0}\right)^n \frac{d\sigma(\tau)}{d\tau}d\tau, \tag{3}$$

or after integrating by parts

$$\varphi(\varepsilon) = \sigma(t) + \int_0^t K\left(\frac{t-\tau}{\tau_0}\right)^{n-1}\sigma(\tau)d\tau, \tag{4}$$

where τ_0 = 1 h or τ_0 = 1 s, i.e., depending on the unit of time used, K and n are material parameters and $\varphi(\varepsilon)$ represent the instantaneous or elastic stress–strain curve, in the limiting case when the stress (strain) rate tends to be infinite. The function was constructed using the CSR test data measured at the highest strain rate, with the assumption of two versions according to the case of imposed stress state or strain state, i.e.,

$$\varepsilon = A\sigma^2 + B\sigma, \tag{5}$$

$$\sigma = \varphi(\varepsilon) = a\varepsilon^2 + b\varepsilon, \tag{6}$$

where A, B, a, b are elastic constants of the material.

The conventional numerical methodology for Volterra-type integrals entails a conversion of the transient compliance to Prony's series [46,47],

$$K\left(\frac{t}{\tau_0}\right)^{n-1} = \sum_{i=1}^{N} k_i e^{-\frac{t}{\tau_i}}, \tag{7}$$

consequently, at the time t, Equation (4) can be written as

$$\varphi(\varepsilon) = \sigma(t) + \sum_{i=1}^{N}\int_0^t k_i e^{-\frac{t-\tau}{\tau_i}}\sigma(\tau)d\tau\,, \tag{8}$$

or

$$\varphi(\varepsilon) = \sigma(t) + \sum_{i=1}^{N} F_i(\sigma,t), \tag{9}$$

the subsequent time step calculation results in a recursive formulation,

$$\varphi(\varepsilon) = \sigma(t+\Delta t) + \sum_{i=1}^{N} F_i(\sigma, t+\Delta t), \tag{10}$$

where

$$\begin{aligned} &\sum_{i=1}^{N} F_i(\sigma, t+\Delta t) \\ &= \sum_{i=1}^{N} e^{-\frac{\Delta t}{\tau_i}} F_i(\sigma, t) \\ &+ \left(\frac{\sigma(t+\Delta t)+\sigma(t)}{2}\right) \sum_{i=1}^{N} k_i \tau_i \left(1 - e^{-\frac{\Delta t}{\tau_i}}\right), \end{aligned} \tag{11}$$

in a more concise format

$$\sum_{i=1}^{N} F_i(\sigma, t+\Delta t) = Y(\sigma, t) + \left(\frac{\sigma(t+\Delta t)+\sigma(t)}{2}\right) Z, \tag{12}$$

if the stress evolution is imposed, then the strain becomes

$$\begin{aligned} \varepsilon(t+\Delta t) = A&\left(\sigma(t+\Delta t) + Y(\sigma, t) + \left(\tfrac{\sigma(t+\Delta t)+\sigma(t)}{2}\right) Z\right)^2 \\ &+B\left(\sigma(t+\Delta t) + Y(\sigma, t) + \left(\tfrac{\sigma(t+\Delta t)+\sigma(t)}{2}\right) Z\right), \end{aligned} \tag{13}$$

alternatively, if the strain evolution is imposed, then the stress becomes,

$$\sigma(t+\Delta t) = \frac{\varphi(\varepsilon(t+\Delta t)) - 2Y(\sigma, t) - Z\sigma(t)}{2+Z}. \tag{14}$$

5. Experimental Results

The eleven distinct tests outlined in Tables 1 and 2 and Figure 1, encompassing creep loading, constant strain rate (CSR), and stress relaxation, were successfully completed. The results for each load case are presented below. They are based on the average of three specimens.

The presentation of experimental results together with theoretical predictions is intended to avoid the repetition of plots. The following section is devoted to a comprehensive presentation of the methodology used to determine the model parameters.

5.1. Creep Tests

The predictive capability of the models was evaluated by extrapolation at creep loading by extending the tests to five decades. The creep test at 6 MPa was run for 86,400 s (equivalent to 24 h) as shown in Figure 2. It is evident that the BM is unable to accurately predict creep strain over timescales exceeding 1000 s. The long-term creep strain rate is influenced by the dashpot associated with the Maxwell element, which assumes a Newtonian viscosity law. The NB-SH and SM curves are in complete overlap throughout the entire plot, becoming almost indistinguishable from one another. This is near the experimental data, but not as close as the Rabotnov model. These models incorporate the creep power law function, which ensure reliable long-term creep extrapolations as observed in many cases [39]. An overview of the multi-step creep loading case results, plotted in Figures 3–5, follows Table 3 containing the mean squared relative error ($MSRE = \frac{1}{n}\sum_{i=1}^{n}\left(\frac{y_i^{exp} - y_i^{pred}}{y_i^{exp}}\right)^2$) for each load step.

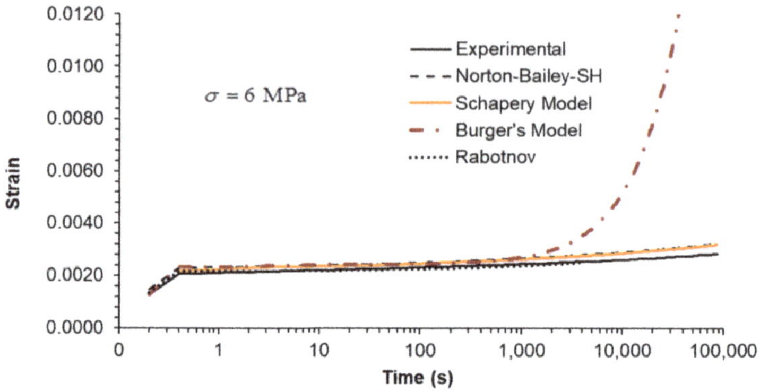

Figure 2. Creep test at 6 MPa during 24 h: experimental data compared against theoretical models (extrapolation from 300 s to 86,400 s).

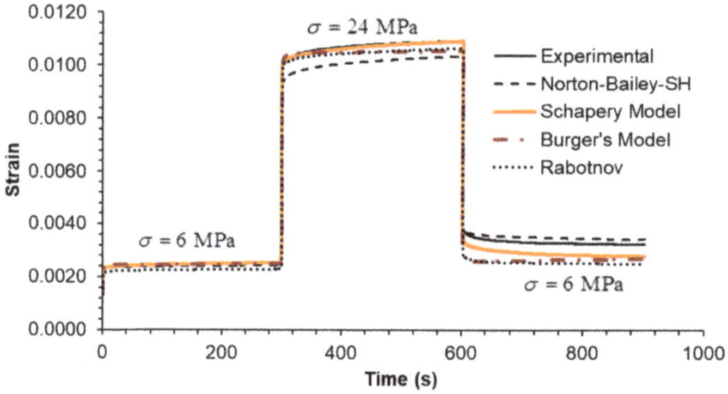

Figure 3. Case A: experimental data compared against theoretical models.

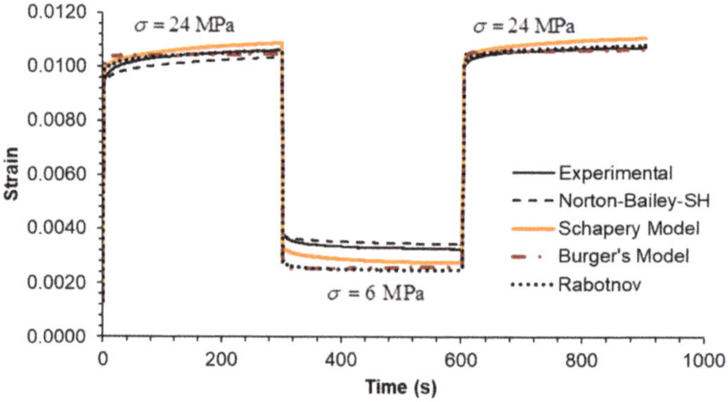

Figure 4. Case B: experimental data compared against theoretical models.

In Case A, all models demonstrate an MSRE value below 5.5% for each step. The largest discrepancies are observed in the final step during the recovery phase, except for the NB-SH model, which accounts for the recovery of the viscoplastic strain. In Case B, all models exhibit an MSRE value lower than 6.2%. The largest discrepancies are observed in the intermediate step 2, except for the NB-SH model, due to the same reason as previously stated. In Case C, all models exhibited MSRE values below 1%, thereby demonstrating comparable performance. The multi-step loading cases A and B showed greater discrepancies from the experimental data than case C. Consequently, the model predictions for the three multi-step loading cases were in reasonable agreement with the experimental data, except for the unloading phases. The exception was the NB-SH model.

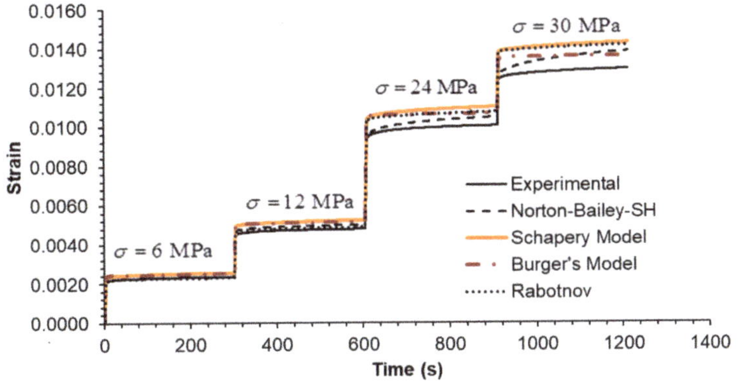

Figure 5. Case C: experimental data compared against theoretical models.

Table 3. Mean squared relative error (MSRE) calculated for all models and multi-step loading cases.

Case	Step	BM	NB-SH	SM	Rabotnov
A	1	0.00012	0.00072	0.00001	0.00684
	2	0.00064	0.00336	0.00002	0.00056
	3	0.04205	0.00319	0.01642	0.05510
B	1	0.00042	0.00065	0.00050	0.00009
	2	0.05453	0.00261	0.02069	0.06175
	3	0.00042	0.00009	0.00082	0.00032
C	1	0.00620	0.00924	0.00614	0.00023
	2	0.00665	0.00178	0.00779	0.00047
	3	0.00548	0.00108	0.00863	0.00504
	4	0.00464	0.00199	0.00957	0.00715

Table 4 contains the MSRE values for the complete loading sequences. The respective Coefficients of Correlation (R^2) between the observed and predicted data are given in Table 5.

Table 4. Mean squared relative error (MSRE) calculated for all models and all loading cases.

Test Type	Case Nr.	BM *	NB-SH	SM	Rabotnov
Creep	1	0.00615	0.00904	0.00661	0.00153
	2	0.01434	0.00242	0.00551	0.02088
	3	0.01838	0.00111	0.00731	0.02065
	4	0.00552	0.00393	0.00752	0.00378
CSR	5	0.00972	0.03956	0.00536	0.00319
	6	0.00189	0.00438	0.00113	0.00029
	7	0.00146	0.00378	0.00176	0.00017
Stress Relax	8	0.00083	0.00155	0.00044	0.00002
	9	0.00024	0.00001	0.00094	0.00022
	10	0.00255	0.00188	0.00021	0.00070
Multi-step Relaxation		0.02912	0.00251	0.02107	0.02199
MSRE Sum		0.06109	0.06766	0.03679	0.07342

* Predictions after 1000 s were not considered for Case 1.

Table 5. Coefficients of correlation (R^2) between the observed and predicted data.

Test Type	Case Nr.	BM *	NB-SH	SM	Rabotnov
Creep	1	0.84169	0.99014	0.99131	0.92453
	2	0.99340	0.99906	0.99743	0.99561
	3	0.99817	0.99833	0.99692	0.99954
	4	0.99958	0.99902	0.99964	0.99969
CSR	5	0.99614	0.99259	0.99923	0.99991
	6	0.99749	0.99364	0.99965	0.99999
	7	0.99850	0.99380	0.99993	0.99998
Stress Relax	8	0.78597	0.97890	0.98601	0.95147
	9	0.82708	0.98150	0.98512	0.95191
	10	0.70199	0.98279	0.98146	0.95144
Multi-step Relaxation		0.99361	0.99977	0.99511	0.99789
Average R^2		0.92124	0.99178	0.99380	0.97927

* Predictions after 1000 s were not considered for Case 1.

5.2. Constant Strain Rate Tests

Three strain rates separated by a factor of ten were used, as shown in Figures 6–8. All MSRE values are below 4%, as documented in Table 4. The NB-SH shows the poorest performance as it is the only model calibrated using only the creep tests at 6 and 24 MPa. All other models show comparable performance with a high degree of correlation with experimental data as observed in Table 5.

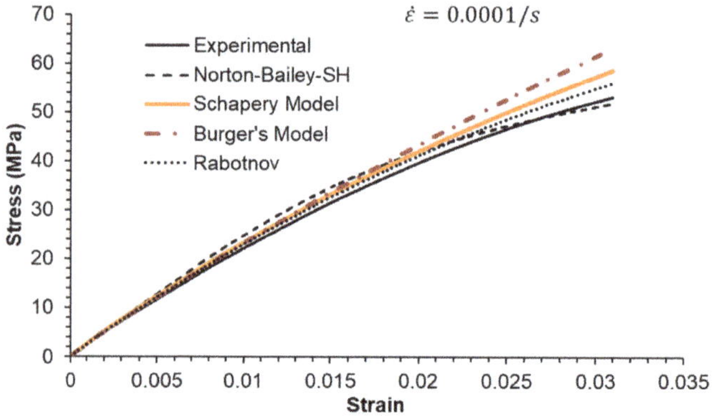

Figure 6. Constant strain rate of 0.0001/s: experimental data compared against theoretical models.

The epoxy resin exhibits a pronounced nonlinear elastic effect, observable above the 0.5% strain threshold. This effect is not included in the NB-SH model as currently formulated. However, it was satisfactorily reproduced as a time-dependent phenomenon.

5.3. Stress Relaxation Tests

Finally, stress relaxation tests were carried out at three different strain levels. As an ideal step strain cannot be applied by the testing machine, it was replaced by applying a constant strain rate ramp until the desired strain level was reached. Therefore, the rise time was included in the model calculations to closely follow the experiments. The inclusion of the rise time in the plots (Figures 9–11) highlights this important detail.

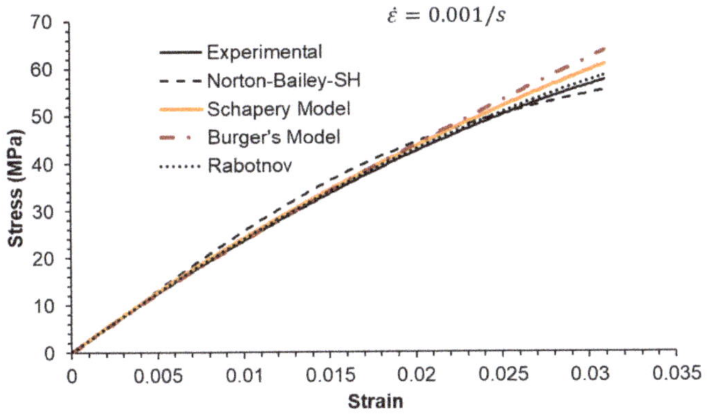

Figure 7. Constant strain rate of 0.001/s: experimental data compared against theoretical models.

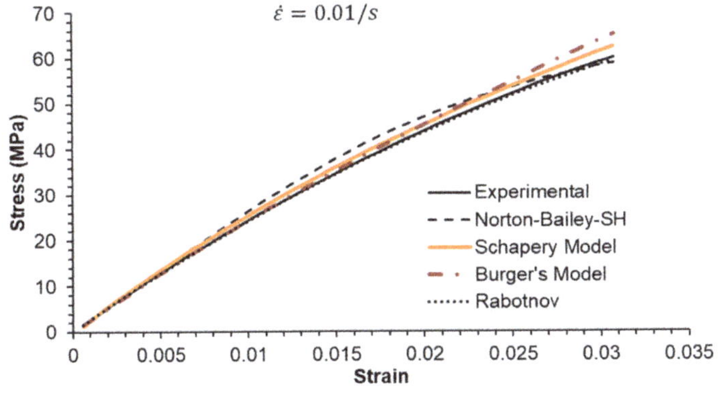

Figure 8. Constant strain rate of 0.01/s: experimental data compared against theoretical models.

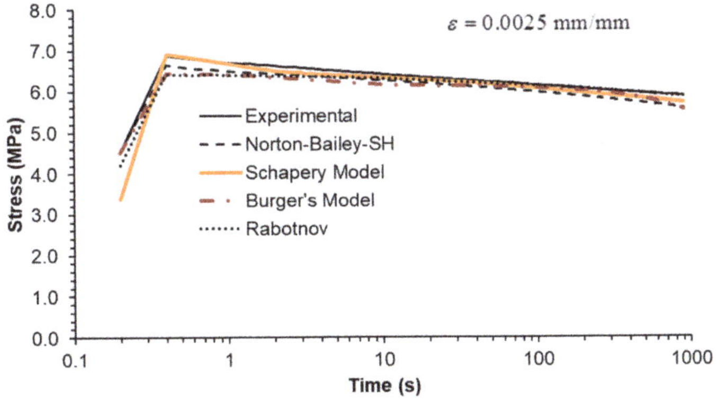

Figure 9. Stress relaxation under a constant strain of 0.0025: experimental data compared against theoretical models.

The model predictions for each strain level are in good agreement with the experimental data, except for the NB-SH model at the higher strain level (Figures 9–11). This is due to the pronounced nonlinear elastic effect not included in the model.

A linear behaviour below the 0.5% strain level is observed after plotting the normalised experimental stress relaxation data versus time (Figure 12). This is consistent with the observations from the CSR experimental data.

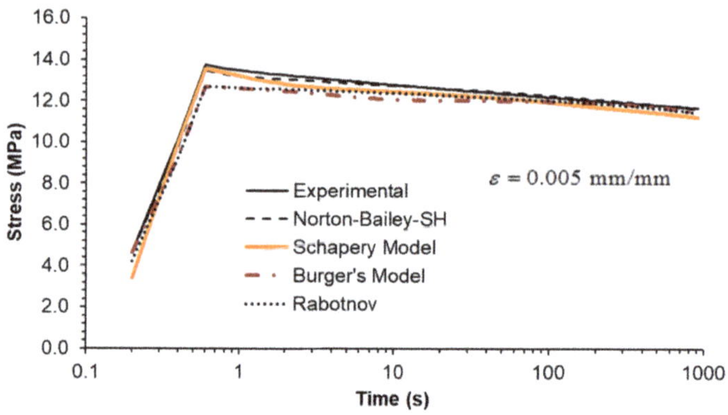

Figure 10. Stress relaxation under a constant strain of 0.005: experimental data compared against theoretical models.

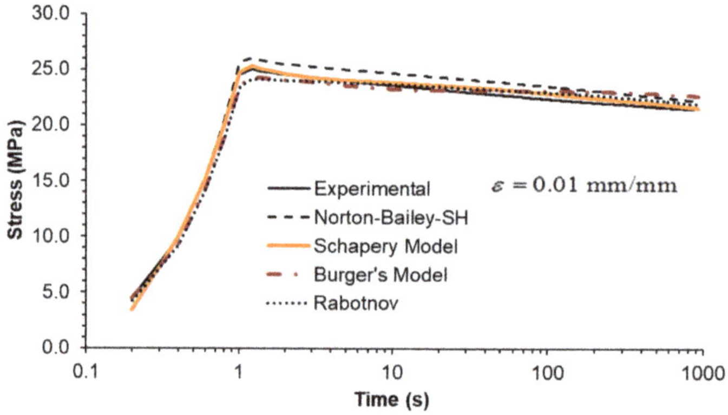

Figure 11. Stress relaxation under a constant strain of 0.01: experimental data compared against theoretical models.

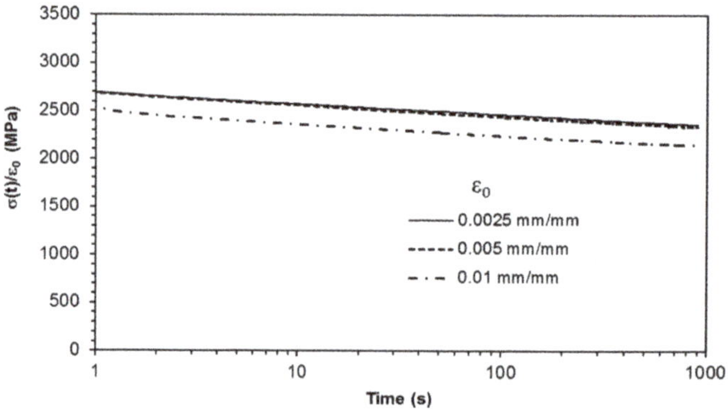

Figure 12. Comparison of normalised stress relaxation measured under different constant strain levels.

The multi-stage relaxation test led to the conclusion that the performance of the NB-SH model was much better than the others (Figure 13). Although it predicts higher stress levels at the last two strain steps, the prediction at the unloading steps is remarkably close to the experimental data. The performance of the other models is quite similar with no significant differences.

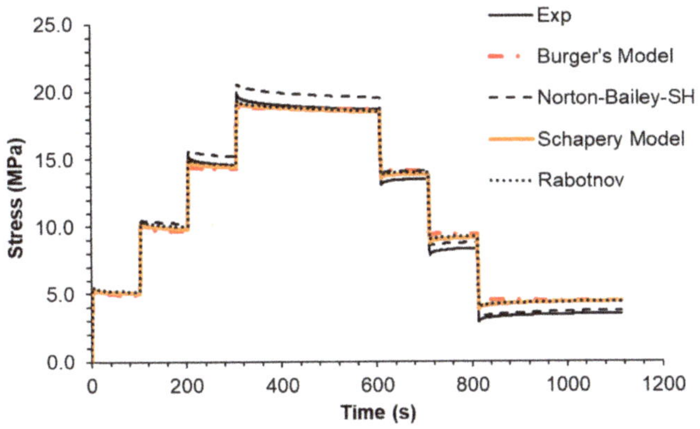

Figure 13. Multi-step stress relaxation test: experimental data compared against theoretical models.

6. Methodology Employed to Determine the Model Parameters

The adopted approach was based on creep strain data measured at 6 and 24 MPa during 300 s, to calibrate the model parameters through the curve fitting of analytical expressions. Since the parameters display stress dependency, linear and exponential stress functions were assumed for SM and BM, respectively. It was observed that the creep curves did not provide enough information about the nonlinear elastic behaviour. As the stress level reached around 60 MPa at constant strain rate, creep strain data measured at 6 and 24 MPa, regardless of time, cannot provide sufficient information about nonlinear elastic behaviour above 24 MPa. Accordingly, the constant strain rate curve (CSR) at the highest rate was employed to calibrate the nonlinear elastic parameters, except for the SB-SH model, which does not encompass the nonlinear elastic behaviour. The assumption was that the time-dependent effects would be negligible at the highest rate.

Estimation of Parameters

The model parameters were determined following the methodology previously outlined. The SM parameters are presented in Table 6. The transient compliance (power law) was converted into a Prony series by imposing seven retardation times, $1/\lambda_i$, i = 1, 2, 3, . . ., 7 which accurately represents the 24 h period. The results are presented in Table 7. The BM parameters are illustrated in Table 8. In the present case, all the BM parameters are stress dependent. Appendix A contains a comprehensive explication of the parameters associated with the SM and BMs. The NB-SH parameters are shown in Table 9. As previously observed, the NB-SH includes the concept of a "positive" viscoplastic strain generated during the loading phase, which is responsible for the viscous force acting in the opposite direction, and a "negative" viscoplastic strain generated during the unloading phase, which opposes the unloading. The same model describes both events with the same parameters, except for one, as indicated. The Rabotnov model parameters are shown in Table 10. As was conducted previously for SM, the power law was converted into a Prony series, as shown in Table 11.

Table 6. Schapery model parameters (power law).

D_0	D_1	n
MPa^{-1}		
3.543×10^{-4}	2.570×10^{-5}	0.17
$g_0 = 5.6433 \times 10^{-3}\sigma + 0.93703$ $g_2 = 0.0065884\sigma + 0.96047$		

Table 7. Prony series calculated after the power law.

	1/s		MPa^{-1}
-	-	D_0	3.5430×10^{-4}
l_1	0.000001	D_1	1.2838×10^{-4}
l_2	0.00001	D_2	6.2690×10^{-5}
l_3	0.0001	D_3	4.1898×10^{-5}
l_4	0.001	D_4	2.8442×10^{-5}
l_5	0.01	D_5	2.0994×10^{-5}
l_6	0.1	D_6	8.5457×10^{-6}
l_7	1	D_7	3.0854×10^{-5}

Table 8. Burger's model parameters.

$R_1 =$	$2.6497 \times 10^3 e^{-3.0452 \times 10^{-3} \sigma}$	(MPa)
$h_1 =$	$1.3702 \times 10^7 e^{7.7016 \times 10^{-2} \sigma}$	(MPa·s)
$R_2 =$	$5.0442 \times 10^4 e^{-1.2602 \times 10^{-2} \sigma}$	(MPa)
$h_2 =$	$2.0177 \times 10^5 e^{-1.2602 \times 10^{-2} \sigma}$	(MPa·s)

Table 9. Norton–Bailey (NB-SH) parameters.

E	*m*	*n*	*A* *
(MPa)	-	-	**(MPa)$^{(1-m)}$**
2772	0.08324	−0.8300	1.5909×10^{-5}

* During unloading, A′ = A/2.

Table 10. Rabotnov's parameters (power law).

A	*B*	*a*	*b*	K	*n*
1/(MPa)^2	**1/(MPa)**	**(MPa)**	**(MPa)**	-	-
−22,952	2662	2.739×10^{-5}	3.386×10^{-4}	0.007000	−0.8000

Table 11. Prony series for Rabotnov's model.

	s		
t_1	2	k_1	6.7199×10^{-3}
t_2	20	k_2	1.4547×10^{-3}
t_3	200	k_3	1.3459×10^{-4}
t_4	2000	k_4	5.5475×10^{-5}

7. General Discussion

The maximum stress level attained at creep and stress relaxation tests was around 40% of the failure tensile stress. This stress level matches the in-service stress range applied to the material, assuming a safety factor of around 2. The nonlinear effects were observed above the 0.5% strain level, which is considered a typical value for polymers [6].

Earlier studies on epoxy resin-based composites at room temperature uncoupled the time-dependent deformation into viscoelastic and viscoplastic strains [28]. While the viscoelastic strain is recoverable during and after the unloading, the viscoplastic strain remains unrecoverable [29,31,36,47]. Although no specific tests were conducted to obtain the viscoplastic strains of this epoxy resin, from the previous studies it should be expected a viscoelastic–viscoplastic response. Still, the applied models being purely viscoelastic (SM, BM, and Rabotnov) or purely viscoplastic (NB-SH), were able to capture all the experimental data in a unified manner. This was more evident for the last model, since it distinguishes the loading and unloading states. This phenomenon has been the subject of detailed description and modelling by Ellyin et al. [12,48]. Another note concerns the BM where the mechanical response of the dashpot in series acts as a viscoplastic component, i.e., its cumulative strain is unrecoverable.

In the case of NB-SH, the model calibration was based on the creep tests only. This epoxy shows nonlinear elastic behaviour, but the NB-SH does not include this type of response.

The quality of model predictions, measured globally by the MSRE (Table 10), allows us to observe the overall differences in model performance. The corresponding Coefficients of Correlation (R^2) confirm the overall good quality of the models' predictions. In summary, SM performs better than the other models. The number of model parameters plays a role in the ability of the model to capture the experimental data. These considerations apply when the model is used to fit the experimental data. However, it is not certain that the model predictions would simply improve by increasing the number of parameters. In the present case, the number of parameters varied from four for NB-SH and Rabotnov to seven for SM and eight for BM. The NB-SH offers a very good compromise between the number of parameters and its performance. In addition, the NB-SH would benefit from the inclusion of the nonlinear elastic effect.

8. Conclusions

Four viscoelastic models were selected to assess its relative performance using experimental data obtained for an epoxy resin. These tests comprise a set of different loading cases, including constant strain rate, creep, and relaxation tests.

A methodology was established to perform the parameters calibration of the viscoelastic models, based on a limited amount of creep experimental data up to 300 s. The remaining data were used to assess the prediction ability of the models.

The observed trends were correctly predicted by all the models. Additionally, the creep tests were extended over five decades, and the models were able to predict correctly the creep strain after 24 h. The unique exception was BM; decidedly, it cannot be used to extrapolate long-term creep strain. Altogether, the four constitutive models describe in a unified manner the time-dependent mechanical response of epoxy resin under various loading modes, i.e., creep, constant strain rate, and stress relaxation.

The quality of predictions was evaluated through the mean squared relative error (MSRE) and Coefficients of Correlation (R^2); SM distinctively performs better than the other models. Alternatively, NB-SH offers a good compromise between the number of parameters and its performance.

Prospects for investigating the performance of these models will include temperature effects and, at a later stage, extending the models to biaxial loading, since in many situations it can be assumed that the polymers are subjected to a plane stress state.

Author Contributions: Conceptualization, R.M.G. and J.L.M.; methodology, R.M.G.; software, R.M.G.; validation, R.M.G. and J.L.M.; formal analysis, R.M.G.; investigation, R.M.G.; resources, R.M.G. and J.L.M.; data curation, R.M.G.; writing—original draft preparation, R.M.G.; writing—review and editing, R.M.G. and J.L.M.; visualisation, R.M.G.; supervision, R.M.G.; project administration, R.M.G. and J.L.M.; funding acquisition, R.M.G. and J.L.M. All authors have read and agreed to the published version of the manuscript.

Funding: This research received no external funding.

Institutional Review Board Statement: Not applicable.

Informed Consent Statement: Not applicable.

Data Availability Statement: The original contributions presented in this study are included in the article. Further inquiries can be directed to the corresponding author.

Acknowledgments: The authors would like to thank Anurag Singh for assisting with experimental testing.

Conflicts of Interest: The authors declare no conflicts of interest.

Appendix A

Appendix A.1. Schapery Model (SM)

The nonlinear viscoelastic strain–stress relationship for the uniaxial loading condition induced by an arbitrary stress history is given by [8]

$$\varepsilon(t) = D_0 g_0 \sigma(t) + g_1 \int_0^t \Delta D(\psi - \psi') \frac{d(g_2\sigma(\tau))}{d\tau} d\tau, \tag{A1}$$

where D_0 is the elastic compliance and the kernel $\Delta D(t)$ is the time dependent compliance, the correspondent reduced times Ψ and Ψ' given by

$$\Psi = \int_0^t \frac{d\tau'}{a_\sigma}, \Psi' = \int_0^\tau \frac{d\tau'}{a_\sigma}, \tag{A2}$$

where g_0, g_1, g_2, a_σ are four stress-dependent nonlinearising parameters.

In the present case, the material model used is a simplification of the Schapery equation by imposing $g_1 = a_\sigma = 1$, and the kernel $\Delta D(t)$ is given by the power law

$$\varepsilon(t) = D_0 g_0 \sigma(t) + \int_0^t D_1 \left(\frac{t-\tau}{\tau_0}\right)^n \frac{d(g_2\sigma(\tau))}{d\tau} d\tau, \tag{A3}$$

where D_1 and the exponent n are material parameters. The creep response of the model, i.e., the response to $\sigma = \sigma_0$ for $t \geq 0$, is

$$\varepsilon(t) = D_0 g_0 \sigma_0 + D_1 g_2 \left(\frac{t}{\tau_0}\right)^n \sigma_0. \tag{A4}$$

For general loading cases the strain history was calculated via Equation (A3), after eliminating the Volterra-type integrals, by converting the transient compliance to Prony's series [46,47]

$$\varepsilon\Delta D(t) = \sum_{i=1}^{N} D_i\left(1 - e^{-\lambda_i t}\right), \tag{A5}$$

where D_i, λ_i are linear viscoelastic parameters. Substituting (A5) into (A3), we obtain

$$\varepsilon(t) = D_0 g_0 \sigma(t) + \sum_{i=1}^{N} \int_0^t D_i\left(1 - e^{-\lambda_i(t-\tau)}\right) \frac{d(g_2\sigma(\tau))}{d\tau} d\tau. \tag{A6}$$

After algebraic manipulations, the following recursive formula is obtained [31]:

$$\varepsilon(t) = g_0 D_0 \sigma(t) + \sum_{i=1}^{N} \varepsilon_i(t), \tag{A7}$$

where

$$\begin{aligned}\varepsilon_i(t_{j+1}) &= e^{-\lambda_i \Delta t} \cdot \varepsilon_i(t_j) + \left[1 - \tfrac{1}{\lambda_i \Delta t}\left(1 - e^{-\lambda_i \Delta t}\right)\right] D_i\, g_2 \sigma(t_{j+1}) \\ &+ \left[\tfrac{1}{\lambda_i \Delta t}\left(1 - e^{-\lambda_i \Delta t}\right) - e^{-\lambda_i \Delta t}\right] D_i\, g_2 \sigma(t_j)\end{aligned}$$

with

$$\Delta t = t_{j+1} - t_j.$$

The total strain given by Equation (A7), can then be written as

$$\varepsilon(t_{j+1}) = g_0 D_0 \sigma(t_{j+1}) + R(t_{j+1}), \tag{A8}$$

where

$$R(t_{j+1}) = g_1 \sum_{i=1}^{N} \varepsilon_i(t_{j+1}). \tag{A9}$$

The previous formulation allows us to calculate the strain under any given stress history. If the strain history is given, then the stress history must be determined. Equation (A8) can be used to do the following:

$$\sigma(t_{j+1}) = \frac{\varepsilon(t_{j+1}) - R(t_{j+1})}{g_0 D_0}. \tag{A10}$$

This formula becomes an implicit equation since $R(t_{j+1})$ depends on the stress $\sigma(t_{j+1})$ level.

Appendix A.2. Burger's Model (BM)

In the elementary mechanical models, a spring for elastic behaviour and a dashpot for viscous behaviour can be combined to obtain differential stress–strain relations. The four-element model, known as Burger's model, has been used to describe the creep behaviour of epoxy resins [25]. Burger's model is a combination of the Maxwell and Kelvin-Voigt models in series, as shown in Figure A1.

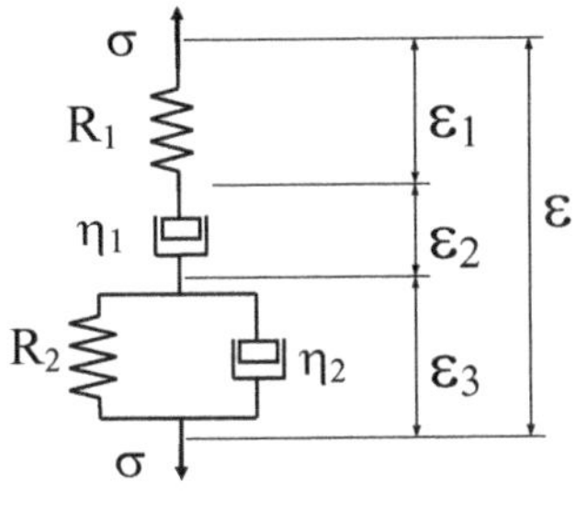

Figure A1. Burger's model.

The total strain is given as

$$\varepsilon = \varepsilon_1 + \varepsilon_2 + \varepsilon_3, \tag{A11}$$

where

$$\varepsilon_1 = \frac{\sigma}{R_1}, \dot{\varepsilon}_2 = \frac{\sigma}{\eta_1} \text{ and } \dot{\varepsilon}_3 + \frac{R_2}{\eta_2}\varepsilon_3 = \frac{\sigma}{\eta_2}. \tag{A12}$$

The dot stands for derivation to t (time). Eliminating the internal variables (ε_1, ε_2, ε_3), the following differential equation is obtained:

$$\sigma + \left(\frac{\eta_1}{R_1} + \frac{\eta_1}{R_2} + \frac{\eta_2}{R_2}\right)\dot{\sigma} + \left(\frac{\eta_1\eta_2}{R_1 R_2}\right)\ddot{\sigma} = \eta_1\dot{\varepsilon} + \frac{\eta_1\eta_2}{R_2}\varepsilon. \tag{A13}$$

The response of the model to the case of creep, i.e., constant load condition $\sigma = \sigma_0$ for $t \geq 0$, is

$$\varepsilon = \frac{\sigma_0}{R_1} + \frac{\sigma_0}{\eta_1}t + \frac{\sigma_0}{R_2}\left(1 - e^{-\frac{R_2}{\eta_2}t}\right). \tag{A14}$$

Yet, the integration for a general loading history can be performed incrementally, using finite differences, through the internal variables (ε_1, ε_2, ε_3)

$$\varepsilon(t_{j+1}) = \varepsilon_1(t_{j+1}) + \varepsilon_2(t_{j+1}) + \varepsilon_3(t_{j+1}). \tag{A15}$$

The total strain can be obtained recursively by using finite differences

$$\begin{gathered}\varepsilon_1(t_{j+1}) = \tfrac{1}{R_1}\sigma(t_{j+1}),\ \varepsilon_2(t_{j+1}) = \tfrac{\Delta t}{\eta_1}\sigma(t_{j+1}) + \varepsilon_2(t_j) \\ \text{and } \varepsilon_3(t_{j+1}) = \frac{\left[\frac{\Delta t}{\eta_2}\sigma(t_{j+1}) + \varepsilon_3(t_j)\right]}{\left(1+\frac{R_2}{\eta_2}\Delta t\right)},\end{gathered} \tag{A16}$$

or

$$\begin{aligned}\varepsilon(t_{j+1}) = {} & \tfrac{1}{R_1}\sigma(t_{j+1}) + \tfrac{\Delta t}{\eta_1}\sigma(t_{j+1}) + \varepsilon_2(t_j) \\ & + \frac{\left[\frac{\Delta t}{\eta_2}\sigma(t_{j+1}) + \varepsilon_3(t_j)\right]}{\left(1+\frac{R_2}{\eta_2}\Delta t\right)}.\end{aligned} \tag{A17}$$

It is possible to invert the calculation, i.e., imposing $\varepsilon(t_{j+1})$ and calculating $\sigma(t_{j+1})$

$$\sigma(t_{j+1}) = \frac{\left[\varepsilon(t_{j+1}) - \varepsilon_2(t_j) - \varepsilon_3(t_j)\frac{\eta_2}{\eta_2+R_2\Delta t}\right]}{\left(\frac{1}{R_1} + \frac{\Delta t}{\eta_1} + \frac{\Delta t}{\eta_2+R_2\Delta t}\right)}. \tag{A18}$$

Burger's model predicts, on a long-term basis, an unlimited strain increases at a constant strain rate. However, the experimental findings for polymers observed continuous strain increase with decreasing strain rate, complying with the Findlay's power law [37].

References

1. Jin, F.L.; Li, X.; Park, S.J. Synthesis and application of epoxy resins: A review. *J. Ind. Eng. Chem.* **2015**, *29*, 1–11. [CrossRef]
2. Brinson, H.F.; Brinson, L.C. *Polymer Engineering Science and Viscoelasticity: An Introduction*; Springer: Berlin/Heidelberg, Germany, 2008; pp. 1–446.
3. Guedes, R.M. Mathematical analysis of energies for viscoelastic materials and energy based failure criteria for creep loading. *Mech. Time-Depend. Mater.* **2004**, *8*, 169–192. [CrossRef]
4. Kontou, E.; Spathis, G. Viscoplastic response and creep failure time prediction of polymers based on the transient network model. *Mech. Time-Depend. Mater.* **2014**, *18*, 373–386. [CrossRef]
5. Lobato, H.; Cernuda, C.; Zulueta, K.; Arriaga, A.; Matxain, J.M.; Burgoa, A. Prediction of long-term creep modulus of thermoplastics using brief tests and interpretable machine learning. *Int. J. Solids Struct.* **2024**, *304*, 113014. [CrossRef]
6. McCrum, N.; Buckley, C.P.; Bucknall, C.B. *Principles of Polymer Engineering*, 2nd ed.; Oxford University Press: Oxford, UK, 1997; p. 126.
7. Knauss, W.G.; Emri, I. Volume change and the nonlinearly thermo-viscoelastic constitution of polymers. *Polym. Eng. Sci.* **1987**, *27*, 86–100. [CrossRef]
8. Schapery, R.A. On the characterization of nonlinear viscoelastic materials. *Polym. Eng. Sci.* **1969**, *9*, 295–310. [CrossRef]
9. Chen, Y.; Xia, Z.; Ellyin, F. Evolution of residual stresses induced during curing processing using a viscoelastic micromechanical model. *J. Compos. Mater.* **2001**, *35*, 522–542. [CrossRef]
10. Yi, S.; Hilton, H.H. Free edge stresses in elastic and viscoelastic composites under uniaxial extension, bending, and twisting loadings. *J. Eng. Mater. Technol. Trans. ASME* **1997**, *119*, 266–272. [CrossRef]
11. Schapery, R.A. Nonlinear viscoelastic solids. *Int. J. Solids Struct.* **2000**, *37*, 359–366. [CrossRef]
12. Ellyin, F.; Xia, Z. Nonlinear viscoelastic constitutive model for thermoset polymers. *J. Eng. Mater. Technol. Trans. ASME* **2006**, *128*, 579–585. [CrossRef]
13. Jamshidi, M.; Shokrieh, M.M. On the Schapery nonlinear viscoelastic model: A review. *Eur. J. Mech. A/Solids* **2024**, *108*, 105403. [CrossRef]
14. Khan, F.; Yeakle, C. Experimental investigation and modeling of non-monotonic creep behavior in polymers. *Int. J. Plast.* **2011**, *27*, 512–521. [CrossRef]
15. Chiu, W.K.; Jones, R. Unified constitutive model for thermoset adhesive, FM73. *Int. J. Adhes. Adhes.* **1995**, *15*, 131–136. [CrossRef]

16. Yu, X.X.; Crocombe, A.D.; Richardson, G. Material modelling for rate-dependent adhesives. *Int. J. Adhes. Adhes.* **2001**, *21*, 197–210. [CrossRef]
17. Scott, M.L.; Elder, D.J.; Feih, S.; Gunnion, A.J.; Liu, X.L.; Thomson, R.S. Engineering solutions for complex composite material behaviour spanning time and temperature scales. *Philos. Mag.* **2010**, *90*, 4153–4174. [CrossRef]
18. Burgers, J.M. *Mechanical Considerations—Model Systems—Phenomenological Theories of Relaxation and of Viscosity*, 2nd ed.; First Report on Viscosity and Plasticity. Prepared by the Committee for the Study of Viscosity of the Academy of Sciences at Amsterdam; Nordemann Publ: New York, NY, USA, 1939.
19. Norton, F.H. *The Creep of Steel at High Temperatures*; McGraw-Hill Book Company, Inc.: New York, NY, USA, 1929.
20. Bailey, R.W. The utilization of creep test data in engineering design. *Proc. Inst. Mech. Eng.* **1935**, *131*, 131–349. [CrossRef]
21. Rabotnov, Y.N. *Elements of Hereditary Solid Mechanics*; Mir Publishers: Moscow, Russia, 1980.
22. Suvorova, J.V.; Ohlson, N.G.; Alexeeva, S.I. An approach to the description of time-dependent materials. *Mater. Des.* **2003**, *24*, 293–297. [CrossRef]
23. Muliana, A.; Rajagopal, K.R.; Wineman, A. On the response of Burgers' fluid and its generalizations with pressure dependent moduli. *Mech. Time-Depend. Mater.* **2013**, *17*, 147–169. [CrossRef]
24. Dey, A.; Basudhar, P.K. *Applicability of Burger Model in Predicting the Response of Viscoelastic Soil Beds*; Geotechnical Special Publication: Reston, VA, USA, 2010.
25. Majda, P.; Skrodzewicz, J. A modified creep model of epoxy adhesive at ambient temperature. *Int. J. Adhes. Adhes.* **2009**, *29*, 396–404. [CrossRef]
26. Zehsaz, M.; Vakili-Tahami, F.; Saeimi-Sadigh, M.A. Creep analysis of adhesively bonded single lap joint using finite element method. *J. Mech. Sci. Technol.* **2014**, *28*, 2743–2748. [CrossRef]
27. Hutchinson, J.M. Physical aging of polymers. *Prog. Polym. Sci.* **1995**, *20*, 703–760. [CrossRef]
28. Pupure, L.; Varna, J.; Joffe, R. Methodology for macro-modeling of bio-based composites with inelastic constituents. *Compos. Sci. Technol.* **2018**, *163*, 41–48. [CrossRef]
29. Oza, A.; Vanderby, R., Jr.; Lakes, R.S. Interrelation of creep and relaxation for nonlinearly viscoelastic materials: Application to ligament and metal. *Rheol. Acta* **2003**, *42*, 557–568. [CrossRef]
30. Touati, D.; Cederbaum, G. On the prediction of stress relaxation from known creep of nonlinear materials. *J. Eng. Mater. Technol. Trans. ASME* **1997**, *119*, 121–124. [CrossRef]
31. Guedes, R.M.; Marques, A.T.; Cardon, A. Analytical and Experimental Evaluation of Nonlinear Viscoelastic-Viscoplastic Composite Laminates under Creep, Creep-Recovery, Relaxation and Ramp Loading. *Mech. Time-Depend. Mater.* **1998**, *2*, 113–128. [CrossRef]
32. Pupure, L.; Pakrastins, L.; Varna, J. Challenges in developing of 3D nonlinear viscoelastic models. *IOP Conf. Ser. Mater. Sci. Eng.* **2021**, *1190*, 012005. [CrossRef]
33. Schapery, R.A. Nonlinear Viscoelastic and Viscoplastic Constitutive Equations Based on Thermodynamics. *Mech. Time-Depend. Mater.* **1997**, *1*, 209–240. [CrossRef]
34. Fothe, T.; Azeufack, U.G.; Kenmeugne, B.; Fogue, M. A one-dimensional elasto-viscoplastic model coupled to damage for the description of creep in wooden materials. *J. Wood Sci.* **2022**, *68*, 8. [CrossRef]
35. Singh, A.; Mitchell, J.K. General stress-strain-time function for soils. *J. Soil Mech. Found. Div.* **1968**, *94*, 21–46. [CrossRef]
36. Merry, S.M.; Bray, J.D. Time-dependent mechanical response of HDPE geomembranes. *J. Geotech. Eng.* **1997**, *123*, 57–65. [CrossRef]
37. Findley, W.N.; Lai, J.S.; Onaran, K. *Creep and Relaxation of Nonlinear Viscoelastic Materials: With an Introduction to Linear Viscoelasticity*; Dover Civil and Mechanical Engineering; Dover Publications: Mineola, NY, USA, 1989.
38. Findley, W.N. 26-Year creep and recovery of poly(vinyl chloride) and polyethylene. *Polym. Eng. Sci.* **1987**, *27*, 582–585. [CrossRef]
39. Scott, D.W.; Lai, J.S.; Zureick, A.H. Creep Behavior of Fiber-Reinforced Polymeric Composites: A Review of the Technical Literature. *J. Reinf. Plast. Compos.* **1995**, *14*, 588–617. [CrossRef]
40. Heymans, N.; Bauwens, J.C. Fractal rheological models and fractional differential equations for viscoelastic behavior. *Rheol. Acta* **1994**, *33*, 210–219. [CrossRef]
41. Glöckle, W.G.; Nonnenmacher, T.F. Fractional Integral Operators and Fox Functions in the Theory of Viscoelasticity. *Macromolecules* **1991**, *24*, 6426–6434. [CrossRef]
42. Guedes, R.M. A viscoelastic model for a biomedical ultra-high molecular weight polyethylene using the time-temperature superposition principle. *Polym. Test.* **2011**, *30*, 294–302. [CrossRef]
43. Guedes, R.M. A practical model to predict the time-dependent behaviour of angle-ply laminates from limited creep data. *Mech. Time-Depend. Mater.* **2023**, *27*, 1043–1067. [CrossRef]
44. Kim, W.; Sun, C.T. Modeling relaxation of a polymeric composite during loading and unloading. *J. Compos. Mater.* **2002**, *36*, 745–755. [CrossRef]
45. Guedes, R.M. Viscoplastic analysis of fiber reinforced polymer matrix composites under various loading conditions. *Polym. Compos.* **2009**, *30*, 1601–1610. [CrossRef]

46. Czyz, J.A.; Szyszkowski, W. An effective method for non-linear viscoelastic structural analysis. *Comput. Struct.* **1990**, *37*, 637–646. [CrossRef]
47. Tuttle, M.E.; Pasricha, A.; Emery, A.F. The Nonlinear Viscoelastic-Viscoplastic Behavior of IM7/5260 Composites Subjected to Cyclic Loading. *J. Compos. Mater.* **1995**, *29*, 2025–2046. [CrossRef]
48. Ellyin, F.; Vaziri, R.; Bigot, L. Predictions of two nonlinear viscoelastic constitutive relations for polymers under multiaxial loadings. *Polym. Eng. Sci.* **2007**, *47*, 593–607. [CrossRef]

materials

MDPI

Article

Features of the Defect Structure of the Compositionally Homogeneous Crystal $LiNbO_3:Er^{3+}$(3.1 wt%) and the Gradient Crystal $LiNbO_3:Er^{3+}$ and Their Manifestation in the IR Transmission Spectra in the Region of Stretching Vibrations of Hydrogen Atoms of OH^--Groups

Nikolay Sidorov [1], Alexander Pyatyshev [2,*], Elena Stroganova [3], Valeriy Galutskiy [3], Andrey Bushunov [4] and Mikhail Tarabrin [4]

1 Tananaev Institute of Chemistry—Subdivision of the Federal Research Centre, Kola Science Centre of the Russian Academy of Sciences (ICT RAS), 184209 Apatity, Russia; n.sidorov@ksc.ru
2 P. N. Lebedev Physical Institute of the Russian Academy of Sciences, 119991 Moscow, Russia
3 Department of Optoelectronics, Kuban State University, Stavropolskaya 149, 350040 Krasnodar, Russia; stroganova@phys.kubsu.ru (E.S.); galutskiy17v@mail.ru (V.G.)
4 Infrared Laser Systems Laboratory, Bauman Moscow State Technical University, 2nd Baumanskaya Str., 5, b. 1, 105005 Moscow, Russia; zendel@yandex.ru (A.B.); tarabrinmike@yandex.ru (M.T.)
* Correspondence: jb_valensia@mail.ru

Abstract: Based on the analysis of the IR transmission spectra in the region of stretching vibrations of hydrogen atoms of OH^--groups, it was established that the oxygen-octahedral MeO_6 clusters (Me-Li^+, Nb^{5+}, vacant octahedron V, impurity ion) of the structure of the compositionally homogeneous crystal $LiNbO_3:Er^{3+}$(3.1 wt%) and the gradient crystal $LiNbO_3:Er^{3+}$(congruent composition by the main components, Er gradient of 0.55 at%/cm) have a shape close to the regular one. In this case, the value of R = [Li]/[Nb] $\approx$ 1, and in the structure of both crystals, there are practically no point defects in Nb_{Li} responsible for the photorefraction effect. By using the IR transmission spectra and Klauer's method, it was found that the volume concentration of OH^--groups in the gradient crystal $LiNbO_3:Er^{3+}$ is almost an order of magnitude lower than in the compositionally homogeneous $LiNbO_3:Er^{3+}$(3.1 wt%) crystal. This fact explains the lower hydrogen conductivity of the gradient crystal $LiNbO_3:Er^{3+}$ and the lower photorefraction effect compared to the compositionally homogeneous $LiNbO_3:Er^{3+}$(3.1 wt%) crystal. The results obtained are important for the development of materials for active nonlinear laser media and for the conversion of laser radiation.

Keywords: lithium niobate; Czochralski method; erbium doping; composition gradient; IR transmission spectrum; OH^--group; Klauer's method; point defect; nonlinear material

1. Introduction

Nonlinear optical lithium niobate single-crystal ($LiNbO_3$) has a number of unique properties, which ensures its wide application as functional elements of modern optoelectronics, telecommunication devices and integrated and laser optics [1–5]. At the same time, for the production of functional elements in industry, compositionally homogeneous $LiNbO_3$ single-crystals of congruent composition (R = [Li]/[Nb] = 0.946), both nominally pure and doped, are mainly used.

In recent years, a new direction has been intensively developing in materials science related to the creation of structures with physical characteristics that change in volume: magnetic [6], ferroelectric [7] and piezoelectric [8]. Along with the predicted change in the ferroelectric characteristics of such systems compared to homogeneous samples (e.g., a change in the phase transition), a number of unique effects were discovered in them. For example, when placing a sample with a composition gradient in the Sawyer–Tower scheme, a significant shift of the hysteresis loop along the polarization axis is observed. Based on this effect, a device was developed that allows for charge amplification up to 150 times under external energy impact. In this case, the type of energy impact can be different—thermal, deformation, or electrical [9,10]. In a number of such systems, a "giant" pyroelectric effect is observed, in which the pyroelectric response is 10–100 times higher than the signal of homogeneous ferroelectrics of the corresponding composition [11,12]. The axial doping gradient of some laser materials allows for a significant reduction in thermal loads and an increase in the output power of laser radiation [13–15]. For LuAG:Ce-LuAG scintillator crystals, the composition gradient allowed us to increase the integral luminescence intensity [16]. The composition gradient allows us to provide optimal conditions for the growth of semiconductor solid solutions [17] and control a number of dielectric and magnetic characteristics of ceramics and crystals [18–21]. Some results in this direction are summarized in the reviews [22,23]. In this case, the directed changes in the physical characteristics of the material are determined by the spatial change in the composition of the material. Such materials are called "gradient materials". Materials with a gradient refractive index are used, in particular, in the creation of film waveguides [24] and sensors [25,26].

To date, studies of the optical properties of gradient $LiNbO_3:Er^{3+}$ crystals have been primarily focused on the analysis of luminescence. In [27], for a double-doped gradient crystal $LiNbO_3:Er^{3+}:Yb^{3+}$, a multidirectional change in the intensity of green (525–585 nm) and IR luminescence (1400–1700 nm) was found depending on the concentration of erbium ions. This difference was explained [28] by various lifetimes and mechanisms of population of the energy levels of the doping ions. It was found that heating of the double-doped gradient crystal $LiNbO_3:Er^{3+}:Yb^{3+}$ in the range of 300–473 K leads to a change in the refractive indices [29] and, unusually, to an increase in the luminescence intensity by 15–20% in the range of 1480–1680 nm. At the same time, for the single-doped gradient crystal $LiNbO_3:Er^{3+}$, a decrease in the luminescence intensity by 30% is observed with increasing temperature [30]. Based on the double-doped gradient crystal $LiNbO_3:Er^{3+}:Yb^{3+}$, a design of proton-exchange waveguides was proposed [31]. In the article [31], a change in the absorption spectrum was discovered for a double-doped crystal $LiNbO_3:Er^{3+}:Yb^{3+}$ with a rare-earth ion concentration gradient, correlating with a change in the composition concentration along the crystal growth axis. However, the configuration of the oxygen-octahedral clusters MeO_6 and the state of defects of the gradient crystals were not analyzed. Meanwhile, the degree of distortion of the MeO_6 clusters and the features of the distribution of the main (Li^+ and Nb^{5+}) and doping cations over the O_6 octahedra determine the position of the Er^{3+} ions and polarons Nb_{Li}^{4+} in the $LiNbO_3:Er^{3+}$ crystal and, accordingly, the intensity of luminescence and laser generation. In particular, the Er^{3+} ion in the $LiNbO_3$ crystal matrix is a very effective emitter in the visible and IR spectral regions [32–34]. For this reason, $LiNbO_3:Er^{3+}$ crystals successfully combine the excellent nonlinear optical properties of the nominally pure $LiNbO_3$ matrix and the laser characteristics of the Er^{3+} ion. Radiation with wavelengths of 980 and 1480 nm is effectively absorbed by the erbium ion, causing excess heat generation inside the crystal. Excess heat generation affects the temperature conditions that ensure radiation focusing in the crystal. Thus, at a significant value of pump radiation, heat will be generated in $LiNbO_3:Er^{3+}$, violating the parameters of the pump radiation propagation medium. Uneven distribution of the thermal field along the length of the element leads to the formation of "thermal lenses", which entails the effect of changing the refractive index and defocusing of the laser radiation, and "thermal shock"—the destruction of the active element structure and, ultimately, the crystal itself. The gradient distribution of Er^{3+} ions in the crystal suggests a change in the ion concentration along the length of the active element and uniform heat generation of excess energy throughout the crystal. Thus, the composition gradient in the $LiNbO_3:Er^{3+}$ crystal allows for compensation of temperature distortions and creates optimal conditions for laser generation and for the conversion of laser radiation [29,30]. In addition, compositionally homogeneous and gradient $LiNbO_3:Er^{3+}$ crystals are characterized by a low photorefraction effect, comparable to that of the congruent crystal $LiNbO_{3cong}$ [32], which is important for the development of laser materials and materials for converting laser radiation.

When growing gradient $LiNbO_3$ crystals (both nominally pure and doped), the composition gradient is realized under highly nonequilibrium conditions in accordance with the incongruent melting of the crystal as a non-stoichiometric phase of variable composition and the features of the phase diagram for a specific composition [27]. It would seem that gradient $LiNbO_3$ crystals, obtained under more nonequilibrium crystallization conditions (compared to compositionally homogeneous crystals), should be characterized by an increased content of various types of point and spatial defects, significantly reducing the quality of the optical material. However, there is reason to believe that this is not always the case. The features of certain chemically active complexes in the melt [35–37], which determine the features of localization in the oxygen-octahedral MeO_6 clusters (Me-Li^+, Nb^{5+}, vacant octahedron V, impurity ion) of the $LiNbO_3$ crystal of the main (Li^+, Nb^{5+}) and doping ions [32] (as well as hydrogen atoms bound to oxygen atoms by hydrogen bonds [38–40]), under certain conditions can lead to a more perfect crystal structure of both doped compositionally inhomogeneous crystals and gradient $LiNbO_3$ crystals. Crystals with a more perfect structure and a lower photorefraction effect are more preferable for the development of functional materials for active nonlinear laser media and for the conversion of laser radiation.

The main structural defects that have a significant effect on the physical characteristics of the $LiNbO_3$ crystal are point defects in the form of basic cations (Li^+, Nb^{5+}) located in the oxygen octahedra O_6, not in their positions and cations of impurity metals [41], as well as microinclusions of impurity phases of other lithium niobates and cluster defects. An important type of defect formed in the $LiNbO_3$ crystal during its growth by the Czochralski method in an air atmosphere is complex defects caused by the presence of hydroxyl OH^--groups in the structure [38–42]. When growing $LiNbO_3$ crystals in an air atmosphere, hydrogen contained in water vapor interacts with oxygen atoms of oxygen-octahedral MeO_6 clusters of oxygen octahedra O_6 of the $LiNbO_3$ crystal structure, forming hydrogen bonds, which leads to the formation of crystal lattice defects in the form of hydroxyl OH^--groups. It should be noted that in the ideal structure of a $LiNbO_3$ crystal of strictly stoichiometric composition (R = 1), in which there are no Nb_{Li} point defects, the placement of hydrogen atoms is impossible [38].

The arrangement of hydrogen atoms in the structure of real $LiNbO_3$ crystals changes depending on the value of the R, concentration and type of dopant [38–41]. In this case, the configuration of oxygen-octahedral clusters of the MeO_6 structure, responsible for spontaneous polarization, nonlinear optical and ferroelectric properties of the crystal, is disrupted. The presence of hydrogen in the structure of the $LiNbO_3$ crystal leads to the formation of complex defects: V_{Li}-OH, Nb_{Li}-OH, Me-OH, Me-OH-Me, etc., and to a change in its nonlinear optical and ferroelectric characteristics, affecting such practically significant properties of the $LiNbO_3$ crystal as photoluminescence, photorefractive thermal fixation, dark conductivity and photorefractive sensitivity [41].

The frequency of the stretching vibration of the hydrogen atom in the hydroxyl OH^--group of the heterodesmic $LiNbO_3$ crystal, due to the smallness of the hydrogen bond compared to the covalent (Nb-O) and electrostatic (Li-O) bonds in the oxygen-octahedral MeO_6 clusters (Me-Nb, Li, vacancy V, impurity ion), is extremely sensitive to the slightest changes in the crystal field, caused, among other things, by the composition gradient. Therefore, the methods of vibrational spectroscopy (IR absorption spectroscopy and Raman spectroscopy) in the region of stretching vibrations of hydrogen atoms of OH^--groups can be very effective for obtaining information on the subtle features of the defect structure of the $LiNbO_3$ crystal. In this case, the IR absorption spectra are much more sensitive to the presence of OH^--groups than the Raman spectra, due to the strong absorption of infrared radiation. This method can be used to study the diffusion processes of hydrogen and deuterium [43,44], to numerically evaluate the efficiency of increasing the volume concentration of hydrogen [45], analyze the depth of ion penetration during diffusion alloying with metals [46], and to analyze concentration thresholds [47].

In this work, comparative studies of the state of defects of the gradient crystal $LiNbO_3$:Er^{3+}(congruent composition by the main components, Er gradient of 0.55 at%/cm) (c$LiNbO_3$:Er^{3+}(0.55 at%/cm)) are performed for the first time using IR transmission spectra in the region of stretching vibrations of hydrogen atoms of hydroxyl OH^--groups. Compositionally homogeneous $LiNbO_3$:Er^{3+}(3.1 wt%) single-crystal was used as a comparison sample. It should be noted that the concentration of 3.1 wt% of Er^{3+} ions in the compositionally homogeneous $LiNbO_3$ crystal was not chosen by chance. It is known that the $LiNbO_3$:Er^{3+} crystal is characterized by concentration thresholds at which the mechanism of Er^{3+} impurity entry into the structure and the physical characteristics of the crystal change abruptly [32,33]. In [32], it was established that for "post-threshold" $LiNbO_3$:Er^{3+} crystals ([Er] > 2.5 wt%), a significant decrease in the Er distribution coefficient is observed, and it approaches unity. Consequently, $LiNbO_3$:Er^{3+} crystals ([Er] > 2.5 wt%), including the $LiNbO_3$:Er^{3+}(3.1 wt%) crystal, will have higher compositional and optical homogeneity than crystals with a lower erbium content, since for the latter, the distribution coefficient is significantly higher than unity.

2. Materials and Methods

To grow a gradient crystal c$LiNbO_3$:Er^{3+}(0.55 at%/cm), the Czochralski method with liquid feeding (double crucible method) was used [27]. A system of platinum crucibles inserted into each other was used: an outer (main crucible) with a diameter of 8 cm and an inner (crucible reactor) with a diameter of 5.2 cm. The system of crucibles moved relative to each other. Setting the speed of movement of the crucible reactor relative to the main crucible is necessary to maintain the concentration gradient of Er^{3+} ions in the $LiNbO_3$ crystal specified before growth. The composition of $LiNbO_3$, congruent in the main components, was preserved during the relative movement of the crucibles with the melt. Before pulling the crystal and fusing the melt, the type of distribution profile of the laser impurity Er^{3+} in the crystal was specified as a function of the coordinate of the crystal along the length. The crystal pulling rate was set before the start of growth, and its value was determined by the relaxation of the melt composition at the crystallization front. The relaxation of the melt composition was determined by the change in the Er^{3+} ion concentration at the crystallization front over time due to the melt being fed into the crucible reactor with melt from the main crucible with a different Er^{3+} concentration. Figure 1 shows a photograph of the grown gradient crystal c$LiNbO_3$:Er^{3+}(0.55 at%/cm).

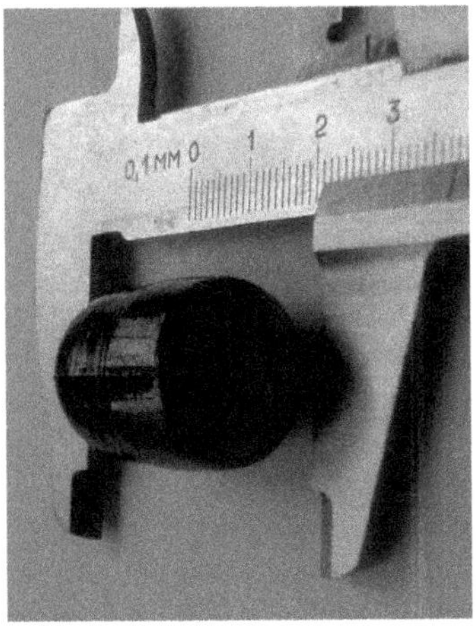

Figure 1. Photo of the grown gradient crystal $cLiNbO_3:Er^{3+}$(0.55 at%/cm).

Niobium oxide (purity 99.97%), erbium oxide (purity 99.99%) and lithium carbonate (purity 99.3%) (all reagents are produced by JSC «Vekton», Saint Petersburg, Russia) taken in the required mass ratios to obtain a compound with the chemical formula $LiNbO_3$ of congruent composition were used to prepare the charge [27]. The charge was tableted at a pressure of 60–120 atmospheres in plexiglass press molds. The resulting tablets were placed in a muffle furnace for solid-phase synthesis for 24 h at a temperature of 900 °C. To grow $LiNbO_3$ crystals with the Er^{3+} concentration profile, the charge was melted in the form of tablets in a crucible system. The grown single-crystal was annealed in a muffle furnace at a temperature of 1200 °C for 5 h. The technology for obtaining gradient crystals $cLiNbO_3:Er^{3+}$(0.55 at%/cm) is presented in more detail in [27]. A plate of about 3 mm thickness was cut from the grown crystal. This plate was cut along the entire length of the crystal. The plate was cut along the Z axis; the X axis was directed perpendicular to the plate. Then, this plate was cut into 9 areas evenly spaced from each other. The size of the areas is justified by the dimensions of the aperture for measuring the Raman and IR transmission spectra. The number of studied areas of the $cLiNbO_3:Er^{3+}$(0.55 at%/cm) reflects the number of measurements required to track the dynamics of the transformation of the properties of Er^{3+} impurity centers with a change in the Er^{3+} concentration in the crystal. With an increase in the step (size) of the crystal samples, as a result of optical measurements, an integral characteristic along the growth direction (Z axis) was realized, while in the X or Y measurement direction, the type and quantitative relationships between the characteristics would change inside the crystalline sample. Therefore, the thickness of the plates in the X axis direction (due to the absence of a concentration gradient in this direction) was related to the step (thickness) of the crystal plate along the Z axis (where the concentration gradient was present). Area 1 corresponded to the initial point of the crystalline boule; the concentration of Er^{3+} ions in it was 3.0 at%. With an increase in the area number, the concentration of Er^{3+} ions decreases with a given gradient of 0.55 at%/cm.

The comparison sample, a compositionally homogeneous $LiNbO_3:Er^{3+}$(3.1 wt%) single-crystal, was grown by the Czochralski method in an air atmosphere in a platinum crucible under conditions of a comparatively small (2–4 °C/cm) axial temperature gradients in the direction of the polar Z axis at constant rotation (16 rpm) and movement (0.8 mm/h) speeds [32]. During crystal growth, the crystallization front was flat. A high-purity, chemically homogeneous, single-phase, granulated lithium niobate batch of congruent composition with high bulk density, developed at the Tananaev Institute of

Chemistry—Subdivision of the Federal Research Centre, Kola Science Centre of the Russian Academy of Sciences (Technical Specification No. 0.027.039), was used. The alloying element Er^{3+} was introduced into the batch immediately before deposition in the form of Er_2O_3 (purity 99.99%, JSC «Solikamsc Magnesium Plant», Solikamsk, Russia). The melt for homogenization of impurity in it before the beginning of crystal growth was maintained for 6 h under conditions of overheating by 150 °C. The grown single-crystal was annealed at a temperature of 1200 °C for 5 h. The heating rates of the batch and cooling of the crystal were the same—50 °C/h. The concentration of erbium ions in the $LiNbO_3{:}Er^{3+}$(3.1 wt%) crystal was determined by analyzing plates cut in the upper (conical) and lower (end) parts of the single-crystal boule, using the emission spectrometry method using a spectrometer ICPE-9000 (Shimadzu, Kyoto, Japan, 2011).

The $LiNbO_3{:}Er^{3+}$(3.1 wt%) single-crystal was monodomainized by high-temperature electrodiffusion annealing with the application of constant electric voltage during the cooling of the samples at a rate of 20 °C/h in the temperature range of 1240–890 °C. The technology for obtaining a compositionally homogeneous $LiNbO_3{:}Er^{3+}$(3.1 wt%) single-crystal, used in this work as a comparison sample, is presented in more detail in [32].

The IR transmission spectra in the region of stretching vibrations of the OH^--groups were recorded using a Bruker VERTEX 70x spectrometer (Bruker, Karlsruhe, Germany) with a spectral resolution of 0.4 cm^{-1}. The measurements were carried out taking into account the reflection of all elements and the absorption of the optical path. The photometric accuracy of the spectrometer used is better than 0.1%. The measurements were carried out in a vacuum at a pressure of 1.78 hPa (in order to eliminate the absorption lines of the atmosphere) and room temperature.

Raman spectra were recorded using a BWS465-785H i-Raman Plus spectrometer (B&W Tek, Plainsboro Township, NJ, USA) using a known technique [48–50]. All spectra were recorded at 293 K.

3. Results

Figures 2 and 3 illustrate the IR transmission spectra in the range of 1000–6000 cm^{-1} of the compositionally homogeneous $LiNbO_3{:}Er^{3+}$(3.1 wt%) single-crystal and the gradient crystal $cLiNbO_3{:}Er^{3+}$(0.55 at%/cm). IR transmission spectra were recorded along the ferroelectric axis of the crystal.

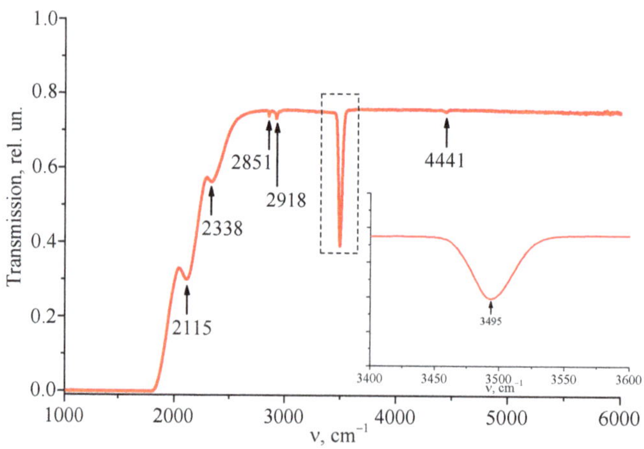

Figure 2. IR transmission spectrum of a compositionally homogeneous $LiNbO_3{:}Er^{3+}$(3.1 wt%) crystal in the region of 1000–6000 cm^{-1}. The inset shows an enlarged view of the IR transmission spectrum in the region of stretching vibrations of OH^--groups.

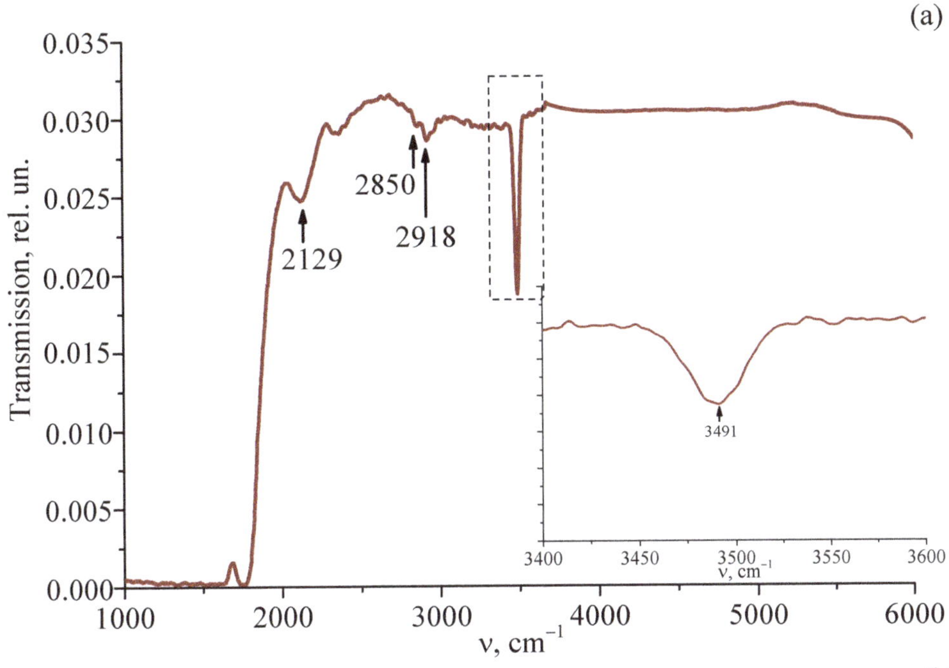

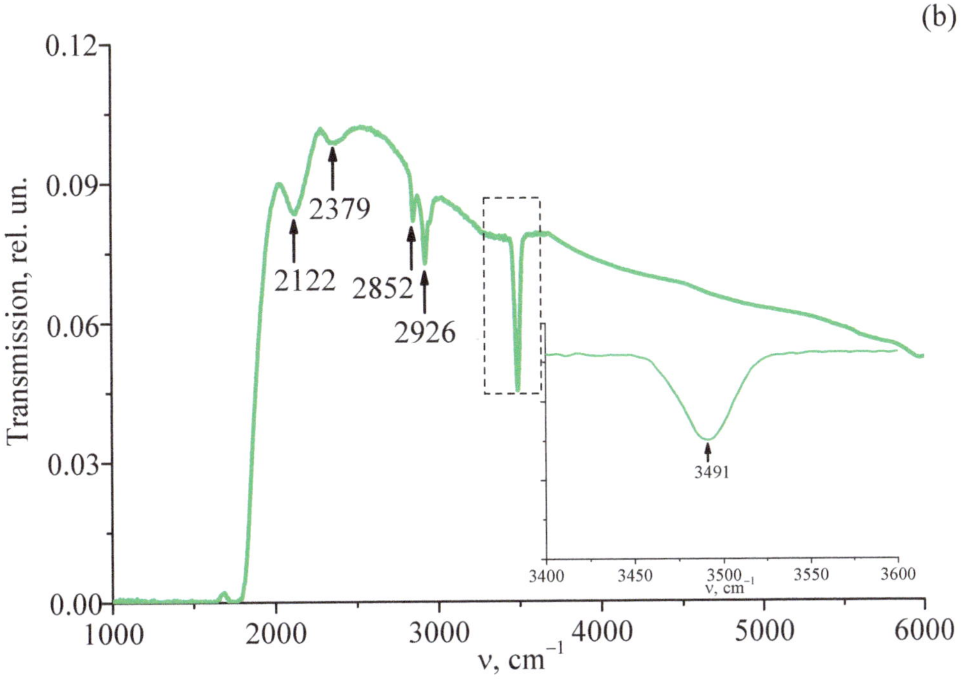

Figure 3. *Cont.*

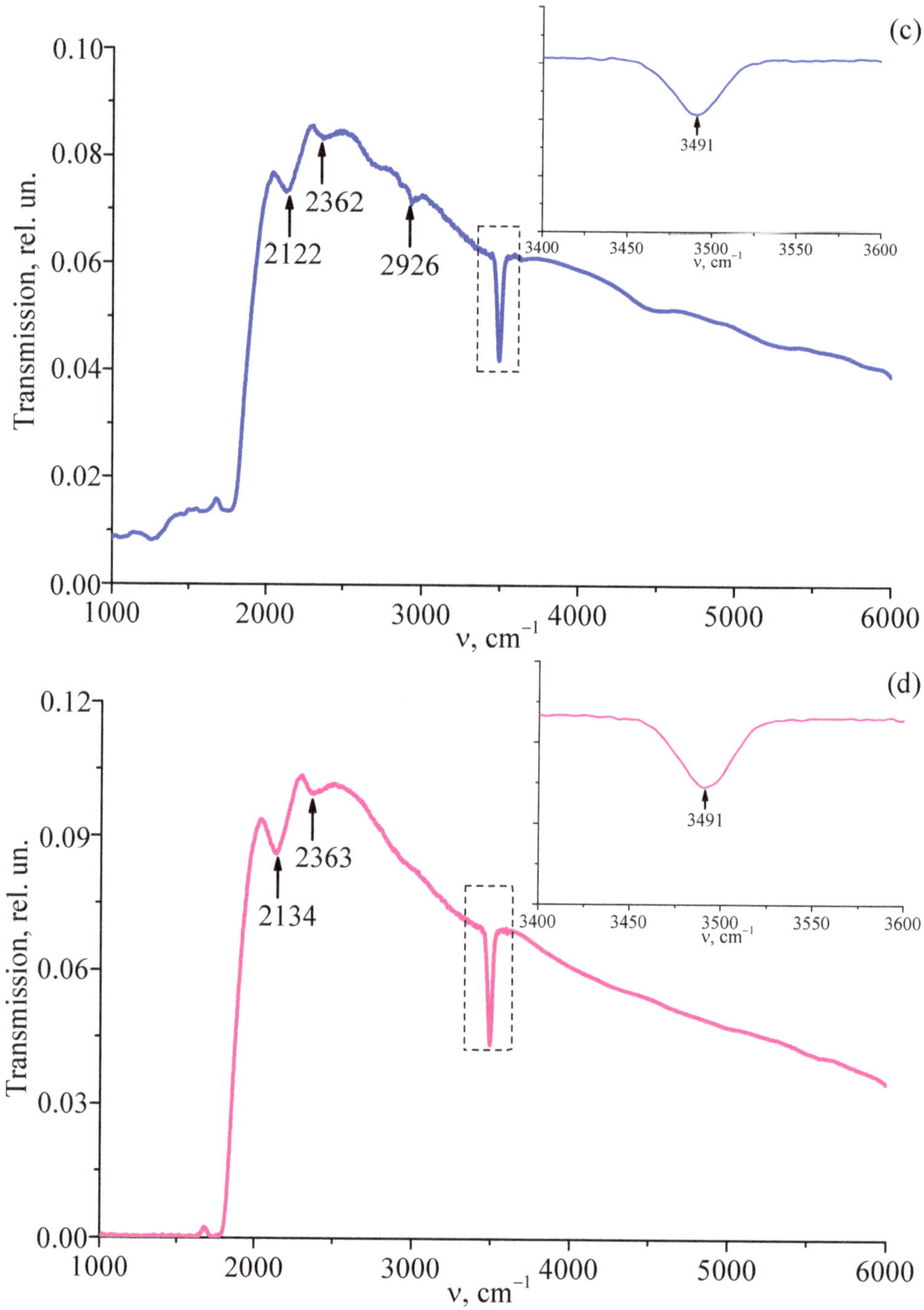

Figure 3. *Cont.*

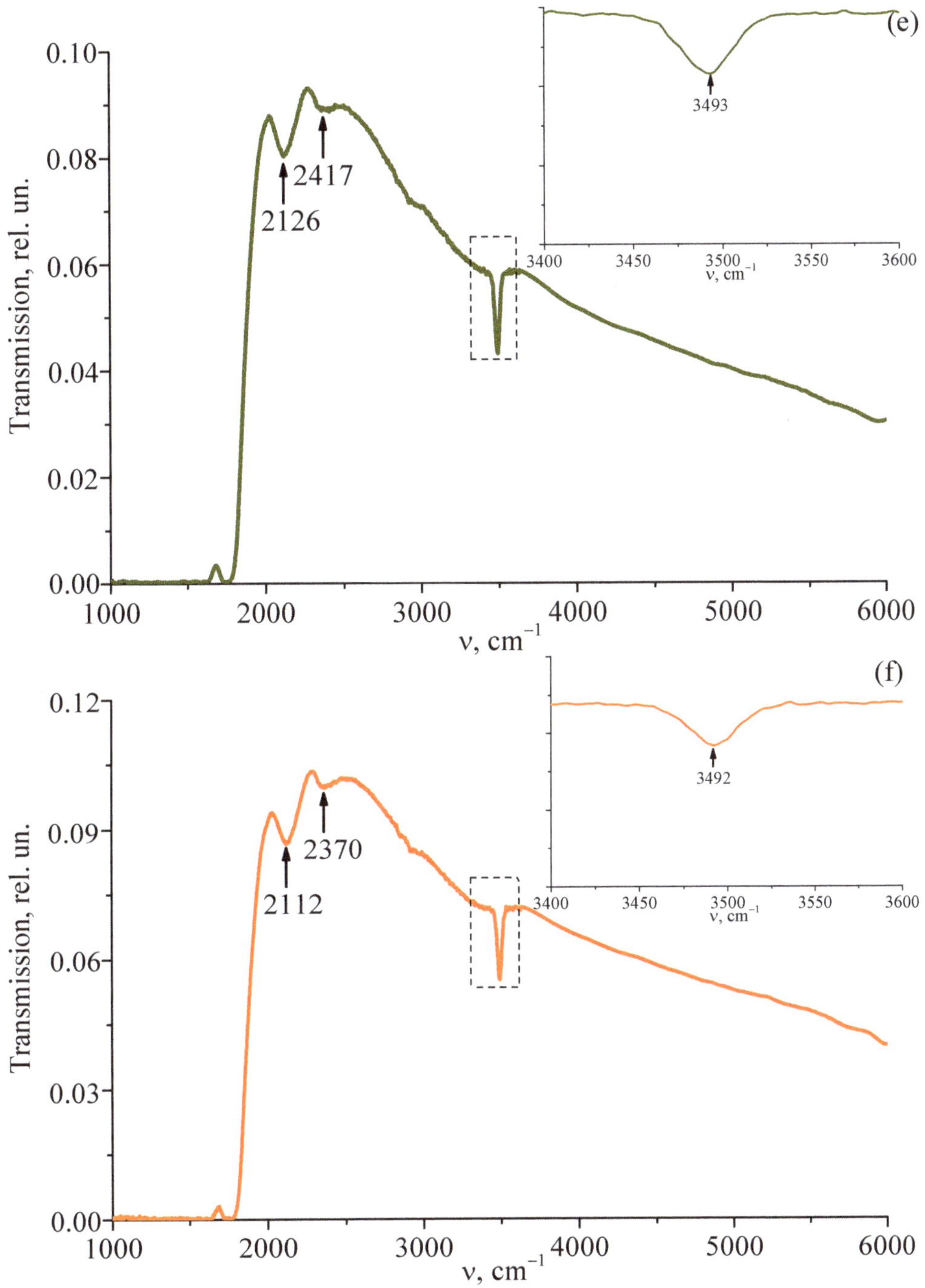

Figure 3. *Cont.*

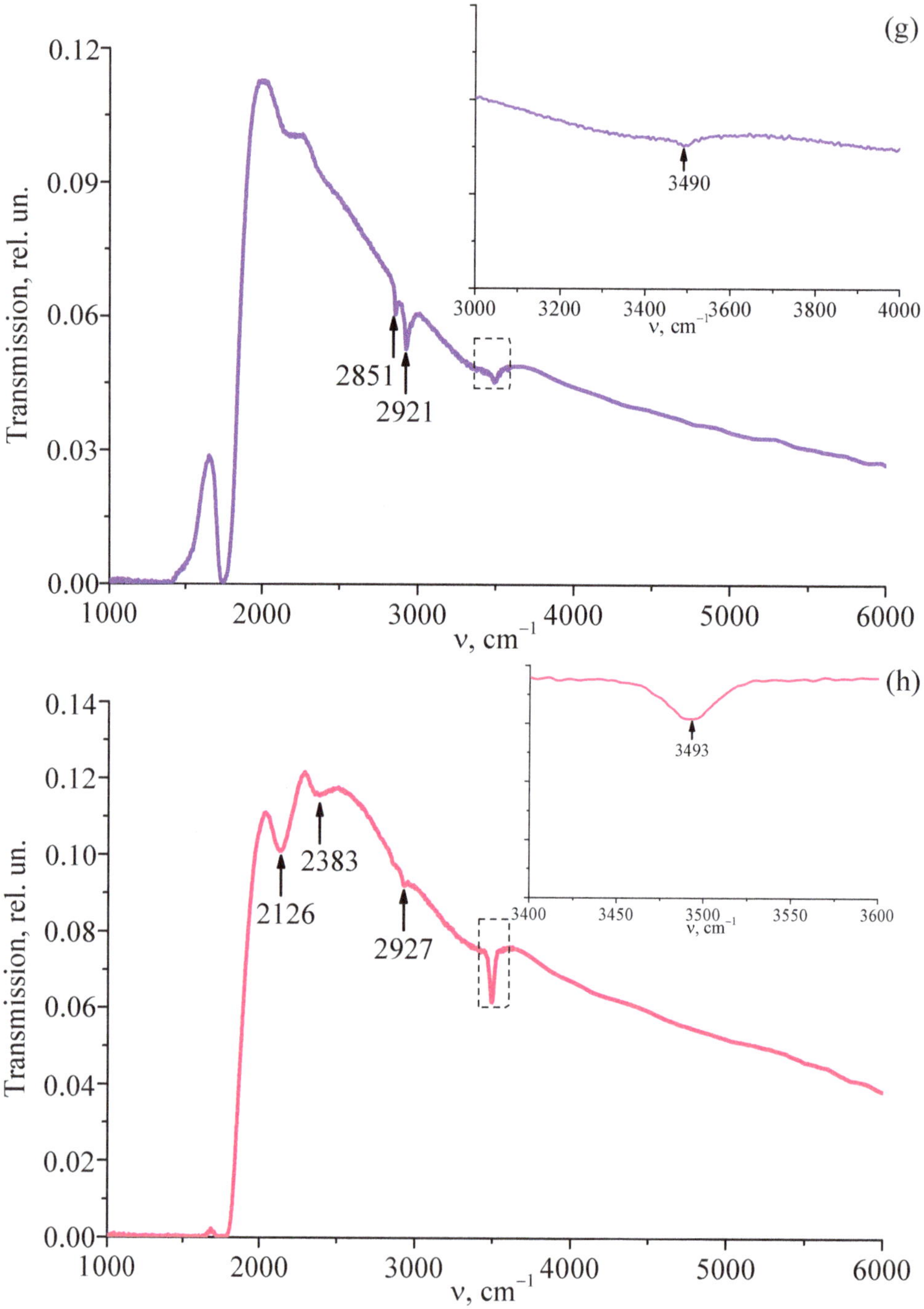

Figure 3. *Cont.*

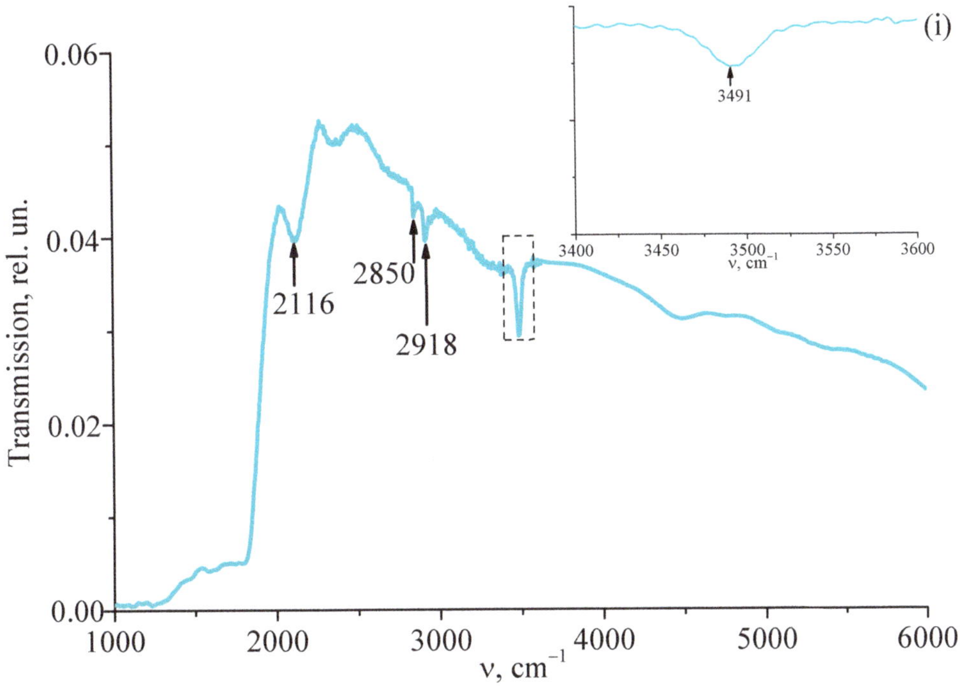

Figure 3. IR transmission spectra from different areas of the gradient crystal $cLiNbO_3:Er^{3+}$(0.55 at%/cm): area 1 (**a**), area 2 (**b**), area 3 (**c**), area 4 (**d**), area 5 (**e**), area 6 (**f**), area 7 (**g**), area 8 (**h**) and area 9 (**i**) in the region of 1000–6000 cm^{-1}. The insets show an enlarged view of the IR transmission spectrum in the region of stretching vibrations of OH^--groups.

As can be seen from Figure 2, several lines are clearly visible in the IR transmission spectrum of the $LiNbO_3:Er^{3+}$(3.1 wt%) crystal. The line with a frequency of 3495 cm^{-1} corresponds to the stretching vibrations of the hydrogen atom of the hydroxyl OH^--group in the cluster defect site Er^{3+} (Er_{Li}^{2+}-OH-Er_{Nb}^{2-}), which consists of one Er_{Li} point defect (Er^{3+} ion in the position of the Li^+ ion) and one Er_{Nb} point defect (Er^{3+} ion in the position of the Nb^{5+} ion) of the ideal $LiNbO_3$ structure of stoichiometric composition [51]. For double-doped crystals, this absorption band shifts by ≈10 cm^{-1} towards lower frequencies [51,52]. The absorption lines with frequencies of 2851 and 2918 cm^{-1} correspond to the symmetric and antisymmetric stretching vibrations of the hydrogen atoms of the C-H bond of the CH_2 group. Compounds with CH_2 groups can be present in microquantities on the crystal surface due to wiping the crystal surface with ethyl alcohol after polishing. The absorption lines with frequencies of 2115 and 2338 cm^{-1} correspond to the absorption bands of the CO and CO_2 groups. Their origin can be associated with the presence of lithium carbonate, from which a batch is prepared for growing lithium niobate crystals.

Figure 3 presents the IR transmission spectra recorded from different areas of the gradient crystal $cLiNbO_3:Er^{3+}$(0.55 at%/cm). IR transmission spectra were recorded along the ferroelectric axis of the crystal. Comparison of Figures 2 and 3 shows that the IR transmission spectra of the compositionally homogeneous $LiNbO_3:Er^{3+}$(3.1 wt%) crystal and the gradient crystal $cLiNbO_3:Er^{3+}$(0.55 at%/cm) differ significantly. The gradient crystal has a much higher absorption in the frequency range of 2000–4000 cm^{-1}. At the same time, a characteristic feature of the compositionally homogeneous $LiNbO_3:Er^{3+}$(3.1 wt%) crystal and the gradient crystal $cLiNbO_3:Er^{3+}$(0.55 at%/cm) studied by us is that the line corresponding to the stretching vibrations of hydrogen ions in the OH^--group is not split into separate components (inserts in Figures 2 and 3), as is observed for many compositionally homogeneous non-stoichiometric $LiNbO_3$ crystals, both nominally pure and doped with a wide range of different metals [38–41].

The IR transmission spectrum also contains a number of lines characteristic of the batch components or technological processes for growing gradient crystals $cLiNbO_3:Er^{3+}$(0.55 at%/cm). The absorption lines with frequencies of 2112–2134 and 2362–2363 cm^{-1} correspond to the absorption bands of the CO and CO_2 groups. Thus, the absorption bands of 2850–2852 and 2916–2927 cm^{-1} correspond to symmetric and antisymmetric stretching vibrations of the hydrogen atoms of the C-H bond of the CH_2 group.

The significant absorption of the gradient crystal $cLiNbO_3:Er^{3+}$(0.55 at%/cm) may be due to the fact that its structure is closer to polycrystalline than to monocrystalline. A less intense absorption band is observed in the region of the stretching vibrations of OH^--groups. This may be due to the design feature of the thermal unit in which the gradient crystal is grown. Despite the wide range studied (1000–6000 cm^{-1}), differences are observed mainly in the region of stretching vibrations of OH^--groups of the studied crystals. This fact indicates a special role of hydroxyl groups in the formation of structural features and properties of $LiNbO_3:Er^{3+}$ crystals of different genesis.

In the IR transmission spectrum and in the Raman spectrum of non-stoichiometric, non-gradient crystals of different compositions, the lines corresponding to the stretching vibrations of the hydrogen atoms of the OH^--groups are observed in the frequency range of 3450–3550 cm^{-1} [38–40]. In this case, the number of observed lines (provided that the oxygen-octahedral clusters of MeO_6 are nonequivalent to each other) should not exceed six (according to the number of OH^--groups in the MeO_6 cluster). The lengths of the O-H stretching bonds and, accordingly, their quasi-elastic constants in such nonequivalent MeO_6 clusters will differ. Therefore, the frequencies of the stretching vibrations of the hydrogen atoms of the OH^--groups will also differ. The more perfect (symmetrical) the MeO_6 cluster, the less the frequencies of the stretching vibrations of the hydrogen atoms of the OH^--groups will differ from each other [38]. In the case of a highly perfect stoichiometric crystal, there is only one position for the hydrogen atom, and its IR transmission spectrum exhibits only one line with a frequency of 3466 cm^{-1} [38]. In the structure of most non-stoichiometric $LiNbO_3$ crystals, there are more than two positions for hydrogen atoms, and their IR absorption spectrum exhibits more than two lines in the frequency range of 3450–3550 cm^{-1}. In particular, in the IR transmission spectrum of the congruent crystal $LiNbO_{3cong}$, according to numerous studies, three lines with frequencies of 3470, 3483 and 3486 cm^{-1} are observed [38–40].

The formation of hydrogen bonds in the $LiNbO_3$ crystal leads to a noticeable change in the wave functions of the outer electron orbitals of oxygen ions and the parameters of its electron polarizability [39–41] and, accordingly, to a noticeable distortion of the entire oxygen-octahedral MeO_6 cluster. The absence of line components in the IR transmission spectrum corresponding to the stretching vibrations of hydrogen atoms in the OH^--groups of various oxygen-octahedral MeO_6 clusters of the crystal structure indicates that the MeO_6 clusters in the structure of the $LiNbO_3:Er^{3+}$(3.1 wt%) crystal and the gradient crystal $cLiNbO_3:Er^{3+}$(0.55 at%/cm), as well as the MeO_6 clusters of the highly perfect crystal of stoichiometric composition [36], have a shape close to regular, that is, they are practically undistorted. The results we obtained are a good confirmation of the results of works [32], in which the method of X-ray structural analysis showed that at erbium concentrations in compositionally homogeneous crystals of $LiNbO_3:Er^{3+}$ of more than 2.66 wt%, a reduction in the difference between the values of long and short distances in the niobium octahedron (NbO_6) is observed. This fact also indicates that the shape of the oxygen-octahedral MeO_6

clusters in the crystal structure tends to the correct shape, characteristic of a highly perfect crystal of stoichiometric composition.

Thus, in the compositionally homogeneous crystal $LiNbO_3:Er^{3+}$(3.1 wt%) and in the gradient crystal $cLiNbO_3:Er^{3+}$(0.55 at%/cm) studied by us, the value of R ≈ 1, i.e., it is close to that for a crystal of stoichiometric composition. The correctness of this conclusion is also confirmed by the absence of a second-order IR transmission spectrum (Figure 2) and a second-order Raman spectrum (Figure 4) in the studied crystal $LiNbO_3:Er^{3+}$(3.1 wt%) in the frequency range of 1000–2000 cm^{-1}. It is widely known that for a highly ordered stoichiometric crystal $LiNbO_{3stoich}$, there is no second-order Raman spectrum in the frequency range of 1000–2000 cm^{-1}. But the second-order spectrum is reliably observed for non-stoichiometric $LiNbO_3$ crystals, both nominally pure and doped [44–47]. What is unusual is that in the gradient crystal $cLiNbO_3:Er^{3+}$(0.55 at%/cm) in the frequency range of 1000–2000 cm^{-1}, the second-order Raman spectrum is reliably observed (Figure 5), but the second-order IR transmission spectrum is absent (Figure 3).

If the value of R ≈ 1 in the $LiNbO_3$ crystal (near-stoichiometric composition), then the crystal structure should be practically free of Nb_{Li} point defects responsible for the photorefraction effect, and the erbium ions will occupy not one, but two positions in the cation sublattice. Moreover, one of the positions is the main one (Er_{Li}) and is populated significantly more than the other position (Er_V). Indeed, according to the data of X-ray structural analysis, in $LiNbO_3:Er^{3+}$ crystals with an erbium concentration of more than 2.66 wt%, studied in [32], the proportion of Nb_{Li} point defects is small. Moreover, Nb_{Li} defects at erbium concentrations of more than 2.66 wt% appear due to the displacement of niobium ions by erbium ions from empty octahedra; that is, there is also a decrease in the concentration of Er_V point defects. From the obtained results, it becomes clear why the $LiNbO_3:Er^{3+}$(3.1 wt%) crystal has a low photorefraction effect: it has a low concentration of Nb_{Li} defects responsible for the photorefraction effect [32].

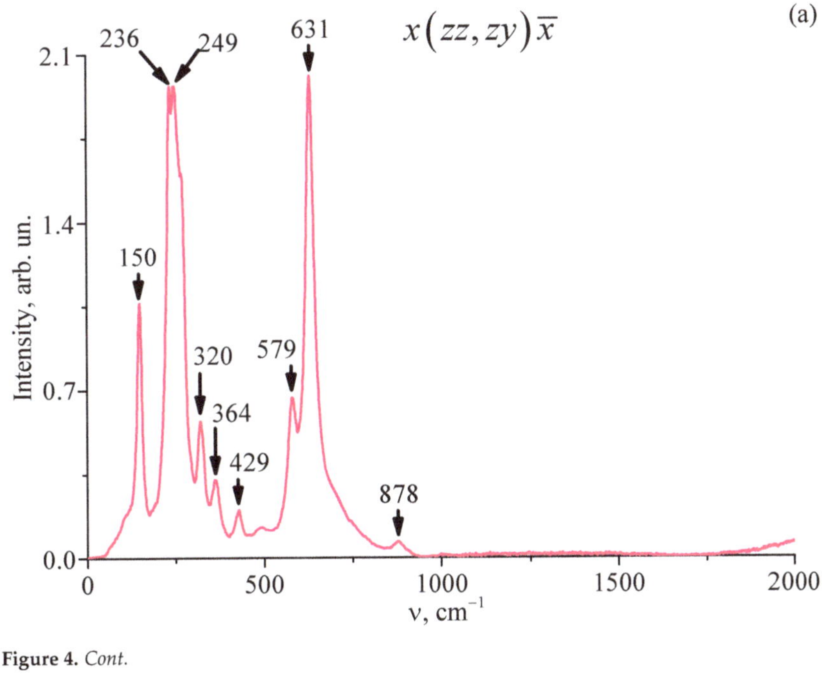

Figure 4. *Cont.*

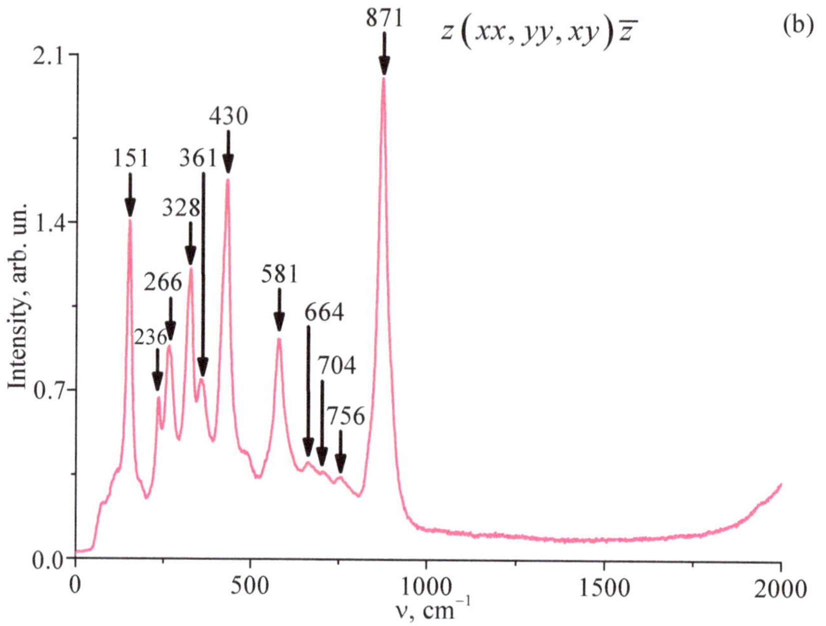

Figure 4. Raman spectra of the $LiNbO_3$:Er^{3+}(3.1 wt%) crystal in two different backscattering geometries upon excitation by radiation with λ_0 = 785 nm. (**a**) $x(zz; zy)\bar{x}$ geometry, (**b**) $z(xx; yy; xy)\bar{z}$ geometry.

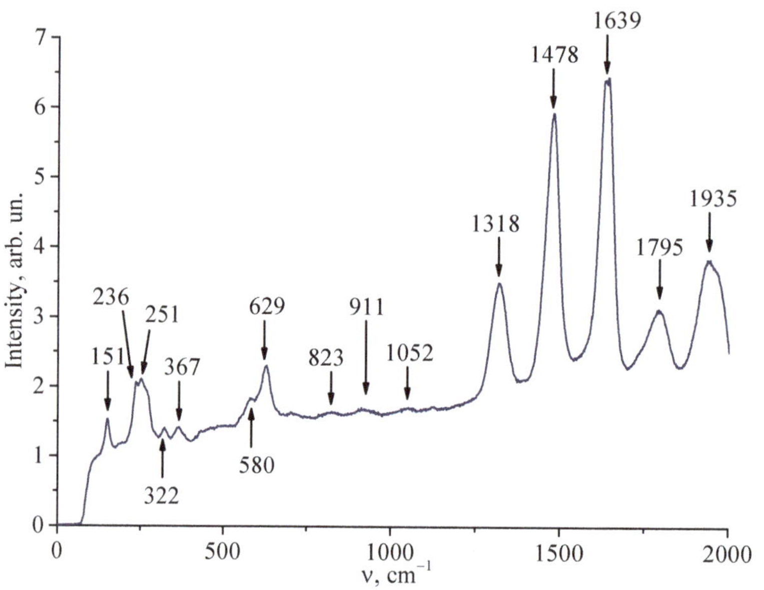

Figure 5. Raman spectrum of the gradient crystal c$LiNbO_3$:Er^{3+}(0.55 at%/cm) in a wide frequency range, including the range of fundamental vibrations of the crystal lattice (first-order spectrum, 100–1000 cm^{-1}) and the second-order spectrum (range 1000–2000 cm^{-1}) upon excitation by radiation with λ_0 = 785 nm.

The contribution to the photorefraction effect in $LiNbO_3$:Er^{3+}(3.1 wt%) crystal and gradient crystal c$LiNbO_3$:Er^{3+}(0.55 at%/cm) can also be made by OH^--groups, since their presence in the crystal increases the electrical conductivity of the crystal, and, accordingly, the photorefraction effect will also increase [36]. Based on the IR transmission spectra, the volume concentration of OH^--groups in the studied crystals can be calculated using Klauer's method [53,54]. The calculation results obtained, taking into account the thickness of the studied gradient samples, are presented in Table 1.

Table 1. Calculated concentration of OH^--groups for different areas of the gradient crystal $cLiNbO_3{:}Er^{3+}$(0.55 at%/cm).

Area	Concentration of the OH^--Groups, cm^{-3}
1	2.62×10^{16}
2	6.51×10^{16}
3	4.21×10^{16}
4	4.98×10^{16}
5	3.23×10^{16}
6	3.42×10^{16}
7	8.33×10^{15}
8	2.86×10^{16}
9	1.90×10^{16}

It is evident from Table 1 that the concentration of OH^--groups in the gradient crystal $cLiNbO_3{:}Er^{3+}$(0.55 at%/cm) does not exceed 7.0×10^{16} cm^{-3}. At the same time, the concentration of OH^--groups in the compositionally homogeneous crystal $LiNbO_3{:}Er^{3+}$(3.1 wt%) is 6.7×10^{17} cm^{-3}, which is almost an order of magnitude greater than in the gradient crystal $cLiNbO_3{:}Er^{3+}$(0.55 at%/cm). The concentration of OH^--groups in the congruent $LiNbO_{3cong}$ and stoichiometric $LiNbO_{3stoich}$ crystals are, respectively, 3.26×10^{17} and 1.58×10^{17} cm^{-3}. This fact indicates a lower conductivity of the gradient crystal $cLiNbO_3{:}Er^{3+}$(0.55 at%/cm) and, accordingly, a lower photorefraction effect, compared to compositionally homogeneous crystals of $LiNbO_3{:}Er^{3+}$(3.1 wt%), $LiNbO_{3cong}$ and $LiNbO_{3stoich}$. It should be noted that in the visible region, direct measurements of the photorefraction effect in the gradient crystal $cLiNbO_3{:}Er^{3+}$(0.55 at%/cm) are significantly complicated due to strong absorption by the crystal, which has a dark color (Figure 1).

The obtained results are of interest for many reasons. Since the presence of OH^--groups play an important role in the formation of the secondary structure and physical characteristics of the crystal, determination of its concentration is necessary to determine the defect structure. In this case, it is much easier to do using IR spectroscopy due to its high sensitivity than using Raman spectroscopy. Hydrogen has its own diffusion coefficient in a single-crystal. Spatial variations in its concentration are possible. In the case of the gradient crystal studied in our work, analysis of the spatial distribution of the OH^--group concentration provides information on the change in the Er_{Li}^{2+}-OH-Er_{Nb}^{2-} defect. By calculating the concentration of OH^--groups, we can select the required section of the gradient crystal $cLiNbO_3{:}Er^{3+}$(0.55 at%/cm) with the required conductivity value. In general, the analysis of changes in the vibrational spectra of the gradient crystal $cLiNbO_3{:}Er^{3+}$(0.55 at%/cm) will allow us to establish concentration thresholds within which the crystal structure is preserved. Numerical estimates of these boundaries will allow us to grow a crystal with the highest possible Er^{3+} concentration required for laser properties and minimal defects that can be redistributed within the crystal during intense heat generation during laser transitions.

4. Conclusions

Based on the analysis of the IR transmission spectra in the region of OH^--group stretching vibrations, it was established for the first time that the oxygen-octahedral MeO_6 clusters in the structure of the $LiNbO_3$:Er^{3+}(3.1 wt%) single-crystal and the gradient crystal $LiNbO_3$:Er^{3+}(congruent composition by the main components, Er gradient of 0.55 at%/cm), as well as the oxygen-octahedral MeO_6 clusters of the highly perfect compositionally homogeneous crystal of stoichiometric composition, have a shape close to the correct one, i.e., they are practically undistorted. The obtained results make it clear why the compositionally homogeneous $LiNbO_3$:Er^{3+}(3.1 wt%) single-crystal and the gradient crystal $LiNbO_3$:Er^{3+}(congruent composition by the main components, Er gradient of 0.55 at%/cm) have a low photorefraction effect: they have a low concentration of Nb_{Li} defects. The correctness of this conclusion is also confirmed by the absence of a second-order vibrational spectrum (IR transmission spectrum and Raman spectrum) of the compositionally homogeneous $LiNbO_3$:Er^{3+}(3.1 wt%) single-crystal in the frequency range of 1000–2000 cm^{-1}, which is reliably manifested in the spectra of non-stoichiometric, compositionally homogeneous $LiNbO_3$ crystals, nominally pure and doped. The gradient crystal $LiNbO_3$:Er^{3+}(congruent composition by the main components, Er gradient of 0.55 at%/cm) does not have a second-order IR transmission spectrum but has a second-order Raman spectrum. The presence of a second-order Raman spectrum can be associated with high microheterogeneity of the gradient crystal $LiNbO_3$:Er^{3+}(congruent composition by the main components, Er gradient of 0.55 at%/cm). By using the IR transmission spectra in the studied crystals, the volume concentration of OH^--groups was calculated using Klauer's method. The concentration of OH^--groups in the gradient crystal $LiNbO_3$:Er^{3+}(congruent composition by the main components, Er gradient of 0.55 at%/cm) does not exceed 7.0×10^{16} cm^{-3}. At the same time, the concentration of OH^--groups in the compositionally homogeneous crystal $LiNbO_3$:Er^{3+}(3.1 wt%) is 6.7×10^{17} cm^{-3}, which is an order of magnitude greater than in the gradient crystal $LiNbO_3$:Er^{3+}(congruent composition by the main components, Er gradient of 0.55 at%/cm). This fact also indicates a lower conductivity of the gradient crystal $LiNbO_3$:Er^{3+}(congruent composition by the main components, Er gradient of 0.55 at%/cm) and, accordingly, a lower photorefraction effect, compared to the compositionally homogeneous crystal $LiNbO_3$:Er^{3+}(3.1 wt%).

Author Contributions: N.S.: conceptualization, writing—supervision; A.P.: formal analysis, data curation, writing—original draft preparation; V.G.: resources, investigation; E.S.: resources, investigation; A.B.: visualization, data curation; M.T.: visualization, data curation. All authors have read and agreed to the published version of the manuscript.

Funding: N.S. acknowledges the Ministry of Science and Higher Education Russian Federation, scientific topic FMEZ-2025-0055. A.P. acknowledges the Russian Science Foundation (Grant No. 19-79-30086-P). V.G. and E.S. acknowledges the Ministry of Science and Higher Education Russian Federation, scientific topic FZEN-2023-0006. The research was carried out using the infrastructure of the Educational Design Center for Opto- and Microelectronics of the Bauman Moscow State Technical University. A.B. and M.T. acknowledge the financial support of the Ministry of Science and Higher Education of the Russian Federation grant (agreement dated 6 March 2024; number 075-02-2024-1519).

Institutional Review Board Statement: Not applicable.

Informed Consent Statement: Not applicable.

Data Availability Statement: The raw data required to reproduce these findings are available from the corresponding author, A.P., upon reasonable request.

Conflicts of Interest: The authors declare no conflicts of interest.

References

1. Boes, A.; Chang, L.; Langrock, C.; Yu, M.; Zhang, M.; Lin, Q.; Lončar, M.; Fejer, M.; Bowers, J.; Mitchell, A. Lithium niobate photonics: Unlocking the electromagnetic spectrum. *Science* **2023**, *379*, eabj4396. [CrossRef] [PubMed]
2. Sánchez-Dena, O.; Villalobos-Mendoza, S.D.; Farías, R.; Fierro-Ruiz, C.D. Lithium niobate single crystals and powders reviewed—Part II. *Crystals* **2020**, *10*, 990. [CrossRef]
3. Wang, L.; Du, H.; Zhang, X.; Chen, F. Optical nonlinearity of thin film lithium niobate: Devices and recent progress. *J. Phys. D Appl. Phys.* **2025**, *58*, 023001. [CrossRef]
4. Wilke, I.; Monahan, J.; Toroghi, S.; Rabiei, P.; Hine, S. Thin-film lithium niobate electro-optic terahertz wave detector. *Sci. Rep.* **2024**, *14*, 4822. [CrossRef]
5. Skvortsov, A.P.; Voronov, M.M.; Pevtsov, A.B.; Starukhin, A.N.; Reznitsky, A.N.; Polgar, K. Luminescent thermometry and up-conversion in $LiNbO_3$:Er crystals. *Opt. Spectrosc.* **2023**, *131*, 1394–1396. [CrossRef]
6. Mantese, J.V.; Micheli, A.L.; Schubring, N.W.; Hayes, R.V.; Srinivasan, G.; Aplay, S.P. Magnetization-graded ferromagnets: The magnetic analogs of semiconductor junction elements. *Appl. Phys. Lett.* **2005**, *87*, 082503. [CrossRef]
7. Ban, Z.-G.; Aplay, S.P.; Mantese, J.V. Fundamentals of graded ferroic materials and devices. *Phys. Rev. B* **2003**, *67*, 184104. [CrossRef]
8. Hauke, T.; Kouvator, A.; Steinhausen, R.; Seifert, W.; Beige, H.; Langhammer, H.; Abicht, H.P. Bending behavior of functionally gradient materials. *Ferroelectrics* **2000**, *238*, 195–202. [CrossRef]
9. Mantese, J.V.; Schubring, N.W.; Micheli, A.L. Polarization-graded ferroelectrics: Transpacitor energy gain. *Appl. Phys. Lett.* **2001**, *79*, 4007. [CrossRef]
10. Mantese, J.V.; Schubring, N.W.; Micheli, A.L. Polarization-graded ferroelectrics: Transpacitor push-pull amplifier. *Appl. Phys. Lett.* **2002**, *80*, 1430–1431. [CrossRef]
11. Jin, F.; Auner, G.W.; Naik, R.; Schubring, N.W.; Mantese, J.V.; Catalan, A.B.; Micheli, A.L. Giant effective pyroelectric coefficients from graded ferroelectric devices. *Appl. Phys. Lett.* **1998**, *73*, 2838–2840. [CrossRef]
12. Schubring, N.W.; Mantese, J.V.; Micheli, A.L.; Catalan, A.B.; Lopez, R.I. Charge pumping and pseudopyroelectric effect in active ferroelectric relaxor-type films. *Phys. Rev. Lett.* **1992**, *68*, 1778–1781. [CrossRef]
13. Dai, Y.; Dong, J.; Zhang, Z.; Zhang, Z.; Kou, H.; Su, L.; Liu, J.; Gao, J.; Dou, R.; Zhang, Q.; et al. Growth of a gradient-doped single-crystal rod with designable distribution of doped ions by a laser-heated pedestal growth method. *Cryst. Growth Des.* **2023**, *23*, 2343–2350. [CrossRef]
14. Tian, F.; Jiang, N.; Liu, Y.; Chen, H.; Xie, T.; Hreniak, D.; Li, J. Fabrication and properties of multistage gradient doping Yb: YAG laser ceramics. *J. Am. Ceram. Soc.* **2023**, *106*, 2309–2316. [CrossRef]
15. Wei, M.E.; Cheng, T.Q.; Dou, R.Q.; Zhang, Q.L.; Jiang, H.H. High-peak-power electro-optically Q-switched laser with a gradient-doped Nd: YAG crystal. *Opt. Lett.* **2021**, *46*, 5016–5018. [CrossRef]
16. Wang, X.; Li, X.; Deng, M.; Zhang, Z.; Yang, F.; Kou, H.; Su, L.; Wu, A.; Chen, J. A couple-free structured LuAG:Ce-LuAG scintillating single crystal fiber grown by a laser-heated pedestal growth method. *Cryst. Growth Des.* **2024**, *24*, 3333–3341. [CrossRef]
17. Aghamaliyev, Z.A. Three-component zone-melting method: Modeling of the concentration-component distribution in single crystals of Ge–Si solid solutions. *Semiconductors* **2021**, *55*, 283–288. [CrossRef]
18. Shut, V.N.; Syrtsov, S.R.; Lobanovskii, L.S.; Yanushkevich, K.I. Crystal structure and magnetic properties of $(Ni_{1-x}Zn_x)Fe_2O_4$ composition graded ceramic. *Phys. Solid State* **2016**, *58*, 1975–1980. [CrossRef]
19. Shut, V.N.; Syrtsov, S.R.; Trublovsky, V.L.; Strukov, B.A. Compositionally graded ceramics based on $Ba_{1-X}Sr_XTiO_3$ solid solutions. *Inorg. Mater.* **2011**, *47*, 87–92. [CrossRef]
20. Shut, V.N.; Kashevich, I.F.; Syrtsov, S.R. Ferroelectric properties of triglycine sulfate crystals with a nonuniform distribution of chromium impurities. *Phys. Solid State* **2008**, *50*, 118–121. [CrossRef]
21. Misirlioglu, I.B.; Alpay, S.P. Compositionally graded ferroelectrics as wide band gap semiconductors: Electrical domain structures and the origin of low dielectric loss. *Acta Mater.* **2017**, *122*, 266–276. [CrossRef]
22. Wu, J.; Yin, K.; Xiao, S.; Wu, Z.; Zhu, Z.; Duan, J.A.; He, J. Laser fabrication of bioinspired gradient surfaces for wettability applications. *Adv. Mater. Interfaces* **2021**, *8*, 2001610. [CrossRef]
23. Matsumoto, Y.; Maruyama, S.; Kaminaga, K. Compositionally graded crystals as a revived approach for new crystal engineering for the exploration of novel functionalities. *CrystEngComm* **2022**, *24*, 2359–2369. [CrossRef]
24. Savotchenko, S.E. Temperature-controlled waveguide properties of the linearly graded-index film in the photorefractive crystal. *Appl. Phys. B* **2023**, *129*, 7. [CrossRef]
25. Dash, D.; Saini, J.; Goyal, A.K.; Massoud, Y. Exponentially index modulated nanophotonic resonator for high-performance sensing applications. *Sci. Rep.* **2023**, *13*, 1431. [CrossRef] [PubMed]
26. Goyal, A.K.; Dash, D.; Saini, J.; Massoud, Y. Theoretical analysis of graded-index topological resonator for improved sensing performance. *Opt. Express* **2024**, *32*, 4102–4110. [CrossRef]

27. Galutskiy, V.V.; Vatlina, M.I.; Stroganova, E.V. Growth of single crystal with a gradient of concentration of impurities by the Czochralski method using additional liquid charging. *J. Cryst. Growth* **2009**, *311*, 1190–1194. [CrossRef]
28. Stroganova, E.V.; Nalbantov, N.N.; Galutsky, V.V.; Yakovenko, N.A. A study of the quantum efficiency of multichannel relaxation in $LiNbO_3$:Yb,Er crystals. *Opt. Spectrosc.* **2016**, *121*, 856–861. [CrossRef]
29. Stroganova, E.V.; Galutskiy, V.V.; Nalbantov, N.N.; Kozin, A.S. Spectral-luminescent properties of gradient-activated $LiNbO_3$ crystals with concentration profiles of Yb^{3+} and Er^{3+} ions. *Optoelectron. Instrument. Proc.* **2017**, *53*, 77–81. [CrossRef]
30. Babenko, I.D.; Galutskiy, V.V.; Ivashko, S.S.; Stroganova, E.V. Temperature dependence of Er^{3+}, Yb^{3+} kinetic spectra in the gradient crystals of lithium niobate. *Opt. Mater.* **2020**, *102*, 109818. [CrossRef]
31. Galutskiy, V.V.; Ponetaeva, I.G.; Puzanovskiy, K.V.; Stroganova, E.V. Formation of channel proton-exchange waveguides in Yb^{3+}, Er^{3+}:PPLN. *Appl. Nanosci.* **2022**, *12*, 3417–3420. [CrossRef]
32. Palatnikov, M.N.; Biryukova, I.V.; Shcherbina, O.B.; Sidorov, N.V.; Makarova, O.V.; Teplyakova, N.A. Growth of $LiNbO_3$:Er crystals and concentration dependences of their properties. *Crystallogr. Rep.* **2016**, *61*, 1031–1038. [CrossRef]
33. Zhang, Z.; Wu, Z.; Zhang, Y.; Liu, S.; Liu, H.; Zheng, D.; Kong, Y.; Xu, J. The doping concentration optimization of Er-doped $LiNbO_3$ crystals for LNOI lasers and amplifiers. *Cryst. Growth Des.* **2024**, *24*, 6838–6844. [CrossRef]
34. Lovisa, L.X.; Nunes, T.B.O.; Tavares, E.C.; Machado, R.C.L.; Dos Santos, L.F.; Bomio, M.R.D.; Motta, F.V. Effects of rare earth ions on structural, morphological and photoluminescent properties of nonstoichiometric $LiNbO_3$. *Appl. Phys. A* **2024**, *130*, 226. [CrossRef]
35. Uda, S.; Tiller, W.A. The dissociation and ionization of $LiNbO_3$ melts. *J. Cryst. Growth* **1992**, *121*, 155–190. [CrossRef]
36. Kimura, H.; Koizumi, H.; Uchida, T.; Uda, S. Influence of impurity doping on the partitioning of intrinsic ionic species during the growth of $LiNbO_3$ crystal from the melt. *J. Cryst. Growth* **2009**, *311*, 1553–1558. [CrossRef]
37. Koh, S.; Uda, S.; Huang, X. Partitioning of ionic species and crystallization electromotive force during the melt growth of $LiNbO_3$ and $Li_2B_4O_7$. *J. Cryst. Growth* **2007**, *306*, 406–412. [CrossRef]
38. Lengyel, K.; Péter, Á.; Kovács, L.; Corradi, G.; Pálfavi, L.; Hebling, J.; Unferdorben, M.; Dravecz, G.; Hajdara, I.; Szaller, Z.; et al. Growth, defect structure, and THz application of stoichiometric lithium niobate. *Appl. Phys. Rev.* **2015**, *2*, 040601. [CrossRef]
39. Cabrera, J.M.; Olivares, J.; Carrascosa, M.; Rams, J.; Müller, R.; Diéguez, E. Hydrogen in lithium niobate. *Adv. Phys.* **1996**, *45*, 349–392. [CrossRef]
40. Kovács, L.; Rebouta, L.; Soarest, J.C.; da Silva, M.F.; Hage-Ali, M.; Stoquert, J.P.; Siffert, P.; Sanz-Garcia, J.A.; Corradi, G.; Szaller, Z.; et al. On the lattice site of trivalent dopants and the structure of Mg^{2+}-OH^--M^{3+} defects in $LiNbO_3$:Mg crystals. *J. Phys. Condens. Matter* **1993**, *5*, 781–794. [CrossRef]
41. Volk, T.; Wöhlecke, M. *Lithium Niobate. Defects, Photorefraction and Ferroelectric Switching*; Springer: Berlin/Heidelberg, Germany, 2008; 250p.
42. Gao, L.; Wexler, R.B.; Fei, R.; Rappe, A.M. Ab initio study of hydrogen niobate $HNbO_3$: Structural, thermodynamic, dielectric, and optical properties. *J. Phys. Chem. C* **2023**, *127*, 5931–5940. [CrossRef]
43. Kofahl, C.; Dörrer, L.; Wulfmeier, H.; Fritze, H.; Ganschow, S.; Schmidt, H. Hydrogen diffusion in $Li(Nb,Ta)O_3$ single crystals probed by infrared spectroscopy and secondary ion mass spectrometry. *Chem. Mater.* **2024**, *36*, 1639–1647. [CrossRef]
44. Kofahl, C.; Dörrer, L.; Fritze, H.; Schmidt, H. Hydrogen diffusion in proton-exchanged congruent lithium niobate during post-annealing. *Solid State Ion.* **2023**, *403*, 116383. [CrossRef]
45. Yatsenko, A.V.; Yagupov, S.V.; Shul'gin, V.F.; Yatsenko, A.A. On an increase in the volume concentration of hydrogen in undoped $LiNbO_3$ crystals of congruent composition. *Tech. Phys.* **2023**, *68*, 629–634. [CrossRef]
46. Sugak, D.; Yakhnevych, U.; Syvorotka, I.I.; Buryy, O.; Popov, A.I.; Ubizskii, S. Optical investigation of the OH^- groups in the $LiNbO_3$ doped by copper. *Integr. Ferroelectr.* **2019**, *196*, 32–38. [CrossRef]
47. Kling, A. Correlation between infrared absorption and lithium sublattice disorder in magnesium-doped lithium niobate. *Materials* **2023**, *16*, 797. [CrossRef]
48. Gorelik, V.S.; Abdurakhmonov, S.D. Overtone Raman scattering in lithium niobate single crystal doped with terbium. *Crystallogr. Rep.* **2022**, *67*, 252–255. [CrossRef]
49. Sidorov, N.; Palatnikov, M.; Pyatyshev, A.; Skrabatun, A. Investigation of the structural perfection of a $LiNbO_3$:Gd^{3+}(0.003):Mg^{2+}(0.65 wt.%) double-doped single crystal using the Raman spectra excited by laser lines in the visible (532 nm) and near-IR (785 nm) regions. *Appl. Sci.* **2023**, *13*, 2348. [CrossRef]
50. Sidorov, N.; Palatnikov, M.; Pyatyshev, A.; Sverbil, P. Second-order Raman scattering in ferroelectric ceramic solid solutions $LiNb_xTa_{1-x}O_3$. *Crystals* **2022**, *12*, 456. [CrossRef]
51. Qian, Y.; Wang, R.; Xing, L.; Xu, Y.; Yang, C.; Liu, X. Influence of Zn^{2+} ions concentration on the optical properties of Zn/Er: $LiNbO_3$ crystals. *Cryst. Res. Technol.* **2011**, *46*, 1137–1142. [CrossRef]
52. Tsuboi, T.; Kaczmarek, S.M.; Nakai, Y.; Huang, W. The effect of Mg^{2+} ions in Er^{3+}, Tm^{3+}-codoped $LiNbO_3$ single crystal. *J. Lumin.* **2014**, *149*, 99–102. [CrossRef]

53. Klauer, S.; Wöhlecke, M.; Kapphan, S. Influence of H-D isotopic substitution on the protonic conductivity of $LiNbO_3$. *Phys. Rev. B* **1992**, *45*, 2786–2799. [CrossRef] [PubMed]
54. Salloum, Y.; Grunsky, O.S.; Man'shina, A.A.; Tver'yanovich, A.S.; Tver'yanovich, Y.S. Investigation of lithium niobate composition by optical spectroscopy methods. *Russ. Chem. Bull.* **2009**, *73*, 2228–2232. [CrossRef]

materials

MDPI

Article

Preparation of Copper Oxide Film at Low Temperature in Basic Conditions on a Copper Substrate

Francesca Irene Barbaccia [1,2], Tilde de Caro [3], Fulvio Federici [3], Alessio Mezzi [3], Lucia Sansone [4], Michele Giordano [4] and Andrea Macchia [2,*]

1 Department of Technological Innovation Engineering, International Telematic University Uninettuno, Corso Vittorio Emanuele II 39, 00186 Rome, Italy
2 Youth in Conservation of Cultural Heritage (YOCOCU APS), Via T. Tasso 108, 00185 Rome, Italy
3 Institute of Nanostructured Materials, National Research Council (ISMN-CNR), Strada Provinciale 35d, 9, 00010 Rome, Italy; tilde.decaro@cnr.it (T.d.C.); fulvio.federici@cnr.it (F.F.)
4 Institute for Polymers, Composites, and Biomaterials, National Research Council (IPCB-CNR), Piazzale Enrico Fermi 1, 80055 Portici, Italy
* Correspondence: andrea.macchia@unical.it

Abstract: Copper is widely used in both its metallic form and as oxide across numerous scientific and industrial domains. The primary copper oxides, cuprite (Cu_2O) and tenorite (CuO), naturally form on the copper surface and play a crucial role in advanced technologies, such as solar cells, lithium batteries, and sensors. Tenorite is appreciated for its optical properties, stability, low toxicity, and reactivity. While copper oxide thin films are traditionally synthesized through thermal treatments and oxidation in alkaline environments, these conventional high-temperature methods not only require significant energy consumption but can also compromise the metal–film interface. This study aims to develop a sustainable alternative approach for forming a homogeneous CuO layer on a copper substrate through environmentally friendly treatments based on low temperature, cost-effective, and time-saving procedures. Three different eco-conscious treatments were investigated: (i) immersion in NaOH basic solution, (ii) exposure to NaOH basic solution vapours, and (iii) a combined treatment involving both immersion and vapour exposure. This green synthesis approach significantly reduces energy consumption compared to traditional thermal methods while maintaining product quality. The surface oxide layer was investigated through an optical microscope (OM), scanning electron microscopy (SEM), spectrocolorimetric analysis, peel-off test, μ-Raman and X-ray photoelectron spectroscopy (XPS) analysis to investigate the surface oxidation state.

Keywords: CuO; copper oxide; tenorite; μ-Raman

1. Introduction

Copper is used in many sectors both as an alloy and in the form of oxide. The applications of this metal extend from artistic production [1–3] to the medical field by exploiting the antibacterial properties of its oxides [4–6], in the industrial sectors as humidity sensors, gas sensors and biosensors [7–9], as well as in electronic and thermal application [10–13]. Additionally, copper is employed in the production of pipelines, wire, sheet, and marine industry components for communication technologies and water purification or desalination systems [14–18].

Copper oxides exist in two primary forms: cuprite Cu_2O (copper I) or cuprous oxide and tenorite CuO (copper II) or cupric oxide. In addition to these, a third copper oxide, paramelaconite Cu_4O_3, has been identified as an intermediate phase between these two structures [19–21]. Copper oxide formation is divided into two steps: the initial ionization of copper (Equation (1)), followed by its oxidation (Equations (2) and (3)):

$$Cu \rightarrow Cu^{n+} + ne^- \quad (1)$$

$$Cu^+ + O^{2-} \rightarrow Cu_2O \quad (2)$$

$$Cu^{2+} + O^{2-} \rightarrow CuO \quad (3)$$

Due to their semiconductive properties, copper oxides are used in the production of solar cells, photo-electro-chemical cells, photocatalysts, lithium-ion batteries, gas sensors, magnetic media, optical devices and thermal control coating in spacecraft [22–24]. Cu_2O has a band gap range of 1.9 and 2.3 eV [20,25,26] whereas CuO, with a band gap of 1.1 and 1.5 eV, exhibits strong sunlight absorption properties, making it a suitable material for energy applications [12,24,27,28].

However, conventional CuO synthesis methods require high-temperature processing, which can induce structural stress at the metal–film interface and compromise film homogeneity [29–31]. Additionally, some techniques involve chemical environments that pose environmental and safety risks. For example, chemical vapour deposition (CVD) and physical vapour deposition (PVD) typically operate at high temperatures, leading to excessive energy consumption and potential thermal deformation of the metallic substrate. Moreover, some synthesis techniques involve volatile organic solvents or toxic reagents, increasing environmental and operational hazards [32–34]. The use of organometallic precursors in CVD, for instance, can release harmful byproducts into the atmosphere. These factors highlight the need for alternative synthesis methods that are both efficient and environmentally friendly [35].

Green synthesis approaches have gained attention in recent years as sustainable alternatives to traditional methods. The eco-friendly approaches prioritize lower energy consumption, non-toxic reagents, and minimal environmental impact while maintaining the effectiveness of oxide layer formation [29,32,36–38]. Pulsed laser synthesis is an example of an eco-friendly method for metal oxide production. Key advantages include the generation of nanomaterials without hazardous chemical reagents, process simplicity, and cost-effectiveness. Potential drawbacks may involve multi-step procedures for film deposition onto desired substrates, or high-energy requests [39–42]. Other studies have demonstrated the feasibility of CuO production using plant extracts, bacterial synthesis or fungal digestion [5,42,43]. Such biological methods not only eliminate the need for hazardous chemicals but also enable fine-tuning of oxide properties by leveraging natural biogenic agents [35,44].

In addition to biological synthesis, alternative chemical methods using alkaline environments have been explored. Studies indicate that exposing copper to the basic solution at moderate temperatures can facilitate CuO layer formation with minimal energy input [32]. The combination of heat and alkaline conditions enables controlled oxidation, ensuring a uniform and stable CuO layer. This approach significantly reduces the environmental footprint of traditional oxidation methods that rely on extreme heat or aggressive chemical agents.

To address these concerns, the present study aims to develop an eco-friendly approach for forming a tenorite (CuO) layer on a copper substrate. The objective is to implement a simple, rapid, and cost-effective procedure that utilizes relatively low temperatures and alkaline environments, thereby reducing the ecological footprint of CuO production while ensuring the structural integrity of the oxidized film. This sustainable approach not only minimizes the environmental risks associated with conventional methods but also promotes energy efficiency and operational safety in copper oxide film synthesis.

By leveraging a combination of immersion and vapour-phase exposure to alkaline solutions, this study seeks to optimize the oxidation process while maintaining a low processing temperature. The effectiveness of different treatment durations and temperatures is assessed to determine the optimal conditions for CuO layer formation. Unlike conventional high-temperature oxidation techniques, which often introduce internal stress and structural defects, this method preserves the mechanical stability of the copper substrate while ensuring a uniform and well-adhered oxide layer.

The oxide layer obtained on the copper surface was characterized using optical microscopy (OM), scanning electron microscope (SEM), peeling test, spectrocolorimetry, micro-Raman spectroscopy (μ-Raman) and X-Ray photoelectron spectroscopy (XPS) [45]. These analyses provide comprehensive insights into the structural, optical, and compositional properties of the CuO layer, confirming its successful formation through an environmentally sustainable approach [42,43].

This research underscores the growing necessity of green chemistry in materials science, emphasizing the importance of low-toxicity and energy-efficient approaches to oxide formation. By implementing sustainable methodologies, this study contributes to the development of advanced copper oxide coatings while minimizing environmental impact.

2. Materials and Methods

Based on Zhang et al. [32] and Neupane et al. [46], research was conducted to avoid a low range of temperatures in combination with NaOH solution to induce the formation of an oxide layer on the surface of copper. The oxide layer was induced using a 10% NaOH (*w*/*v*) solution. NaOH was purchased from Sodax (Sodax, Italy) and solubilized in distilled water.

Before the oxidation treatment, samples of pure copper 99.99% size 30 × 10 × 0.5 mm, were polished with 600, 800 and 1200 SiC paper, degreased with acetone, rinsed with ethanol, washed with distilled water and left to dry at room temperature.

A basic attack for copper oxide formation was carried out using the following methods (Table 1) based on the evidence reported in the literature [47–49].

Table 1. Description of the selected treatment method, related to the sample code.

Sample Code	Method Description
Im	Method A: immersing copper samples for 30 min in a 10% NaOH solution, heated at different temperatures, respectively, 20, 40, 60, and 80 °C. Solutions were magnetically stirred at 500 rpm.
ImVap	Method B: immersing the sample for 1 min in a 10% NaOH solution, which was stirred at 500 rpm and heated to 80 °C, followed by exposure to the vapour of the same solution at different times of exposition: 3, 5, 10 and 20 min.
Vap	Method C: by contact with a strongly basic solution (pH 14), involved treating the sample by exposure only to vapours of 10% NaOH solution, magnetically stirred at 500 rpm and heated to 80 °C. Exposure time to the vapour was conducted for 5, 10 and 20 min.

The experimentation procedure and thermal oxidation were conducted in a closed glass vessel put into a silica oil bath. Silica oil was heated on an IKA RCT basic hotplate, and the temperature was controlled using an IKA ETS-D5 thermometer (IKA-Werke GmbH, Staufen im Breisgau, Germany).

Optical surface observation was performed using stereomicroscope Leica M125C acquired with DMC4500 USB (Leica, Wetzlar, Germany) digital camera and scanning electron microscopic (SEM) conducted using SEM Tescam-VEGA 3 (Microcontrolnt, Milan, Italy) at higher magnifications to investigate the morphology and homogeneity of the oxide layer. SEM analyses were performed operating with an accelerating voltage beam at 30 kV under a high vacuum at 12 mm working distance (WD). Imaging was performed in secondary electrons (SE) mode. Spectrocolorimetric analysis was conducted using a Y3060 3nh spectrophotometer (3nh, Guangzhou, China) equipped with an 8 mm aperture lens, enabling acquisition in SCE (Specular Component Excluded) mode. Spectra were collected in the visible region (400–700 nm), under D65 CIE Standard illuminant, with a Standard observer at a 10° angle. The colour was defined using CIELAB colour space coordinates. CIELAB parameters consist in the expression of lightness (L*) limited between 0 and 100 values, where zero is no colour and 100 corresponds to maximum brightness; colour variations from red to green (a*) and colour variations between blue and yellow (b*). An average of ten measurements were taken for each surface sample.

To determine the molecular structure of copper oxides, micro-Raman analysis was performed with a Renishaw RM 2000 Raman spectrometer (Renishaw, Pianezza, Italy), equipped with a CCD detector. The measurements were conducted in backscattering geometry. The Raman spectrometer is based on the Leica optical microscope. The objective has 50× magnification with 0.75 numerical aperture. The excitation of Raman scattering was performed with the 785 nm wavelength diode laser. The laser power was 1.58 mW at the sample. Spectra were collected with 1 s accumulation and 10 repetitions. Normalization 0 to 1 was conducted using Origin 8.5 software, and it was conducted on the values of the spectra collected to compare the intensities of the different spectra acquired on samples. Oxide layer cohesion to the metallic substrate was measured using the peeling test. The test was performed using an Instron 3343 dynamometer, (Instron, Italy) with a fixed angle of 90° and a controlled speed of 10 mm/min. The analysis was repeated on ten different specimens, and for each, the detachment force was recorded along a length of 50 mm. The Maximum Peel Value (MPV) was calculated as the highest force value recorded during the test, while the mean and standard deviation were determined to evaluate the reproducibility of the method [50,51].

XPS analysis was carried out by an ESCALAB 250Xi (Thermo Fisher Scientific Ltd., East Grinstead, UK), equipped with a monochromatic X-ray source (Al Kα—hν = 1486.6 eV) and 6-channeltron detection device. The measurements were performed operating in a UHV system at a base pressure of 1×10^{-10} mbar and using a constant pass energy of 50 eV. The samples were fixed to the sample holder by a metallic clip, in order to ensure the electrical contact, while the binding energy (BE) scale was calibrated positioning the Fermi level at BE = 0 eV. All spectra were collected and processed by Avantage software v5.9 (Thermo Fisher Scientific Ltd., East Grinstead, UK).

3. Results

Reported results investigate the impact of alkaline treatment using 10% NaOH solution and the role of temperature on the chemical and physical transformations of copper surfaces. The results provide insights into the morphological and chemical characteristics of the surface, to highlight the effectiveness of different methods in achieving desired outcomes.

Figures 1 and 2 show macroscopic images and reflectance spectra, respectively, of samples treated using method A (Im). Reflectance spectra (Figure 2) of samples treated at 20 °C and 40 °C exhibit a dominant wavelength at 650 nm. However, samples treated at 60 °C and 80 °C, corresponding to the surface changes observed in Figure 1, show a decrease in overall reflectance with increasing temperature and a sharp decrease in reflectance near 570 nm. The pure bulk Cu_2O has is band gap of about 2.1 eV. It corresponds to about 570 nm wavelength. The defects as well as strains may shift the absorption edge value. Starting from this value, a sharp decrease in reflectivity begins to occur. This latter feature is due to the increased presence of oxides on the surface [52]. The enhanced formation of oxides leads to a darkening of the surface. Specifically, the sample immersed in the solution heated at 20 °C possesses an orange-brown hue similar to the elemental copper and, according to reflectance analysis (Figure 2), this sample shows the lightest hue among other samples. The presence of black areas on the surface indicates the growth of an inhomogeneous oxide layer. Samples immersed in solutions heated at 40 and 60 °C develop a more homogeneous oxidation layer while retaining an orange-brown hue. However, they appear darker compared to the sample treated at 20 °C. The sample immersed in the solution heated at 80 °C shows a dark, purplish-brown surface. The hue observed in samples treated at 20, 40 and 60 °C is characterized by varying degrees of orange-red-brown hue, suggesting the formation of a cuprite layer [53] on the surface accompanied by tenorite or paramelaconite, as indicated by a decrease in reflectance. The sample treated by immersion at 80 °C was prevalently characterized by a darkish appearance, suggesting an increasing possibility of tenorite formation.

Figure 1. Surface hue of samples treated by immersion in NaOH 10% solution heated at different temperatures.

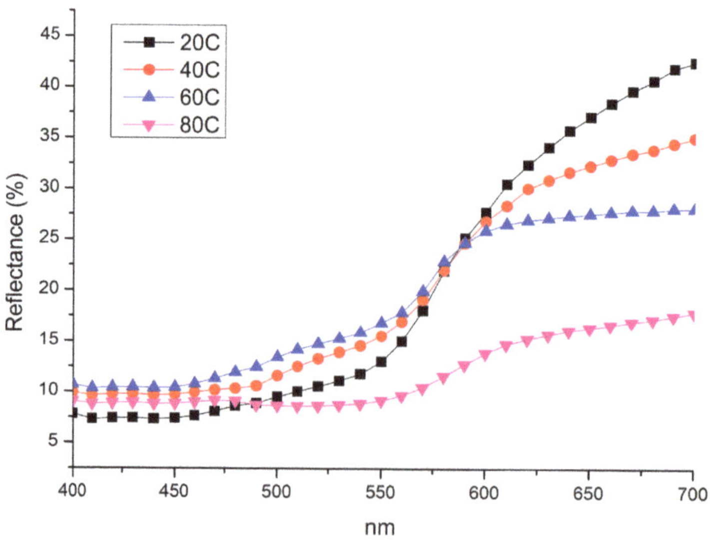

Figure 2. Reflectance spectra collected on samples treated by immersion in NaOH 10% at 20 °C (20C); 40 °C (40C); 60 °C (60C); and 80 °C (80C).

μ-Raman analyses were performed on extended darker areas on the sample surface to investigate the nature of Cu-oxide film (Figures 1 and 2). All compared spectra were normalized from 0 to 1 (minimum–maximum intensity value normalization). Spectra acquired on sample obtained at 80 °C (Figure 3) is characterized by peaks at 151, 516, and 611 cm^{-1}. Peaks at 151 and 516 cm^{-1} are attributable to Cu_2O [54,55] while at 611 cm^{-1} present in the sample spectrum can be traced back to the Cu_2O peak positioned at 634 cm^{-1}. Shift from 634 to 611 cm^{-1} associated with the enlarged peak shape suggests a poor crystallization of the Cu_2O phase [56,57].

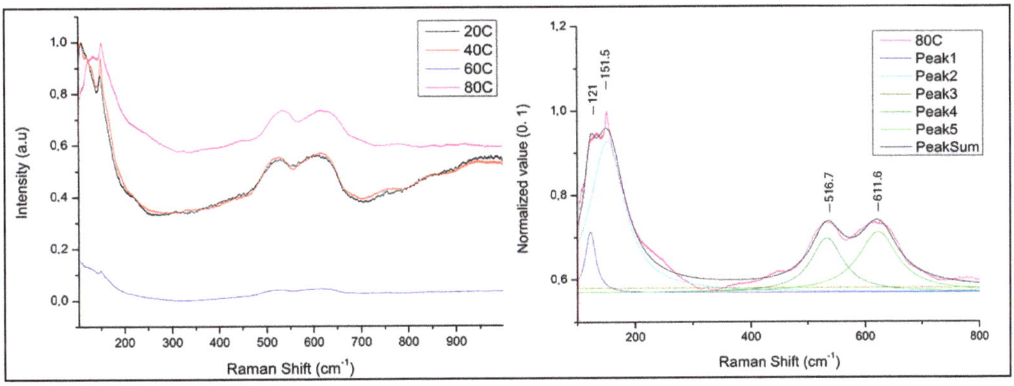

Figure 3. μ-Raman spectrum of the sample treated by immersion in NaOH 10% solution heated at different temperatures (**left**) and deconvolution of 80 °C treated sample (**right**).

The results obtained using method B are presented below. The reflectance spectra (Figure 4) indicate that longer exposure times promote the formation of a darker patina. The spectrum recorded at 3 min exhibits spectral features at 425, 475, and 650 nm. At 5 and 10 min, the spectra display an increasingly linear trend, which becomes more pronounced at 20 min, where the surface develops a nearly colourless appearance, ultimately resulting in a deep black coloration.

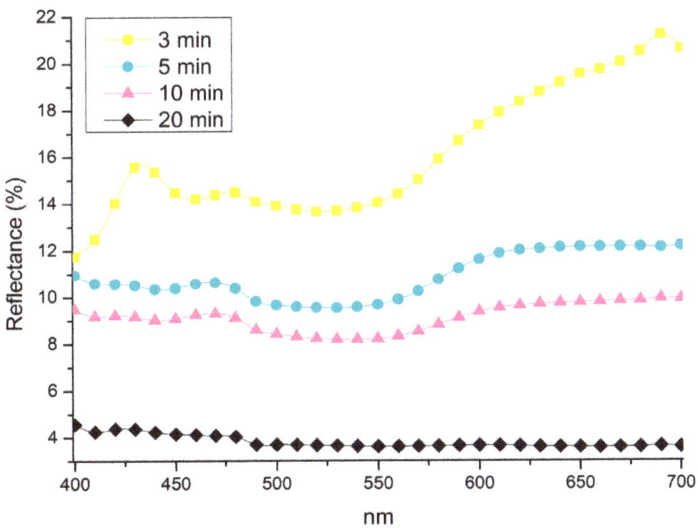

Figure 4. Spectra collected from samples treated by 1 min immersion and exposed to vapour for 3 min (3 min); 5 min (5 min); 10 min (10 min); and 20 min (20 min).

To define the composition of darker oxide, μ-Raman spectra were collected from treated samples with method B at 5, 10 and 20 vapour exposition time. Samples exposed to vapour for 3 min were excluded due to the absence of blackish hue and relative tenorite formation, as confirmed by μ-Raman and colorimetric results obtained from method A. All spectra from the selected treated samples with method B are characterized by the presence of similar main peaks (Figure 5), although the ratio of peak intensities is different in relationship to the exposition time. The main peaks are at 121, 156, 301, 515, and 606 cm^{-1}. The highest intensity of the peaks at 301 and 606 cm^{-1} is assigned to the sample treated for 1 min of immersion and 10 min exposure to NaOH 10% solution. Peaks at 301 cm^{-1} are due to a shift in the main characteristic CuO peaks at 297–298 cm^{-1} [32,36,58]. The peak at 515 cm^{-1} is attributable to the intrinsic violation mechanism of selection rules in Cu_2O phonon frequencies during structure orientation passage into paramelaconite Cu_4O_3 [59], but it is also attributed to CuO [60,61]. Support for attributing this peak to paramelaconite is reported in the study of Jagadish et al. [62] where the peak at 532 cm^{-1} is caused by Raman-allowed paramelaconite crystals' mode. The peak at 606 cm^{-1} can be attributed to a Cu-O stretching mode [59] shifted from 602 cm^{-1}, and similar to the previous samples. The variation in the shape of the latter peak indicates changes in the crystal structure induced by the temperature dependence of formation [32,60–62].

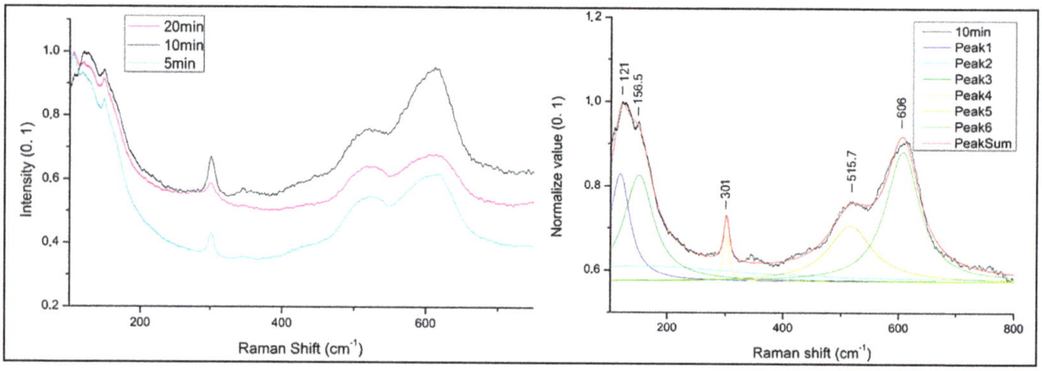

Figure 5. Comparison of μ-Raman spectra collected by sample treated with 1 min immersion and different exposition time to solution vapours: 5 min exposition to vapour (5 min); 10 min exposition to vapour (10 min) and 20 min exposition to vapour (20 min) (**left**); Deconvolution of μ-Raman spectra collected from the sample (10 min) treated with 1 min immersion and 10 min exposition to vapour solution, heated at 80 °C and magnetically stirred (**right**).

To investigate whether the formation of the copper oxide layer by exposure to the vapours of NaOH solution also occurs in the absence of immersion of the sample for 1 min in the same solution, an analysis was carried out by exposing the sample only to the vapours, without prior immersion. Reflectance spectra were collected, followed by optical microscopy (OM) imaging to assess the surface appearance. Finally, μ-Raman spectra were acquired from three different samples: one exposed exclusively to vapours (Vap), one subjected to immersion followed by 10 min of vapour exposure (10 min), and one treated solely by immersion in the 80 °C NaOH solution (80C).

Images collected through OM observations (Figure 6) reveal distinct oxidation patterns resulting from the three tested methods. The sample exposed exclusively to NaOH vapours exhibits an orange-brown coloration with non-homogeneous red oxidation spots. More pronounced oxidation is observed in the samples treated by immersion, which display a combination of orange-brown regions interspersed with larger blackened areas, indicating a more advanced oxidation process. The most uniform and intense oxidation is found in the sample subjected to both immersion and subsequent vapour exposure, where the surface is entirely covered by a dark-coloured patina, suggesting the formation of a more continuous and well-developed oxide layer.

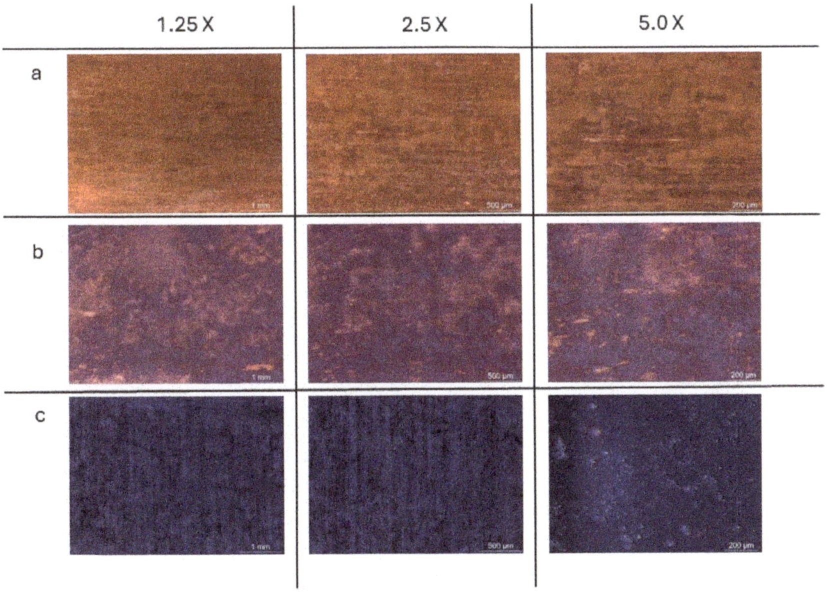

Figure 6. OM images of surface samples from samples treated with vapour exposition treatment (**a**); immersion treatment (**b**) and combining immersion and vapour exposition treatment (**c**), observed at 1.25×, 2.5× and 5.0× magnification.

A comparison of the reflectance spectra analysis of the samples corroborates the OM observations, indicating a higher reflectance percentage and lighter coloration in the sample exposed exclusively to NaOH vapours. In contrast, samples treated by only immersion or by the combined immersion–vapour process exhibit lower reflectance values, corresponding to a darker surface appearance. Among these, the sample subjected to both immersion and vapour exposure presents the lowest reflectance and the most intense dark coloration, suggesting a more advanced oxidation process and a higher concentration of CuO (Figure 7). Table 2 presents the CIE $L^*a^*b^*$ colour coordinate values for samples treated with the three methods. According to OM analysis and reflectance spectra, samples exposed only to vapours exhibited the highest lightness (L*) and yellowness (b*), indicating minimal surface alteration (i.e., oxide formation). Increasing the treatment temperature and combining the methodologies, all the colours shift toward a darker and cooler hue.

μ-Raman spectra collected on the three selected samples Vap, 80C and 10 min (Figure 8), show a peak attributed to tenorite (301, 515 and 606 cm^{-1}), only in the sample treated by immersion and subsequent exposure to the vapours of the 10% NaOH solution heated to 80C (method B). In the spectra collected on the other two samples, only the peaks attributed to cuprite (121, 151, 516 and 606 cm^{-1}) are present [55].

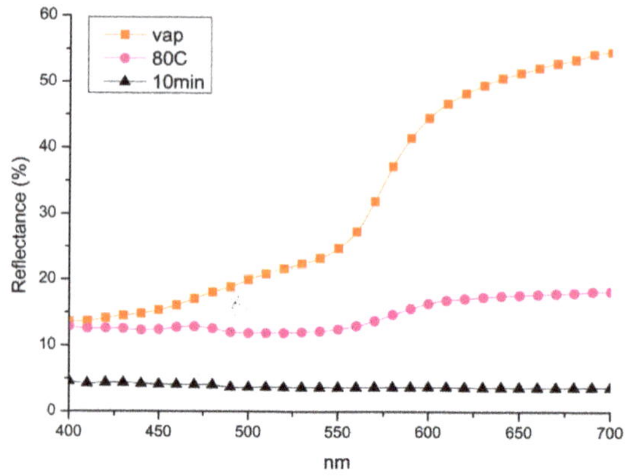

Figure 7. Comparison of reflectance data collected from samples treated only by vapours exposition (Vap), samples treated only by immersion treatment (80C) and samples treated by combining immersion and exposition treatments (10 min).

Table 2. Colour space CIELAB values obtained on samples treated with the three methods.

Method	Sample Code	L*	a*	b*
A: Im	20 °C	48.1	22.1	25.1
	40 °C	50.2	16.0	20.0
	60 °C	49.5	10.9	18.5
	80 °C	38.8	10.1	4.9
B: ImVap	3 min	45.7	6.3	1.3
	5 min	38.4	4.5	−0.2
	10 min	35.4	3.2	−1.4
	20 min	22.5	0.9	−2.7
C: Vap	Vap	52.3	21.2	28.3

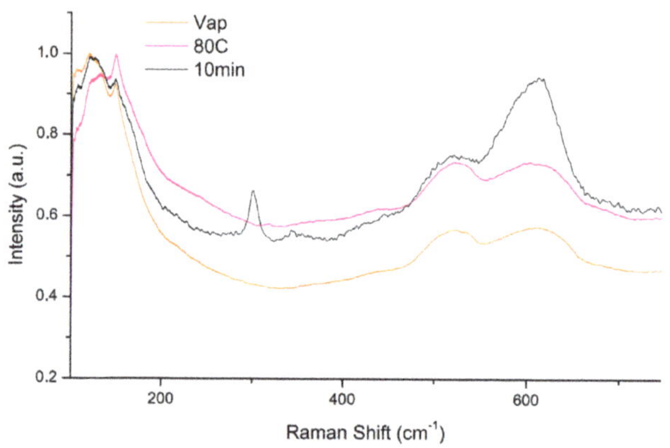

Figure 8. Comparison of micro-Raman spectra collected by sample treated with vapour exposition (Vap), immersion in solution heated at 80 °C (80C) and 1 min immersion and 10 min vapour exposition (10 min) of 10% NaOH solution.

XPS studies were performed to identify the surface chemical composition of the sample, including the oxidation state of copper on the three different treatment methods, and to confirm the formation of CuO on the surface of the sample. The obtained results are listed in Tables 3–5 where it can be seen that all the samples were characterized by the presence of Cu, C, O and, with the exception of the one exposed to vapour of NaOH. The peak-fitting analysis evidence that C 1 s signal included several contributions found at BE = 285.0, 286.4 eV and 288.4 eV assigned to C–C bond, C–O or C = O bond and carboxylic groups, respectively. In the case of the sample immersed in the NaOH solution, the C 1 s signal also showed the contribution of the carbonate, at BE = 289.1 eV [62].

Table 3. Cu-Vap (Vap) XPS sample results.

Name	Peak BE (eV)	FWHM (eV)	Area (P) CPS.eV	Atomic %	Assignment	Auger Parameter α' (eV)
C1s-1	285.0	1.69	17,108.94	61.8	C–C	
C1s-2	286.4	1.69	1448.38	5.2	C–O, C = O	
C1s-3	288.4	1.69	1397.89	5.1	-COOH(R)	
Cu2p3-1	932.5	1.68	15,757.20	4.4	Cu(I)	1849.1
Cu2p3-2	935.2	2.39	8147.80	2.3	$Cu(OH)_2$	
O1s-1	531.9	1.88	8969.87	11.8	hydroxides	
O1s-2	530.5	1.88	5539.83	7.3	Oxides	
O1s-3	533.5	1.88	1610.63	2.1	H_2O	

Table 4. Cu-Im (80C) XPS sample results.

Name	Peak BE (eV)	FWHM (eV)	Area (P) CPS.eV	Atomic %	Assignment	Auger Parameter α' (eV)
C1s-1	285.0	1.57	20,268.04	33.4	C–C	
C1s-2	289.1	1.57	6638.69	10.9	Carbonates	
Cu2p3	932.2	1.56	11,748.05	1.5	Cu(I)	1849.2
Na1s	1071.6	1.84	64,061.97	18.0	Carbonates	
O1s	531.3	1.86	47,907.67	28.8	Carbonates	

Table 5. Cu-ImVap (10 min) XPS sample results.

Name	Peak BE (eV)	FWHM (eV)	Area (P) CPS.eV	Atomic %	Assignment	Auger Parameter α' (eV)
C1s-1	284.8	1.53	15,548.59	39.4	C–C	
C1s-2	286.5	1.53	2145.08	5.4	C–O, C = O	
C1s-3	288.4	1.53	1419.61	3.6	-COOH(R)	
Cu2p3-1	932.3	1.51	52,744.65	10.3	Cu(I)	1849.6
Cu2p3-3	935.6	2.72	9783.49	1.9	$Cu(OH)_2$	
Cu2p3-2	934.2	2.06	23,691.50	4.6	CuO	
Na1s	1071.5	1.77	4462.76	1.9	Carbonates	
O1s-1	530.0	1.56	19,970.00	18.4	Oxides	
O1s-2	531.5	1.56	11,611.16	10.7	Hydroxides	
O1s-3	533.1	1.56	3936.54	3.6	H_2O	

As it concerns the Cu 2p signal, Figure 9 shows that the shape of the signal changes depending on the corrosive treatment. When the sample was immersed in the NaOH solution, a single peak, positioned at BE = 932.2 eV, was found. This value could be ambiguous because it is characteristic of both metallic and Cu^{+1}. Auger parameter can be used to discern this issue. This parameter is calculated by applying the following Formula (4):

$$\alpha' = \mathrm{BE}\ (\mathrm{Cu}\ 2p3/2) + \mathrm{KE}\ (\mathrm{Cu\ LMM}) \tag{4}$$

where BE (Cu 2p3/2) is the BE of the core level Cu 2p3/2, while KE (Cu LMM) is the kinetic energy of the photo-induced Auger transition. The obtained value $\alpha' = 1849.1 \div 1849.6$ eV suggested that Cu was oxidized to +1, rather than being metallic. However, in the other samples, an adjunctive component was found, as a consequence of these treatments, at BE = 934.2 eV and 935.6 eV, was assigned to CuO and $Cu(OH)_2$, respectively, characterized also by their typical satellite peaks located at higher BE (Figure 9) [63].

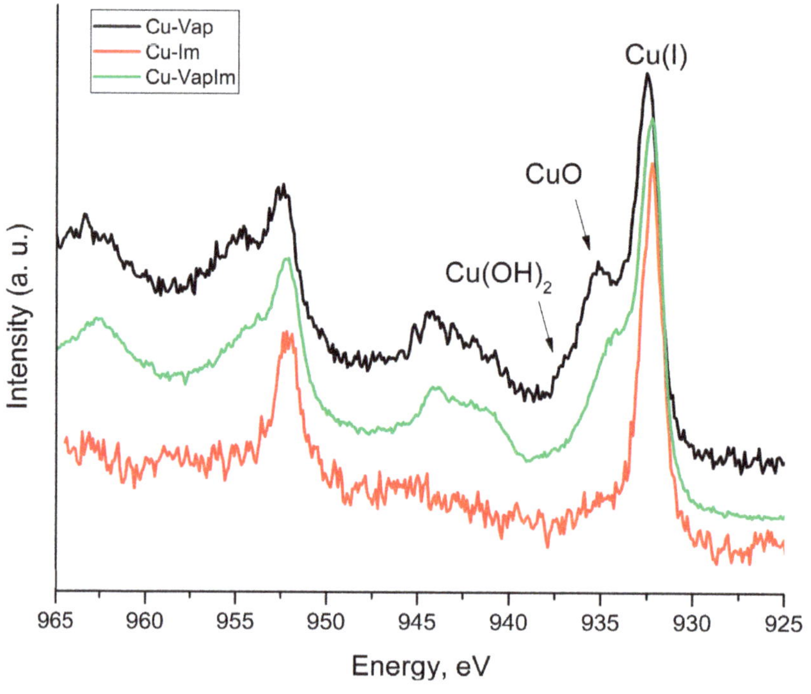

Figure 9. 80C (Cu-Im), Vap (Cu-Vap) and 10 min (Cu-VapIm) XPS spectra.

Scanning electron microscopy (SEM) (Figure 10) and peeling test (Table 6) were conducted exclusively on samples obtained using method B, as this approach yielded the most promising results in terms of oxide layer uniformity, morphology, and adhesion to the substrate.

The images acquired through secondary electron (SEM-SE), with 10 μm resolution, reveal a highly homogeneous and uniform oxide layer on the sample surface (Figure 10c). The presence of scratches, due to the initial polishing of the sample surface with SiC paper, still remains visible.

SEM, which provides higher-resolution topographical information, confirms the formation of a continuous and homogeneous CuO layer, indicating that the inhomogeneities observed by OM are likely due to optical magnification rather than actual surface irregularities.

They are widely used due to their simplicity and speed, especially in characterizing thin films and determining the adhesion strength of materials [50,64]. In Table 6, the peeling test data for three samples prepared using method B are reported. The results indicate a moderate adhesion between the oxide layer and the copper substrate, with Maximum Peel Values (MPV) ranging from 0.0436 N/mm to 0.0523 N/mm. While these values suggest a relatively consistent adhesion across the tested samples, they remain lower than those typically reported for CuO films deposited via methods such as PVD or CVD.

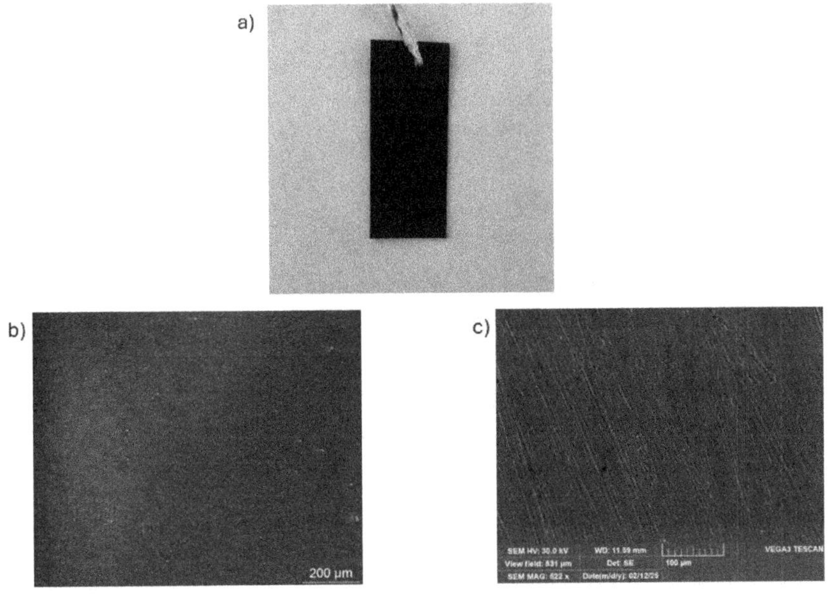

Figure 10. Observation of (**a**) oxide layer on general images, (**b**) optical microscope 5×, and (**c**) SEM magnification 10 µm resolution.

Table 6. Peeling test.

Samples	Maximum Peel Value	Standard Deviation
1	0.0523	0
2	0.0436	$9.43 \times 10^{-0.5}$
3	0.0457	$2.36 \times 10^{-0.5}$
4	0.0578	$2.36 \times 10^{-0.5}$
5	0.0449	$4.71 \times 10^{-0.5}$
6	0.0493	$2.36 \times 10^{-0.5}$
7	0.0513	0
8	0.0445	$4.71 \times 10^{-0.5}$
9	0.0519	$2.36 \times 10^{-0.5}$
10	0.0439	$2.36 \times 10^{-0.5}$

4. Discussion

The aim of this study was to develop a more environmentally friendly approach for the formation of a tenorite (CuO) layer on a metallic copper substrate. The proposed method prioritizes sustainability while also being simple, fast, non-toxic, and cost-effective, ensuring both eco-compatibility and time efficiency.

By leveraging the known effectiveness of thermal treatment and exposure to alkaline environments, three different procedures were tested to enhance cost efficiency, reduce environmental impact and procedural simplicity. The first method involved immersing the copper sample in a heated alkaline NaOH solution at varying temperatures. Temperature increments of 20 °C were applied within the range of boiling water temperatures to assess the role of heat in oxide formation. Optical microscopy and spectrocolorimetric analysis confirmed that immersion in a basic solution, combined with thermal treatment (heating range 20–80 °C), promotes the development of a dark oxide layer on the copper surface. A comparison of data collected at identical immersion time (30 min) revealed that only the highest temperature tested resulted in the formation of a visibly dark oxidation layer. Tenorite, characterized by its bluish-grey-black appearance, was expected in these conditions. μ-Raman analysis did not detect tenorite on the immersion-treated sample, instead indicating the presence of cuprite (Cu_2O). These results suggest that a temperature close to 100 °C is necessary but not solely sufficient to catalyze the formation of tenorite on the copper surface. To further enhance oxide formation, the second experimental approach combined immersion in a heated 10% NaOH solution with subsequent exposure to its vapours for varying durations. This combined treatment yielded promising results. Both reflectance analysis and microscopic observations indicated the formation of a dark-coloured oxide layer consistent with tenorite. μ-Raman analysis, however, revealed the presence of peaks associated with both paramelaconite (Cu_4O_3) and tenorite. Paramelaconite appears to be an intermediate crystalline phase between cuprite and tenorite, as suggested by overlapping spectral features. Specifically, the presence of a peak around 298–300 cm^{-1} suggests a structural relationship between paramelaconite and tenorite, with the former resembling CuO but lacking monoclinic distortion and exhibiting partial oxygen charge removal [25,65]. Previous studies indicate that annealing treatments can induce the transformation of paramelaconite into tenorite by stabilizing its crystalline phase [62]. A third treatment was conducted as a comparative parameter, where the copper sample was exposed only to the vapours of a 10% NaOH solution heated to 80 °C under magnetic stirring. This treatment, performed at relatively low temperatures and for a short duration, proved insufficient for inducing tenorite formation on the copper surface without additional processing. The obtained oxide layer with method B, identified as the most promising procedure to induce the formation of tenorite on the copper surface, was examined by SEM to assess its homogeneity and morphology and by peeling test to assess the adhesion between the formed oxide layer and substrate. The obtained results indicate that this method yields an excellent outcome in terms of uniformity, morphology, and adhesive characteristics of the oxide layer.

Compared to CuO films obtained through wet chemical synthesis or sol–gel methods, which typically exhibit adhesion values below 0.1 N/mm due to weaker interfacial bonding and higher porosity, the present results demonstrate comparable performance.

However, the adhesion of the CuO films produced in this study remains lower than that generally observed for films deposited via physical vapour deposition (PVD) or chemical vapour deposition (CVD) [66,67].

Although adhesion values for PVD- and CVD-grown CuO films are reported to be higher, precise numerical values vary significantly depending on deposition parameters such as temperature, pressure, and substrate preparation. Some studies indicate that optimized deposition conditions can lead to strong film–substrate bonding, particularly when high-energy deposition processes promote interfacial diffusion and chemical bonding.

This suggests that while the proposed low-temperature oxidation method offers an energy-efficient and environmentally friendly alternative, further optimization may be necessary to enhance film adhesion. Strategies such as post-synthesis annealing or surface pretreatment of the copper substrate, both of which have been shown to significantly improve adhesion in oxide coatings, could be explored to strengthen the interfacial bonding between the oxide layer and the substrate. Furthermore, this study demonstrates that modifying alkaline treatment parameters, such as a combination of methodology, extending exposure time to NaOH vapours or incorporating surfactants to enhance surface wetting, can yield a homogeneous and mechanically robust oxide film even at lower temperatures, thereby achieving comparable outcomes to those obtained with more conventional methods.

X-ray photoelectron spectroscopy (XPS) analysis confirmed that the immersion-plus-vapour treatment successfully induced the formation of a thin CuO layer. However, the detection of additional chemical compounds suggests that the process requires refinement. These secondary compounds likely contributed to the spectral shifts observed in μ-Raman analysis and the imperfect black coloration of the sample surface. It is reasonable to infer that their presence interfered with the formation of a well-defined crystalline CuO phase. Further optimization of the treatment parameters, such as exposure time and temperature control, may enhance the purity and crystallinity of the resulting tenorite layer.

5. Conclusions

The three methods selected to induce the formation of tenorite on a metallic copper substrate reflect the most used treatments for the formation of oxides on metal surfaces. The challenge brought by this study was to exploit these treatments by using them to induce the formation of a tenorite layer on the surface of metallic copper using low-cost, environmentally safe, low-temperature, and time-saving experimental conditions. The results demonstrate how this study was successful in achieving this requirement. Even using temperatures below 100 °C in varying the parameters of the basic environment and a combination of immersion treatment and vapour exposure, it is possible to obtain a CuO layer in a brief time and with low costs. Further research on the uniformity and adherence of the produced CuO layer may be carried out.

Author Contributions: Conceptualization, F.I.B. and A.M. (Andrea Macchia); Methodology, T.d.C. and F.F.; Software, A.M. (Alessio Mezzi); Validation, L.S. and M.G.; Formal analysis, F.I.B., T.d.C. and A.M. (Alessio Mezzi); Investigation, F.I.B.; Resources, F.F.; Data curation, F.I.B.; Writing—original draft, F.I.B.; Writing—review & editing, T.d.C., L.S. and A.M. (Andrea Macchia); Visualization, A.M. (Alessio Mezzi); Supervision, M.G. and F.F.; Project administration, F.I.B. All authors have read and agreed to the published version of the manuscript.

Funding: This research received no external funding.

Institutional Review Board Statement: Not applicable.

Informed Consent Statement: Not applicable.

Data Availability Statement: The original contributions presented in the study are included in the article. Further inquiries can be directed to the corresponding author.

Conflicts of Interest: The authors declare no conflicts of interest.

References

1. Buccolieri, G.; Castellano, A.; Serra, A.; Zavarise, G.; Palmiero, E.; Buccolieri, A. Archaeometric analysis of patinas of the outdoor copper statue Sant'Oronzo (Lecce, Italy) preparatory to the restoration. *Microchem. J.* **2020**, *154*, 104538. [CrossRef]
2. Porcaro, M.; Barbaro, B.; Canovaro, C.; Artioli, G.; Lucarelli, C.; Lugli, F.; Depalmas, A.; Brunetti, A. A multi analytical characterization of a small bronze figurine from Gran Carro site (Bolsena Lake, Italy). *J. Archaeol. Sci. Rep.* **2023**, *51*, 104230. [CrossRef]
3. General-Toro, P.; Bordalo, R.; Moreira, P.R.; Vieira, E.; Brunetti, A.; Iannaccone, R.; Bottaini, C. Art Casting in Portuguese 19th Century Industrial Foundries: A Multi-Analytical Study of an Emblematic Copper-Based Alloy Monument. *Heritage* **2021**, *4*, 3050–3064. [CrossRef]
4. Merkl, P.; Long, S.; McInerney, G.M.; Sotiriou, G.A. Antiviral Activity of Silver, Copper Oxide and Zinc Oxide Nanoparticle Coatings against SARS-CoV-2. *Nanomaterials* **2021**, *11*, 1312. [CrossRef]
5. Waris, A.; Din, M.; Ali, A.; Ali, M.; Afridi, S.; Baset, A.; Khan, A.U. A comprehensive review of green synthesis of copper oxide nanoparticles and their diverse biomedical applications. *Inorg. Chem. Commun.* **2021**, *123*, 108369. [CrossRef]
6. Verma, N.; Kumar, N. Synthesis and Biomedical Applications of Copper Oxide Nanoparticles: An Expanding Horizon. *ACS Biomater. Sci. Eng.* **2019**, *5*, 1170–1188. [CrossRef]
7. Steinhauer, S. Gas Sensors Based on Copper Oxide Nanomaterials: A Review. *Chemosensors* **2021**, *9*, 51. [CrossRef]

8. Bao, J.; Huang, T.; Wang, Z.; Yang, H.; Geng, X.; Xu, G.; Samalo, M.; Sakinati, M.; Huo, D.; Hou, C. 3D graphene/copper oxide nano-flowers based acetylcholinesterase biosensor for sensitive detection of organophosphate pesticides. *Sens. Actuators B Chem.* **2019**, *279*, 95–101. [CrossRef]
9. Di Tocco, A.; Robledo, S.; Osuna, Y.; Sandoval-Cortez, J.; Granero, A.; Vettorazzi, N.; Martínez, J.; Segura, E.; Iliná, A.; Zon, M.; et al. Development of an electrochemical biosensor for the determination of triglycerides in serum samples based on a lipase/magnetite-chitosan/copper oxide nanoparticles/multiwalled carbon nanotubes/pectin composite. *Talanta* **2018**, *190*, 30–37. [CrossRef]
10. Majumdar, D.; Ghosh, S. Recent advancements of copper oxide based nanomaterials for supercapacitor applications. *J. Energy Storage* **2021**, *34*, 101995. [CrossRef]
11. Geuli, O.; Mandler, D. The synergistic effect of benzotriazole and trimethylsiloxysilicate towards corrosion protection of printed Cu-based electronics. *Corros. Sci.* **2018**, *143*, 329–336. [CrossRef]
12. Vikraman, D.; Park, H.J.; Kim, S.-I.; Thaiyan, M. Magnetic, structural and optical behavior of cupric oxide layers for solar cells. *J. Alloys Compd.* **2016**, *686*, 616–627. [CrossRef]
13. Song, P.; Shen, S.; Li, C.-C.; Guo, X.-Y.; Wen, Y.; Yang, H.-F. Insight in layer-by-layer assembly of cysteamine and l-cysteine on the copper surface by electrochemistry and Raman spectroscopy. *Appl. Surf. Sci.* **2015**, *328*, 86–94. [CrossRef]
14. Fateh, A.; Aliofkhazraei, M.; Rezvanian, A.R. Review of corrosive environments for copper and its corrosion inhibitors. *Arab. J. Chem.* **2020**, *13*, 481–544. [CrossRef]
15. Jamil, F.; Hassan, F.; Shoeibi, S.; Khiadani, M. Application of advanced energy storage materials in direct solar desalination: A state of art review. *Renew. Sustain. Energy Rev.* **2023**, *186*, 113663. [CrossRef]
16. Ontiveros, M.A.; Quintero, Y.; Llanquilef, A.; Morel, M.; Martínez, L.A.; García, A.G.; Garcia, A. Anti-Biofouling and Desalination Properties of Thin Film Composite Reverse Osmosis Membranes Modified with Copper and Iron Nanoparticles. *Materials* **2019**, *12*, 2081. [CrossRef]
17. RMiller, J.; Adeleye, A.S.; Page, H.M.; Kui, L.; Lenihan, H.S.; Keller, A.A. Nano and traditional copper and zinc antifouling coatings: Metal release and impact on marine sessile invertebrate communities. *J. Nanoparticle Res.* **2020**, *22*, 129. [CrossRef]
18. Farhadi, S.; Siadatnasab, F. Copper(I) sulfide (Cu_2S) nanoparticles from Cu(II) diethyldithiocarbamate: Synthesis, characterization and its application in ultrasound-assisted catalytic degradation of organic dye pollutants. *Mater. Res. Bull.* **2016**, *83*, 345–353. [CrossRef]
19. Oral, A.Y.; Menşur, E.; Aslan, M.H.; Başaran, E. The preparation of copper(II) oxide thin films and the study of their microstructures and optical properties. *Mater. Chem. Phys.* **2004**, *83*, 140–144. [CrossRef]
20. Pierson, J.F.; Thobor-Keck, A.; Billard, A. Cuprite, paramelaconite and tenorite films deposited by reactive magnetron sputtering. *Appl. Surf. Sci.* **2003**, *210*, 359–367. [CrossRef]
21. Gattinoni, C.; Michaelides, A. Atomistic details of oxide surfaces and surface oxidation: The example of copper and its oxides. *Surf. Sci. Rep.* **2015**, *70*, 424–447. [CrossRef]
22. Jayatissa, A.H.; Guo, K.; Jayasuriya, A.C. Fabrication of cuprous and cupric oxide thin films by heat treatment. *Appl. Surf. Sci.* **2009**, *255*, 9474–9479. [CrossRef]
23. Prasanth, D.; Sibin, K.P.; Barshilia, H.C. Optical properties of sputter deposited nanocrystalline CuO thin films. *Thin Solid Film.* **2019**, *673*, 78–85. [CrossRef]
24. Dhineshbabu, N.R.; Rajendran, V.; Nithyavathy, N.; Vetumperumal, R. Study of structural and optical properties of cupric oxide nanoparticles. *Appl. Nanosci.* **2016**, *6*, 933–939. [CrossRef]
25. Debbichi, L.; de Lucas, M.C.M.; Pierson, J.F.; Krüger, P. Vibrational Properties of CuO and Cu_4O_3 from First-Principles Calculations, and Raman and Infrared Spectroscopy. *J. Phys. Chem. C* **2012**, *116*, 10232–10237. [CrossRef]
26. John, A.O. The Inter-play of the Opto-Electrical Properties of Cuprite and Tenorite Semiconductors for Solar Cell Application. *Am. J. Nano Res. Appl.* **2017**, *5*, 81. [CrossRef]
27. Al-Kuhaili, M.F. Characterization of copper oxide thin films deposited by the thermal evaporation of cuprous oxide (Cu_2O). *Vacuum* **2008**, *82*, 623–629. [CrossRef]
28. Zhang, W.; Ding, S.; Yang, Z.; Liu, A.; Qian, Y.; Tang, S.; Yang, S. Growth of novel nanostructured copper oxide (CuO) films on copper foil. *J. Cryst. Growth* **2006**, *291*, 479–484. [CrossRef]
29. Karapetyan, A.; Reymers, A.; Giorgio, S.; Fauquet, C.; Sajti, L.; Nitsche, S.; Nersesyan, M.; Gevorgyan, V.; Marine, W. Cuprous oxide thin films prepared by thermal oxidation of copper layer. Morphological and optical properties. *J. Lumin.* **2015**, *159*, 325–332. [CrossRef]
30. Tang, C.; Sun, F.; Chen, Z.; Chen, D.; Liu, Z. Improved thermal oxidation growth of non-flaking CuO nanorod arrays on Si substrate from Cu film and their nanoscale electrical properties for electronic devices. *Ceram. Int.* **2019**, *45*, 14562–14567. [CrossRef]

31. Castrejón-Sánchez, V.-H.; Solís, A.C.; López, R.; Encarnación-Gomez, C.; Morales, F.M.; Vargas, O.S.; Mastache-Mastache, J.E.; Sánchez, G.V. Thermal oxidation of copper over a broad temperature range: Towards the formation of cupric oxide (CuO). *Mater. Res. Express* **2019**, *6*, 075909. [CrossRef]
32. Zhang, Q.; Zhang, K.; Xu, D.; Yang, G.; Huang, H.; Nie, F.; Liu, C.; Yang, S. CuO nanostructures: Synthesis, characterization, growth mechanisms, fundamental properties, and applications. *Prog. Mater. Sci.* **2014**, *60*, 208–337. [CrossRef]
33. Lillo-Ramiro, J.; Guerrero-Villalba, J.M.; Mota-González, M.d.L.; Tostado, F.S.A.; Gutiérrez-Heredia, G.; Mejía-Silva, I.; Castillo, A.C. Optical and microstructural characteristics of CuO thin films by sol gel process and introducing in non-enzymatic glucose biosensor applications. *Optik* **2021**, *229*, 166238. [CrossRef]
34. Yu, J.; Tian, H.; Huang, M.; Xu, X. Facile Synthesis of Ag NP Films via Evaporation-Induced Self-Assembly and the BA-Sensing Properties. *Foods* **2023**, *12*, 1285. [CrossRef] [PubMed]
35. Reddy, K.R. Green synthesis, morphological and optical studies of CuO nanoparticles. *J. Mol. Struct.* **2017**, *1150*, 553–557. [CrossRef]
36. Sahu, K.; Choudhary, S.; Khan, S.A.; Pandey, A.; Mohapatra, S. Thermal evolution of morphological, structural, optical and photocatalytic properties of CuO thin films. *Nano-Struct. Nano-Objects* **2019**, *17*, 92–102. [CrossRef]
37. Kumar, S.K.; Murugesan, S.; Suresh, S. Preparation and characterization of CuO nanostructures on copper substrate as selective solar absorbers. *Mater. Chem. Phys.* **2014**, *143*, 1209–1214. [CrossRef]
38. Toe, C.Y.; Zheng, Z.; Wu, H.; Scott, J.; Amal, R.; Ng, Y.H. Photocorrosion of Cuprous Oxide in Hydrogen Production: Rationalising Self-Oxidation or Self-Reduction. *Angew. Chem. Int. Ed.* **2018**, *57*, 13613–13617. [CrossRef]
39. ElFaham, M.M.; Mostafa, A.M.; Toghan, A. Facile synthesis of Cu_2O nanoparticles using pulsed laser ablation method for optoelectronic applications. *Colloids Surf. A Physicochem. Eng. Asp.* **2021**, *630*, 127562. [CrossRef]
40. Theerthagiri, J.; Karuppasamy, K.; Lee, S.J.; Shwetharani, R.; Kim, H.-S.; Pasha, S.K.K.; Ashokkumar, M.; Choi, M.Y. Fundamentals and comprehensive insights on pulsed laser synthesis of advanced materials for diverse photo- and electrocatalytic applications. *Light Sci. Appl.* **2022**, *11*, 250. [CrossRef]
41. Begildayeva, T.; Lee, S.J.; Yu, Y.; Park, J.; Kim, T.H.; Theerthagiri, J.; Ahn, A.; Jung, H.J.; Choi, M.Y. Production of copper nanoparticles exhibiting various morphologies via pulsed laser ablation in different solvents and their catalytic activity for reduction of toxic nitroaromatic compounds. *J. Hazard. Mater.* **2021**, *409*, 124412. [CrossRef] [PubMed]
42. Letchumanan, D.; Sok, S.P.M.; Ibrahim, S.; Nagoor, N.H.; Arshad, N.M. Arshad. Plant-Based Biosynthesis of Copper/Copper Oxide Nanoparticles: An Update on Their Applications in Biomedicine, Mechanisms, and Toxicity. *Biomolecules* **2021**, *11*, 564. [CrossRef]
43. Yadav, S.; Jain, A.; Malhotra, P. A review on the sustainable routes for the synthesis and applications of cuprous oxide nanoparticles and their nanocomposites. *Green Chem.* **2019**, *21*, 937–955. [CrossRef]
44. Surendhiran, S.; Gowthambabu, V.; Balamurugan, A.; Sudha, M.; Kumar, V.S.; Suresh, K. Rapid green synthesis of CuO nanoparticles and evaluation of its photocatalytic and electrochemical corrosion inhibition performance. *Mater. Today Proc.* **2021**, *47*, 1011–1016. [CrossRef]
45. Vivas, L.; Chi-Duran, I.; Enríquez, J.; Barraza, N.; Singh, D.P. Ascorbic acid based controlled growth of various Cu and Cu_2O nanostructures. *Mater. Res. Express* **2019**, *6*, 065033. [CrossRef]
46. Neupane, M.P.; Kim, Y.K.; Park, I.S.; Kim, K.A.; Lee, M.H.; Bae, T.S. Temperature driven morphological changes of hydrothermally prepared copper oxide nanoparticles. *Surf. Interface Anal.* **2009**, *41*, 259–263. [CrossRef]
47. Mikić, D.; Ćurković, H.O. Protection of Patinated Bronze with Long-Chain Phosphonic Acid/Organic Coating Combined System. *Materials* **2023**, *16*, 1660. [CrossRef]
48. Fan, J.; Tang, D.; Wang, D. Spontaneous growth of CuO nanoflakes and microflowers on copper in alkaline solutions. *J. Alloys Compd.* **2017**, *704*, 624–630. [CrossRef]
49. Aromaa, J.; Kekkonen, M.; Mousapour, M.; Jokilaakso, A.; Lundström, M. The Oxidation of Copper in Air at Temperatures up to 100 °C. *Corros. Mater. Degrad.* **2021**, *2*, 625–640. [CrossRef]
50. Shan, J.; Shan, C.-H.; Huang, C.; Wu, Y.-P.; Lia, Y.-K.; Chen, W.-J. Study of Microstructure and Mechanical Properties of Electrodeposited Cu on Silicon Heterojunction Solar Cells. *Metals* **2023**, *13*, 1223. [CrossRef]
51. Bartlett, M.D.; Case, S.W.; Kinloch, A.J.; Dillard, D.A. Peel tests for quantifying adhesion and toughness: A review. *Prog. Mater. Sci.* **2023**, *137*, 101086. [CrossRef]
52. Arslan, B.; Demirci, G.; Erdoğan, M.; Karakaya, İ. Formation and characterization of infrared absorbing copper oxide surfaces. *Appl. Surf. Sci.* **2017**, *402*, 218–224. [CrossRef]
53. Ghaedi, A.M.; Karamipour, S.; Vafaei, A.; Baneshi, M.M.; Kiarostami, V. Optimization and modeling of simultaneous ultrasound-assisted adsorption of ternary dyes using copper oxide nanoparticles immobilized on activated carbon using response surface methodology and artificial neural network. *Ultrason. Sonochem.* **2019**, *51*, 264–280. [CrossRef] [PubMed]
54. Ning, W.; Xia, C.; Xiaolan, C.; Yanjun, X.; Lin, G. Porous cuprite films: Facile solution deposition and their application for nitrite sensing. *Analyst* **2010**, *135*, 2106. [CrossRef]

55. De Caro, T.; Angelini, E.; Sebar, L.E. Number 1, 234-240 ACTA IMEKO. 2021. Available online: www.imeko.org (accessed on 22 February 2025).
56. Ospitali, F.; Chiavari, C.; Martini, C.; Bernardi, E.; Passarini, F.; Robbiola, L. The characterization of Sn-based corrosion products in ancient bronzes: A Raman approach. *J. Raman Spectrosc.* **2012**, *43*, 1596–1603. [CrossRef]
57. Htay, M.T.; Okamura, M.; Yoshizawa, R.; Hashimoto, Y.; Ito, K. Synthesis of a cuprite thin film by oxidation of a Cu metal precursor utilizing ultrasonically generated water vapor. *Thin Solid Film.* **2014**, *556*, 211–215. [CrossRef]
58. Montoya, N.; Montagna, E.; Lee, Y.; Doménech-Carbó, M.T.; Doménech-Carbó, A. Raman spectroscopy characterization of 10-cash productions from the late Chinese emperors to the Republic. *J. Raman Spectrosc.* **2017**, *48*, 1337–1345. [CrossRef]
59. Purusottam-Reddy, B.; Sivajee-Ganesh, K.; Jayanth-Babu, K.; Hussain, O.M.; Julien, C.M. Microstructure and supercapacitive properties of rf-sputtered copper oxide thin films: Influence of O2/Ar ratio. *Ionics* **2015**, *21*, 2319–2328. [CrossRef]
60. Moumen, A.; Hartiti, B.; Thevenin, P.; Siadat, M. Synthesis and characterization of CuO thin films grown by chemical spray pyrolysis. *Opt. Quantum Electron.* **2017**, *49*, 70. [CrossRef]
61. Murali, D.S.; Aryasomayajula, S. Thermal conversion of Cu4O3 into CuO and Cu_2O and the electrical properties of magnetron sputtered Cu4O3 thin films. *Appl. Phys. A* **2018**, *124*, 279. [CrossRef]
62. Jagadish, K.A.; Kekuda, D. Thermal annealing effect on phase evolution, physical properties of DC sputtered copper oxide thin films and transport behavior of ITO/CuO/Al Schottky diodes. *Appl. Phys. A* **2024**, *130*, 315. [CrossRef]
63. Li, J.; Li, J.; Jiang, Q.; Jiang, Q.; Xing, X.; Xing, X.; Sun, C.; Sun, C.; Wang, Y.; Wang, Y.; et al. Surface Structure Reformulation from CuO to $Cu/Cu(OH)_2$ for Highly Efficient Nitrate Reduction to Ammonia. *Adv. Sci.* **2024**, *11*, 2404194. [CrossRef]
64. Toader, G.; Diacon, A.; Rusen, E.; Mangalagiu, I.I.; Alexandru, M.; Zorilă, F.L.; Mocanu, A.; Boldeiu, A.; Gavrilă, A.M.; Trică, B.; et al. Peelable Alginate Films Reinforced by Carbon Nanofibers Decorated with Antimicrobial Nanoparticles for Immediate Biological Decontamination of Surfaces. *Nanomaterials* **2023**, *13*, 2775. [CrossRef] [PubMed]
65. Djurek, D.; Prester, M.; Drobac, D.; Ivanda, M.; Vojta, D. Magnetic properties of nanoscaled paramelaconite $Cu_4O_3{-}x$ (x = 0.0 and 0.5). *J. Magn. Magn. Mater.* **2015**, *373*, 183–187. [CrossRef]
66. Dolai, S.; Dey, R.; Das, S.; Hussain, S.; Bhar, R.; Pal, A.K. Cupric oxide (CuO) thin films prepared by reactive d.c. magnetron sputtering technique for photovoltaic application. *J. Alloys Compd.* **2017**, *724*, 456–464. [CrossRef]
67. Nitta, R.; Kubota, Y.; Kishi, T.; Matsushita, N. Fabrication of nanostructured CuO thin films with controllable optical band gaps using a mist spin spray technique at 90 °C. *Thin Solid Film.* **2022**, *762*, 139555. [CrossRef]

materials

MDPI

Article

Experimental Analysis of Gravitational Vortex Turbine Made from Natural Fibers

María Varga [1], Laura Velásquez [1,*], Ainhoa Rubio-Clemente [1,2], Bladimir Ramón Valencia [3] and Edwin Chica [1]

1 Grupo de Energía Alternativa, Facultad de Ingeniería, Universidad de Antioquia, Calle 70 No. 52-21, Medellín 050010, Colombia; mjose.vargas@udea.edu.co (M.V.); ainhoa.rubioc@udea.edu.co (A.R.-C.); edwin.chica@udea.edu.co (E.C.)

2 Escuela Ambiental, Facultad de Ingeniería, Universidad de Antioquia, Calle 70 No. 52-21, Medellín 050010, Colombia

3 Grupo de Investigación de Ingeniería Mecánica de la Universidad de Pamplona-GIMUP, Facultad de Ingeniería y Arquitectur, Universidad de Pamplona, Km 1 Vía Bucaramanga Ciudad Universitaria, Pamplona 543050, Colombia; hbladimir@unipamplona.edu.co

* Correspondence: lisabel.velasquez@udea.edu.co

Abstract: The use of natural fibers in hydro turbine rotors promotes sustainability by offering biodegradable, renewable materials with a lower carbon footprint. This study compares the hydrodynamic performance of two rotors in a gravitational vortex turbine: Rotor 1, 3D-printed with polylactic acid (PLA), and Rotor 2, made from fique fiber and epoxy resin using manual molding. To compare the rotors, experimental tests were conducted on a laboratory-scale setup, where the behavior of both rotors was evaluated under different flow regimes. Rotor 1 achieved 61.01% efficiency at an angular velocity (ω) 160 RPM, while Rotor 2 reached only 19.03% at ω of 165 RPM. The lower performance of Rotor 2 was due to dynamic imbalances and mechanical vibrations, leading to energy losses. These challenges highlight the limitations of manual molding in achieving precise rotor geometry and balance. To improve natural fiber rotor viability, optimizing manufacturing techniques is crucial to enhance dynamic balance and minimize vibrations. Advancements in fabrication could bridge the performance gap between natural and synthetic materials, making bio-based rotors more competitive. This study emphasizes the potential of natural fibers in sustainable energy and the need to refine production methods to maximize efficiency and reliability. Addressing these challenges will help integrate eco-friendly rotors into hydro turbine technologies.

Keywords: gravitational vortex turbine; renewable energy; natural fiber; PLA; 3D printing; energy efficiency

1. Introduction

The rapid increase in global energy consumption has led to a strong dependence on fossil fuels, further exacerbating environmental issues such as climate change, air pollution, and ecosystem degradation [1,2]. As energy demand continues to rise, the urgency to diversify the energy matrix and invest in renewable sources becomes even more critical. However, many traditional solutions, such as large hydroelectric plants, require expensive infrastructure and can have negative environmental impacts, including the disruption of aquatic ecosystems and the displacement of communities [3,4]. In this scenario, it is essential to develop more accessible and sustainable technologies that allow for the efficient use of natural resources while minimizing ecological impact. Additionally, addressing these challenges requires innovative policies, improved energy storage solutions, and decentralized renewable energy systems to ensure a more resilient and environmentally friendly energy future.

Gravitational vortex turbines represent an alternative in the field of hydrokinetic energy generation, as they enable the transformation of water energy into electricity without the need for dams or reservoirs [5,6]. These turbines operate by creating a vortex within a circulation and discharge chamber, harnessing the kinetic energy of the water flow to drive a rotor connected to a generator. Their compact design and ability to function with low water flow rates make them an ideal solution for rural communities or regions with limited water resources. The power generation system consists of an inlet channel, a circulation chamber, a discharge chamber, the rotor, the electric generator, and the regulation and control system [7].

Despite advancements in the design and optimization of gravitational vortex turbines, most current studies have focused on improving their hydrodynamic and mechanical performance through both numerical and experimental research [8–11]. However, these developments have primarily relied on conventional materials such as metals and synthetic polymers, due to their favorable mechanical properties and compatibility with existing manufacturing processes. In laboratory-scale models, the use of 3D printing filaments and recycled polymers has been reported, which not only facilitates the fabrication of complex geometries with high precision but also promotes a more sustainable approach. In practical applications, steel and concrete structures have been employed, where steel allows for faster construction and greater recyclability, while reinforced concrete, though more cost-effective, requires extended curing times [5,12,13].

To date, only one study has been identified that combines fique natural fibers with gravitational vortex turbines. The authors of [14] evaluated a biocomposite material made from unsaturated polyester resin reinforced with fique fiber as a sustainable and cost-effective alternative for constructing rotor blades of a gravitational vortex turbine. Using the manual impregnation method, laminates with three layers of fique fiber were produced, yielding a lightweight material with promising mechanical properties. Compared to conventional materials like aluminum and stainless steel, the biocomposite demonstrated lower density and environmental advantages. Mechanical tests (tensile and bending) confirmed its suitability, and a functional rotor prototype was successfully manufactured, highlighting the biocomposite's potential for application in hydrokinetic systems.

The use of materials such as biopolymers or natural fiber-reinforced composites presents a significant opportunity to reduce environmental impact and enhance the functionality of these systems. Integrating these materials into the development of gravitational vortex turbines could optimize their energy efficiency throughout their entire lifecycle, from production to end-of-life disposal. This opens new avenues for research in the field of sustainable materials, enabling the creation of more efficient and environmentally responsible technologies for renewable energy generation. Furthermore, the implementation of natural materials in the fabrication of turbine rotors could minimize their environmental footprint and promote the use of local resources, thereby contributing to a more sustainable and accessible energy transition. However, their implementation faces critical challenges, including inferior mechanical properties, susceptibility to moisture, complex manufacturing processes, and difficulties in achieving optimal dynamic balance. Additionally, the variability in the properties of natural materials and their potentially higher costs may limit their scalability and competitiveness. Overcoming these obstacles requires advances in material treatment, optimization of manufacturing processes, and the development of strategies for sustainable end-of-life management.

The manufacturing of rotors for gravitational vortex turbines using biopolymers and natural fibers can be approached through various processes, each with its own advantages and challenges. Three-dimensional printing stands out for its high precision and ability to create complex geometries, reducing material waste [15,16]; however, it may be limited by the mechanical strength of biopolymers and the costs associated with large-scale production. This process can be enhanced by incorporating natural fibers, such as fique, hemp, or flax, into biopolymer matrices like polylactic acid (PLA), enabling the creation of sustainable and customized components. Despite its potential, challenges such as the lower mechanical strength of natural fibers, the need to adjust printing parameters to avoid thermal degradation, and the higher costs of producing composite filaments must be addressed. On the other hand, compression molding is a scalable and efficient option for natural fiber-reinforced composites, although it requires high-precision molds that can increase initial costs and pose challenges in achieving optimal dynamic balance. Injection molding offers high repeatability and precision, making it ideal for uniform geometries, but it is restricted to materials with suitable flow properties and involves high equipment and mold costs. The manual molding process provides flexibility in using different fibers and resins, being low-cost for prototypes, but its precision and consistency are lower, which can result in dynamic imbalances. A hybrid approach, combining techniques such as 3D printing for critical parts and compression molding for final assembly, leverages the strengths of multiple processes, though it increases complexity and costs. Finally, post-manufacturing treatments, such as precision machining or the application of coatings, improve rotor quality and durability but add time and production costs. Overall, the choice of process will depend on factors such as production volume, required material properties, and available resources, making it crucial to explore innovative approaches that maximize the performance and sustainability of these materials. Integrating natural fibers into advanced manufacturing processes like 3D printing not only enhances sustainability but also opens new possibilities for creating efficient and eco-friendly energy systems.

This study presents the experimental characterization of a rotor for a gravitational vortex turbine fabricated using natural fibers, aiming to evaluate its performance and viability as an eco-friendly alternative for electric power generation. The importance and novelty of this article lie in the use of natural fibers as a manufacturing material, as the rotor design has already been validated previously. Key parameters such as system efficiency, structural resistance of the materials, and flow behavior are analyzed to assess its potential for applications in resource-limited areas. Additionally, the performance of the natural fiber rotor is compared to that of a 3D-printed rotor made from PLA, a biodegradable thermoplastic, providing insights into the advantages and limitations of each material in terms of energy generation and sustainability.

2. Materials and Methods

2.1. Importance of Natural Fibers in the Manufacturing of Turbine Rotors

The use of natural fibers in the manufacturing of hydro turbine rotors represents a significant step toward more sustainable energy solutions. Unlike synthetic materials such as fiberglass or carbon fiber, natural fibers such as flax, hemp, or jute are biodegradable, renewable, and have a lower carbon footprint. Their production requires less energy and generates fewer emissions, contributing to a more environmentally friendly manufacturing process. In addition, natural fiber composites offer a favorable strength-to-weight ratio, making them a competitive alternative for the reinforcement of modern composite materials [17]. Compared to synthetic fibers and fiberglass, they overcome various limitations by offering improved mechanical, thermal, and physical properties, which can enhance turbine efficiency while reducing mechanical stress on other components [18].

Furthermore, by minimizing reliance on petroleum-based materials, the adoption of natural fibers helps reduce environmental pollution and long-term waste accumulation. From a socio-economic perspective, incorporating locally sourced natural fibers supports rural economies and promotes sustainable agricultural practices, creating a circular economy model. Furthermore, these materials typically require fewer harmful chemicals during processing, lowering health risks for workers and reducing pollution in aquatic ecosystems. By integrating natural fibers into hydro turbine rotors, it is possible to achieve an optimal balance between performance, durability, and environmental responsibility, paving the way for more sustainable hydrokinetic energy systems.

A clear example of their implementation is the development of wind turbine blades made from composites based on sisal fibers, which not only reduces the carbon footprint of their production but also facilitates their recycling at the end of their useful life. Similarly, the automotive industry has begun to integrate natural fibers into the manufacture of structural components and interior panels, achieving lighter and more energy-efficient vehicles [19].

According to Amzil et al., the integration of hemp fibers into the manufacture of wind turbine blades has shown significant economic and environmental advantages, while maintaining established design requirements. These findings highlight their potential as a viable alternative to traditional glass fiber-reinforced composite materials. The studies indicate that this modification allowed for a reduction in both density and cost compared to a reference model composed exclusively of glass fiber and epoxy resin [20].

The study conducted by Rua et al. investigated the use of fique fibers as reinforcement in a composite material with epoxy resin, allowing for exploring new limits in the dynamic behavior of these materials. The results indicated that the composite material had lower stiffness compared to pure epoxy resin; but, at the same time, it showed better performance under stress conditions. In addition, its fracture was not brittle, evidencing greater flexibility and a better response to deformation under dynamic loading. Its impact resistance under static loading conditions was also highlighted [21].

In Colombia, the fique industry is widely recognized, with an annual production close to 30,000 tons, making this fiber an attractive alternative for the development of new composite materials. The viability and application of these composites depend largely on economic factors, where weight reduction, cost reduction, and the availability of sustainable raw materials play a key role. Based on these findings, fique fiber was selected as a reinforcing material in this research.

For the matrix, a polyester resin from the Poliescol brand was used. It is thermosetting with a density of 1.15 g/mL, has gel time of 6 to 9 min, has a viscous pink appearance, and the recommended hardener is methyl ethyl ketone peroxide (MEK peroxide) mixed with 1% by weight. The polyester resin has an elasticity modulus of 3.6 GPa, a tensile strength of 45 Mpa, and an elongation of 2.4%. The fique fiber, also called Furcraea Andina, is an endemic species native to the South American Andes, which has an asymmetric 0/90° braiding configuration, with double fiber in the warp direction. The fique has a modulus of 8.2 to 9.1 GPa, a tensile strength of 132.4 Mpa, and an elongation of 9.8%. This fiber was analyzed using various characterization techniques to evaluate its morphology, chemical structure, and thermal properties. Morphological analysis by scanning electron microscopy (SEM) reveals that fique fibers have a compact surface, without fibrillation, and are separated by approximately 10 µm. They present roughness, typical of raw natural fibers, with some impurities and waxy material. In its cross-section, cell walls formed by microfibrils and a central lumen are observed, with a wall thickness between 1 and 10 µm and a lumen diameter of 0.1 to 20 µm. Fique reflects an irregular and fibrous structure characteristic of natural fibers. In the analysis of fique fiber by Fourier Transform Infrared

Spectroscopy (FTIR), the presence of hydroxyl groups, C-H in cellulose and hemicellulose, was observed. In addition, carbonyl groups (C=O) and lignin (C=C) confirmed the lignocellulosic composition of the fiber [22].

The characterization of the fique fiber composite with a polyester resin matrix focused on the interaction between the fiber and the polymeric matrix, evaluating its mechanical, thermal, and interface behavior. The morphological analysis of the fiber–matrix interface by SEM observed the interface between the fiber and the polyester resin. The fiber, with its rough surface and compact structure, shows a good interaction with the resin, which translates into greater compatibility and anchoring between both components. The resin seems to effectively impregnate the fique fibers, taking advantage of the surface roughness for better adhesion. In the mechanical properties, the results indicate that the composite presents an improvement in mechanical resistance compared to natural fibers without a matrix, thanks to the reinforcement provided by the polyester resin. The thermogravimetric analysis of the composites showed that the presence of polyester resin affects the thermal behavior of the material. The resin contributes to the thermal stabilization of the composite, although it may slightly alter the degradation temperature range compared to pure fique fibers. This type of composite, which combines natural fibers with polymeric matrices, presents a promising path towards sustainability in engineering materials, especially in industries seeking green solutions without compromising mechanical performance.

2.2. Geometry of the Gravitational Vortex Turbine Rotor

Gravitational vortex turbines are devices designed for hydrokinetic energy conversion through the generation of a stable vortex within a basin or vortex chamber. Their operating principle is based on inducing a rotational flow of water inside a cylindrical or conical tank, allowing the transformation of the fluid's kinetic energy into mechanical energy via a rotor [23]. Water enters the vortex chamber through a tangential inlet, generating rotational movement within the fluid. The basin's geometry influences the formation of a downward helical flow, characteristic of the gravitational vortex. A rotor, positioned at the center of this vortex, is driven by the motion of the water. The captured mechanical energy is transferred to an electric generator, enabling its conversion into usable power. These turbines exhibit high efficiency in low-flow and low-head environments, making them suitable for renewable energy generation in rivers and water bodies with limited hydraulic conditions [10].

The efficiency of gravitational vortex turbines is typically calculated by comparing the mechanical power output of the turbine (P_{out}) to the hydraulic power available in the flowing water (P). The hydraulic power is determined using the equation $P = \rho g Q H$, where ρ is the density of water, g is the acceleration due to gravity, Q is the volumetric flow rate, and H is the net head. The mechanical power output is usually obtained from measurements of torque (T) and rotational speed (ω). The overall efficiency (η) is then calculated using Equation (1) [8,10]:

$$\eta = \frac{P_{out}}{P} = \frac{T\omega}{\rho g Q H} \quad (1)$$

Figure 1 presents the turbine under study, designed for efficient energy conversion from the gravitational vortex. The rotor, optimized by Velásquez et al. [8], features an advanced geometry that enhances performance by maximizing momentum transfer and minimizing energy losses. The curved blade configuration contributes to vortex stability, reinforcing its structure and optimizing turbine efficiency. Additionally, the blades are strategically inclined toward the rotor's center, promoting vortex concentration in the central region, which enhances fluid interaction and improves energy conversion.

Figure 1. Rotor design.

The rotor design incorporates specific inlet and outlet angles that govern fluid interaction with the blades, directly influencing velocity and momentum transfer—see Figure 2. The blade inlet angle (α) is set at 16°, allowing water to enter smoothly while maintaining vortex stability, while the blade outlet angle (ϕ) is 90°, ensuring an efficient redirection of flow. Similarly, the absolute velocity inlet angle (θ) is 40°, and the absolute velocity outlet angle (β) is 43.8°, optimizing hydrodynamic performance by minimizing energy dissipation and improving momentum transfer to the rotor. The relative velocity twist angle (λ) of 68.8° further enhances fluid guidance along the blade surface, contributing to a stable and efficient flow pattern.

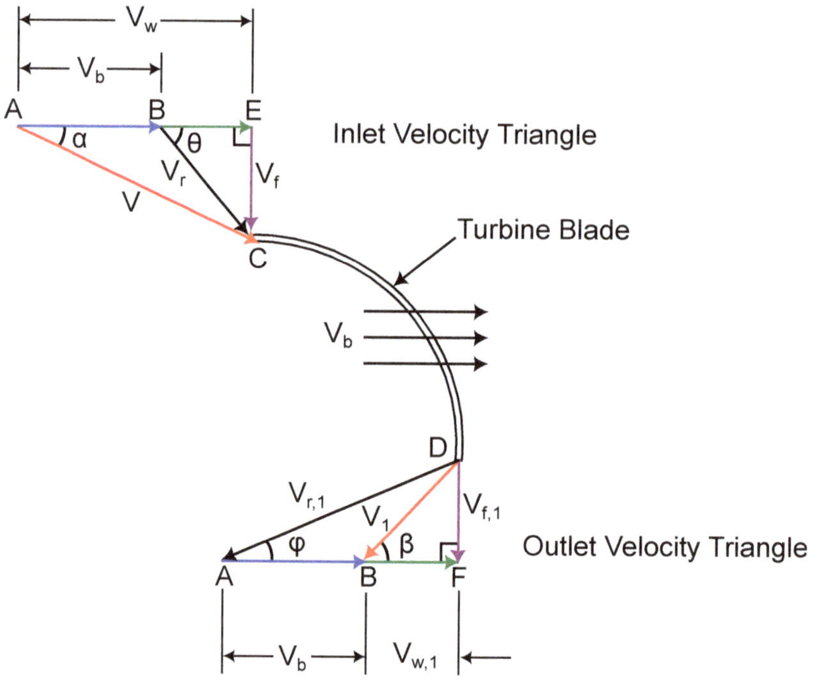

Figure 2. Velocity triangle.

Additionally, the rotor has a blade height (L) of 200 mm, a diameter of 292 mm, and consists of six blades, each with a thickness of 3 mm. These blades are joined to a central shaft with a diameter of 19.5 mm. Figure 3 and Table 1 present the main dimensions of the rotor under study. These geometric parameters influence vortex formation and the effective interaction between the fluid and the rotor, ensuring a balance between sustaining a strong vortex structure and maximizing energy extraction.

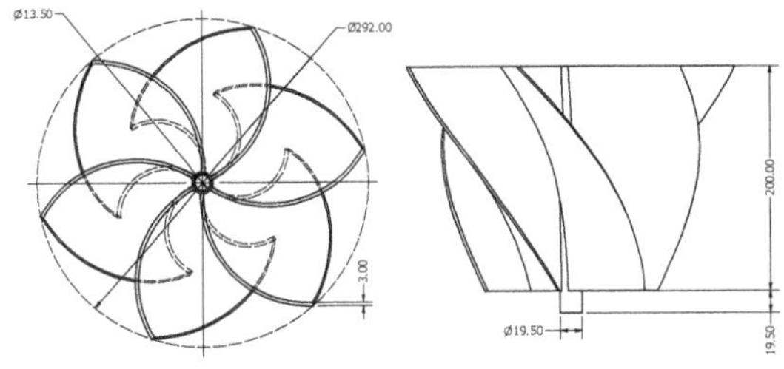

Figure 3. Dimension of the rotor.

Table 1. Main geometric parameters of the rotor.

Symbol	Variable	Value
α	Blade inlet angle [Deg]	16.00
ϕ	Blade outlet angle [Deg]	90.00
θ	Water inlet angle [Deg]	40.00
β	Water outlet angle [Deg]	43.80
λ	Twist angle [Deg]	68.80
L	Blade height [mm]	200.00
D	Rotor diameter [mm]	292.00
n	Number of blades	6.00
t	Blade thickness [mm]	3.00

2.3. Manufacturing Process of Turbine Rotors Using 3D Printing

The manufacturing process of turbine rotors using 3D printing begins with the creation of a digital 3D model using computer-aided design (CAD) software Autodesk Inventor professional 2025, where the rotor's geometry is optimized to meet performance criteria such as flow efficiency, strength, and durability. Once the design is finalized, an appropriate material is chosen for 3D printing, which may include thermoplastics, metal alloys, or composite filaments depending on the desired properties of the rotor, such as flexibility, strength, or resistance to corrosion. Using additive manufacturing, the rotor is then fabricated layer by layer, with the printer extruding or melting the chosen material through a nozzle to build the rotor's structure from the bottom up. This method offers precise control over the geometry of the rotor, allowing for complex and detailed shapes that would be challenging to achieve through traditional manufacturing techniques [24,25].

After printing, the rotor typically undergoes post-processing to enhance its surface finish and mechanical properties, including the removal of support structures, sanding, polishing, and curing the material for maximum strength. In the case of metal 3D printing, additional heat treatment may be used. Once the rotor is completed, it is assembled with other turbine components and tested under controlled conditions to assess its performance, mechanical strength, and efficiency. Based on the results, adjustments can be made to the design or manufacturing process to optimize the rotor's performance, durability, and functionality. This process of 3D printing turbine rotors offers several advantages, such as reduced material waste, faster production times, and the ability to create highly customized and efficient designs for hydrokinetic and wind energy systems. The printing strategy of the rotor was first a 3D modeling in CAD software (Autodesk Inventor software 2025e) and then exported in STL format. PrusaSlicer was used to generate the G-code, optimizing printing parameters. The printing method was fused deposition molding (FDM), which is the standard technology in Ender 3 printers (Creality, Shenzhen, China). This method allows the rotor to be manufactured layer by layer, achieving a functional structure for initial tests. The rotor was printed using a PEGASUS Lite 3D printer (Creality, Shenzhen, China), which features a build volume of 300 × 300 × 300 mm. PLA was used for printing, with printing parameters with layer height 0.2 mm (for a balance between precision and printing time), 100% infill, printing speed 60 mm/s, extruder temperature 210 °C, bed temperature 60 °C (to improve adhesion), and support on the base. Figure 2 presents the printing parameters used in the software, as well as the total printing time and the amount of filament consumed. The total printing time was 14 h and 52 min, with 176.16 m of filament consumed, equivalent to 538.11 g. The printing parameters are presented in Figure 4.

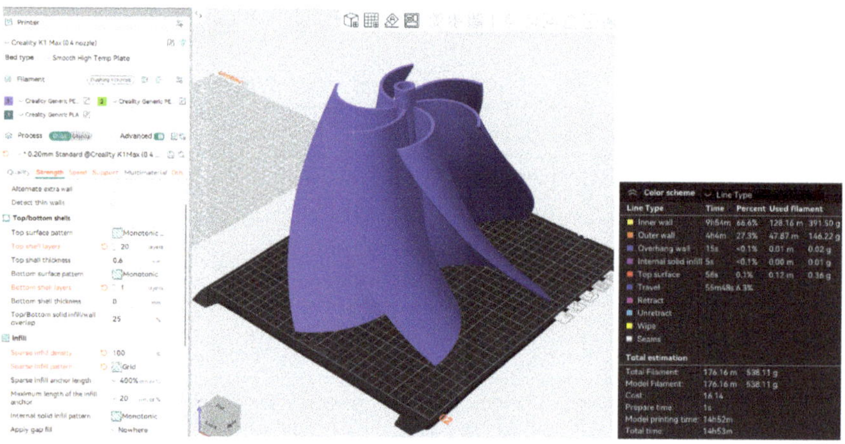

Figure 4. Printing parameters.

2.4. *Manufacturing Process of Natural Fiber Turbine Rotor*

The manufacturing process for turbine rotors using natural fibers involves several key steps. First, the appropriate fique fiber material is selected and prepared. This preparation includes cleaning the selected fiber to remove impurities, ensuring that it will be suitable for integration into the composite material. Next, the natural fibers are combined with resins, either biocompatible or synthetic, to create a fiber composite material. The fibers are impregnated with resin using manual processes. The choice of resin depends on the performance needs of the rotor, with options including thermoplastic or thermosetting resins, which contribute to the mechanical strength of the material. Once the composite is prepared, it is molded into the desired blade shape using vacuum infusion, die casting, or hand lamination techniques. During molding, the resin hardens, providing structural integrity and strength to the composite. Once the blade is molded, it undergoes a curing process to ensure that the resin hardens completely, solidifying the fibers and allowing the rotor to meet the required mechanical properties. Once cured, the blades undergo finishing processes, which involve smoothing the surfaces and removing imperfections. The blades are then assembled onto the central shaft of the rotor, ensuring proper alignment and secure attachment. Finally, the rotor is tested under simulated operating conditions to assess its energy efficiency. The rotor is also checked to ensure it meets dynamic balancing and safety standards. During this stage, any necessary adjustments are made to optimize the rotor's performance. The use of natural fibers in hydroelectric turbine rotors not only improves the sustainability of the energy system, but also offers better performance and durability, contributing to the development of more environmentally friendly hydrokinetic energy solutions. The process began with the design of the blades using Autodesk Inventor software, where geometric precision was pursued to meet hydraulic efficiency and structural stability requirements, such as blade thickness, among others. The rotor was modeled with six symmetrical blades with specific curvatures that maximize the energy conversion of the gravitational flow. The CAD model incorporated key geometric details, such as the aerodynamic profile and the length of the blades, which are decisive for the performance of the runner. Once the optimal design was finalized, only one blade was left and exported in high-resolution STL (Stereolithography) format for use in subsequent stages. The base model was then 3D printed using PLA filament. Printing was accomplished layer by layer, adjusting parameters such as nozzle temperature of 210 °C and layer speed of 30 mm/s, to obtain an accurate reproduction of the original design. The printed model, as seen in Figure 5, served as a template to replicate the geometry of the blades, taking advantage of the properties of PLA to capture fine details and ensure accuracy in the following stages.

Figure 5. Model of the blade of the gravitational vortex turbine.

The next step was to manufacture the mold from the printed model. To achieve this, the piece was placed on a support structure and a mixture of polyester resin and fiberglass was applied. A fiberglass cloth of suitable size was arranged to cover the model and was subsequently impregnated with resin using a brush, ensuring uniform coverage. Once the material had solidified, the base model was removed, obtaining a precise cavity that reproduced the shape of the blade. For the manufacture of the blades, a combination of polyester resin and two layers of fique fibers as reinforcement supplied were used. The fibers were previously cut and prepared to fit the mold. Each layer was manually impregnated with polyester resin, guaranteeing complete saturation and the elimination of air bubbles. The layers were arranged following a weave pattern to optimize mechanical strength and structural uniformity. After lamination, the piece was left to cure at room temperature according to the resin manufacturer's specifications. After curing, the blades were carefully removed to avoid damage to the edges and subjected to a manual polishing process with fine-grained abrasives as shown in Figure 6; this improved the surface quality and aerodynamics of the parts. Additionally, individual weight control was performed to ensure uniformity, a crucial aspect to prevent imbalances and vibrations during operation. In cases where weight variations were detected, adjustments were made by fine sanding until a homogeneous distribution of mass was achieved in all the parts.

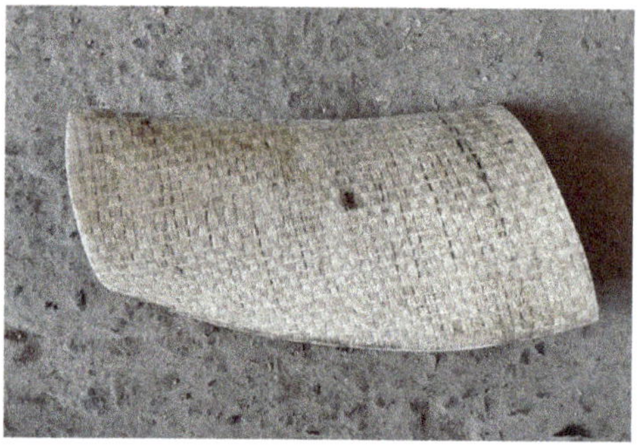

Figure 6. Blade manufactured with good finishes.

The central shaft was manufactured exclusively from polyester resin, molded into a cylindrical structure that provided the necessary rigidity and durability. For the assembly of the impeller, a special support was designed using 3D printing in PLA as seen in Figure 7, which allowed the blades to be precisely aligned around the shaft and maintain their position during the joining process.

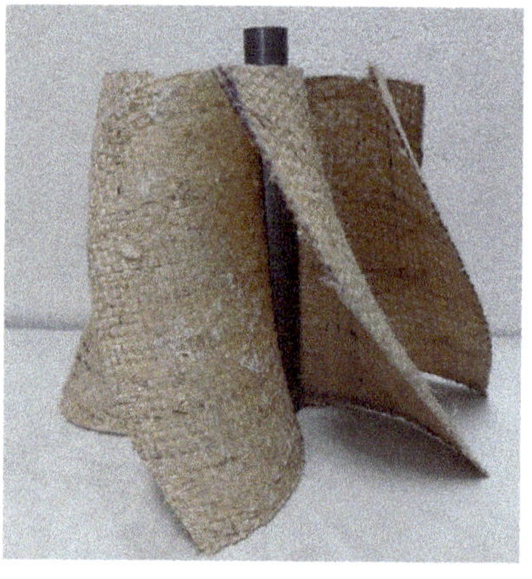

Figure 7. Fitting the blades to the shaft mold.

Finally, the blades were assembled and fixed to the shaft made of polyester resin, ensuring a firm and stable connection. The assembly was then left to cure completely to ensure optimal adhesion between the components and improve structural strength. The result was a solid and durable connection, as seen in Figure 8.

Figure 8. Final gravitational vortex turbine rotor.

The use of natural fibers such as fique in the construction of turbine blades presents several limitations. One of the main challenges is the natural variability of the fibers, as their properties can vary significantly depending on their origin and the treatments they undergo, which affects the reproducibility of the rotor. Additionally, while fique fibers have good tensile strength, they exhibit limited resistance to fatigue and erosion. In conditions of turbulent flow, erosion, or exposure to moisture, the fibers tend to degrade more quickly, reducing their durability. Furthermore, the high moisture absorption of fique fibers can compromise the structural integrity of the turbine. Moisture absorption can degrade the matrix or induce swelling, leading to a reduction in the dimensional and mechanical stability of the blades. Finally, fique composites tend to have a higher surface roughness compared to other materials if appropriate compaction techniques are not employed, which may negatively impact the hydraulic efficiency of the rotor.

2.5. Setup and Experimental Configuration for Rotor Testing

The test bench shown in Figure 9 is designed to evaluate the performance of a gravitational vortex turbine. The test bench consists of several components. The pumping system (1) includes an electric motor coupled to a centrifugal pump that generates the required water flow to feed the test channel. A flow control system (2) is installed at the pump outlet to regulate water velocity before entering the test section. The lower tank (3) collects the water after passing through the turbine, allowing its recirculation. The supply pipe (4) transports water from the lower tank to the feeding tank (6), where the flow stabilizes before entering the test section. A flow meter (5) measures the water flow entering the system, ensuring precise control of the test conditions. The spiral inlet channel (7) directs water toward the turbine, creating a vortex flow that enhances the gravitational vortex effect. The measurement and control system (8) records key parameters such as torque and rotational speed. The discharge cone (9) facilitates water exit from the turbine, ensuring proper flow toward the lower tank while minimizing unwanted turbulence. The support structure (10) holds the feeding system and ensures the stability of the test channel. A loading motor (11) is coupled to the turbine shaft to apply a controlled mechanical load during the tests, while a torque sensor (12) measures the torque generated by the turbine. The adjustable support (13) allows modifications to the position of the measurement and control system to accommodate different experimental configurations. A vertical shaft (14) connects the turbine rotor to the measurement and control system, transmitting rotational motion. Finally, the gravitational vortex turbine (15) is submerged at the bottom of the system, harnessing the vortex-induced flow to convert water energy into mechanical energy.

The experimental tests of the gravitational vortex turbine in the described test system follow a structured procedure. First, a specific flow rate is set, causing the fluid to rise through the pipe and fill the feeding tank. Once the tank reaches a certain level, water flows into the inlet channel, where it moves towards the discharge cone. Inside the cone, a water vortex forms, driving the rotation of the turbine. The measurement and control system records the turbine's rotational velocity and torque. The sensor, through a connection to a computer, provides real-time values of torque and rotational speed under the different testing conditions.

Next, the motor coupled to the control system is powered, applying a load to the rotor and gradually slowing its rotation. As the current and voltage supplied to the rotor increase, the load intensifies, further reducing the rotor's speed. This process is repeated incrementally until the rotor comes to a complete stop. The procedure is conducted multiple times to obtain the turbine's performance curve. The tests were performed at two flow rates, 2.5 L/s and 3.0 L/s, and for two different rotors: Rotor 1, made of PLA, and Rotor 2, made of natural fiber.

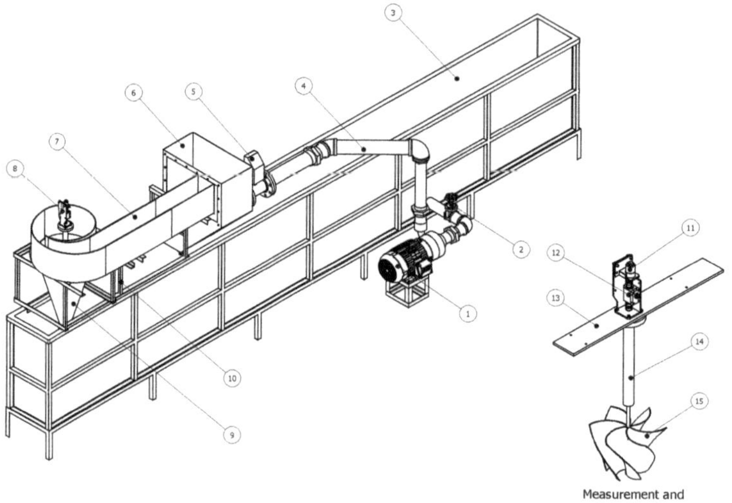

Figure 9. Experimental bench: (1) Centrifugal pump, (2) flow control system, (3) lower tank, (4) supply pipe, (5) flow meter, (6) feeding tank, (7) spiral inlet channel, (8) measurement and control system, (9) discharge cone, (10) support structure, (11) loading motor, (12) torque sensor, (13) adjustable support, (14) vertical shaft, and (15) rotor.

Possible sources of measurement errors in the experimental setup include sensor inaccuracies, signal noise, mechanical misalignments, and flow instabilities. Torque and rotational speed measurements may be affected by slight miscalibrations of the sensor or variations in the connection between the rotor and the measuring shaft. Additionally, fluctuations in the flow rate due to turbulence or inconsistent valve regulation can introduce variability in the turbine's performance data. To minimize these errors, all sensors were calibrated prior to testing, and measurements were taken multiple times to ensure repeatability. The flow was regulated using a constant head in the feeding tank to reduce instabilities, and the mechanical components were carefully aligned and checked for looseness or asymmetry. Data were recorded in real time and averaged over short intervals to smooth out transient variations and improve the reliability of the results.

3. Results and Discussion

3.1. Flow Visualization and Vortex–Rotor Interaction

Figures 10 and 11 present Rotors 1 and 2 operating inside the cone, highlighting the interaction between the gravitational vortex and the rotors. The rotors are partially submerged in the swirling water flow. The vortex forms a well-defined central core, with high-speed water spiraling downwards around the rotor. The transparency of the water surface allows for the visualization of flow patterns and turbulence generated by the rotating fluid. At first glance, the vortex behavior around both rotors seems similar, suggesting that the general flow structure is maintained regardless of the rotor material. However, further analysis would be required to determine subtle differences in energy extraction and efficiency between the two configurations.

Figure 10. Vortex formation around Rotor 1 in a gravitational vortex turbine.

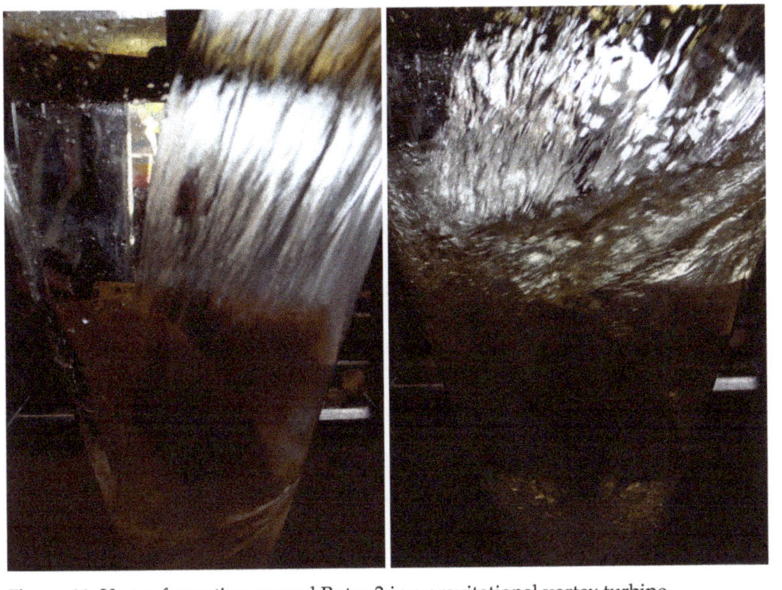

Figure 11. Vortex formation around Rotor 2 in a gravitational vortex turbine.

3.2. Experimental Performance

Figure 12 presents the experimental performance for both rotors, showing the relationship between angular velocity and efficiency for the tested flow rates of 2.5 L/s and 3.0 L/s. The curves illustrate how efficiency varies with increasing angular velocity for each rotor and operating condition. Efficiency was calculated using Equation (1), based on the torque and rotational speed values provided by the measurement and control system. For both tests, the net head H used in the calculations was 42.2 cm.

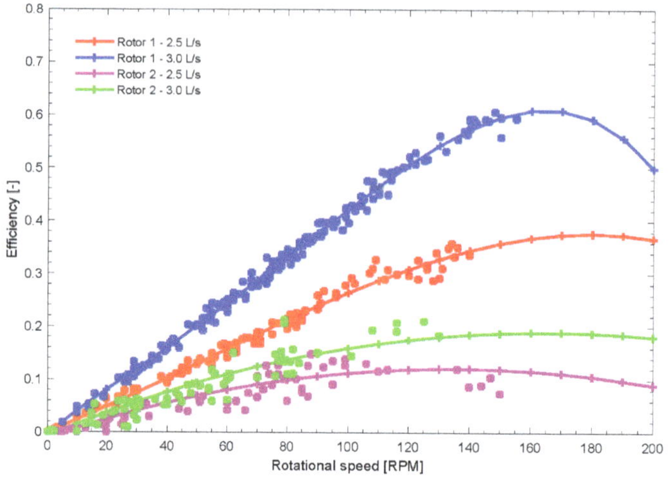

Figure 12. Experimental results.

Rotor 1, made of PLA, exhibits superior performance compared to Rotor 2, made of natural fiber, under both flow rate conditions. At 2.5 L/s, Rotor 1 achieves a maximum efficiency of approximately 0.35 at around 140 RPM. When the flow rate increases to 3.0 L/s, its performance improves significantly, reaching a peak efficiency close to 0.65 at approximately 160 RPM. In contrast, Rotor 2 demonstrates lower efficiency values across all conditions. At 2.5 L/s, its efficiency remains below 0.15, with a maximum reached near 120 RPM. When the flow rate increases to 3.0 L/s, the efficiency improves slightly but remains under 0.20, peaking around 140 RPM.

3.3. *Impact of Manufacturing Quality on Performance*

The natural fiber rotor (920 g) is significantly heavier than the PLA rotor (538.11 g). This difference affects its moment of inertia, meaning that more energy is required to start and stop it. A heavier rotor may take longer to reach its optimal speed, which could be reflected in performance curves with a smoother slope at the beginning. When increasing the load on the control system, the heavier rotor may better resist deceleration due to its higher inertia. However, if the material's efficiency is low, it may also generate less power. The lower efficiency of the natural fiber rotor may be attributed to its greater weight, which implies higher resistance to motion. If the natural fiber rotor has higher inertia, it may require a stronger flow to achieve the same rotational speed as the PLA rotor. Additionally, the PLA rotor was fabricated in 14 h, whereas the natural fiber rotor required three days for production. This disparity in manufacturing time could result in variations in surface finish and geometric precision, ultimately influencing hydrodynamic efficiency. A rougher surface finish on the natural fiber rotor, potentially attributable to the fabrication process, may induce increased frictional drag, thereby diminishing its hydrodynamic performance. Conversely, the smooth surface of the PLA rotor can mitigate frictional forces and enhance fluid flow characteristics, contributing to improved efficiency. Furthermore, the dimensional accuracy afforded by additive manufacturing techniques, such as 3D printing, facilitates the production of more uniform and optimized rotor geometries, enhancing overall performance [26]. In contrast, the increased surface roughness of the fique fiber rotor may promote turbulence and energy dissipation, resulting in reduced efficiency.

Moreover, the inherent variability in the fiber's morphology can lead to inconsistencies in performance metrics. These differences in weight, inertia, and surface characteristics highlight the complex interplay between material properties and rotor performance.

3.4. Dynamic Balance of the Rotor

Beyond these factors, another crucial aspect influencing the efficiency of gravitational vortex turbines is the dynamic balance of the rotor. Regardless of the material used, an unbalanced rotor can introduce additional mechanical losses, vibrations, and structural stresses that further hinder energy conversion. The imbalance of a rotor is inherent such that the rotor's rotation axis is not coincident with the geometric axis, resulting in rotational inertial force motivated when rotating [27]. Therefore, ensuring proper rotor balancing is essential for optimizing performance and maintaining the long-term stability of the system.

The importance of having a dynamically balanced rotor in gravitational vortex turbines lies in its direct impact on the efficiency and stability of the energy capture process. An unbalanced rotor generates vibrations and uneven forces that can negatively affect system performance, increasing mechanical losses and reducing the efficient conversion of water's kinetic energy into usable power [28,29]. Proper rotor balancing minimizes unwanted oscillations that could compromise the turbine's structural integrity and accelerate the wear of mechanical components [27]. Additionally, it reduces hydrodynamic resistance and improves operational stability, allowing the blades to maintain an optimal angle of attack relative to the water flow. This results in higher energy extraction efficiency, lower maintenance requirements, and an extended system lifespan. In the context of gravitational vortex turbines, where flow velocities can vary and generate fluctuating dynamic loads, a balanced rotor also enhances the turbine's reliability. This is particularly crucial in remote applications or decentralized energy systems, where maintenance access may be limited. Therefore, ensuring precise dynamic balancing of the rotor not only optimizes energy efficiency but also enhances the sustainability and economic viability of gravitational vortex turbines, guaranteeing stable and long-term operation in various aquatic environments. In the case of gravitational vortex turbines, the manufacturing process of Rotor 2 did not ensure proper dynamic balancing, which may have led to vibrations and negatively impacted its energy conversion performance. As a result, before conducting experimental tests, it is necessary to perform a dynamic balancing process to correct any existing imbalance. However, this correction entails additional costs in both manufacturing and assembly, potentially affecting the economic feasibility of the design. This aspect was overlooked in the original study, suggesting that the differences observed in efficiency when comparing the two rotors may be directly attributed to the lack of proper balancing in Rotor 2. Therefore, future research should incorporate a more detailed analysis of dynamic balancing as a critical factor in the design and manufacturing of gravitational vortex turbine rotors.

The dynamic balancing of a rotor made from natural fibers using a manual molding process requires a systematic approach to minimize vibrations and optimize performance. First, an initial static balancing is performed by placing the rotor on a low-friction support to identify and correct any significant mass asymmetries by adding or removing material [30]. Next, for dynamic balancing, the rotor is mounted on a test rig equipped with vibration sensors and a high-speed rotational system. By gradually increasing the rotation speed, an accelerometer or laser vibrometer is used to detect imbalance-induced vibrations. Based on the collected data, small counterweights or resin adjustments are strategically applied to the rotor until the vibrations are minimized [27,31].

Given that the manual molding process may introduce inconsistencies in material distribution, achieving a precise balance is essential to enhance the turbine's efficiency, reduce mechanical wear, and ensure long-term operational stability. Additionally, the manual molding process presents several disadvantages that can affect the quality and performance of the rotor in a gravitational vortex turbine. One of the main limitations is the difficulty in precisely controlling the applied material, leading to variations in blade thickness and uniformity. Factors such as applied pressure, resin distribution, and fiber arrangement may vary in each fabrication, making it challenging to achieve consistent rotor geometry. These inconsistencies can negatively impact the rotor's aerodynamic and hydrodynamic properties, affecting its dynamic balance and overall energy conversion efficiency [32], especially when compared to automated manufacturing methods like injection molding or vacuum infusion.

To address this issue, several strategies can be implemented:

- Use of high-precision rigid molds: Constructing molds from aluminum or reinforced polymers to ensure better dimensional stability in the blades.
- Application of compaction techniques: Utilizing rollers or vacuum systems during resin curing to achieve a more uniform material distribution and prevent thickness inconsistencies.
- Thickness quality control: Inspecting components with calipers or 3D scanners to detect variations and make necessary corrections before dynamic balancing.
- Post-processing through machining or sanding: Manually adjusting the final blade thickness after fabrication to improve geometric precision and minimize imbalances. As part of future research, these measures will be implemented to refine the fabrication process and improve the performance of the gravitational vortex turbine rotors. By addressing these manufacturing challenges, the study aims to enhance the efficiency, durability, and feasibility of using natural fiber-based rotors in hydrokinetic applications.

4. Conclusions

The manufacturing of the rotor of the gravitational vortex turbine using composite materials and natural fibers represents a significant advance in engineering, demonstrating that it is possible to develop new sustainable materials applicable to hydraulic turbines. This study marks a first step towards optimizing the manufacturing process, allowing for improving rotor efficiency, optimizing material use, and reducing costs compared to traditional methods. Rotor 1, made of PLA, outperforms Rotor 2, made of natural fiber, in terms of efficiency. At a flow rate of 3.0 L/s, Rotor 1 achieves a peak efficiency of approximately 0.65 at around 160 RPM. In contrast, under the same flow conditions, Rotor 2 exhibits significantly lower efficiency, remaining below 0.20 and peaking at approximately 140 RPM. While the efficiency metrics of fique fiber rotors may show a performance deficit compared to their PLA counterparts, their inherent material properties offer distinct advantages in specific operational contexts. In such scenarios, the trade-off between efficiency and durability favors the use of fique fiber, ensuring prolonged and reliable performance in challenging environments. The selection of the natural fiber-based biocomposite was presented as a viable alternative for rotor manufacturing. Furthermore, proper curing of the polyester resin shaft guaranteed a firm and stable connection between the blades and the shaft, contributing to the structural durability of the assembly. Reinforcement with natural fibers allowed a significant reduction in weight compared to conventional materials, while maintaining adequate mechanical performance. This feature is especially beneficial in renewable energy applications, where structural efficiency is a key factor.

Nevertheless, several limitations and challenges were encountered during this study. The use of the manual rolling method, while accessible and cost-effective, posed significant difficulties in achieving uniform surface quality and consistent adhesion between layers, which are critical for maintaining a homogeneous blade thickness. Additionally, ensuring dynamic balance in the rotor proved challenging, directly affecting the turbine's operational performance and lifespan. Vibrations resulting from imbalance not only reduced efficiency but also risked inducing structural damage over time. The handmade manufacturing process also introduced variability in the rotor dimensions, influencing the reproducibility of results. Future research should explore alternative fabrication techniques, such as vacuum infusion, to enhance the mechanical properties, surface finish, and durability of the rotors made with natural fibers. Additionally, optimizing fiber orientation and resin curing parameters could further improve structural integrity and overall turbine performance.

Author Contributions: Methodology, M.V. and B.R.V.; Validation, L.V.; Formal analysis, L.V. and B.R.V.; Investigation, M.V.; Writing—original draft, L.V.; Writing—review & editing, A.R.-C. and E.C.; Supervision, A.R.-C. and E.C.; Project administration, E.C.; Funding acquisition, E.C. All authors have read and agreed to the published version of the manuscript.

Funding: This research was funded by Universidad de Antioquia (Contract No. 2022-52624).

Institutional Review Board Statement: Not applicable.

Informed Consent Statement: Not applicable.

Data Availability Statement: The original contributions presented in this study are included in the article. Further inquiries can be directed to the corresponding author.

Conflicts of Interest: The authors declare no conflict of interest.

Abbreviations

PLA	Polylactic acid
MEK	Methyl ethyl ketone
SEM	Scanning electron microscopy
FTIR	Fourier transform infrared
CAD	Computer-aided design
FDM	Fused deposition molding
STL	Stereolithography
P_{out}	Mechanical power output
P	Hydraulic power
ρ	Density of water
g	Gravity
Q	Volumetric flow rate
H	Net head
T	Torque
η	Efficiency
ω	Angular velocity
α	Blade inlet angle
ϕ	Blade outlet angle
θ	Absolute velocity inlet angle
β	Absolute velocity outlet angle
λ	Relative velocity twist angle
L	Blade height
D	Rotor diameter
n	Number of blade
t	Blade thickness

References

1. Singh, S. Energy crisis and climate change: Global concerns and their solutions. In *Energy: Crises, Challenges and Solutions*; John Wiley & Sons Ltd.: London, UK, 2021; pp. 1–17.
2. Lak, S.Z.; Rezaei, J.; Rahimpour, M.R. Health and pollution challenges of fossil fuels utilization. *Encycl. Renew. Energy Sustain. Environ.* **2024**, *8*, 155.
3. Rahman, A.; Farrok, O.; Haque, M.M. Environmental impact of renewable energy source based electrical power plants: Solar, wind, hydroelectric, biomass, geothermal, tidal, ocean, and osmotic. *Renew. Sustain. Energy Rev.* **2022**, *161*, 112279. [CrossRef]
4. Singh, A.; Mandavkar, R.; Singh, S.; Prabhu Bhembre, R.; Jain, D.; Rufuss, D.D.W. Socio-environmental impacts of hydro power technology-a review. In Proceedings of the ASME Power Conference. American Society of Mechanical Engineers, Virtual, Online, 20–22 July 2021; Volume 85109, p. V001T09A003.
5. Maika, N.; Lin, W.; Khatamifar, M. A review of gravitational water vortex hydro turbine systems for hydropower generation. *Energies* **2023**, *16*, 5394. [CrossRef]
6. Velásquez, L.; Rubio-Clemente, A.; Posada, A.; Chica, E. Gravitational water vortex hydraulic turbine implementation in Colombia: Hydropower potential and prospects. *Rev. UIS Ing.* **2023**, *22*, 39–54. [CrossRef]
7. Velásquez, L.; Posada, A.; Chica, E. Optimization of the basin and inlet channel of a gravitational water vortex hydraulic turbine using the response surface methodology. *Renew. Energy* **2022**, *187*, 508–521. [CrossRef]
8. Velásquez, L.; Romero-Menco, F.; Rubio-Clemente, A.; Posada, A.; Chica, E. Numerical optimization and experimental validation of the runner of a gravitational water vortex hydraulic turbine with a spiral inlet channel and a conical basin. *Renew. Energy* **2024**, *220*, 119676. [CrossRef]
9. Betancour, J.; Romero-Menco, F.; Velásquez, L.; Rubio-Clemente, A.; Chica, E. Design and optimization of a runner for a gravitational vortex turbine using the response surface methodology and experimental tests. *Renew. Energy* **2023**, *210*, 306–320. [CrossRef]
10. Edirisinghe, D.S.; Yang, H.S.; Gunawardane, S.; Lee, Y.H. Enhancing the performance of gravitational water vortex turbine by flow simulation analysis. *Renew. Energy* **2022**, *194*, 163–180. [CrossRef]
11. Vinayakumar, B.; Antony, R.; Binson, V.; Youhan, S. Experimental and numerical study on gravitational water vortex power plant for small water bodies. *E-Prime Electr. Eng. Electron. Energy* **2024**, *7*, 100460. [CrossRef]
12. Zamora-Juárez, M.Á.; Guerra-Cobián, V.H.; Ferriño-Fierro, A.L.; Bruster-Flores, J.L.; Fonseca Ortiz, C.R.; López-Rebollar, B.M. Assessment of a prototype of gravitational water vortex turbine: experimental validation of efficiency. *Clean Technol. Environ. Policy* **2024**, *26*, 691–711. [CrossRef]
13. Sedai, A.; Singh, G.; Dhakal, R.; Kumal, B.B.; Ghimire, N.; Yadav, B.K. Technical and economic analysis of site implementations of gravitational water vortex power plant. In Proceedings of the 2021 IEEE International Conference on Intelligent Systems, Smart and Green Technologies (ICISSGT), Indianapolis, IN, USA, 1 November 2021; pp. 128–133.
14. Villamizar, J.; Ramon, B.; Vargas, M.; Selection of a biocomposite material as an alternative for the manufacture of the rotor of a gravitational vortex turbine. In Proceedings of the 2nd International Conference on Renewable Energies and Power Quality (ICREPQ'24), Bilbao, Spain, 26–28 June 2024; pp. 106–111.
15. Javaid, M.; Haleem, A.; Singh, R.P.; Suman, R.; Rab, S. Role of additive manufacturing applications towards environmental sustainability. *Adv. Ind. Eng. Polym. Res.* **2021**, *4*, 312–322. [CrossRef]
16. Vanaei, H.R.; Khelladi, S.; Tcharkhtchi, A. 3D Printing as a Multidisciplinary Field. In *Industrial Strategies and Solutions for 3D Printing: Applications and Optimization*; John Wiley & Sons Ltd.: London, UK, 2024; pp. 1–24.
17. Balasubramanian, M.; Saravanan, R.; Sathish, T. Exploring natural plant fiber choices and treatment methods for contemporary composites: A comprehensive review. *Results Eng.* **2024**, *24*, 103270. [CrossRef]
18. Castro, D.; Pertuz, A.; León-Becerra, J. Mechanical behavior analysis of a vertical axis wind turbine blade made with fique-epoxy composite using FEM. *Procedia Comput. Sci.* **2022**, *203*, 310–317. [CrossRef]
19. Miliket, T.A.; Ageze, M.B.; Tigabu, M.T.; Zeleke, M.A. Experimental characterizations of hybrid natural fiber-reinforced composite for wind turbine blades. *Heliyon* **2022**, *8*, e09092. [CrossRef]
20. Amzil, L.; Fertahi, S.; Raffak, T.; Mouhib, T. Structural performance analysis of hemp fiber-reinforced hybrid composites in wind turbine blade manufacturing. *Structures* **2023**, *58*, 105373. [CrossRef]
21. Rua, J.; Buchely, M.F.; Monteiro, S.N.; Echeverri, G.I.; Colorado, H.A. Impact behavior of laminated composites built with fique fibers and epoxy resin: A mechanical analysis using impact and flexural behavior. *J. Mater. Res. Technol.* **2021**, *14*, 428–438. [CrossRef]
22. Pabón Rojas, J.J.; Ramón Valencia, B.A.; Bolívar Osorio, F.J.; Ramirez, D. The role of fiber-matrix compatibility in vacuum processed natural fiber/epoxy biocomposites. *Cellulose* **2021**, *28*, 7845–7857. [CrossRef]
23. Velásquez, L.; Rengifo, J.P.; Urrego, J.; Rubio-Clemente, A.; Chica, E. Experimental Assessment of Hydrodynamic Behavior in a Gravitational Vortex Turbine with Different Inlet Channel and Discharge Basin Configurations. *Energies* **2024**, *17*, 5773. [CrossRef]

24. Praveena, B.; Lokesh, N.; Buradi, A.; Santhosh, N.; Praveena, B.; Vignesh, R. A comprehensive review of emerging additive manufacturing (3D printing technology): Methods, materials, applications, challenges, trends and future potential. *Mater. Today Proc.* **2022**, *52*, 1309–1313.
25. Srinivasan, D.; Meignanamoorthy, M.; Ravichandran, M.; Mohanavel, V.; Alagarsamy, S.; Chanakyan, C.; Sakthivelu, S.; Karthick, A.; Prabhu, T.R.; Rajkumar, S. 3D printing manufacturing techniques, materials, and applications: An overview. *Adv. Mater. Sci. Eng.* **2021**, *2021*, 5756563. [CrossRef]
26. Sinha, A.; Swain, B.; Behera, A.; Mallick, P.; Samal, S.K.; Vishwanatha, H.; Behera, A. A review on the processing of aero-turbine blade using 3D print techniques. *J. Manuf. Mater. Process.* **2022**, *6*, 16. [CrossRef]
27. Li, L.; Cao, S.; Li, J.; Nie, R.; Hou, L. Review of rotor balancing methods. *Machines* **2021**, *9*, 89. [CrossRef]
28. Deng, W.; Tong, M.; Zheng, Q.; Ren, X.; Yang, Y. Investigation on transient dynamic balancing of the power turbine rotor and its application. *Adv. Mech. Eng.* **2021**, *13*, 16878140211007325. [CrossRef]
29. Zhao, S.; Ren, X.; Liu, Y.; Lu, K.; Fu, C.; Yang, Y. A Dynamic-Balancing Testing System Designed for Flexible Rotor. *Shock Vib.* **2021**, *2021*, 9346947. [CrossRef]
30. MacCamhaoil, M. *Static and Dynamic Balancing of Rigid Rotors*; Bruel & Kjaer: Naerum, Denmark, 2016; pp. 0276–12.
31. Smolík, L.; Dyk, Š.; Rendl, J. Role of dynamic unbalance in dynamics of turbocharger rotors. *Int. J. Mech. Sci.* **2023**, *249*, 108237. [CrossRef]
32. Chen, Y.; Griffith, D.T. Blade mass imbalance identification and estimation for three-bladed wind turbine rotor based on modal analysis. *Mech. Syst. Signal Process.* **2023**, *197*, 110341. [CrossRef]

materials

MDPI

Article

Study of the Impact of Epoxidized Soybean Oil on the Characteristics of Wood-Polymer Composites

Andrii Kulikov [1,*], Dmytro Kryvolapov [2], Kostyantyn Sukhyy [2], Oleksandr Yeromin [2], Marcel Fedak [1], Olena Prokopenko [2], Iryna Sukha [2], Angelo Musaio [3] and Tomas Hrebik [1]

[1] Faculty of Manufacturing Technologies with the Seat in Presov, Technical University of Kosice, Bayerova 1, 08001 Presov, Slovakia; marcel.fedak@tuke.sk (M.F.); tomas.hrebik@tuke.sk (T.H.)
[2] Department of Ecology, Heat Transfer and Labor Protection, Faculty of Mechanical Engineering and Environmental Protection, Ukrainian State University of Science and Technologies, 49000 Dnipro, Ukraine; dmipro706@gmail.com (D.K.); ksukhyy@gmail.com (K.S.); aoeremin@gmail.com (O.Y.); eprok777@ukr.net (O.P.); irinasuha3@gmail.com (I.S.)
[3] Rectorate, Università degli Studi di Genova, 16167 Genova, Italy; angelo.musaio@unige.it
* Correspondence: andrii.kulikov@tuke.sk

Abstract: The effect of epoxidized soybean oil on the operational, technological, and physical and mechanical properties of composites based on high-density recycled polyethylene filled with wood floor was investigated. It has been shown that the introduction of epoxidized soybean oil in the amount of 0.5 wt.% into the wood-polymer composite (WPC) improves the physical, mechanical, and operational properties of the material: the Charpy impact strength (on notched samples) increases from 7.5 kJ/m^2 to 20.0 kJ/m^2, the bending strength increases from 31.6 MPa to 50.8 MPa, and the coefficient of linear thermal expansion decreases by 15%. With a further increase in the content of epoxidized soybean oil in the composite, its water absorption and technological shrinkage decrease, but its physical and mechanical properties deteriorate.

Keywords: water absorption; wood-polymer composite; epoxidized soybean oil; high-density polyethylene; physical and mechanical properties

1. Introduction

Soybean is an industrial crop primarily grown for soybean oil and protein production, ranking among the highest-yielding oilseed crops worldwide. Today, economically viable industrial applications of soybean oil include paper coatings, alkyd resins, and oleochemical products such as epoxidized soybean oil. Vegetable oils can be modified by converting double bonds into more reactive groups, such as epoxides, acrylates, and hydroxyls. Among these reactions, epoxidation is preferred due to the high reactivity of the oxirane ring and the wide range of potential applications for the resulting products [1]. Given the focus on energy savings and environmental concerns, it is essential to develop low-carbon, eco-friendly, renewable biomass-based materials to replace petrochemical materials, particularly in the construction industry and building materials sector. In recent years, biological materials have gained popularity in material production due to their cost competitiveness, renewability, and abundant availability—especially for manufacturing wood adhesives. Additionally, research has been conducted on bio-based construction materials, such as composites made by mixing straw and lime paste as an aggregate, hempcrete, and composite materials from hemp flax. These bio-based composite materials have demonstrated acceptable mechanical properties, making them suitable for use in structural and construction applications. Natural vegetable oils, particularly soybean oil,

https://doi.org/10.3390/ma18112455

have high practical value due to their fatty acid and triglyceride content exceeding 99%. Utilizing discarded or recycled soybean oil in the production of building materials represents an environmentally friendly and sustainable approach. The synthesis of epoxidized soybean oil (ESO) through an epoxidation reaction enhances the reactivity of soybean oil. ESO serves as an eco-friendly cross-linking agent, containing multiple epoxide groups within its molecules, allowing it to rapidly cross-link with reactants to improve the strength and water resistance of materials. Bio-based resins derived from epoxidized soybean oil have been successfully applied in the production of grafted fibers, coatings, adhesives, and multifunctional composite materials. Therefore, the use of ESO-based composite materials as a substitute for petrochemical-based materials can contribute to energy savings and environmental protection [2].

The development of bio-based thermoset materials requires selecting an oil with an unsaturated structure that contains highly reactive functional groups, such as carbon-carbon double bonds. Due to its natural appeal, the advancement of bio-based materials has garnered significant societal interest, with soybean oil derivatives emerging as a focal point. In traditional wood–plastic composites, more expensive coupling agents are often used, such as maleinized polyolefins. In contrast, epoxidized soybean oil can serve as a cost-effective alternative, helping to reduce overall production costs while maintaining the material's performance characteristics. Among them, epoxidized soybean oil (ESO) stands out as the most extensively studied raw material for synthesizing epoxy resins (EPs). The preference for it is due to its high content of unsaturated fatty acids, which allow the efficient incorporation of epoxy groups, as well as its biodegradability and renewable nature, making it an environmentally friendly alternative to petroleum-based precursors. Additionally, ESO exhibits good compatibility with various curing agents, enabling the formulation of epoxy resins with desirable mechanical and thermal properties. Its relatively low cost compared to conventional epoxy precursors further enhances its appeal for sustainable material development. An alternative approach involves subjecting ESO to an acrylation process, resulting in acrylated epoxidized soybean oil (AESO). While this material forms a thermoset through radical polymerization, it exhibits suboptimal mechanical properties and low glass transition temperatures [3]. Epoxidized soybean oil (ESO) contains multiple epoxide groups in its molecular structure, allowing it to effectively cross-link with reactants, thereby enhancing the mechanical strength and water resistance of polymers. ESO is considered an environmentally friendly material [4]. Wood-polymer composites are thermoplastic polymers containing a certain amount of wood in the form of flour or short fibers [5]. These materials can be easily processed using typical woodworking procedures and can also be processed into products by extrusion or injection molding [6]. They are typically used as a wood substitute (e.g., for fencing, flooring, and decking). Their use is convenient in humid environments or in contact with water: the hydrophobic polymer insulates and protects hydrophilic wood fibers, which increases the durability and reduces maintenance interventions [7]. A WPC is also used in acoustics [8] and in the automotive industry [9].

A WPC is used mainly for exterior work. In particular, this material is used to make terrace floors, siding, decorative fences, fence systems, steps, universal profiles, various accessories and components for exterior decoration. An important limitation of the polymers used in WPCs is the requirements for processing conditions (temperature, pressure) that do not lead to the thermal decomposition of the wood filler. Wood decomposes at a temperature of about 220 °C, so general-purpose polymers are commonly used for WPC production: polyethylene [10], polypropylene [11], polyvinyl chloride [12], and polystyrene [13]. The main processes for processing wood-polymer composites are extrusion, injection molding, and compression molding or thermoforming (pressing). New manufacturing processes for WPCs include additive manufacturing using fused layer modeling and laser sintering [14–17]. As for the properties of finished products, each method has its own set of parameters, advantages, and disadvantages [18,19]. The main advantages of wood as a filler are the reduced costs and improved environmental performance of the resulting composite. The former is quite obvious, since wood particles are usually inexpensive materials that are often obtained from wood or industrial waste. The latter is also obvious, since the corresponding fraction, sometimes reaching 60 or 70 wt.%, of non-biodegradable material of fossil origin is replaced by an environmentally friendly component [20,21]. But natural fibers as fillers have their drawbacks. Wood fibers are hydrophilic due to the hydroxyl groups contained in the molecular chains of cellulose and hemicellulose. And most thermoplastics used for WPCs are hydrophobic. Wood fibers agglomerate during compounding due to physical (hydrogen) bonds. As a result, wood can be poorly dispersed in polymer matrices, and thus its reinforcing effect can be impaired. The mechanical properties of composites can decrease when the aggregation of filler particles occurs [6,21]. To improve the adhesion at the polymer–wood filler interface, aprotectants are used, the functional groups of which interact with the surface functional groups of the wood filler [22]. Many apretizing agents have been used in WPCs: silicates, titanates, organic acids, chlorotriazines, acid anhydrides, isocyanates, acrylates, amides, imides, silanes, polymeric compounds, etc. [23,24].

The aim of this study is to investigate the effect of using epoxidized soybean oil as a binder modifier on the physical, mechanical, and operational properties of a WPC based on recycled high-density polyethylene. This additive was selected due to its high reactivity, which allows it to effectively interact with the polymer matrix and improve adhesion between components. Additionally, ESO is widely available and cost-effective, making it a practical alternative to other bio-based oils. Its biodegradability and renewable nature align with sustainability goals. Moreover, ESO was used to investigate its properties as an organic material with the aim of reducing the CO_2 footprint of the WPC.

2. Materials and Methods

A composite with the following composition was used as the object of study [25]:

- High-density recycled polyethylene—30 wt.%;
- Wood floor—54 wt.%;
- Calcite—11 wt.%;
- Polyethylene wax—1.1 wt.%;
- Stearic acid—1.1 wt.%;
- UV stabilizer and colorants.

The content of epoxidized soybean oil (epoxy number 6.6%) in the composition was varied from 0 to 3 wt.%. This range was selected based on preliminary considerations regarding ESO's compatibility with the polymer matrix and its potential effects on composite properties. Higher concentrations could lead to phase separation, negatively affecting mechanical integrity and adhesion within the composite. Additionally, previous studies on bio-based modifiers suggest that beyond a certain threshold, further increases in concentration do not yield proportional improvements and may even compromise structural stability. Thus, the chosen range allows for a systematic evaluation of ESO's influence while maintaining processability and cost-effectiveness in the development of WPC composition.

The physical and mechanical properties of the used recycled high-density polyethylene are given in Table 1.

Table 1. Physical and mechanical properties of recycled high-density polyethylene.

Indicator	Value
Appearance	Dark green granules 3–5 mm in size
Density, kg/m^3	956
Vick softening point, °C	130
Melt flow index, g/10 min	1.0
Curl flow, mm	370
Flexural strength, MPa	50
Charpy impact strength (on unnotched samples), kJ/m^2	samples do not collapse
Tensile yield strength, MPa	22.3
Elongation at break, %	570

It should be noted that the used recycled HDPE is first recycled, meaning it has undergone only one recycling cycle. Its properties remain comparable to those of virgin HDPE, while offering a lower cost and greater environmental benefits.

In the current work, oak wood flour of the M200 brand according to DEST 16361-87 was used. The main characteristics of wood flour of the M200 brand are given in Table 2.

Table 2. Quality characteristics of wood flour grade M200.

Name of the Indicator	Value
Moisture, %, not more	5
Mass fraction of ash, %, not more	1.0
Mass fraction of colored impurities, %, not more	4.0
Mass fraction of metallomagnetic impurities, %, not more	0.0016
Bulk density, kg/m^3	100 to 140
Mass fraction of residue on the grid 0.18, %, not more	5.0
Mass fraction of residue on the grid 0.25, %, not more	0.4
Mass fraction of resins and oils, %, not more	5.0

As a mineral filler, untreated microcalcite of the MMC-1 brand produced by ChAT "Novgorod-Seversky Building Materials Plant", Ukraine, was used. As lubricants, stearic acid of the 1860 brand produced by PT. Dua Kuda, Indonesia, and polyethylene wax of the CWN-105 brand produced by Liebenzell AG Group Company, Valencia, Spain were used.

The main stages of preparation of samples of wood-polymer composite included the following stages: preparation of raw materials, weighing of components, mixing, granulation, extrusion of terrace board, cooling and cutting, and finishing.

The preparatory stage includes the quality control of raw materials and the preliminary drying of components (if necessary). Quality control involved the visual inspection of the raw materials to ensure uniform granule size, the absence of contaminants, and consistent coloration. Additionally, for recycled HDPE, key parameters such as density and the melt flow index were verified against supplier specifications to confirm compliance with expected material characteristics. Then, on electronic scales (3rd accuracy class), the components were weighed and dosed into appropriate containers with a total mass of 150 kg. After that, using a screw feeder, the raw material was fed into a blade mixer model skywinswhl 600/1200 (Figure 1), where at a blade speed of 800 rpm, a mechanical mixture was prepared for 45–60 min until a temperature of 100 °C was reached to ensure the removal of residual moisture. Epoxidized soybean oil (ESO) was added last during the loading of components into the mixer. This ensured its uniform distribution throughout the polymer–wood flour mixture. The controlled mixing time and temperature facilitated the effective dispersion of ESO, optimizing its role as a binder modifier before further processing.

Figure 1. Paddle mixer SKY WIN SWHL 600/1200.

Next, the mixture was fed into a twin-screw granulator of the skywin swmsz-3 brand (Figure 2), where the composite material was homogenized at a temperature of 180 °C and screw speeds of 250–350 rpm. The composite material in the form of granules was loaded into big bags and transported to the decking board extrusion line.

Figure 2. Twin-screw extruder SKY WIN SWMSZ-3.

Extrusion of the terrace board took place on a twin-screw extruder manufactured by KraussMaffei (Munich, Germany), model KMD-60, with the following specified parameters: temperatures in all zones of 125 °C, an adapter temperature of 140 °C, a molding head temperature of 145 °C in circumference, a screw rotation speed of 10 rpm, and an extruder main drive load of 80%, which is provided by the rotation of the raw material dispenser. The vacuum was not created in the degassing zone. After the extrudate exited the head, the board entered the calibrator and cooling bath, where its final molding took place at a water temperature of 14–16 °C. Then, the formed and cooled board entered the cutting device, where the boards were cut to a certain size. To improve the appearance and functional properties, the decking board was subjected to surface treatment by grinding and tufting. Samples for research were taken from the produced boards at the beginning, middle, and end of the batch.

The technological process of producing the WPC consisted of the following stages: preparation of components; drying and quality control of raw materials; preparation of a mechanical mixture in a paddle mixer (SKY WIN SWHL 600/1200) at a speed of 800 rpm; granulation of the resulting mass in a twin-screw extruder (sky win swmsz-3) at a temperature of 160 to 2000 °C; and extrusion of a decking board from which samples for research were cut.

In the mechanical testing of samples, technological and operational properties were determined according to standard methods of research and testing of plastics. Statistical processing of the experimental data was performed by regression-correlation analysis. Measurements of the properties of each sample were repeated five times, with a confidence interval of 0.95. The scheme of the analysis and measurement process is depicted in Figure 3.

Experimental data were obtained according to existing standard methods and modern research methods:

- Charpy impact strength was determined according to ISO 179:2017, using notched specimens of 80 × 10 × 5 mm, with an impact energy of 2 J and a test temperature of 23 °C.
- Flexural strength was measured according to ISO 178:2019, using the three-point bending test, with specimens of 160 × 10 × 5 mm, a test speed of 2 mm/min, and a span length of 64 mm.
- Water absorption was determined according to ISO 62:2008, by immersing specimens in distilled water at 23 °C for 24 h, followed by weighing with an accuracy of 0.001 g.
- The coefficient of linear thermal expansion was measured according to ASTM D 696, using a dilatometer within a temperature range of −15 °C to 80 °C.
- Technological shrinkage was determined according to DIN 16901, by measuring the dimensional change in molded specimens in the transverse direction after 24 h of conditioning at 23 °C and 50% relative humidity.
- Density was determined according to ISO 1183-1, using the buoyancy method, with analytical balances and isopropyl alcohol as the immersion medium.

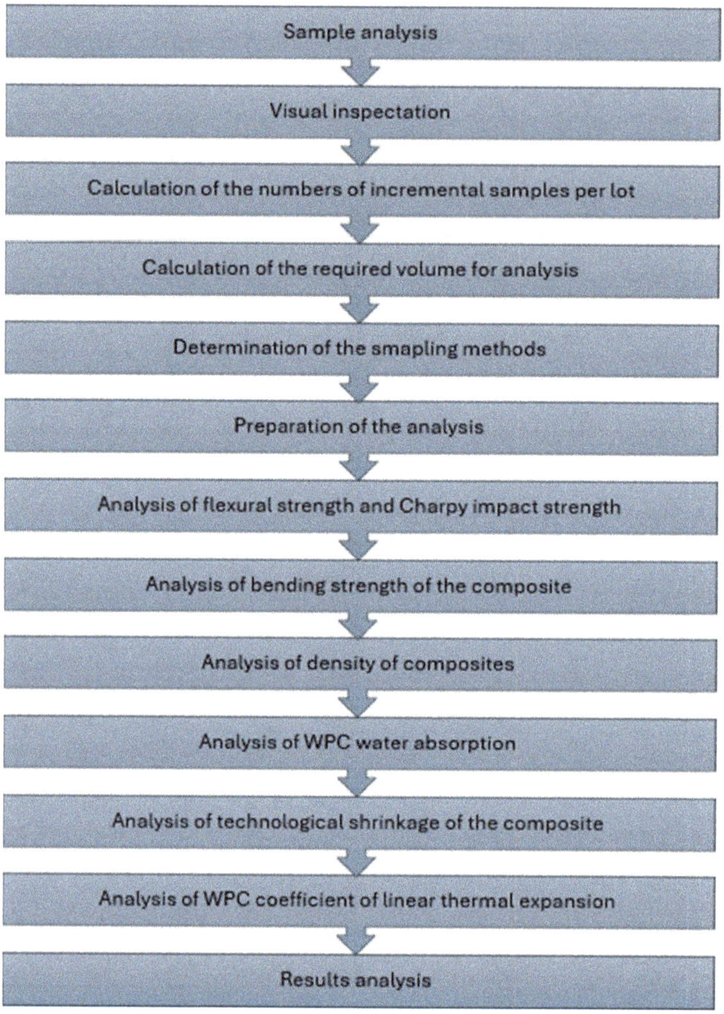

Figure 3. The scheme of the analysis and measurement process.

3. Results

When determining the properties of a WPC material for outdoor use, physical and mechanical parameters are among the most informative. Previous studies have shown that for a WPC as a finishing material for facades and terraces, bending strength and Charpy impact strength are crucial. Figures 4 and 5, respectively, show the effect of the content of epoxy soybean oil on the flexural strength and Charpy impact strength (on notched samples) of wood-polymer composites. Figures 4–9 present the standard deviation of the measured values depending on the concentration of epoxy soybean oil.

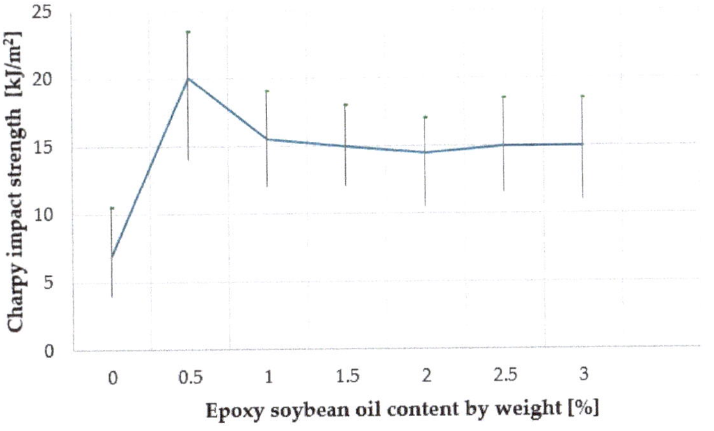

Figure 4. Dependence of the Charpy impact strength (on specimens with a cut) of the composite on the content of epoxy soybean oil.

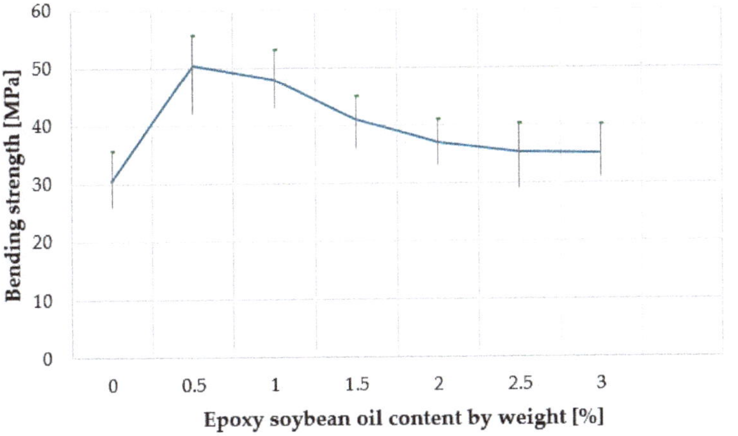

Figure 5. Dependence of the bending strength of the composite on the content of epoxy soybean oil.

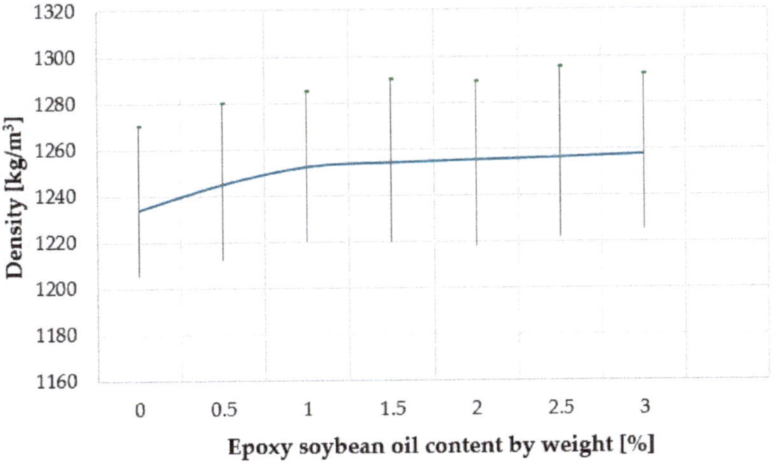

Figure 6. Dependence of the density of composites on the content of epoxidized soybean oil.

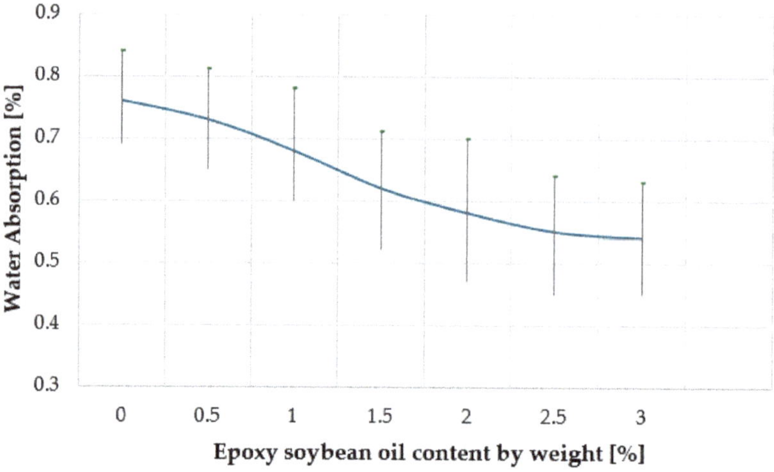

Figure 7. Dependence of WPC water absorption on the content of epoxidized soybean oil.

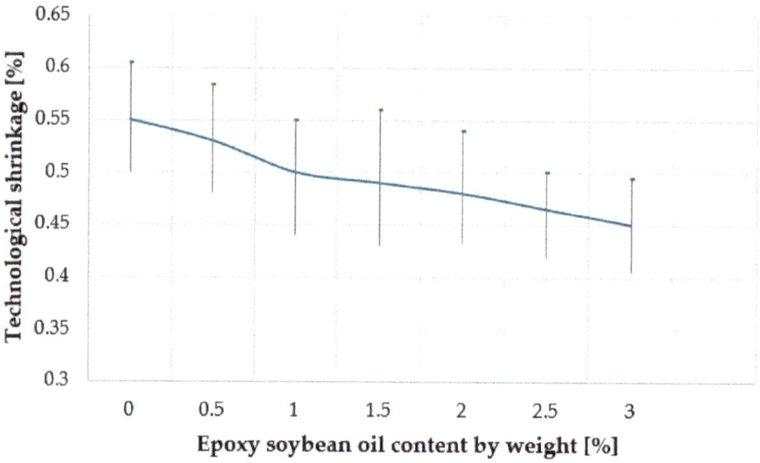

Figure 8. Dependence of the technological shrinkage of the composite on the content of epoxy soybean oil.

Analyzing the data of Figures 4 and 5, it can be stated that at the content of epoxidized soybean oil of 0.5 wt.% in the composition, the mechanical properties of the wood polymer composite improve: the Charpy impact strength (on notched samples) increases from 7.5 kJ/m^2 to 20.0 kJ/m^2 (almost three times), and the bending strength increases from 31.6 MPa to 50.8 MPa (by 60%). With a further increase in the content of epoxidized soybean oil to 3.0 wt.%, the mechanical properties of the composite deteriorate: the bending strength decreases to 34.2 MPa, and the Charpy impact strength (on notched samples) decreases to 15.2 kJ/m^2. A statistical analysis was performed to assess the significance of these differences. The results indicate that the observed variations in impact strength and bending strength are statistically significant at the 0.95 confidence level, confirming the reliability of the trends presented in the study.

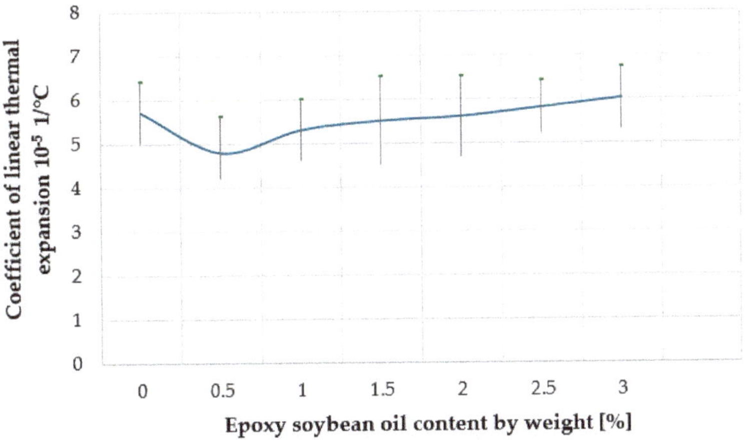

Figure 9. Dependence of the coefficient of linear thermal expansion of the composite on the content of epoxy soybean oil.

This strengthening mechanism can be explained by the fact that at low concentrations, epoxidized soybean oil acts as a plasticizer, improving the mobility of polymer chains and contributing to a more efficient distribution of stresses under load [18]. This leads to an increase in impact strength and flexural strength. However, when the concentration of epoxidized soybean oil increases to 1.5 wt.%, it can start to act as a defect, forming microscopic inclusions that become stress concentrators [19]. This leads to a decrease in flexural strength and impact strength, since the energy required to fracture the material is reduced due to premature crack formation in stress concentration areas. This effect is typical of excessive amounts of plasticizers in polymer composites [20].

In other words, this strengthening mechanism can be explained by the fact that the epoxy groups contained in epoxidized soybean oil interact with the hydroxyl groups of wood, and the nonpolar fatty acid moieties contained in soybean oil have good compatibility with polyethylene, thereby increasing the compatibility between the polymer and the wood. When the amount of epoxidized soybean oil in the composite increases, the polymer matrix plasticizes because the relatively small molecules of epoxidized soybean oil diffuse into the polymer, pushing out larger polyethylene molecules, surrounding them with a monolayer, and shielding the polar groups, resulting in a decrease in flexural strength and Charpy impact toughness.

This mechanism of reinforcement can be explained by the fact that the epoxy groups contained in epoxy soybean oil interact with the hydroxyl groups of wood, and the non-polar fatty acid fragments contained in soybean oil are compatible with polyethylene, and therefore, the compatibility between the polymer and wood increases. With an increase in the content of epoxy soybean oil in the composite, due to the fact that relatively small oil molecules diffuse into the polymer, push macromolecules apart, surround them with a monomolecular layer, and screen the polar groups, the polymer matrix plasticizes, and, as a result, the bending strength and Charpy impact strength decrease. Many manufacturers consider the density of composite building materials as a factor determining the weight of the profile (in terms of transportation costs), the ease of the installation of decks, and as a factor determining the consumption of raw materials. The density of the composite material also largely determines the service life and performance characteristics of products [18]. The decomposition of polymer and lignin from wood fibers during processing leads to the formation of volatile organic compounds, and thus to the porosity of the composite and a decrease in density. Figure 6 shows the dependence of WPC density on the content of epoxidized soybean oil.

As can be seen from Figure 6, with an increase in the content of epoxy soybean oil, the density of the composite increases from 1234 kg/m^3 to 1258 kg/m^3, indicating a more ordered structure. That is, epoxy soybean oil acts as a dispersant that prevents the agglomeration of wood particles during processing. The observed density increase aligns with the expected effect of ESO improving compatibility between the polymer and the wood, rather than a direct contribution from its own mass. Given that ESO has a density close to HDPE (~1000 kg/m^3) and its evaporation is minimal at processing temperatures, its addition primarily influences particle distribution and matrix uniformity, rather than significantly altering the overall density.

Increased water absorption of WPC products can significantly reduce their durability, making them ineffective for exterior decoration. The main reasons for the increase in WPC water absorption are their porosity and low adhesion at the polymer–wood filler interface. The dependence of the water absorption of composites on the content of epoxied soybean oil is shown in Figure 7.

From the data in Figure 7, it can be concluded that with an increase in the content of epoxy soybean oil, the water absorption of wood-polymer composites decreases from 0.76% (in the absence of epoxy soybean oil in the composite) to 0.54% (with an epoxy soybean oil content of 3 wt.%). This positive trend can be explained, firstly, by the more ordered structure of the WPC (as indicated by the increase in density), and secondly, by the hydrophobic nature of epoxied soybean oil.

Technological shrinkage is a phenomenon that occurs during the production of products and is associated with their dimensional changes due to temperature, humidity, or other processing factors. In the case of wood-polymer composites, technological shrinkage can occur during the molding process or during the cooling of the material after production. shrinkage can affect the dimensional accuracy of the product and its appearance. In this work, we investigated the effect of the content of epoxy soybean oil on the technological shrinkage of the composite (Figure 8).

As can be seen from Figure 8, with an increase in the content of epoxidized soybean oil in the WPC to 3% by weight, the technological shrinkage decreases from 0.55% to 0.45%. This behavior is explained primarily by an increase in the density of the composite material, which leads to a decrease in the reorientation of polymer macromolecules during the cooling of the extrudate. Technological shrinkage was measured in the transverse direction to the extrusion flow. In WPC processing, shrinkage is typically anisotropic, with differences between the extrusion direction and the transverse direction. The alignment of polymer chains and wood fibers during extrusion often results in greater shrinkage along the extrusion direction, while transverse shrinkage is generally lower. In injection molding, shrinkage may be more uniform, but it depends on cooling conditions and pressure distribution. Lower technological shrinkage at a high ESO content has important implications for the production of large WPC profiles. Reduced shrinkage in the transverse direction helps minimize dimensional variations, making it easier to control the final shape and size of extruded products. Additionally, lower shrinkage reduces internal stresses, which can improve the long-term stability and durability of large WPC structures.

Thermal expansion and contraction is a universal phenomenon and is observed for all solids. The higher the ambient temperature, the higher the amplitude of vibrations of atoms and molecules of materials, which increases the effective volume of objects and, as a result, their size in all directions. The linear expansion–compression coefficient (k) can be calculated using the following formula: $k = \Delta L/(L \times \Delta T)$, where ΔL is the expansion or contraction of the material in the temperature range ΔT, and L is the initial size of the material at the beginning of the temperature range. The dependence of the linear coefficient of thermal expansion of the WPC on the content of epoxied soybean oil in the composite is shown in Figure 9.

Analyzing the data in Figure 9, it can be concluded that when 0.5% by weight of epoxy soybean oil is added to the composite, the coefficient of linear thermal expansion decreases by 15%. However, with a further increase in the content of epoxidized soybean oil in the composite, the coefficient of linear thermal expansion increases. This behavior can be explained by the fact that at low concentrations, ESO exhibits a weak plasticizing effect, while simultaneously improving the compatibility of wood particles with polyethylene and preventing their agglomeration. However, at higher concentrations, ESO begins to excessively plasticize the polymer matrix, increasing the mobility of macromolecules. This excessive plasticizing effect leads to structural weakening, as confirmed by the decline in mechanical properties at concentrations of 1.5 wt.% and above.

In order to verify the mechanism of interaction between the epoxy groups of epoxidized soybean oil and the hydroxyl groups of wood, IR spectra of epoxidized soybean oil, the WPC without epoxidized soybean oil, and the WPC with 0.5 wt.% epoxidized soybean oil were obtained (Figure 9) [26].

Analyzing the IR spectra (Figure 10), it is possible to see a shift in the peak at 1423.03 cm^{-1}, which is in the spectrum of the pure composite, by 2.05 cm^{-1} in the long-wave direction, to 1420.98 cm^{-1}, after the introduction of 0.5 wt.% epoxidized soybean oil. This band can be attributed to the deformation vibrations of the R-O-H group. Such a shift can be explained as a result of blocking interactions between hydroxyl groups of cellulose macromolecules in the surface layers of wood particles due to the reaction between epoxy and hydroxyl groups with the formation of ether alcohol fragments. At the temperatures used to process the composite mass into a board, this is entirely possible. In this way, the hydrophobic molecules of the epoxidized soybean oil chemically bond to the wood surface and improve its thermodynamic affinity with polyethylene [27,28].

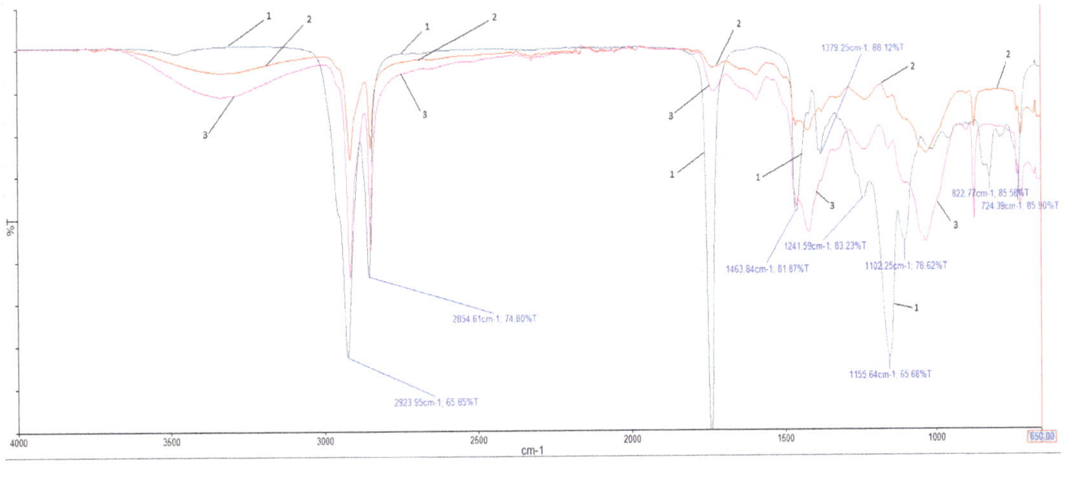

Figure 10. IR spectra of epoxidized soybean oil (1), WPC without epoxidized soybean oil (2), and WPC with epoxidized soybean oil content of 0.5 wt.% (3).

4. Discussion

Their effectiveness as an alternative to traditional materials is well established. With the global trend toward using natural compounds, wood-polymer composites (WPCs) not only offer technological advantages over natural materials, but also consist primarily of renewable components [29].

However, most WPCs combine natural lignocellulosic fibers with plastics derived from fossil fuels, which raises concerns regarding the sustainability and carbon footprint. Additionally, the recycling of WPCs remains limited, as the extrusion process creates strong interfacial bonds, making it difficult to separate renewable and non-renewable components after processing.

Since terrace decking represents the largest share of WPC products, the incorporation of epoxidized soybean oil (ESO) as a finishing agent in composites based on recycled polymers and wood flour enables the production of cost-effective, high-quality decking boards with enhanced physical, mechanical, and operational properties. The use of recycled polymers in WPC formulations can significantly reduce environmental impact, lowering dependence on virgin fossil-based plastics (Table 3). Further research into biodegradable polymer matrices and advanced recycling techniques could improve the sustainability of WPCs, addressing these concerns.

Table 3. Comparative characteristics of pine and WPC.

Parameters	WPC	Pine
Appearance	Smooth, no knots	Rough, knots present
Susceptibility to rot and mold	Not susceptible	Susceptible
Susceptibility to cracking	Not susceptible	Prone
Need for special care	Does not require	Requires
Resistance to UV radiation	Resistant	Not resistant
Bending strength, MPa	50.8	49.5
Water absorption, %	0.72	185
Charpy impact strength, kJ/m^2	20.0	35.0
Coefficient of linear thermal expansion, 1/°C	4.8×10^{-5}	29×10^{-6}
Average service life, years	20–50	10–20

Thus, wood-polymer composite flooring boards have a number of advantages over natural wood boards, which makes them suitable for use in the construction and decoration of exterior structures. The results of many studies show the effect of using a WPC or a combination of WPC/wood compared to classic materials [30,31]. However, it should be noted that WPC decking boards are generally more expensive than pressure-treated pine boards, with prices sometimes twice as high. Despite the higher initial cost, WPC offers greater durability, resistance to moisture, insects, and decay, and requires minimal maintenance, unlike pine, which may need regular sealing, staining, or replacement due to weathering.

5. Conclusions

According to the results of the studies on the effect of epoxidized soybean oil on the properties of a WPC based on high-density recycled polyethylene, it was found that the use of epoxy soybean oil in wood-polymer composites reduces technological shrinkage and water absorption, which makes it possible to use these composites to produce products with more accurate dimensions that operate in high-humidity conditions.

The rational content of epoxidized soybean oil, which provides the highest mechanical and operational performance in the WPC, is 0.5 wt.%, but with a further increase in the content of epoxidized soybean oil, its physical, mechanical and operational properties deteriorate.

Since soybean oil is a plant-based product, its use in wood-polymer composites has a positive environmental impact and allows for the production of WPC boards at a lower cost compared to conventional WPC formulations. This cost reduction is achieved through improved processing efficiency and reduced reliance on synthetic additives, making an ESO-modified WPC a more economical alternative within the WPC market.

The ESO-modified composites could be processed at lower temperatures, which can reduce energy consumption and even mitigate equipment wear over time. These advantages highlight the eco-friendly nature of ESO and its potential to lower manufacturing costs while maintaining or improving material performance. The results have important implications for industrial applications of WPCs made with recycled polymers. Improved dimensional stability means that extruded or molded WPC products will retain their intended shape and dimensions more reliably, easing installation and improving fit in construction uses. The reduction in water absorption observed at optimal ESO levels translates to greater durability in outdoor or humid environments, as the composites will be less prone to moisture-induced swelling, decay, or freeze–thaw damage over time. Furthermore, the enhanced thermal stability and lower melting temperature of ESO-containing composites suggest easier processing and the possibility of faster production cycles or the use of lower processing temperatures in existing industrial equipment. Together, these improvements can broaden the range of applications for WPC materials—for example, in decking, cladding, or outdoor furniture—by improving their service life and reliability. Properties such as moisture resistance and dimensional stability were measured in accelerated tests, but real-world factors like prolonged UV exposure, seasonal temperature fluctuations, and biological attack (mold or microbes) were not fully simulated. Future research should address the above limitations to fully realize the potential of ESO in wood-polymer composites. Long-term performance testing is especially critical: extended outdoor exposure trials in diverse climates would clarify how ESO-modified WPCs endure years of weathering, including UV radiation, rain/humidity, and temperature cycling. Such studies will determine if the initial improvements in water resistance and dimensional stability are retained over the service life, or if issues like plasticizer leaching or the biodegradation of the ESO emerge after prolonged exposure.

Author Contributions: Methodology, O.P.; Validation, I.S.; Formal analysis, O.Y. and A.M.; Investigation, K.S.; Data curation, Resources, D.K.; Data curation, T.H.; Data curation, Writing—review & editing A.K.; Validation, Funding acquisition, M.F. All authors have read and agreed to the published version of the manuscript.

Funding: This article was supported by the Cultural and Educational Grant Agency of the Ministry of Education, Research, Development, and Youth of the Slovak Republic, through the project KEGA 024TUKE-4/2024 and the project VEGA 1/0723/25.

Institutional Review Board Statement: Not applicable.

Informed Consent Statement: Not applicable.

Data Availability Statement: The original contributions presented in this study are included in the article. Further inquiries can be directed to the corresponding author.

Conflicts of Interest: The authors declare no conflict of interest.

References

1. Işik, M.; Ahmetli, G. Nanocomposites based on MWCNT and nanoclay: Effect of acrylated epoxidized soybean oil on curing and composite properties. *Ind. Crops Prod.* **2024**, *221*, 119421. [CrossRef]
2. Wang, W.; Luo, R.; Deng, S.; Essawy, H.; Lee, S.H.; Lum, W.C.; Zhou, X.; Zhang, J. High performance epoxy soybean oil-based composite reinforced by wood powder. *Ind. Crops Prod.* **2024**, *211*, 118195. [CrossRef]
3. Lascano, D.; Gomez-Caturla, J.; Garcia-Sanoguera, D.; Garcia-Garcia, D.; Ivorra-Martinez, J. Optimizing biobased thermoset resins by incorporating cinnamon derivative into acrylated epoxidized soybean oil. *Mater. Des.* **2024**, *243*, 113084. [CrossRef]
4. Jiang, J.; Wang, W.; Lee, S.H.; Lum, W.C.; Essawy, H.; Du, G.; Zhou, X.; Zhang, J. Flame retardancy, photostability, and thermal properties of vanillin-epoxidized soybean oil composites reinforced with walnut shell. *Ind. Crops Prod.* **2024**, *218*, 118912. [CrossRef]
5. Adekomaya, O.; Jamiru, T.; Sadiku, R.; Huan, Z. A review on the sustainability of natural fiber in matrix reinforcement—A practical perspective. *J. Reinf. Plast. Compos.* **2016**, *35*, 3–7. [CrossRef]
6. Gardner, D.J.; Han, Y.; Wang, L. Wood-plastic composite technology. *Curr. For. Rep.* **2015**, *1*, 139–150. [CrossRef]
7. Fortini, A.; Mazzanti, V. Combined effect of water uptake and temperature on wood polymer composites. *J. Appl. Polym. Sci.* **2018**, *135*, 46674. [CrossRef]
8. Santoni, A.; Bonfiglio, P.; Fausti, P.; Marescotti, C.; Mazzanti, V.; Pompoli, F. Characterization and vibro-acoustic modeling of wood composite panels. *Materials* **2020**, *13*, 1897. [CrossRef]
9. Isa, A.; Toyoda, T.; Suzuki, S.; Kojima, Y.; Ito, H.; Makise, R.; Okamoto, M. The effects of wet-milled wood flour on the mechanical properties of wood flour/polypropylene composites. *J. Wood Chem. Technol.* **2014**, *34*, 20–30. [CrossRef]
10. Adhikary, K.B.; Pang, S.; Staiger, M.P. Dimensional stability and mechanical behavior of wood-plastic composites based on recycled and virgin high-density polyethylene (HDPE). *Compos. Part B Eng.* **2008**, *39*, 807–815. [CrossRef]
11. Espert, A.; Vilaplana, F.; Karlsson, S. Comparison of water absorption in natural cellulosic fibers from wood and one-year crops in polypropylene composites and its influence on their mechanical properties. *Compos. Part A Appl. Sci. Manuf.* **2004**, *35*, 1267–1276. [CrossRef]
12. Jiang, H.; Kamdem, D.P. Development of poly(vinyl chloride)/wood composites. A literature review. *J. Vinyl Addit. Technol.* **2004**, *10*, 59–69. [CrossRef]
13. Poletto, M.; Dettenborn, J.; Zeni, M.; Zattera, A.J. Characterization of composites based on expanded polystyrene wastes and wood flour. *Waste Manag.* **2011**, *31*, 779–784. [CrossRef] [PubMed]
14. Jiang, K.; Guo, Y.; Bourell, D.L. Study on the Microstructure and Binding Mechanisms of Selective Laser Sintered Wood Plastic Composite. 2013. Available online: https://repositories.lib.utexas.edu/items/107429a1-4939-404c-bd8c-ef8dbcb17592 (accessed on 31 May 2015).
15. Sukhyy, K.; Belyanovskaya, E.; Nosova, A.; Sukha, I.; Sukhyy, M.; Huang, Y.; Kochergin, Y.; Hryhorenko, T. Dynamic Mechanical Properties of Epoxy Composites Modified with Polysulphide Rubber. *Chem. Technol.* **2022**, *16*, 432–439.
16. Gephardt, A. *Understanding Additive Manufacturing*; Hanser Publications: Cincinnati, OH, USA, 2012; p. 164.
17. Sukhyy, K.M.; Belyanovskaya, E.A.; Nosova, A.N.; Huang Yu Hryhorenko, T.I. Influence of Concentration of Thiokol, Amount of the Hardener and Filler on Properties of Epoxy-Polysulphide Composites. *J. Chem. Technol.* **2021**, *29*, 531–539.
18. Wilczyński, K.; Buziak, K.; Lewandowski, A.; Nastaj, A.; Wilczyński, K.J. Rheological basics for modeling of extrusion process of wood polymer composites. *Polymers* **2021**, *13*, 622. [CrossRef] [PubMed]
19. Sukhyy, K.M.; Belyanovskaya, E.A.; Nosova, A.N.; Sukhyy, M.K.; Huang, Y.; Kocherhin, Y.; Hryhorenko, T. Properties of epoxy-thiokol materials based on the products of the preliminary reaction of thioetherification. *Vopr. Khimii Khimicheskoi Tekhnologii* **2021**, 128–136. [CrossRef]
20. Mazzanti, V.; Mollica, F.; El Kissi, N. Rheological and mechanical characterization of polypropylene-based wood plastic composites. *Polym. Compos.* **2016**, *37*, 3460–3473. [CrossRef]
21. Sukhyy, K.M.; Belyanovskaya, E.A. *Polymer-Inorganic Nanostructured Composites Based on Amorphous Silica, Layered Silicates, and Polyionenes*; IGI Global: San Francisco, CA, USA, 2023; p. 303.
22. Li, X.; Tabil, L.G.; Panigrahi, S. Chemical treatments of natural fiber for use in natural fiber-reinforced composites: A review. *J. Polym. Environ.* **2007**, *15*, 25–33. [CrossRef]
23. Klesov, A.A. *Wood-Polymer Composites*; Chmel, A., Translator; Scientific Basis and Technologies: St. Petersburg, Russia, 2010; 736p.
24. Rimar, M.; Kulikova, O.; Kulikov, A.; Fedak, M. Energy Treatment of Solid Municipal Waste in Combination with Biomass by Decentralized Method with the Respect to the Negative Effects on the Environment. *Sustainability* **2021**, *13*, 4405. [CrossRef]
25. Kryvolapov, D.S.; Sukhyy, K.M.; Bashtanyk, P.I.; Tretyakov, A.O.; Bezrukavy, V.A. Influence of stearic acid on the properties of wood-polymer composites. *Vopr. Khimii I Khimicheskoi Tekhnologii* **2023**, *5*, 50–54. [CrossRef]
26. Smith, A.B. Plasticizers in Polymers Properties and Applications. *J. Appl. Polym. Sci.* **2010**, *125*, 1234–1245.
27. Jones, C.D. Stress Concentration in Composite Materials. *Compos. Sci. Technol.* **2015**, *71*, 567–578.
28. Brown, E.F. Effect of Plasticizer Content on Mechanical Properties of Polymers. *Polym. Eng. Sci.* **2005**, *45*, 789–800.

29. Boztoprak, Y. Physical, Mechanical, and Flammability Properties of Wood–Plastic Composites (WPC) Containing Beech-Wood Flour and Flame-Retardant Additives. *Polymers* **2024**, *16*, 2944. [CrossRef] [PubMed]
30. Canto-Borges, L.C.; Cruz, J.C.; Nahuat-Sansores, J.R.; Domínguez-Lepe, J.A.; Gurrola, M.P. Sustainable WPC Production: A Novel Method Using Recycled High-Density Polyethylene and Wood Veneer. *Recycling* **2024**, *9*, 1. [CrossRef]
31. Alnajjar, H.M.; Nagiub, H.M.; Shaker, N.O.; Ahmed, H.; Farag, A.A.; Kandil, U.F. Enhancement of Recycled WPC with Epoxy Nanocomposite Coats. *Egypt. J. Chem.* **2019**, *62*, 555–563. [CrossRef]

materials

MDPI

Article

Mechanical Characteristics of 26H2MF and St12T Steels Under Torsion at Elevated Temperatures

Waldemar Dudda

Faculty of Technical Sciences, University of Warmia and Mazury, Oczapowskiego 11, 10-719 Olsztyn, Poland; dudda@uwm.edu.pl

Abstract: The concept of "material effort" appears in continuum mechanics wherever the response of a material to the currently existing state of loads and boundary conditions loses its previous, predictable character. However, within the material, which still descriptively remains a continuous medium, new physical structures appear and new previously unused physical features of the continuum are activated. The literature is dominated by a simplified way of thinking, which assumes that all these states can be characterized and described by one and the same measure of effort—for metals it is the Huber–Mises–Hencky equivalent stress. Quantitatively, perhaps 90% of the literature is dedicated to this equivalent stress. The remaining authors, as well as the author of this paper, assume that there is no single universal measure of effort that would "fit" all operating conditions of materials. Each state of the structure's operation may have its own autonomous measure of effort, which expresses the degree of threat from a specific destruction mechanism. In the current energy sector, we are increasingly dealing with "low-cycle thermal fatigue states". This is related to the fact that large, difficult-to-predict renewable energy sources have been added. Professional energy based on coal and gas units must perform many (even about 100 per year) starts and stops, and this applies not only to the hot state, but often also to the cold state. The question arises as to the allowable shortening of start and stop times that would not to lead to dangerous material effort, and whether there are necessary data and strength characteristics for heat-resistant steels that allow their effort to be determined not only in simple states, but also in complex stress states. Do these data allow for the description of the material's yield surface? In a previous publication, the author presented the results of tension and compression tests at elevated temperatures for two heat-resistant steels: St12T and 26H2MF. The aim of the current work is to determine the properties and strength characteristics of these steels in a pure torsion test at elevated temperatures. This allows for the analysis of the strength of power turbine components operating primarily on torsion and for determining which of the two tested steels is more resistant to high temperatures. In addition, the properties determined in all three tests (tension, compression, torsion) will allow the determination of the yield surface of these steels at elevated temperatures. They are necessary for the strength analysis of turbine elements in start-up and shutdown cycles, in states changing from cold to hot and vice versa. A modified testing machine was used for pure torsion tests. It allowed for the determination of the sample's torsion moment as a function of its torsion angle. The experiments were carried out at temperatures of 20 °C, 200 °C, 400 °C, 600 °C, and 800 °C for St12T steel and at temperatures of 20 °C, 200 °C, 400 °C, 550 °C, and 800 °C for 26H2MF steel. Characteristics were drawn up for each sample and compared on a common graph corresponding to the given steel. Based on the methods and relationships from the theory of strength, the yield stress and torsional strength were determined. The yield stress of St12T steel at 600 °C was 319.3 MPa and the torsional strength was 394.4 MPa. For 26H2MH steel at 550 °C, the yield stress was 311.4 and the torsional strength was 382.8 MPa. St12T steel was therefore more resistant to high temperatures than 26H2MF. The combined data from the tension, compression, and torsion tests allowed us to determine the asymmetry and plasticity coefficients, which allowed us to model the yield surface according to the Burzyński criterion as a function of temperature. The obtained results also allowed us to determine the parameters of the Drucker-Prager model and two of the three parameters of the Willam-Warnke and Menetrey-Willam models. The research results are a valuable contribution to the design and diagnostics of power turbine components.

Keywords: torsional strength; yield stress; stress hypothesis; yield surfaces; heat-resistant steels

1. Introduction

This paper refers largely to the mechanics of thermal stress, which is the basis for describing such phenomena as low-cycle thermal fatigue, thermo-mechanical destruction, thermo-plastic cracking, high-temperature cyclic creep, etc. These and other unrecognized phenomena occurring in structures made of modern heat-resistant steels are crucial for properly determining the safety of the device, its heating and cooling rate, service life and, in general, resistance to thermo-mechanical load cycles. The data presented in the paper allow for the improvement of strength hypotheses, especially in the context of heat-resistant materials. These materials show a significant disparity between the corresponding mechanical properties (yield stress or strength limits) in tension and compression, especially at elevated temperatures. Currently, the Huber–Mises–Hencky (HMH) hypothesis is most often used to determine the stress state, based on the assumption of the symmetry of critical parameters in tension and compression.

In relation to materials not showing this symmetry of properties, the HMH hypothesis underestimates (in compression areas) and overestimates (in tension areas) the stress state. However, the original combined Huber hypothesis from 1904 [1] took into account this asymmetry, as it consisted in decomposing the strain energy Φ into the volumetric energy Φ_v and the distortional strain energy Φ_f. However, in the description of the stress in the compressive stress area, only the shear energy is used, while in the case of the tension area, the sum of the shear and volumetric energy is used. In turn, the Beltrami stress measure from 1885 [2] always uses both energy components, and the von Mises hypothesis from 1913 [3] always uses only the distortional strain energy. During the First International Congress of Applied Mechanics in Delft (1924), Huber was persuaded by Mises and Hencky to reject Beltrami's combined hypothesis in favor of an indefinitely cylindrical surface (only distortional strain energy), which he finally did in 1930. The original division of the energy hypotheses is presented in Table 1 and illustrated in Figure 1.

Table 1. The original division of energy hypotheses.

Author, Date of Hypothesis	Hypothesis Condition	Figure
Beltrami, 1885	$\Phi \leq K$ always	Figure 1a
Huber, 1904	$\Phi_v + \Phi_f \leq K$ when $\sigma_1 + \sigma_1 + \sigma_3 > 0$ $\Phi_f \leq K$ when $\sigma_1 + \sigma_1 + \sigma_3 < 0$	Figure 1b
Mises, 1913 HMH, 1930	$\Phi_f \leq K$ always	Figure 1a,b
	K—critical energy value determined experimentally	

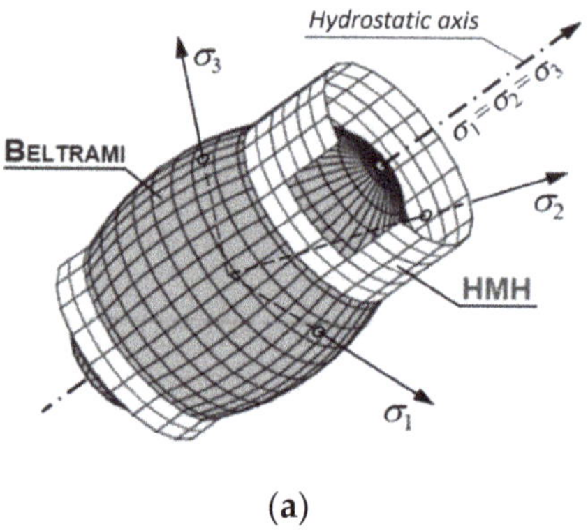

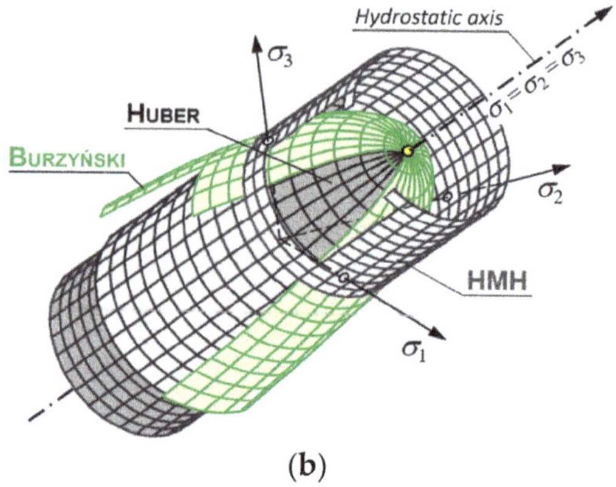

Figure 1. Drawings of boundary surfaces: Beltrami and HMH (**a**); Huber, Huber–Mises–Hencky (HMH), and Burzyński (**b**); σ_1, σ_2, σ_3—principal stresses.

Many contemporary researchers emphasize the advantage of combined hypotheses [4–8], among them is the Burzyński hypothesis, which is an extension of the original combined hypothesis of Huber from 1904.

Burzyński formulates the equation of his hypothesis in such a way that it describes both the area of tensile and compressive stresses. This allows for taking into account materials showing differences between properties in tension and compression. In addition, he made his measure of effort dependent on properties in torsion. The construction of the hypothesis itself is such that with the same properties in tension and compression and with an appropriate value in torsion, it smoothly passes into the Huber–Mises–Hencky cylinder. The issue with this hypothesis and its use in relation to the numerical analysis of turbine blades has been widely described in works [9–12].

There are many equations describing anisotropic materials, such as the Drucker-Prager, Willam-Warnke, and Menetry–Willam models. However, they are most often used for materials such as concrete and other cohesive-friction materials (rocks, soil, and ceramics) [13–16].

The Burzyński model was corrected by Życzkowski in 1981 [17]. In 1999 Życzkowski, while analyzing discontinuous bifurcations of the Burzyński condition, demonstrated the parameters of this model through which we obtain, for example, the Beltrami-Schleicher ellipsoid, the Tor paraboloid, or the Drucker-Prager cone [18]. In all these cases we obtain circular sections in the deviatoric plane. In 2011 Pęcherski modifies the Burzyński hypothesis by introducing the third invariant of the stress tensor deviator, linking it with the Lode angle. This allowed modeling of non-circularity in the deviatoric section. Pęcherski et al. analyzed the range of variability of the modified model parameters and calibrated their values for the Inconel 718 alloy. They graphically presented the small depth of the plastic surface corrugation of this alloy both in the deviatoric cross-section and in the cross-section containing the hydrostatic axis [19].

A separate aspect in determining the strength of power turbine components is the estimation of the strength of components that are essentially subject to torsion. There is a lack of such studies in the literature, especially in the context of determining the yield stress and strength during torsion at elevated temperatures. For example, Khoddam and Hodgson conducted studies to determine the strain hardening indices during torsion using the Fields and Backofen method [20]. Zhang et al. studied the phenomenon of recrystallization and microstructure evolution in duplex stainless steel SAF2205 during torsional deformation at elevated temperatures [21], and Bhanuse and Jankar developed a double-twist torsion technique. They used 316 stainless steel as a sample material for experimental evaluation of recrystallization behavior and determination of the non-recrystallization temperature [22]. Hot torsion tests have been used to analyze and physically model the flow process and microstructure evolution of materials and alloys, e.g., API-X70 alloy steel [23], or to simulate thermo-mechanical processing of Al-Cu-Mg alloy [24]. There are works in which the behavior of Nb-Ti API 5L X80 alloy steel during plastic forming is assessed on the basis of torsion tests [25] or the parameters of hot plastic forming are determined for steels such as 30Cr25Ni32Mo3 steel [26] and X22CrNi17 steel [27]. Aiello et al. used hot torsion tests to induce cracks in 9Cr steel to assess the suitability of the material for nuclear reactor components [28]. Vuppala et al. presented an extended method for determining flow curves under shear loading using torsion tests. This method was used to determine strain rate constants for 16MnCrS5 steel [29]. Ballard et al. determined only the shear stress of a sample subjected to torsion as a function of temperature to understand the behavior of ordinary structural steel during fire conditions [30]. Hjorth investigated the effect of torsion at elevated temperatures on the change in the microstructure of AA6082 aluminum [31]. Zhou and Clode used the torsion test to improve the test data in the verification of the constitutive equations of AA5252 aluminum [32]. Li et al. proposed a high-temperature dynamic shear test method for Ti-1023 titanium alloy and investigated the microstructural evolution of the tested material at different temperatures [33].

Initial and extended research results aimed at determining the characteristics and material properties in tension and compression of these two steels were presented in [9,34,35].

The aim of the presented work is to determine the properties of heat-resistant steels St12T and 26H2MF, such as torsional yield stress and torsional strength, as well as strength characteristics in pure torsion and at elevated temperatures. This makes it possible to determine which of the two tested steels is more resistant to high temperatures. The results of the torsion test in combination with the results of tension and compression are intended to determine the yield surface of these steels. This is of fundamental importance in the design process and the analysis of the causes of turbine failures.

2. Tested Materials

Two of the most frequently dedicated steels for turbine components are ST12T and 26H2MF steels. These steels are similar in most alloying elements but contain different amounts of chromium (Table 2).

Table 2. Chemical composition of analyzed steels [34].

Steel		Chemical Composition (%)							
Grade	Signature	C	Si	V	Cr	Mn	Ni	Cu	Mo
26H2MF	24CrMoV55	0.23	0.52	0.25	1.54	0.30	0.12	0.17	0.60
St12T	X22CrMoV12-1	0.16	0.37	0.24	11.10	0.44	0.42	0.13	0.96

St12T steel belongs to the martensitic class of heat-resistant steels, with high corrosion resistance in steam medium while maintaining creep resistance at the same level as steels with lower chromium content. St12T steel is used for elements of rotors and blades of steam turbines operating at temperatures up to 600 °C and even 615 °C. On the other hand, 26H2MF steel is a ferritic steel resistant to creep, abrasion, and mechanical wear. It is used in the power equipment of turbines and boilers operating at temperatures up to 540 °C. It is most often used to make shafts, pin bolts, and nuts of boiler fittings [36,37].

Considering the scope of application, it was decided that static torsion tests will be carried out for the two heat-resistant steels at the temperatures listed in Table 3.

Table 3. Torsion test temperatures of analyzed steels.

Steel	Temperature [°C]				
St12T	20	200	400	600	800
26H2MF	20	200	400	550	800

3. Experimental Method

3.1. Sample Shape and Test Procedure

To perform torsion tests, samples were prepared in the shape shown in Figure 2. Three samples were prepared from each of the tested materials for testing at a given temperature in the range of up to 600 °C (Table 3), and there was one sample for testing at 800 °C. Individual samples were numbered. The measuring lengths of all samples were $L_0 = 46$ mm and they were made with an accuracy of ±0.2 mm. The measuring diameters of each sample were measured with a micrometer and are included together with the testing temperatures in Table 4.

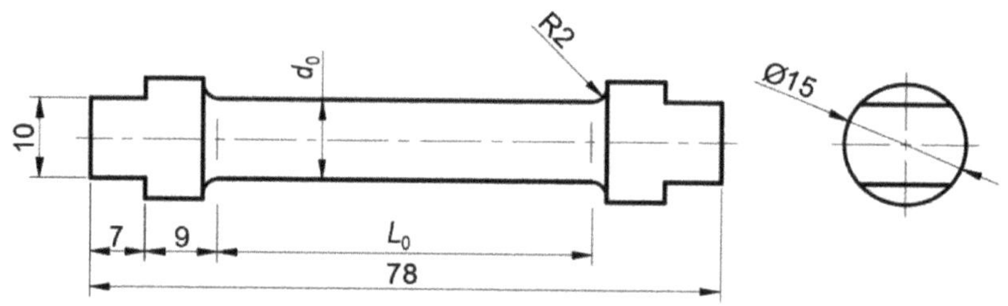

Figure 2. Shape and dimensions of torsion test samples at elevated temperatures.

Table 4. Dimensions of St12T and 26H2MF steel samples.

St12T			26H2MF		
Sample No.	Diameter d_0 [mm]	Temperature [°C]	Sample No.	Diameter d_0 [mm]	Temperature [°C]
01	9.94		01	10.13	
03	9.93	20	03	10.01	20
04	10.05		04	10.02	
31	9.98		31	9.91	
33	9.87	200	33	10.04	200
34	9.95		34	9.92	
41	10.01		41	10.02	
43	10.06	400	43	10.02	400
44	10.00		44	10.16	
61	10.05		61	10.07	
63	10.01	600	63	10.03	550
64	9.88		64	10.04	
81	9.93	800	81	10.06	800

The Heckert EUS-20 universal testing machine was adapted to perform the torsion test. The diagram of the test stand is shown in Figure 3. The screwed sample N is mounted in holders D and E. The sample holder D is mounted via bearings G in the upper crossbeam B of the machine. This allows the free rotation of the holder D together with its lever F around the vertical axis.

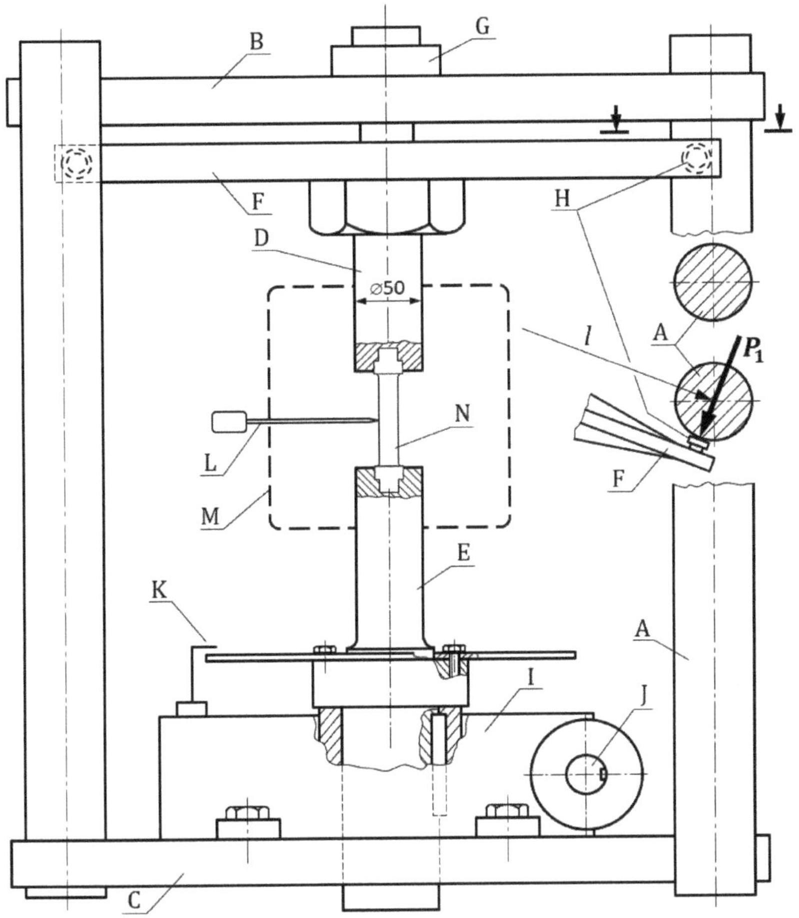

Figure 3. Schematic diagram of the stand for the torsion test of samples at elevated temperatures: machine columns (A), upper crossbeam (B), lower crossbeam (C), upper sample holder (D), lower sample holder (E), upper sample holder lever (F), lever bearing (G), strain gauges (H), worm gear (I), worm gear input shaft coupled with the main angle gauge (J), auxiliary angle gauge (K), thermocouple (L), heating chamber (M), sample (N).

At the ends of lever F, symmetrically with respect to the axis of rotation, strain gauge sensors H are mounted, in contact with columns A of the machine. Simultaneous contact of these sensors with the columns is ensured by adjustment nuts. The signal from the strain gauges is sent via the L-Card E440 analog-to-digital converter to a computer with Power Graph 3.3.8 software for data collection and pre-processing. In order to convert the deformation indicated by the sensor into a force value, calibration is performed. It consisted in recording the readings of the strain sensor when it was loaded with forces of 40, 80, 120, 160, and 200 N, and then determining the average conversion factor value in this range. For this purpose, an AXIS FA200 force meter was used, with a range of 0 ÷ 200 N and an accuracy of 0.05 N. The coefficient determined as a result of this calibration allows the readings of the strain gauge to be converted into the values of forces P_1 and P_2 with an accuracy of 0.2 N. The torque moment of the sample is then calculated from the relationship:

$$M_s = \frac{1}{2}(P_1 + P_2)l, \quad (1)$$

where l is the distance between the points of contact of the gauges with the columns.

Using a keyway connection, the sample holder E was mounted in the socket of the worm gear, which was attached to the lower crossbeam C of the machine. The worm gear drive shaft was driven by an electric motor via a belt drive. The belt transmission ratio was selected so that the engine speed (adjusted by the inverter) could ensure the sample torsion in the range of 0.03–0.3 rad/min. The speed of the belt drive shaft was controlled by a tachometer. The worm gear drive causes the lower sample holder E to rotate. Then, with the upper holder D blocked by lever F, the sample is twisted. The auxiliary angle gauge disc K is attached to the holder E, while the main angle gauge J is attached to the input shaft of the gear. Twisting the sample by 1° requires the worm to rotate by 60°, which significantly improves the accuracy of the angle of rotation reading. After the sample is attached to the holders and the clearances are eliminated, the readings of the main angle gauge and dynamometers are reset to zero. We then closed the heating chamber M and heat the sample to the desired temperature. The accuracy of the temperature measurement with the thermocouple L was ±1 °C, and the accuracy of the temperature control was ±5 °C. After the sample is heated, the heater control module switches to the set temperature maintenance mode. A total of 10 min after reaching the set temperature, the torsion process was started at a speed of 0.03 rad/min, recording the torsion angle φ and the corresponding readings of the dynamometers P_1 and P_2. After the sample had clearly plasticized, the speed was increased to 0.3 rad/min. and torsion was continued until the sample was destroyed. After recording the data and calculating the torsion moment according to the relationship (1), a torsion diagram $M_s = f(\varphi)$ was prepared.

3.2. Method of Determining the Yield Stress and Torsional Strength

Using the enlarged fragment of the graph, the conventional yield strengths $R^T_{0.3s}$ are determined graphically. A straight line B is drawn on the graph (Figure 4), parallel to the proportional part of the torsion curve $f(\varphi)$, from the place corresponding to the conventional permanent shear strain $\gamma = \gamma_{0.3} = 0.003$ [rad]. On the sample surface, the strain $\gamma_{0.3}$ corresponds to the torsion angle $\varphi_{0.3}$ such that

$$tg\gamma_{0.3} = r_0\ \varphi_{0.3}/L_0. \quad (2)$$

For small angles it is correct to assume $tg\gamma_{0.3} = \gamma_{0.3}$, then

$$\varphi_{0.3} = 0.003L_0/r_0 = 0.006L_0/d_0, \quad (3)$$

where r_0 and d_0 are the radius and diameter of the initial cross-section of the sample, and L_0 is the measuring length of the sample (Figure 2, Table 4).

Then, the straight line B is plotted for the torsion angle $\varphi_{0.3}$ given by the relationship (3). The ordinate C (Figure 4) of the intersection point of the straight line B with the torsion curve $M_s = f(\varphi)$ determines the moment $M^T_{0.3s}$, corresponding to the conventional yield strength in torsion. Finally, the value of the yield strength is calculated from the equation

$$R^T_{0.3s} = M^T_{0.3s}/W_0, \quad (4)$$

where W_0 is the cross-section index of the sample in torsion. In the case of samples with a circular cross-section, it is calculated from the formula

$$W_0 = \pi d_0^3/16. \tag{5}$$

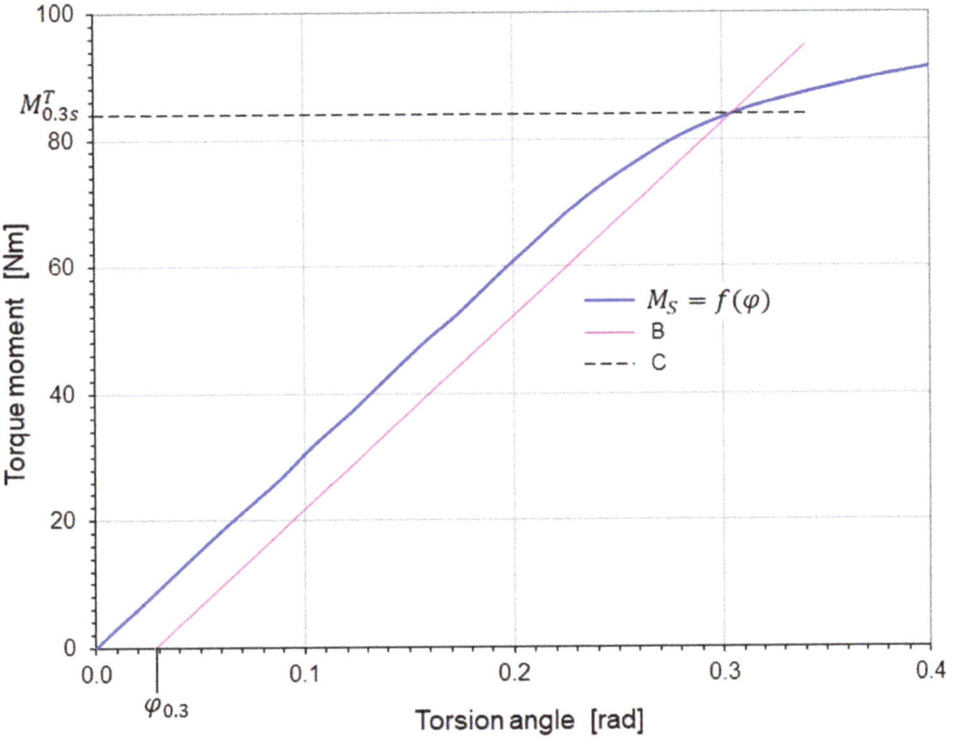

Figure 4. Graphical method for determining of the moment $M^T_{0.3s}$.

The torsional strength of a material for clearly plastic materials, whose cross-section undergoes complete plasticization, is calculated from the relationship

$$R^T_s = \frac{3}{4} M^{max}_s / W_0, \tag{6}$$

where M^{max}_s is the maximum torsional moment recorded during the torsion test of a given sample, until its failure.

4. Experimental Results

4.1. Torsion Diagrams for Heat-Resistant Steels St12T and 26H2MF

Torsion tests were carried out on samples, the specifications and test temperatures of which are listed in Table 4. The fractures of individual samples together with the fracture planes are shown in Figures 5 and 6. Their torsion diagrams $M_s = f(\varphi)$ at variable temperatures are shown in Figures 7 and 8.

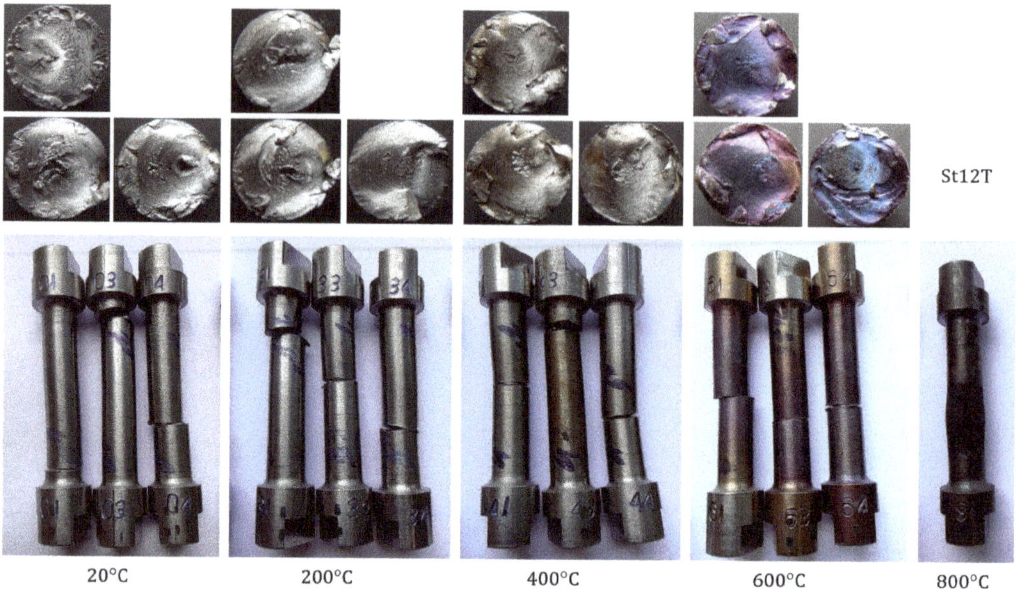

Figure 5. Samples of St12T steel after torsion tests.

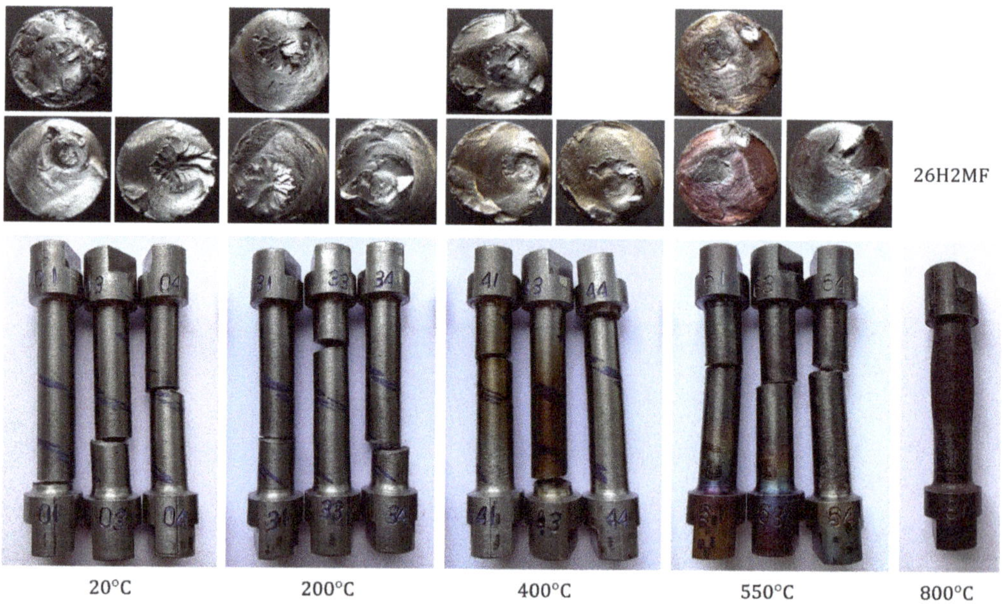

Figure 6. Samples of 26H2MF steel after torsion tests.

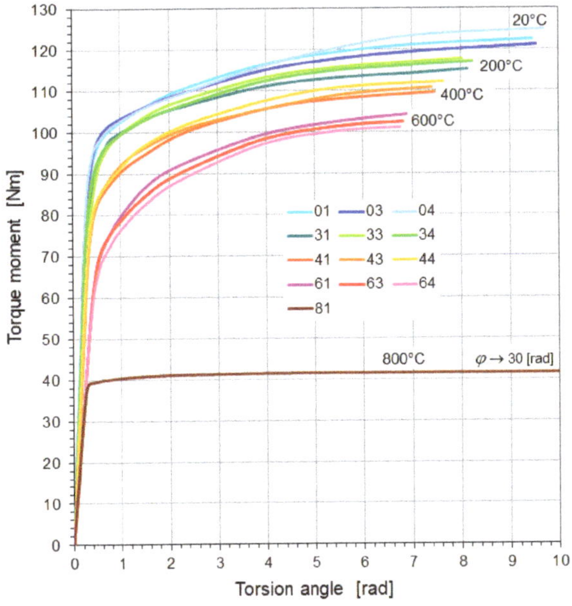

Figure 7. Torsion diagrams of St12T steel samples at elevated temperatures.

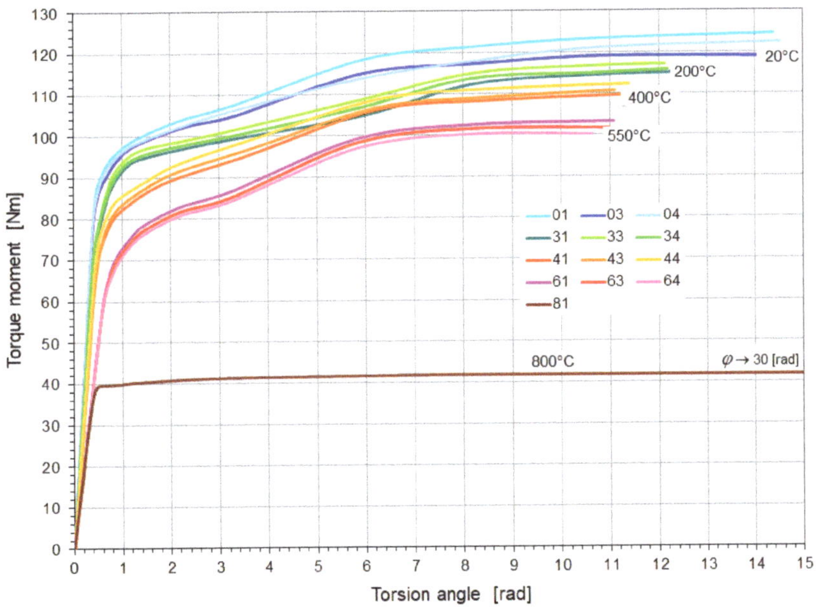

Figure 8. Torsion diagrams of 26H2MF steel samples at elevated temperatures.

All samples and their failure planes shown in Figures 5 and 6 show no visible defects; therefore, the tests can be considered valid.

The characteristics presented in Figures 7 and 8 show that in the temperature range of 20–550 °C, 26H2MF steel shows approximately 50% larger plastic torsion angles before destruction compared to St12T steel. However, at torsion angles of approximately 4 [rad] before failure, the 26H2MF steel no longer showed any tendency to hardening, which is clearly visible especially in the case of samples tested at 550 °C. At 800 °C, both steels did not show hardening and became perfectly plastic in torsion. They were not destroyed, but their base (measuring) part swelled and shortened (Figures 5 and 6). In both cases, the tests were stopped at a torsion angle of about 30 [rad] and with a practically unchanged torsional moment (Figures 7 and 8).

4.2. Torsion Strength Properties of Heat-Resistant Steels St12T and 26H2MF

The values of torsional strength R_s^T and yield stress $R_{0.3s}^T$ determined for individual samples, as well as their average values at a given temperature, are given in Tables 5 and 6, respectively. The change in the average values of the determined properties of both heat-resistant steels with temperature is shown in Figure 9.

Table 5. Yield stress $R_{0.3s}^T$ and torsion strength R_s^T of St12T steel.

Temperature [°C]	Sample No.	$R_{0.3s}^T$ [MPa]		R_s^T [MPa]	
		Samples	Average Value	Samples	Average Value
20	01	435.6	436.4	476.4	473.0
	03	439.5		472.4	
	04	434.0		470.0	
200	31	404.8	401.8	442.7	454.1
	33	397.3		466.8	
	34	403.3		452.9	
400	41	368.1	367.5	412.4	425.9
	43	365.2		421.2	
	44	369.3		444.0	
600	61	321.1	319.3	392.1	394.4
	63	319.9		390.3	
	64	316.9		400.8	
800	81	197.7	197.7	-	-

Table 6. Yield stress $R_{0.3s}^T$ and torsion strength R_s^T of 26H2MF steel.

Temperature [°C]	Sample No.	$R_{0.3s}^T$ [MPa]		R_s^T [MPa]	
		Samples	Average Value	Samples	Average Value
20	01	421.4	417.6	457.5	458.5
	03	416.4		453.2	
	04	415.1		464.7	
200	31	376.8	377.5	451.3	448.7
	33	377.5		441.6	
	34	378.3		453.1	

Table 6. *Cont.*

Temperature [°C]	Sample No.	$R^T_{0.3s}$ [MPa]		R^T_s [MPa]	
		Samples	Average Value	Samples	Average Value
400	41	342.3	346.1	416.1	414.8
	43	346.3		419.9	
	44	349.6		408.3	
550	61	314.2	311.4	386.0	382.8
	63	307.9		384.2	
	64	312.0		378.2	
800	81	190.1	190.1	-	-

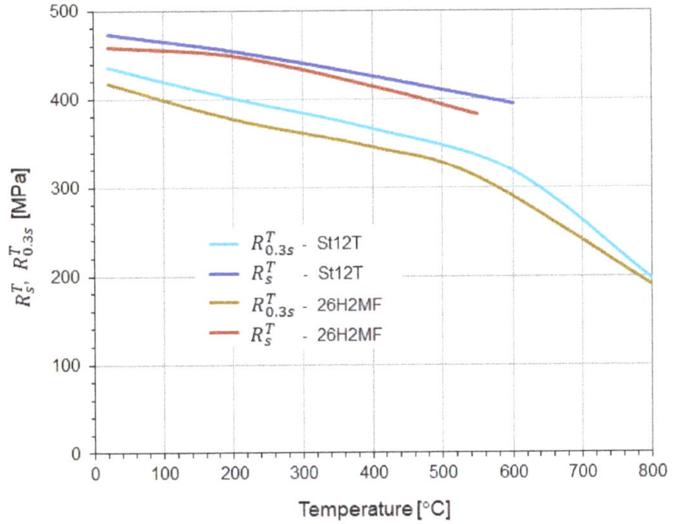

Figure 9. Average values of torsional strength R^T_s and yield strength $R^T_{0.3s}$ at elevated temperatures in steel St12T and 26H2MF.

The determined average value of the yield stress of St12T steel (Table 5) at a temperature of 20 °C was 436.4 MPa, and it decreased to 319.3 MPa at a temperature of 600 °C. In the case of 26H2MF steel, this limit (Table 6) was on average 417.6 MPa at 20 °C, and an increase in temperature to 550 °C caused a decrease in this limit to 311.4 MPa. In the case of both tested steels, a further increase in temperature to 80 °C caused a significant decrease in the yield stress value. In the case of St12T steel, this limit decreased to 197.7 MPa, and for 26H2MF steel to 190.1 MPa.

The temperature change from 20 °C to 600 °C for St12T steel and to 550 °C for 26H2MF steel (Tables 5 and 6) caused a decrease in the torsional strength value from 473 MPa to 394.4 MPa (St12T) and from 458.5 MPa to 382.8 MPa (26H2MF), respectively. In the case of testing samples of both steels at temperatures of 800 °C, it was not possible to destroy the samples (Figures 5 and 6), and at a torsion angle of 30 [rad] the tests were stopped. Hence, the torsional strength could not be formally determined at this temperature.

St12T steel is more resistant to temperature compared to 26H2MF steel. The values of its yield stress and torsional strength are higher than for 26H2MF steel at each of the temperatures: 20, 200, and 400 °C. Moreover, the values of both these properties are higher for St12T steel at 600 °C than for 26H2MF steel at 550 °C.

5. Using Torsion Test Results to Determine the Yield Surface

5.1. Burzynski's Criterion

The results of the torsion test (Tables 5 and 6) and the results of the tension and compression tests [9,12,34,35] were used to extend the Burzyński criterion to thermal stress. The critical stress parameters in his C5 hypothesis [38] for tension k_r, compression k_c, and torsion k_s were related to the yield stress:

$$k_r = R^T_{0.2r},\ k_c = R^T_{0.2c},\ k_s = R^T_{0.3s}, \tag{7}$$

where $R^T_{0.2r}$ is the conventional yield stress determined in the uniaxial tension test, $R^T_{0.2c}$ is the conventional yield stress determined in the uniaxial compression test, and $R^T_{0.3s}$ is the conventional yield stress determined in the torsion test.

All three of these boundaries are defined at variable temperatures. Two equivalent parameters of the Burzyński hypothesis were introduced: the coefficient of asymmetry of the elastic region $\varkappa_T$ and the coefficient of plasticity ν_T, in the form

$$\varkappa_T = \frac{R^T_{0.2c}}{R^T_{0.2r}}, \tag{8}$$

$$\nu_T = \frac{R^T_{0.2r}\,R^T_{0.2c}}{2\left(R^T_{0.3s}\right)^2} - 1\,. \tag{9}$$

Then the hypothesis defining the surface of the onset of plasticity will take the form

$$\begin{aligned}&\sigma_1^2 + \sigma_2^2 + \sigma_3^2 - 2\nu_T(\sigma_1\sigma_2 + \sigma_2\sigma_3 + \sigma_3\sigma_1) + \\ &(\varkappa_T - 1)R^T_{0.2r}(\sigma_1 + \sigma_2 + \sigma_3) - \varkappa_T\left(R^T_{0.2r}\right)^2 = 0.\end{aligned} \tag{10}$$

In order to apply the above thermal stress hypothesis in practice for two heat-resistant steels (St12T and 26H2MF), the coefficients $\varkappa_T$ and ν_T, corresponding to these steels, were determined. The values of these coefficients, calculated in accordance with Equations (8) and (9), are given in Table 7.

Table 7. Coefficients of plasticity and asymmetry at variable temperatures [9,12,34,35].

Steel	Temperature [°C]	Yield Strength [MPa]			Coefficient [-]	
		$R^T_{0.2r}$	$R^T_{0.2c}$	$R^T_{0.3s}$	$\varkappa_T$	ν_T
St12T	20	720.3	786.0	436.37	1.091	0.4866
	200	656.7	731.7	401.76	1.114	0.4883
	400	608.0	666.3	367.51	1.108	0.4997
	600	487.0	639.3	319.28	1.313	0.5271
	800	160.0	290.0	197.65	1.813	1.1033
26H2MF	20	667.3	779.0	417.60	1.167	0.4905
	200	610.7	693.3	377.50	1.135	0.4855
	400	567.0	628.0	346.06	1.108	0.4866
	550	507.3	573.3	311.37	1.130	0.5001
	700	388.5	460.0	238.60 *	1.184	0.5696

* Value obtained from linear interpolation between temperatures 550 and 800 °C.

The values in Table 7 allowed for the analysis of the influence of individual parameters on the geometry of the yield surface defined by the revised hypothesis (10). Figure 10 presents the evolution of the yield surface for the tested heat-resistant steels.

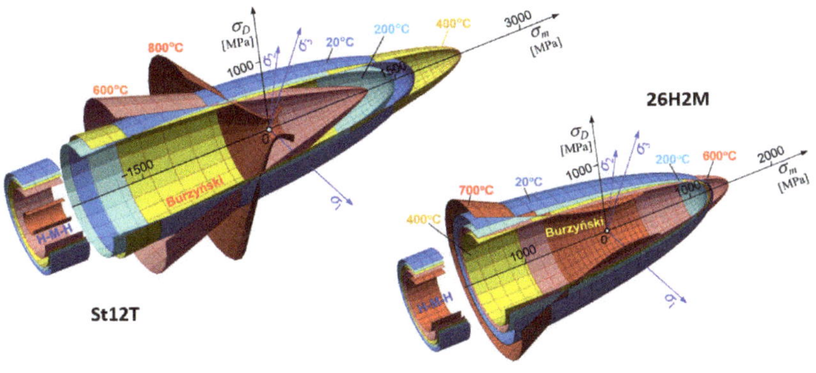

Figure 10. Evolution of the Burzynski and H-M-H yield surfaces of ST12T and 26H2MF steels with temperature.

5.2. Drucker-Prager Model

The experimental data given in Table 7 can be used to determine the yield surface using the Drucker-Prager model [13], which can be written in terms of the principal stresses as

$$\sqrt{\frac{1}{6}[(\sigma_1-\sigma_2)^2+(\sigma_2-\sigma_3)^2+(\sigma_3-\sigma_1)^2]} = A + B\,(\sigma_1+\sigma_2+\sigma_3) \tag{11}$$

where parameters A and B are given by equations

$$A = \frac{2}{\sqrt{3}}\left(\frac{R_{0.2r}^T\,R_{0.2c}^T}{R_{0.2r}^T + R_{0.2c}^T}\right) \;;\; B = \frac{1}{\sqrt{3}}\left(\frac{R_{0.2r}^T - R_{0.2c}^T}{R_{0.2r}^T + R_{0.2c}^T}\right) \tag{12}$$

Different uniaxial yield stresses in tension and in compression are predicted by the Drucker-Prager model. Their asymmetry coefficient is defined in the same way as in Burzyński's case and can be expressed by the parameter B:

$$\varkappa_T = \frac{R_{0.2c}^T}{R_{0.2r}^T} = \frac{1-\sqrt{3}B}{1+\sqrt{3}B} \tag{13}$$

However, this criterion was introduced to describe plastic deformations of soils. It and its many variants have been applied to rock, concrete, polymers, foams, and other pressure-dependent materials. Therefore, in order to assess the usefulness of this criterion for the strength analysis of heat-resistant steel elements, additional experiments should be performed in complex stress states.

5.3. Willam-Warnke and Menetrey-Willam Models

The Willam-Warnke and Menetrey-Willam yield criteria are functions that are used to predict when failure will occur in concrete and other cohesive-frictional materials such as rock, soil, ceramics, and fiber-reinforced concrete [14–16].

Before using the Willam-Warnke yield criterion to predict failure, the following must be determined: σ_t—the uniaxial tensile yield stress, σ_c—the uniaxial compressive yield stress, and σ_b—the equi-biaxial compressive yield stress.

In the case of the Menetrey-Willam model, the following parameters must be previously determined: f_t—the uniaxial tensile yield stress, f_c—the uniaxial compressive yield stress, and e—the parameter defining the roundness of the damage surface. The failure surface has sharp corners if $e = 0.5$, and is fully circular around the hydrostatic axis if $e = 1$ [15]. Alternatively, this model can be based on the parameters $\overline{f}_t$, $\overline{f}_c$, and $\overline{f}_{bc}$, which denote the actual uniaxial compressive strength, uniaxial tensile strength, and biaxial compressive strength, respectively [16].

The description of the yield surface of the tested heat-resistant steels using the Willam-Warnke and Menetrey-Willam models would require additional experiments, consisting of determining the yield stress under equi-biaxial compression at variable temperatures. Therefore, σ_b, f_b, or $\overline{f}_{bc}$ should be determined, while based on Table 7 we have $f_t = \overline{f}_t = R^T_{0.2r}$ and $f_c = \overline{f}_c = R^T_{0.2c}$. Then, in order to determine its applicability, the obtained model should be validated in a complex stress state.

6. Summary and Conclusions

The paper analyses the mechanical properties of two heat-resistant steels (St12T and 26H2MF) subjected to static torsion. Their yield and strength limits were determined when torsion occurred at different temperatures, and the data were presented in tables and graphs.

The conducted research is a response to the lack of experimental data on the behaviour of materials under torsion, which is important when designing power turbine components.

Particular emphasis was placed on comparing the results in the context of the asymmetry of mechanical properties in tension, compression, and torsion, which is crucial for validating the stress hypotheses. The research results were used to determine the parameters of the Burzyński hypothesis, extended by the effect of temperature, which enabled the presentation of the evolution of the yield surface geometry and further use of the results in numerical analyses and FEM simulations.

In addition, the use of the data contained in Table 7 to determine the parameters of the Drucker-Prager model is presented and the limitations in the use of these data in relation to the Willam-Warnke and Menetrey-Willam models are discussed. It was indicated that in this case it would be necessary to perform tests of samples under biaxial compression.

Practical conclusions from the pure torsion test of both tested heat-resistant steels:

1. The increase in temperature from 20 to 600 °C for St12T steel caused a decrease in the yield strength from 436.4 MPa to 319.3 MPa, while the torsional strength decreased from 473.0 MPa to 394.4 MPa.
2. The increase in temperature from 20 to 550 °C for 26H2MF steel caused a decrease in the yield strength from 417.6 MPa to 311.4 MPa, while the torsional strength decreased from 458.5 MPa to 382.8 MPa.
3. At 800 °C, St12T steel showed a yield strength of 197.7 MPa, while 26H2MF steel showed this limit at the level of 190.1 MPa.
4. Both steels tested at 800 °C were not destroyed despite the twisting of the samples to 30 radians, but they were shortened and swollen (Figures 5 and 6), and after exceeding the yield point, the materials of these steels showed features of ideal plasticity in torsion $M_s = f(\varphi) = const.$ (no hardening—Figures 7 and 8).
5. St12T steel has higher resistance to torsion at high temperatures than 26H2MF steel. Both its yield point of 394.4 MPa and torsional strength of 394.4 MPa at temperatures of 600 °C are higher than the yield point of 311.4 MPa and torsional strength of 382.8 MPa of 26H2MF steel determined at a temperature of 550 °C.

The research results presented in this paper offer new empirical data necessary in the modern design of energy machines.

Funding: This research received no external funding.

Institutional Review Board Statement: The study did not require ethical approval.

Informed Consent Statement: Not applicable.

Data Availability Statement: The original contributions presented in this study are included in the article. Further inquiries can be directed to the corresponding author.

Conflicts of Interest: The authors declare no conflict of interest.

References

1. Huber, M.T. Specific work of strain as a measure of material effort. *Arch. Mech.* **2004**, *56*, 173–190.
2. Beltrami, E. Sulle condizioni di resistenza dei corpi elastici. *Il Nuovo C.* **1885**, *18*, 145–155. [CrossRef]
3. Mises, R. Mechanik der festen Korper im Plastisch Deformablen Zustand. *Nachr. Ges. Wiss. Gott.* **1913**, *232*, 582–620.
4. Altenbach, H.; Bolchoun, A.; Kolupaev, V.A. Phenomenological Yield and Failure Criteria. In *Plasticity of Pressure-Sensitive Materials*; Altenbach, H., Öchsner, A., Eds.; Springer: Berlin/Heidelberg, Germany, 2014; pp. 49–152.
5. Altenbach, H.; Eremeyev, V.A.; Lebedev, L.P. On the existence of solution in the linear elasticity with surface stresses. *ZAMM* **2010**, *90*, 231–240. [CrossRef]
6. Kolupaev, V.A.; Altenbach, H.; Bolchoun, A. Comparison of strength criteria based on the measurements on concrete. *J. Eng. Mech.* **2018**, *144*, 04018028. [CrossRef]
7. Kolupaev, V.A. *Equivalent Stress Concept for Limit State Analysis*; Springer International Publishing: Berlin/Heidelberg, Germany, 2018.
8. Altenbach, H. Strength hypotheses—A never ending story. *Czas. Tech. Politech. Krak.* **2010**, *107*, 5–15.
9. Dudda, W.; Ziółkowski, P.J.; Ziółkowski, P.; Bryk, M.; Badur, J. A Concept of Thermal Effort for Heat-Induced Metal Plasticity. *Materials* **2024**, *17*, 4824. [CrossRef]
10. Ochrymiuk, T.; Dudda, W.; Froissart, F.; Badur, J. Principles of Stress-Strength Modelling of the Highly Thermally Loaded Materials—The Influence of an Effect of Strength Differential on the Material Effort. *Materials* **2021**, *14*, 7449. [CrossRef]
11. Bryk, M.; Banaszkiewicz, M.; Kowalczyk, T.; Dudda, W.; Ziółkowski, P. Slowly-closing valve behaviour during steam machine accelerated start-up. *Case Stud. Therm. Eng.* **2022**, *39*, 102457. [CrossRef]
12. Dudda, W.; Kraszewski, B. A theoretical validation of Burzyński hypothesis for a stress-strain analysis of heat-resistant steel. *Case Stud. Therm. Eng.* **2021**, *23*, 100806. [CrossRef]
13. Drucker, D.C.; Prager, W. Soil mechanics and plastic analysis for limit design. *Q. Appl. Math.* **1952**, *10*, 157–165. [CrossRef]
14. Willam, K.J.; Warnke, E.P. Constitutive models for the triaxial behavior of concrete. *Proc. Int. Assoc. Bridge Struct. Eng.* **1975**, *19*, 1–30.
15. Menetrey, P.; Willam, K.J. Triaxial failure criterion for concrete and its generalization. *ACI Struct. J.* **1995**, *92*, 311–318.
16. Dmitriev, A.; Novozhilov, Y.; Mikhalyuk, D.; Lalin, V. Calibration and validation of the Menetrey-Willam constitutive model for concrete. *Constr. Unique Build. Struct.* **2020**, *88*, 8804.
17. Życzkowski, M. *Combined Loadings in the Theory of Plasticity*; PWN: Warszawa, Poland, 1981.
18. Życzkowski, M. Discontinuous bifurcations in the case of the Burzyński-Torre yield condition. *Acta Mech.* **1999**, *132*, 19–35. [CrossRef]
19. Nowak, M.; Ostrowska–Maciejewska, J.; Pęcherski, R.B.; Szeptyński, P. Yeld condition accounting for third invariant of stress tensor deviator. *Eng. Trans.* **2011**, *59*, 273–281.
20. Khoddam, S.; Hodgson, P. Post processing of the hot torsion test results using a multi-dimensional modelling approach. *Mater. Des.* **2010**, *31*, 2578–2584. [CrossRef]
21. Zhang, J.; Yao, K.; Chen, H.; Zhu, G.; Li, F. Microstructure evolution of SAF2205 duplex stainless steel during torsion deformation at elevated temperature. *Rare Met. Mater. Eng.* **2018**, *47*, 1689–1694.
22. Bhanuse, M.M.; Jankar, P.D. Behavior of Steel under Elevated Temperature—Experimental work. *Int. J. Adv. Sci. Tech. Res.* **2013**, *6*, 700–705.
23. Mirzakhani, B.; Salehi, M.T.; Arabi, H.; Khoddam, S.; Seyedein, S.H.; Aboutalebi, M.R. Predication of temperatur̹e distribution in the hot torsion test specimen. *Int. J. Eng. Sci.* **2008**, *19*, 45–50.
24. Carreño, F.; Cepeda-Jiménez, C.M.; Peñalba, F.; Carsí, M.; Ruano, O.A. Simulation of hot rolling processing of an Al-Cu-Mg alloy by torsion tests. *Mater. Sci. Forum* **2012**, *706–709*, 277–282. [CrossRef]
25. Martins, F.H.; Machado, M.L. Estudo do comportamento termomecânico do aço API 5L X80 microligado ao Nb-Ti através de ensaios de torção a quente. *Rev. Esc. Minas* **2015**, *68*, 195–200. [CrossRef]
26. Ebrahimi, G.R.; Keshmiri, H.; Arabshahi, H. Mechanical Characteristics of Superaustenitic Stainless Steel Type 30Cr25Ni32Mo3 at Elevated Temperatures. *Mater. Sci. Appl.* **2010**, *1*, 323–328.

27. Gusel, L. Obtaining the stress-strain behaviour of stainless steel at elevated temperatures. *Mater. Sci.* **2019**, *2*, 47–50.
28. Aiello, G.; Caes, C.; Lamagnere, P.; Martin, A.; Sauzay, M. Tension–torsion ratcheting tests on 9Cr steel at high temperature. *Nucl. Eng. Des.* **2015**, *284*, 207–214.
29. Vuppala, A.; Brüggemann, H.; Bailly, D.; Scharifi, E. An Inverse Piecewise Flow Curve Determination Method for Torsion Tests at Elevated Temperature. *Metals* **2025**, *15*, 219. [CrossRef]
30. Ballard, T.J.; Speer, J.G.; Findley, K.O.; De Moor, E. Double twist torsion testing to determine the non recrystallization temperature. *Sci. Rep.* **2021**, *11*, 1495. [CrossRef]
31. Hjorth, A. The Effect of Deformation by Torsion and Subsequent Heat Treatment on the Microstructure of AA6082. *Mater. Sci. Eng.* **2018**, *3–4*, 1–48.
32. Zhou, M.; Clode, M.P. Thermal analysis of the torsion test under hot-working conditions. *Comput. Mater. Sci.* **1998**, *9*, 411–419. [CrossRef]
33. Li, L.; Wang, Z.; Ma, W. Experimental Study on the High Temperature Impact Torsional Behavior of Ti-1023 Alloy. *Materials* **2022**, *15*, 3847. [CrossRef]
34. Dudda, W. Influence of High Temperatures on the Mechanical Characteristics of 26H2MF and ST12T STEELS. *Mater. Sci.* **2019**, *55*, 435–439. [CrossRef]
35. Dudda, W. Mechanical Characteristics of 26H2MF and St12T steels under compression at elevated temperatures. *Strength Mater.* **2020**, *52*, 325–328. [CrossRef]
36. Ennis, P.J.; Czyrska-Filemonowicz, A. Recent advances in creep-resistant steels for power plant applications. *Sadhana* **2003**, *28*, 709–730. [CrossRef]
37. Klotz, U.E.; Solenthaler, C.; Uggowitzer, P.J. Martensitic-austenitic 9–12% Cr steels—Alloy design, microstructural stability, and mechanical properties. *Mater. Sci. Eng. A* **2008**, *476*, 186–194. [CrossRef]
38. Burzyński, W. Selected passages from Wodzimierz Burzyński´s doctoral dissertation: Study on material effort hypotheses. *Eng. Trans.* **2009**, *57*, 185–215.

materials

MDPI

Article

Stiffness and Density Relationships in Additively Manufactured Structures: A Virial Theorem-Based Approach

Tomáš Stejskal [1], Silvia Maláková [1,*], Marcela Lascsáková [2] and Peter Frankovský [3]

1 Department of Engineeringfor Design of Machines and Transport Equipment, Faculty of Mechanical Engineering, Technical University of Kosice, Letna No. 9, 042 00 Kosice, Slovakia; tomas.stejskal@tuke.sk
2 Department of Applied Mathematics and Informatics, Faculty of Mechanical Engineering, Technical University of Kosice, Letna No. 9, 042 00 Kosice, Slovakia; marcela.lascsakova@tuke.sk
3 Department of Applied Mechanics and Mechanical Engineering, Faculty of Mechanical Engineering, Technical University of Kosice, Letna No. 9, 042 00 Kosice, Slovakia; peter.frankovsky@tuke.sk
* Correspondence: silvia.malakova@tuke.sk; Tel.: +421-55-603-2372

Abstract: Topological optimization uses two main optimization conditions aimed at achieving the maximum stiffness at minimum weight of the loaded object, while not exceeding the allowable stress. This process naturally creates complex structures with varying degrees of density. There is a certain regularity between the density of the structure and stiffness, with the optimal density being related to the golden ratio. This study contributes to materials modeling and their characterization by introducing a mathematical theory related to the virial theorem as a predictive framework for understanding stiffness–density relationships in additively manufactured structures. The definition of virial stability and the methodology for deriving this stability from the kinetic and potential components of a random signal are introduced. The proposed virial-based model offers a generalizable tool for materials characterization, applicable not only to topological optimization but also to broader areas of materials science and advanced manufacturing.

Keywords: virial stability; signal energy; materials modeling; structure stiffness; additive production technologies

1. Introduction

Nowadays, additive manufacturing technologies are increasingly being promoted. Three-dimensional printing allows the creation of complex-shaped structures of products that have optimized parameters. The purpose of topological optimization is to create the desired shapes of components while maintaining the highest possible stiffness at the minimum weight. These two optimization criteria have become the basis of simulation methods for the design of modern machine structures. The first optimization condition is an even distribution of stress in the structure. The second condition is a reduction in weight. A characteristic feature of topological optimization is that products resemble various natural patterns in shape. This connection will be justified in the following text.

The density of structures has a significant impact on the stability of the system under extreme loads. Optimization in terms of structural density is often focused on resistance to damage or absorption of impact energy [1–4]. The issue of the stability of various biomimicry structures has been addressed in works [5–8]. These works are focused on finding optimal structures, often by empirical methods. It can be assumed that there could be a third optimization condition for the ideal choice of the density of structures, which would also show the highest degree of stability and rigidity. A theoretical optimization condition is sought.

In this paper, an adequate mathematical theory is derived to support the results obtained by simulation. The thermodynamic theory of gases was used to find the necessary connections.

2. Materials and Methods

It is known that the total energy of any mechanical system is given by the sum of the potential and kinetic components. In general, the ratio of these components can be arbitrary. However, if we look at a complex system with random processes, the virial theorem applies, and the ratio of total potential energy to total kinetic energy, simultaneously from all components of the system, is 1:2. This result is surprising, because the cause of this phenomenon is not clear at first glance. The virial theorem was derived for the thermodynamic state of a gas, first published by the German researcher Rudolf Clausius. This discovery follows a certain historical sequence, which is linked to the second half of the 19th century.

Clausius's most famous statement about the second law of thermodynamics was published in German in 1854 [9] and in English in 1856 [10]. However, the concept of entropy was not defined until several years later. In 1865, Clausius defined entropy [11], and in 1870 he stated the virial theorem, which applied to heat [12].

The second law of thermodynamics states that the direction of energy propagation is always given by the direction of equalization of different energy potentials and never the other way around. The second law of thermodynamics can validate that the entropy of isolated systems left to spontaneous evolution cannot decrease, because they always tend towards a state of thermodynamic equilibrium, where the entropy is highest for a given internal energy [13]. The increase in the combined entropy of a system and its surroundings accounts for the irreversibility of natural processes, often referred to by the concept of the arrow of time [14,15].

The second law of thermodynamics corresponds from a mathematical point of view to the central limit theorem. In probability theory, the central limit theorem states that under appropriate conditions, the distribution of a normalized version of the sample mean converges to the standard normal distribution. The oldest version of this theorem is the de Moivre–Laplace theorem, according to which the normal distribution can be used as an approximation to the binomial distribution [16,17].

The binomial distribution is the basis for the principle of energy propagation in a cloud of gas particles. This consideration provides a basis for investigating whether another statistical tool does not inherently describe other thermodynamic quantities without physical connections. Such a phenomenon was found in the form of the virial theorem. This theorem is presented with a physical example; it is also possible to construct its mathematical form. For comparison, we first present the physical model of the virial theorem.

Physical Model of the Virial Theorem

The definition of the second law of thermodynamics implies that the total energy in a closed system tends to be distributed evenly throughout the entire volume. If this were not the case, then there would naturally have to be different energy potentials, and the entropy would not reach its maximum value. It is a random process in which the overall ratio of average potential and average kinetic energy stabilizes. Based on this, the virial theorem is a direct consequence of the second law of thermodynamics.

A set of particles of different masses is determined in a potential field by position vectors $\mathbf{r}_i$ on which forces caused by potential energy and by kinetic energy simultaneously act. Thus, in general, a system of particles and applied forces corresponds to the dot product of momentum and position. Then, the potential quantity G, which generally includes all types of forces, can be defined as

$$G = \sum_i \mathbf{p_i r_i} \,, \tag{1}$$

where $\mathbf{p_i} = m_i \mathbf{v_i}$ is the momentum of the i-th component of the system, and $\mathbf{r_i}$ is the position vector of the i-th component measured from the origin. The law of conservation of momentum in a closed system applies. The time derivate separates the potential E_p and the kinetic energy E_k of components [18]:

$$\frac{dG}{dt} = \sum_i \mathbf{p_i}\frac{d\mathbf{r_i}}{dt} + \sum_i \frac{d\mathbf{p_i}}{dt}\mathbf{r_i} \,, \tag{2}$$

$$\frac{dG}{dt} = \sum_i m_i \mathbf{v_i}^2 + \sum_i \mathbf{F_i r_i} \,, \tag{3}$$

$$\frac{dG}{dt} = 2E_k + E_p \,. \tag{4}$$

Equation (4) states that for a very large number of components, an average value can be considered, which determines the average value of both kinetic and potential energy over a sufficiently long time interval. The local components of the energy may change, but the total energy from all components does not change with time τ. To distinguish them from the usual energy notation, average energies are indicated by angle brackets. The energy state can be expressed as follows:

$$\frac{1}{\tau}\int_0^\tau \frac{dG}{dt} \equiv \langle \frac{dG}{dt} \rangle = 2\langle \frac{1}{2}\sum_i m_i \mathbf{v_i}^2 \rangle + \langle \sum_i \mathbf{F_i r_i} \rangle = \frac{1}{\tau}[G(\tau) - G(0)] \,. \tag{5}$$

The function on the right-hand side of the equation approaches zero for a sufficiently large τ, then

$$2\langle E_k \rangle + \langle E_p \rangle = 0 \,. \tag{6}$$

The resulting ratio of energy components can be expressed using a virial proportion in the form

$$\frac{|\langle E_p \rangle|}{|-\langle E_k \rangle|} = 2 \,. \tag{7}$$

The virial theorem allows one to calculate potential energy from knowledge of kinetic energy alone and vice versa.

3. Results

The virial theorem can be derived based on mathematical assumptions. This model is suitable for the design of material structures.

3.1. Mathematical Model of the Virial Theorem

The mathematical model is based only on mathematical assumptions. The starting point is mathematical statistics represented by the binomial distribution of random events. The binomial distribution can describe random physical phenomena, such as the random motion of gas molecules, but it is inherently a concept, independent of the physical world. The mathematical approximation of the binomial distribution to a continuous function is the well-known normal probability distribution. The normal probability distribution for the general parameters of standard deviation σ and mean μ is defined by a probability density of the form

$$g(x) = \frac{1}{\sigma\sqrt{2\pi}} e^{-\frac{(x-\mu)^2}{2\sigma^2}} \,, \qquad g(x) \in (0,1) \,, \tag{8}$$

where x is the parameter of the random event. Random phenomena can be of a different spatial nature. The simplest manifestation of randomness is a sequence of independent values describing some phenomenon. It could also be a random signal. In a random signal, time is not understood as a continuous physical quantity but as a sequence of normally distributed random events that change the value of the signal. In this way, physical time can be discretized and perceived as a mathematical object without physical meaning. This means that a time unit, for example a second, is given by the order of values. The randomness principle causes the probability distribution of values around the mean value of the signal to be naturally normal (Figure 1).

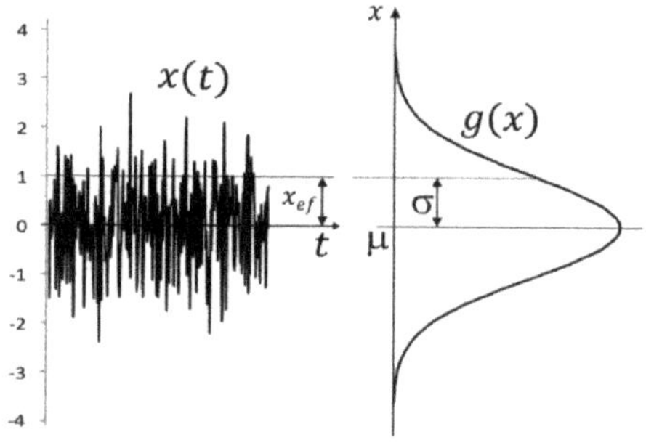

Figure 1. Normal distribution of a white noise signal.

A random signal has certain specific properties that partly express the relationship to physical models. It is about the power and energy signal.

The power signal is officially defined in the form

$$P_s = \lim_{T\to\infty} \frac{1}{T} \int_{-T/2}^{T/2} x^2(t)dt = x_{ef}^2\,, \tag{9}$$

where $x(t)$ is an arbitrary signal without specifying a physical dimension, x_{ef} is the rms value of the signal, and T is the period of the signal. The rms value of the signal has a normal distribution for white noise, and it is equal to the standard deviation of the distribution.

The energy signal is defined as follows:

$$E_s = \int_{-\infty}^{\infty} x^2(t)dt\,. \tag{10}$$

In a steady, stationary signal, the average signal power does not change over a sufficiently long period of time. The above results are presented in signal processing theory [19,20].

In general, signal energy increases continuously over time. Therefore, to characterize the process, the probability of the signal $g(x)$ occurring depending on the deviation x is used. This probability is proportional to the average power of the signal.

Both the power and signal energy parameters depend on the square of the deviation. Analogously, it can be considered in relation to the normal distribution that there is a power or energy distribution of a random phenomenon in the form of a quadratic function.

The exponentiation of the normal distribution slims down the resulting distribution by a constant ratio $1/\sqrt{2}$.

$$g^2(x) = \frac{1}{\sigma^2 2\pi} e^{-\frac{(x-\mu)^2}{\sigma^2}} = \frac{1}{\sigma 2\sqrt{\pi}} \frac{1}{\left(\frac{\sigma}{\sqrt{2}}\right)\sqrt{2\pi}} e^{-\frac{(x-\mu)^2}{2\left(\frac{\sigma}{\sqrt{2}}\right)^2}}. \tag{11}$$

We get a new normal distribution with standard deviation $\sigma/\sqrt{2}$. The expression before the exponent only changes the height of the division but does not change the shape.

Let the distance of a point from the mean value of a random signal express the potential energy of the signal. Then the shape of the probability distribution of the potential energy as a function of the deflection x can be obtained by exponentiating the normal distribution. This exponentiating is similar in nature to determining the efficiency of a mechanical system. The total efficiency is given by the product of the efficiencies of the individual components. In this case, it is the probability of reaching a given deviation and the probability of the energy value at a given deviation. Both probabilities have the same distribution. To highlight the probabilistic property of the normal distribution, in the following text we will work with the standard normal distribution $g(x)_{0,1}$. With the standard normal distribution, the displayed curve clearly expresses the probability of the phenomenon. Two identical standard normal distributions are multiplied together. The probability distribution of potential energy is then of the form

$$E_{sp} = k \int_{-\infty}^{\infty} g^2(x)_{0,1} dx = \int_{-\infty}^{\infty} P_{sp}(x) dx = 1\,, \tag{12}$$

where E_{sp} is the area proportional to the total potential energy of the signal or the average power from the potential energy, $P_{sp}(x)$ is the probability of the power of the potential energy of the signal depending on the deviation, $g(x)_{0,1}$ is the standard normal distribution of the signal deviation, and $k \approx 35.4461$ is the conversion coefficient that ensures a unit area under the curve $P_{sp}(x)$ (Figure 2). This probability corresponds in the time domain to the power of the potential energy signal.

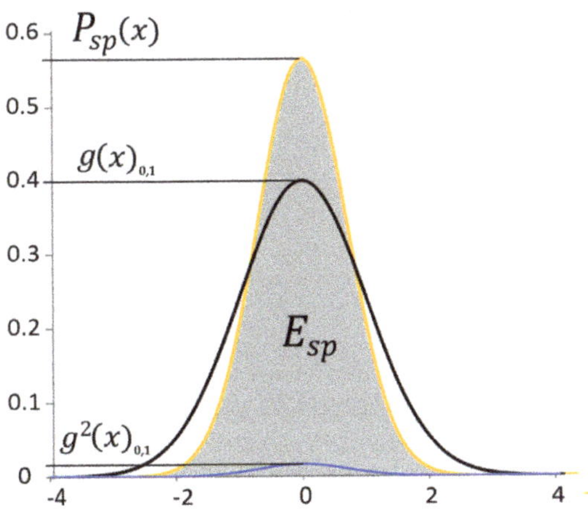

Figure 2. The standard normal distribution of a random signal and the distribution of the potential energy signal.

In mechanical systems such as a mechanical oscillator, kinetic energy is the derivative of the change in potential energy. In such a simple case, when it comes to the harmonic function of the oscillator's deflection, the sum of two squared harmonic functions gives a constant. This result of the sum of the potential and kinetic energy corresponds to the total energy, which does not change based on the conservation law of energy. Analogously, it can be established in relation to the normal distribution that the kinetic energy of the signal will have a random distribution obtained by the derivative of the normal distribution (Figure 3), which we will subsequently square.

The procedure for deriving the kinetic energy of a signal is designed based on the analogy with a mechanical oscillator.

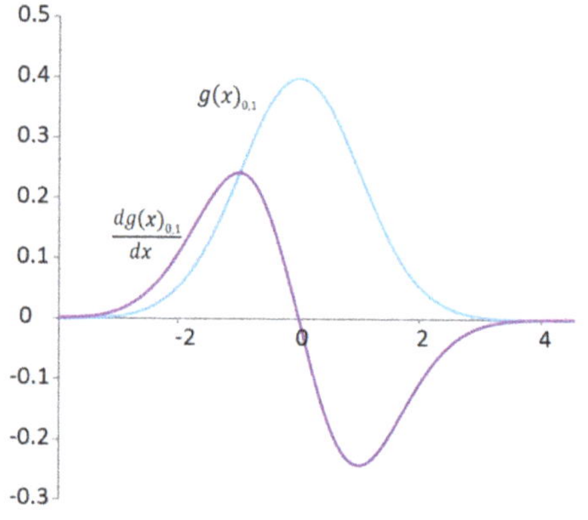

Figure 3. Standard normal distribution of a random deflection signal and distribution of the kinetic energy signal.

By exponentiating the derivative of the normal distribution, the distribution of the average kinetic energy output is obtained (Figure 4). Using Equation (12), the area corresponding to the kinetic energy can be determined:

$$E_{sk} = k \int_{-\infty}^{\infty} \left(\frac{dg(x)_{0,1}}{dx} \right)^2 dx = \int_{-\infty}^{\infty} P_{sk}(x) dx = 0.5\,, \tag{13}$$

where E_{sk} is the area proportional to the total kinetic energy of the signal or the average power from the kinetic energy, $k \approx 3.54461$ is the conversion coefficient, and $P_{sk}(x)$ is the probability of the power of the kinetic energy of the signal depending on the deflection. That is to say, it is the kinetic power signal. This probability corresponds in the time domain to the power of the kinetic signal energy. In this case, the same conversion coefficient as for potential energy must be used.

Remark 1. *The value of the conversion coefficient k does not matter. If any coefficient is used, the ratio of areas under the curves does not change.*

The virial proportion is the ratio of the potential and kinetic energy of the entire random signal equal to two. The virial theorem, from a mathematical point of view, arises by comparing the areas of probability distributions that correspond to the total potential and total kinetic energy of a signal.

$$\frac{E_{sp}}{E_{sk}} = 2\,. \tag{14}$$

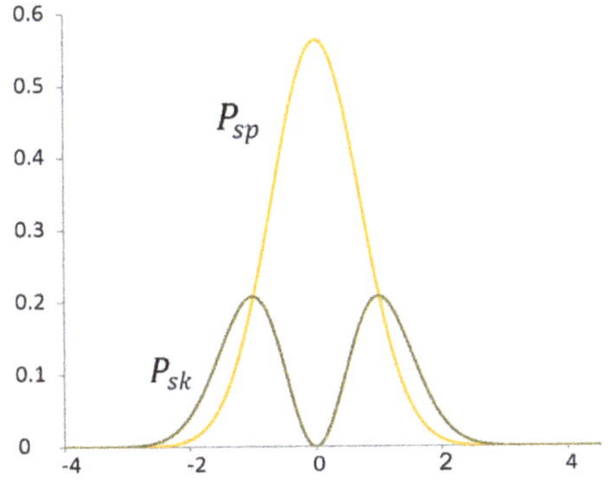

Figure 4. Power distribution of potential and kinetic energy signals.

This result is a specific property of the normal distribution, which is a mathematical concept. The result supports the validity of the shape of kinetic energy but is not direct evidence of it. An area ratio equal to two can theoretically exhibit different curve shapes. On the other hand, the derivative of the normal distribution is the natural shape of the curve. In the following text, it is shown that this shape of the curve is consistent with the Maxwell–Boltzmann distribution of the velocity of molecules of an ideal gas, which supports the correctness of the assumption.

Remark 2. *The area calculation process can be verified numerically. It is therefore appropriate to show numerically that with an increasing accuracy of curve integration, the result of the virial proportion converges to the number two (Table 1).*

Table 1. Convergence of the virial proportion, which is dependent on the integration step size.

Step Size	Virial Proportion
0.5	2.0631
0.2	2.0100
0.1	2.0025
0.05	2.0006

3.2. Concept of the Virial Theorem for Mechanical Systems with Random Motion

The concept of random molecular motion can also be transferred to compact mechanical systems. Mechanical systems can be understood as various connected oscillators. Mutually independent oscillators that have the same oscillation period creates a random system. The signal will be given by the deviation of the mass point from the equilibrium position. These are mechanical oscillators that are unsteady and have a random distribution of deviations over time. This physical model can be rewritten into a mathematical model by recording each oscillator at a given random position as a sequence of random deflections. The result will be a signal with an arbitrary probability distribution.

For completely independent oscillators, an arbitrary probability distribution can be proposed. In practice, there are only dependent oscillators that together form a complex system. In this case, we are dealing with coupled oscillators, where one oscillator is always the driving oscillator, and the other is the driven oscillator. During the energy transfer, there is a quarter phase shift of the oscillations between the driving and driven oscillators. After all the energy has been relocated from one oscillator to the other, the roles are reversed. In the long term, with a large number of oscillators, the same principle can be considered as in the collisions of gas molecules. Then the energy distribution of the oscillators involved in creating the maximum potential energy will have a normal distribution, and the virial theorem can be applied. The theory of the energy transfer of coupled oscillators is well described in the relevant literature [21–24].

It is worth noting that simple systems such as an isolated oscillator exhibit a virial ratio of 1:1, because there is no influence of randomness in such a system.

3.3. Comparison of the Maxwell–Boltzmann Distribution with the Kinetic Power Signal

A closed system with material objects is subject to gravitational potential energy. The role of potential energy is to hold material objects in a given space. Mass particles must also have kinetic energy. Otherwise, all objects would immediately fall into a common center of gravity, and the motion would cease.

In the case of a gas observed on Earth, only kinetic energy is dominant, which is directly dependent on the square of the molecular velocity. To prevent molecules from escaping, a closed space is necessary. The material walls of an enclosed space represent potential energy. The manifestation of potential energy is then expressed by the gas pressure.

Maxwell was the first to formulate the probability distribution of the velocity of a complex mechanical system [25,26]. The random movement of molecules creates pressure when they collide with a wall. At any moment, some molecules increase the force exerted, while others decrease the force and leave the wall. There are also differences in the speed of impact and the magnitude of the force impulse. We assume that the principle of the central limit theorem applies, and, based on many partial influences, none of which is dominant, a normal distribution of the contribution of individual molecules to the total pressure is realized. Based on this, the potential energy of molecules has a normal distribution.

For the molecules' speed, two opposing conditions can be defined, determining the final state of the system:

1. The distribution of molecular speeds follow the normal distribution $g(x)$, where zero speed is the most probable. At zero speed, the potential energy of the molecule is at its maximum. This phenomenon occurs at the moment of collision of two molecules, when maximum pressure is created.
2. The distribution of molecular velocity is subject to a quadratic function $f(x) = x^2$, whose minimum is at zero velocity, which is a characteristic of kinetic energy as a function of velocity's growth. At the moment of collision of two molecules, the kinetic energy of the molecule under consideration is naturally zero.

The resulting state of the probability distribution of the velocity of molecules in one orientation is given by the product of the above contradictory conditions. The result corresponds to the Maxwell–Boltzmann distribution of molecular velocity.

The parametric expression of the probability density of the Maxwell–Boltzmann distribution is given in the form

$$m(x) = \frac{1}{a^3}\sqrt{\frac{2}{\pi}}x^2 e^{-\frac{1}{2a^2}x^2}, \qquad x \in (0, \infty), \tag{15}$$

where a is the partition parameter depending on the temperature of the ideal gas. After substituting physical units, we have

$$x = v \tag{16}$$

$$a = \sqrt{\frac{k_B T}{m}} \tag{17}$$

where v is the velocity of the molecules, k_B is the Boltzmann constant, T is the temperature of the gas, and m is the mass of the molecules.

The kinetic energy performance probability curve P_{sk} is very close in shape to the Maxwell–Boltzmann distribution $m(x)$ (Figure 5). A numerical comparison can be made by slimming down the Maxwell–Boltzmann distribution $m(x)_{mod}$ using the standard deviation $\sigma = 1/\sqrt{2}$. The obtained distribution is adjusted in height by multiplying it with a correction coefficient $z = 10$, thus obtaining a good overlap with the kinetic energy power probability curve. The value of the correction coefficient allows one to achieve the best overlap of the two curves, but at the same time it does not affect the shape of the curves. During the analysis, the emphasis is placed on achieving a good shape matching of the curves.

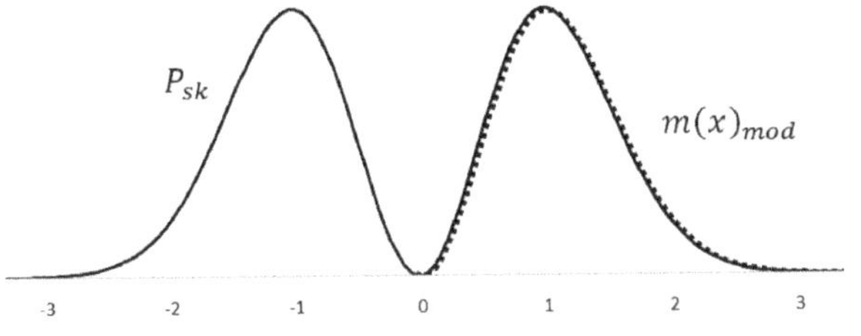

Figure 5. Overlapping of the Maxwell–Boltzmann distribution and the square of the derivative normal distribution.

The resulting correlation of the curve shape is quite high. For 82 numerical values that correspond to the 3 sigma interval, the correlation coefficient is $\rho = 0.9987$. The good shape match supports the theory that the kinetic component of the signal energy can also well describe the probability distribution of the molecular velocity.

3.4. Virial Stability of a Random System

Any mechanical system with cyclic or chaotic motion can theoretically have any ratio of kinetic and potential energy. For a random system, the virial theorem was derived, according to which although the components may have different ratios of kinetic and potential energy, the average ratio of all components of the system in the long term will be 2:1. This is a law that speaks about a property of the whole, but it says nothing about the statistical distribution of kinetic energy of this property in space. From the analysis it is clear that in the chaotic motion of gas molecules all states lie between two extreme states. At the zero value of the distribution, there is an extreme probability in which the molecules are not moving and have a maximum potential energy. At the other extreme of the distribution, molecules move at maximum speed, have maximum kinetic energy, and minimum potential energy. Statistically, only a minimal number of molecules are found at both extremes. Most molecules are in a state between these two extremes.

Definition 1. *The virial stability of a random system is the point at which the mode of the kinetic energy distribution of the system has exactly the equality of the probabilities of the potential and kinetic energy outputs.*

Every uncontrolled mechanical system tries to assume the most stable position of its components. The virial theorem expresses this thesis in mathematical language. The most probable state is given by virial stability. The disruption of virial stability can be imagined as the moment when a warm volume of gas mixes with a cold one. Both gas volumes were in virial stability before mixing. A new virial stability occurs only after the temperatures of the mixed gases equalize.

3.5. Proportionality of the Virial Theorem

For the proportionality analysis of the normal distribution, it is convenient to compare significant points with the derivative curve in one graph (Figure 6). The ratio of the peaks of the maximum values of random distributions is quite close to the ratio of the golden ratio $1 : \Phi$ for $\Phi \approx 1.618$.

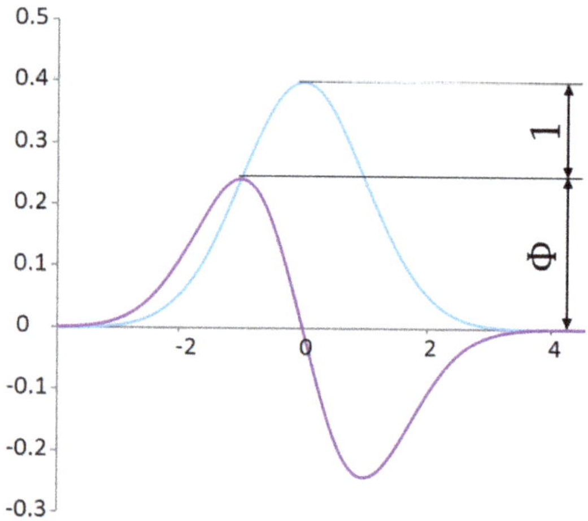

Figure 6. The proportionality between the standard normal distribution and its derivative.

The point that determines the division in the golden ratio is not only the mode of the power distribution of the kinetic energy of the signal but also the intersection of the power distribution with the potential energy of the signal (Figure 7). This dividing point has a specific property. At this point, the increment of the original function is equal to the derivative of the original function. This point expresses virial stability.

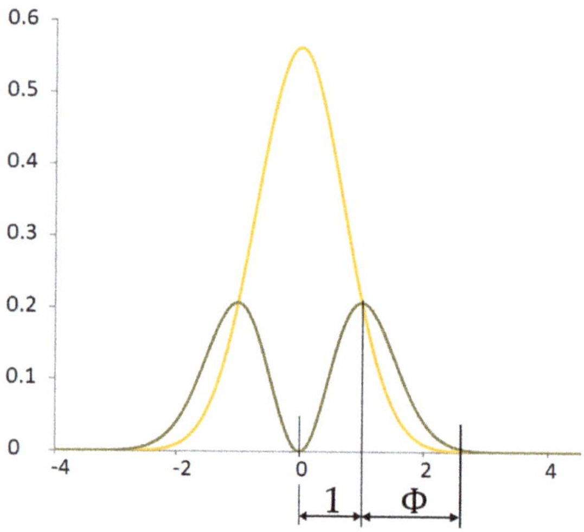

Figure 7. Proportionality between probability distribution of potential and kinetic power signals.

It can be stated that at the point of virial stability, the component of the system is equally far from the state of complete immobility as from the state of maximum mobility.

If the least probable states of the system represent extreme states, then the most probable state lies in the golden ratio between the two extremes.

3.6. Design of Random Material Structures for 3D Printing

Homogeneous structures essentially have two variable parameters.

The first parameter is the arrangement of the components by size. This arrangement corresponds to a geometric series of component sizes. The series can be modified by changing the quotient of the series or by omitting some components.

The second parameter is the rate of components. From an energetic point of view, the most likely relationship is an inverse relationship between the size of a component and its rate in the structure.

Nature always creates the most efficient structures. These structures have the form of a fractal. This is a natural law that is connected to the second law of thermodynamics. The essence is that the distribution of energy to create a structure is influenced by the size of the component and its rate. The analysis shows that the distribution of dynamic elements is ideally in the form of the kinetic power of the signal. This is essentially the derivative of the square of the normal distribution. If the horizontal axis is the size of the components, then the vertical axis will present the ideal rate of elements in the structure. The highest rate is at the point of virial stability.

It can be assumed that deviation from these ideal parameters will create structures with worse properties under load. At a constant volume of mass, which is the structure, this will be reflected in a reduced stiffness compared to a structure with ideal parameters.

The inspiration for this claim is an analysis of a cross-section of sourdough bread (Figure 8).

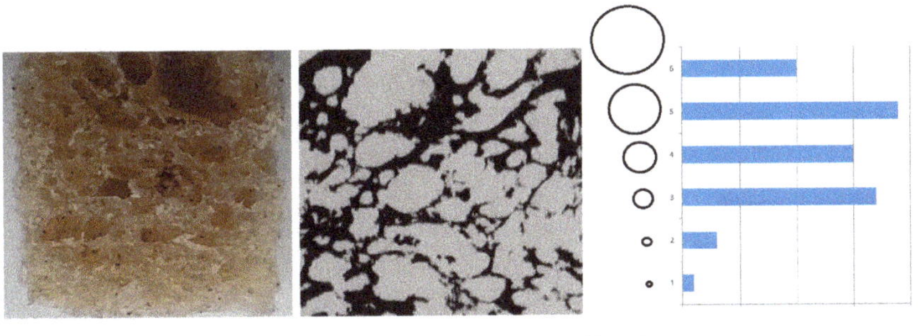

Figure 8. Analysis of the number of holes in a cross-section of bread.

Such a random structure can be both analyzed and designed for 3D printing objects.

4. Conclusions

The novelty of this paper is its reference to the mathematical description of the virial, which is independent of physics. This suggests that it is a general principle that applies in various fields of technology. The principle generally also applies to the structure of materials with a random distribution of building elements. In the process of manufacturing the structure, dynamic forces act, which are related to kinetic energy. The ideal distribution of the size and frequency of the elements of the structure will be subject to this law. This can be advantageously used in the design of 3D printing as an optimization condition.

Author Contributions: Conceptualization, T.S.; methodology, T.S.; resources, T.S.; writing—original draft preparation, T.S.; supervision, M.L. and P.F.; formal analysis, M.L.; writing—review and editing, S.M. and M.L.; project administration, S.M. funding acquisition, S.M. and P.F. All authors have read and agreed to the published version of the manuscript.

Funding: This work was funded by the Ministry of Education, Science, Research and Sport of the Slovak Republic within projects VEGA 1/0152/24 and KEGA 037TUKE-4/2024.

Institutional Review Board Statement: Not applicable.

Informed Consent Statement: Not applicable.

Data Availability Statement: The original contributions presented in this study are included in the article. Further inquiries can be directed to the corresponding author.

Acknowledgments: This work was supported by the Slovak Research and Development Agency under Contract nos. VEGA 1/0152/24, VEGA 1/0243/23 and VEGA 1/0191/24 and Contract nos. KEGA 037TUKE-4/2024, KEGA 004TUKE-4/2024 and KEGA 039TUKE-4/2024.

Conflicts of Interest: The authors declare no conflicts of interest.

References

1. Nasrullah, A.I.H.; Santosa, S.P.; Dirgantara, T. Design and Optimization of Crashworthy Components Based on Lattice Structure Configuration. *Structures* **2020**, *26*, 969–981. [CrossRef]
2. Mueller, J.; Matlack, K.H.; Shea, K.; Daraio, C. Energy Absorption Properties of Periodic and Stochastic 3D Lattice Materials. *Adv. Theory Simul.* **2019**, *2*, 1900081. [CrossRef]
3. Wang, P.; Bian, Y.; Yang, F.; Fan, H.; Zheng, B. Mechanical Properties and Energy Absorption of FCC Lattice Structures with Different Orientation Angles. *Acta Mech.* **2020**, *231*, 3129–3144. [CrossRef]
4. Bernard, A.R.; ElSayed, M.S.A. Crashworthiness of 3D Lattice Topologies under Dynamic Loading: A Comprehensive Study. *Materials* **2024**, *17*, 1597. [CrossRef] [PubMed]
5. Benyus, J.M. *Biomimicry: Innovation Inspired by Nature*; Harper Perennial: New York, NY, USA, 2002.
6. Yaraghi, N.A.; Kisailus, D. Biomimetic structural materials: Inspiration from design and assembly. *Annu. Rev. Phys. Chem.* **2018**, *69*, 23–57. [CrossRef] [PubMed]
7. Sreeramagiri, S.V.S. Biomimetics: Applications in structural design. *Int. J. Innov. Res. Sci. Eng. Technol. (IJIRSET)* **2021**, *10*, 8899–8908.
8. Hu, Z. Biomimetic Design and Topology Optimization of Discontinuous Carbon Fiber-Reinforced Composite Lattice Structures. *Biomimetics* **2023**, *8*, 148. [CrossRef] [PubMed]
9. Clausius, R. Uebereineveränderte Form des zweiten Hauptsatzes der mechanischen Wärmetheorie. *Ann. Phys.* **1854**, *169*, 481–506. [CrossRef]
10. Clausius, R.X. On a modified form of the second fundamental theorem in the mechanical theory of heat. *Lond. Edinb. Dublin Philos. Mag. J. Sci.* **1856**, *12*, 81–98. [CrossRef]
11. Clausius, R. Ueberverschiedenefür die Anwendungbequeme Formen der Hauptgleichungen der mechanischen Wärmetheorie. *Ann. Phys.* **1865**, *201*, 353–400. [CrossRef]
12. Clausius, R. Uebereinen auf die Wärme anwendbaren mechanischen Satz. *Ann. Phys.* **1870**, *217*, 124–130. [CrossRef]
13. Halliwell, H.F.; Nymburg, S.C. Statements of First and Second Laws of Thermodynamics. *Sci. Prog.* **1958**, *46*, 450–458.
14. Scott, D.S. The arrow of time. *Int. J. Hydrogen Energy* **2003**, *2*, 147–149. [CrossRef]
15. Albrecht, A. From Eternity to Here: The Quest for the Ultimate Theory of Time. *Phys. Today* **2010**, *63*, 54–55. [CrossRef]
16. Johnson, N.L.; Kemp, A.W.; Kotz, S. *Univariate Discrete Distributions*; Wiley Series in Probability and Statistics; Wiley: Hoboken, NJ, USA, 2005; ISBN 9780471272465.
17. Walker, H.M. De Moivre on the Law of Normal Probability. 2006. Available online: https://www.york.ac.uk/depts/maths/histstat/demoivre.pdf (accessed on 19 May 2025).
18. Cline, D. *Variational Principles in Classical Mechanics*; University of Rochester River Campus Librarie: Rochester, NY, USA, 2017.
19. Bendat, J.S.; Piersol, A.G. *Random Data: Analysis and Measurement Procedures*; John Wiley & Sons: Hoboken, NJ, USA, 2011.
20. Li, H.; Hu, Y.; Wang, S. A Novel Blind Signal Detector Based on the Entropy of the Power Spectrum Subband Energy Ratio. *Entropy* **2021**, *23*, 448. [CrossRef] [PubMed]
21. Stejskal, T. Coupled pendulums—Consequences of an experiment inconsistent with the theory. *Acta Tech. CSAV* **2015**, *60*, 193–207.
22. Aronson, D.G.; Ermentrout, G.B.; Kopell, N. Amplitude response of coupled oscillators. *Phys. D Nonlinear Phenom.* **1990**, *41*, 403–449. [CrossRef]
23. Makarov, D. Coupled Harmoni Oscillator in a System of Free Particles. *Mathematics* **2022**, *10*, 294. [CrossRef]
24. Urzúa, A.R.; Ramos-Prieto, I.; Fernández-Guasti, M.; Moya-Cessa, H.M. Solution to the Time-Dependent Coupled Harmonic Oscillators Hamiltonian with Arbitrary Interactions. *Quantum Rep.* **2019**, *1*, 82–90. [CrossRef]

25. Maxwell, J.C., V. Illustrations of the dynamical theory of gases—Part I. On the motions and collisions of perfectly elastic spheres. *Lond. Edinb. Dublin Philos. Mag. J. Sci.* **1860**, *19*, 19–32. [CrossRef]
26. Maxwell, J.C., II. Illustrations of the dynamical theory of gases. *Lond. Edinb. Dublin Philos. Mag. J. Sci.* **1860**, *20*, 21–37. [CrossRef]

MDPI AG
Grosspeteranlage 5
4052 Basel
Switzerland
Tel.: +41 61 683 77 34

Materials Editorial Office
E-mail: materials@mdpi.com
www.mdpi.com/journal/materials

www.ingramcontent.com/pod-product-compliance
Lightning Source LLC
Chambersburg PA
CBHW061558220926
51850CB00005B/366
* 9 7 8 3 7 2 5 8 7 0 9 8 1 *